TRIGONOMETRY
FIFTH EDITION

TRIGONOMETRY
FIFTH EDITION

CHARLES P. MCKEAGUE
Cuesta College

MARK D. TURNER
Cuesta College

THOMSON
™
BROOKS/COLE

Australia • Canada • Mexico • Singapore • Spain
United Kingdon • Unites States

Publisher: *Robert W. Pirtle*

Acquisitions Editor: *John-Paul Ramin*

Assistant Editor: *Lisa Chow*

Editorial Assistant: *Darlene Amidon-Brent*

Technology Project Manager: *Christopher Delgado*

Marketing Manager: *Karin Sandberg*

Marketing Assistant: *Jennifer Gee*

Advertising Project Manager: *Bryan Vann*

Project Manager, Editorial Production: *Andy Marinkovich*

Print/Media Buyer: *Jessica Reed*

Permissions Editor: *Beth Zuber*

Production Service: *Graphic World Publishing Services*

Text Designer: *Kathleen Cunningham*

Photo Researcher: *Kathleen Olson, Kofoto*

Cover Designer: *Lisa Henry*

Cover Image: *J.A. Kraulis/Wonderfile*

Compositor: *Graphic World, Inc.*

Printer: *R.R. Donnelley (Willard)*

Cover Printer: *Phoenix Color Corp.*

For more information about our products, contact us at:

Thomson Learning Academic Resource Center

1-800-423-0563

For permission to use material from this text, contact us by:

Phone: 1-800-730-2214   **Fax:** 1-800-730-2215

Web: http://www.thomsonrights.com

Library of Congress Control Number: 2003111537

Student Edition: ISBN 0-534-40392-1

Instructor's Edition: ISBN 0-534-40401-4

Brooks/Cole/Thomson Learning

10 Davis Drive

Belmont, CA 94002-3098

USA

Asia

Thomson Learning

5 Shenton Way #01-01

UIC Building

Singapore 068808

Australia

Thomson Learning

102 Dodds Street

Southbank, Victoria 3006

Australia

Canada

Nelson

1120 Birchmount Road

Toronto, Ontario M1K 5G4

Canada

Europe/Middle East/Africa

Thomson Learning

High Holborn House

50/51 Bedford Row

London WC1R 4LR

United Kingdom

Latin America

Thomson Learning

Seneca, 53

Colonia Polanco

11560 Mexico D.F.

Mexico

IDENTITIES AND FORMULAS

Basic Identities [5.1]

	Basic Identities	Common Equivalent Forms
Reciprocal	$\csc \theta = \dfrac{1}{\sin \theta}$	$\sin \theta = \dfrac{1}{\csc \theta}$
	$\sec \theta = \dfrac{1}{\cos \theta}$	$\cos \theta = \dfrac{1}{\sec \theta}$
	$\cot \theta = \dfrac{1}{\tan \theta}$	$\tan \theta = \dfrac{1}{\cot \theta}$
Ratio	$\tan \theta = \dfrac{\sin \theta}{\cos \theta}$	
	$\cot \theta = \dfrac{\cos \theta}{\sin \theta}$	
Pythagorean	$\cos^2 \theta + \sin^2 \theta = 1$	$\sin^2 \theta = 1 - \cos^2 \theta$
		$\sin \theta = \pm\sqrt{1 - \cos^2 \theta}$
		$\cos^2 \theta = 1 - \sin^2 \theta$
		$\cos \theta = \pm\sqrt{1 - \sin^2 \theta}$
	$1 + \tan^2 \theta = \sec^2 \theta$	
	$1 + \cot^2 \theta = \csc^2 \theta$	

Sum and Difference Formulas [5.2]

$$\sin (A + B) = \sin A \cos B + \cos A \sin B$$
$$\sin (A - B) = \sin A \cos B - \cos A \sin B$$
$$\cos (A + B) = \cos A \cos B - \sin A \sin B$$
$$\cos (A - B) = \cos A \cos B + \sin A \sin B$$
$$\tan (A + B) = \frac{\tan A + \tan B}{1 - \tan A \tan B}$$
$$\tan (A - B) = \frac{\tan A - \tan B}{1 + \tan A \tan B}$$

Double-Angle Formulas [5.3]

$$\sin 2A = 2 \sin A \cos A$$

$$
\begin{aligned}
\cos 2A &= \cos^2 A - \sin^2 A && \text{First form} \\
&= 2 \cos^2 A - 1 && \text{Second form} \\
&= 1 - 2 \sin^2 A && \text{Third form}
\end{aligned}
$$

$$\tan 2A = \frac{2 \tan A}{1 - \tan^2 A}$$

Cofunction Theorem [2.1]

$$\sin x = \cos (90° - x)$$
$$\cos x = \sin (90° - x)$$
$$\tan x = \cot (90° - x)$$

Half-Angle Formulas [5.4]

$$\sin \frac{A}{2} = \pm\sqrt{\frac{1 - \cos A}{2}}$$

$$\cos \frac{A}{2} = \pm\sqrt{\frac{1 + \cos A}{2}}$$

$$\tan \frac{A}{2} = \frac{1 - \cos A}{\sin A} = \frac{\sin A}{1 + \cos A}$$

Even/Odd Functions [3.3]

$$\cos (-\theta) = \cos \theta \qquad \text{Even}$$
$$\left.\begin{aligned} \sin (-\theta) &= -\sin \theta \\ \tan (-\theta) &= -\tan \theta \end{aligned}\right\} \quad \text{Odd}$$

Sum to Product Formulas [5.5]

$$\sin \alpha + \sin \beta = 2 \sin \frac{\alpha + \beta}{2} \cos \frac{\alpha - \beta}{2}$$

$$\sin \alpha - \sin \beta = 2 \cos \frac{\alpha + \beta}{2} \sin \frac{\alpha - \beta}{2}$$

$$\cos \alpha + \cos \beta = 2 \cos \frac{\alpha + \beta}{2} \cos \frac{\alpha - \beta}{2}$$

$$\cos \alpha - \cos \beta = -2 \sin \frac{\alpha + \beta}{2} \sin \frac{\alpha - \beta}{2}$$

Product to Sum Formulas [5.5]

$$\sin A \cos B = \frac{1}{2} [\sin (A + B) + \sin (A - B)]$$

$$\cos A \sin B = \frac{1}{2} [\sin (A + B) - \sin (A - B)]$$

$$\cos A \cos B = \frac{1}{2} [\cos (A + B) + \cos (A - B)]$$

$$\sin A \sin B = \frac{1}{2} [\cos (A - B) - \cos (A + B)]$$

Pythagorean Theorem [1.1]

$$c^2 = a^2 + b^2$$

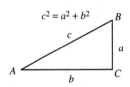

The Law of Sines [7.1]

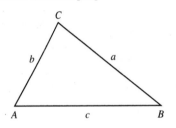

$$\frac{\sin A}{a} = \frac{\sin B}{b} = \frac{\sin C}{c}$$

or, equivalently,

$$\frac{a}{\sin A} = \frac{b}{\sin B} = \frac{c}{\sin C}$$

The Law of Cosines [7.3]

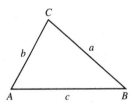

$$a^2 = b^2 + c^2 - 2bc \cos A$$
$$b^2 = a^2 + c^2 - 2ac \cos B$$
$$c^2 = a^2 + b^2 - 2ab \cos C$$

or, equivalently,

$$\cos A = \frac{b^2 + c^2 - a^2}{2bc}$$

$$\cos B = \frac{a^2 + c^2 - b^2}{2ac}$$

$$\cos C = \frac{a^2 + b^2 - c^2}{2ab}$$

BRIEF CONTENTS

CONTENTS

PREFACE TO THE INSTRUCTOR

This fifth edition of *Trigonometry* retains the same format and style as the previous editions. It is a standard right-triangle approach to trigonometry. Each section is written so that it can be discussed in a 45- to 50-minute class session. The focus of the textbook is on understanding the definitions and principles of trigonometry and their applications to problem solving. Exact values of the trigonometric functions are emphasized throughout the textbook.

The text covers all the material usually taught in trigonometry. In addition, there is an appendix on functions and inverse functions, and a second appendix on exponential and logarithmic functions. The appendix sections can be used as a review of topics that students may already be familiar with, or they can be used to provide thorough instruction for students encountering these concepts for the first time.

There are numerous calculator notes placed throughout the text to help students calculate values when appropriate. Most of the calculator-related material has been revised in this edition. As there are many different models of graphing calculators, and each model has its own set of commands, we have tried to avoid an overuse of specific key icons or command names. If you are using a graphing calculator in your classroom, the text-specific Graphing Calculator Manual found at the McKeague book companion Web site provides more detailed guidance on calculator keystrokes for popular models.

NEW TO THIS EDITION

Content Changes The material on heading has been moved from Section 2.5 to Section 7.2. The concept of work is introduced in Section 2.5, and later revisited in Section 7.6. Amplitude is now covered in Section 4.1 to provide more opportunities for students to practice graphing the basic trigonometric functions in the problem sets. The material on inverse trigonometric functions has been expanded, including more detailed coverage of graphs of the inverse functions and additional examples. A new appendix provides a review of functions and inverse functions that is of sufficient depth for a first encounter with these concepts. Section 7.5 on vectors has been split into two sections, with the dot product now covered separately in Section 7.6. We have added $r\text{cis}\,\theta$ notation for complex numbers in Chapter 8.

Group Projects Each chapter now concludes with a group project, which involves some interesting problem or application that relates to or extends the ideas introduced in the chapter. Many of the projects emphasize the connection of mathematics with other disciplines or illustrate real-life situations in which trigonometry is used. The projects are designed to be used in class with groups of three or four students each, but the problems could also be given as individual assignments for students wanting an additional challenge.

Research Projects A research project is also offered at the end of each chapter. The research projects ask students to investigate an historical topic or personage that is in some way connected to the material in the chapter; these projects are intended to promote an appreciation for the rich history behind trigonometry. Students may find the InfoTrac College Edition database or the Internet to be helpful resources in doing their research.

Getting Ready for Class Located before each problem set, Getting Ready for Class sections require written responses from students, and can be answered by reading the preceding section. They are to be done before the students come to class.

More Applications Many new application problems have been added to the problem sets, including problems involving cable cars, work, and cycling, to help motivate students and stimulate their interest in trigonometry.

Graphing Calculator Exercises For those of you who are using graphing calculators in your classes, new exercises that require graphing calculators have been added to some of the problem sets. These exercises are clearly marked with a special icon ( ), and may easily be omitted if you are not using this technology in your classroom.

CONTINUING FEATURES

Chapter Introductions Each chapter opens with an introduction in which a real-world application, historical example, or link between topics is used to stimulate interest in the chapter. Many of these introductions are expanded on later in the chapter and then carried through to topics found later in the book. Many sections open in a similar fashion.

Study Skills Study Skills sections are found in the first six chapter openings and help students become organized and efficient with their time.

Using Technology Using Technology sections throughout the book show how graphing calculator technology can be used to enhance the topics covered. These sections have been completely revised to provide more direct and meaningful connections with the material. All graphing calculator material is optional, but even if you are not using graphing calculator technology in your classroom, these segments can provide additional insight into the standard trigonometric procedures and problem solving found in the section.

Three Definitions All three definitions for the trigonometric functions are contained in the text. The point-on-the-terminal-side definition is contained in Section 1.3. The right triangle definition is in Section 2.1. Circular functions are given in Section 3.3.

Applications Application problems are titled according to subject for easy reference. We have found that students are more likely to put some time and effort into trying application problems if they do not have to work an overwhelming number of them at one time and if they work on them every day. For this reason, a few application problems are included toward the end of almost every problem set in the book.

Review Problems Each problem set, beginning with Chapter 2, contains a few review problems. Where appropriate, the review problems cover material that will be needed in the next section. Otherwise, they cover material from the previous chapter. Continual review will help students retain what they have learned in previous chapters and reinforce main ideas.

Extending the Concepts Scattered throughout the text, Extending the Concepts problems involve students in researching, writing, and group work on topics such as English, history, religion, and map-making. These problems address the AMATYC and NCTM goals of increased writing and group activities. Some also require research on the World Wide Web.

Chapter Summaries Each chapter summary lists the new properties and definitions found in the chapter. The margins in the chapter summaries contain examples that illustrate the topics being reviewed.

Chapter Tests Every chapter ends with a chapter test that contains a representative sample of the problems covered in the chapter. Answers to both odd and even problems are given in the back of the book.

SUPPLEMENTS TO THE TEXTBOOK

FOR THE INSTRUCTOR

Annotated Instructor's Edition This special version of the complete student text contains a Resource Integration Guide as well as answers to all the exercises in the text.

Test Bank The test bank includes eight tests per chapter as well as three final exams. The tests are made up of a combination of multiple-choice, free response, true/false, and fill-in-the-blank questions.

Complete Solutions Manual Written by Ross Rueger, College of the Sequoias, the complete solutions manual provides worked-out solutions to all of the problems in the text.

BCA Instructor Version With a balance of efficiency and high performance, simplicity and versatility, Brooks/Cole Assessment gives you the power to transform the learning and teaching experience. BCA Instructor Version is made up of two components: BCA Testing and BCA Tutorial. BCA Testing is a revolutionary, Internet-ready, text-specific testing suite that allows instructors to customize exams and track student progress in an accessible, browser-based format. BCA offers full algorithmic generation of problems and free response mathematics. BCA Tutorial is a text-specific, interactive tutorial software program, that is delivered via the Web (at http://bca.brookscole.com) and is offered in both student and instructor versions. Like BCA Testing, it is browser-based, making it an intuitive mathematical guide even for students with little technological proficiency. So sophisticated, it's simple, BCA Tutorial allows students to work with real math notation in real time, providing instant analysis and feedback. The tracking program built into the instructor version of the software enables instructors to carefully monitor student progress.

WebTutor ToolBox for WebCT and WebTutor ToolBox for Blackboard Preloaded with content and available free via pincode when packaged with this text, *WebTutor ToolBox for WebCT and WebTutor ToolBox for Blackboard* pairs all the content of this text's rich Book Companion Web Site with all the sophisticated course management functionality of a WebCT and Blackboard product. You can assign materials (including online quizzes) and have the results flow *automatically* to your gradebook. ToolBox is ready to use as soon as you log on—or, you can customize its preloaded content by uploading images and other resources, adding Web links, or creating your own practice materials.

FOR THE STUDENT

A Digital Video Companion (DVC) This single CD-ROM containing over 6 hours of instructional video instruction is packaged with each book. The foundation of the DVC is the video lessons. There is one video lesson for each section of the book. Each video lesson is 8 to 12 minutes long. The 🔧 icon is used in the problem sets in the text to denote problems worked on the Digital Video Companion. On the CD, the problems worked during each video lesson are listed next to the viewing screen, so that students can work them ahead of time, if they choose. To help students evaluate their progress, each section contains a 10-question Web quiz (the results of which can be e-mailed to the instructor), and each chapter contains a chapter test, with answers to each problem on each test.

The DVC also includes MathCue Tutorial software, as well as Graphing Calculator Tutorials.

MathCue This computer software package of tutorials has problems that correspond to every section of the text. The software presents problems to solve and tutors students by displaying annotated, step-by-step solutions. Students may view partial solutions to get started on a problem, see a continuous record of progress, and back up to review missed problems. Student scores can also be printed. Available for Windows and Macintosh.

MathCue Practice This algorithm-based software allows students to generate large numbers of practice problems keyed to problem types from each section of the book. Practice scores students' performance and saves students' scores session to session. Available for Windows and Macintosh.

Graphing Calculator Tutorials These tutorials provide detailed, step-by-step instruction for each Using Technology segment as well as selected examples in the text. Full keystroke sequences and screen images are given for the TI-83, TI-83 Plus, and TI-86 graphing calculators. The 🔧 icon next to a Using Technology segment or example indicates that a corresponding calculator tutorial is available on the DVC.

Text-Specific Videotapes These text-specific videotape sets, available at no charge to qualified adopters of the text, feature 10- to 20-minute problem-solving lessons that cover each section of every chapter. On the videotapes we work selected problems from the text. The 📼 icon is used in the problem sets in the text to denote problems worked on the text-specific videotapes.

BCA Student Version A complete content mastery resource offered online or on CD-ROM, *BCA Tutorial* allows students to work with real math notation in real time, with unlimited practice problems, instant analysis and feedback, and streaming video to illustrate key concepts. And live, online, text-specific tutorial help is just a click away with *vMentor,* accessed seamlessly through BCA Tutorial.

Student Solutions Manual Written by Judy Barclay, Cuesta College, the Student Solutions Manual provides worked-out solutions to the odd-numbered problems from the problem sets and all the problems in the chapter tests in the text.

Web Site http://mathematics.brookscole.com
When you adopt a Thomson-Brooks/Cole mathematics text, you and your students will have access to a variety of teaching and learning resources. This Web site features everything from book-specific resources to newsgroups. It's a great way to make teaching and learning an interactive and intriguing experience.

ACKNOWLEDGMENTS

A project of this size cannot be handled by two people. We had lots of help. To begin, John-Paul Ramin, our editor at Brooks/Cole Publishing, was very helpful and encouraging in all parts of the revision process. He is a pleasure to work with. Carol O'Connell of Graphic World Publishing Services handled the production part of this revision, and has done an outstanding job of moving this project from manuscript to the book you have in front of you. Andy Marinkovich, production project manager, helped get the manuscript into production. Lisa Chow coordinated the production of the supplemental material for the book, and Darlene Amidon-Brent coordinated the review of the manuscript. Tammy Fisher helped edit the videotapes that accompany the text. John Van Eps from Cal Poly State University, Kate Pawlik, and Ann Ostberg did an excellent job with proofreading and accuracy checking. Judy Barclay, Cuesta College, produced the outstanding Student's Solutions Manual that accompanies the book, and Ross Rueger, College of the Sequoias, produced the Complete Solutions Manual. Patrick McKeague produced the first Digital Video Companion that accompanied this book, and it has proven to be an outstanding tool for students and instructors alike. Our thanks to all these people; this book would not have been possible without them.

Thanks also to Diane McKeague, Amy Jacobs, and Anne Turner for their continued encouragement

Finally, we are grateful to the following instructors for their suggestions and comments on this revision. Some reviewed the entire manuscript, while others were asked to evaluate the development of specific topics, or the overall sequence of topics. Our thanks go to the people listed below.

Yossef Balas, *University of Southern Mississippi*

Jeffrey E. Cohen, *El Camino College*

John Garlow, *Tarrant County College*

Geoffrey Hagopian, *College of the Desert*

Christy MacBride-Hart, *Santa Ana College*

Lynn D. Hearn, *Valencia Community College*

Marta L. Hidden, *Orange Coast College*

Gene W. Majors, *Fullerton College*

Marianna E. McClymonds, *Phoenix College*

Michael Montano, *Riverside Community College*

Doris Louise Sutton, *Central Texas College*

Thomas J. Tegtmeyer, *Truman State University*

Barbara J. Tucker, *Tarrant County College*

Charles P. McKeague
Mark D. Turner
October 2003

A NOTE FROM CHARLES McKEAGUE

I am extremely pleased to have Mark Turner as my coauthor on this edition of Trigonometry. Mark has helped me with my books for the past few years, and now he is taking the next step and signing on as coauthor. He is an outstanding teacher and, as you will see, an excellent writer. If you have attended any of the talks he has given recently, then you are also aware that he is an innovative instructor, bringing many new ideas on technology into the classroom. I am very pleased to have Mark as my coauthor on this book.

PREFACE TO THE STUDENT

Trigonometry can be a very enjoyable subject to study. You will find that there are many interesting and useful problems that trigonometry can be used to solve. However, many trigonometry students are apprehensive at first because they are worried that they will not understand the topics we cover. When we present a new topic that they do not grasp completely, they think something is wrong with them for not understanding it. On the other hand, some students are excited about the course from the beginning. They are not worried about understanding trigonometry and, in fact, expect to find some topics difficult.

What is the difference between these two types of students?

Those who are excited about the course know from experience (as you do) that a certain amount of confusion is associated with most new topics in mathematics. They don't worry about it because they also know that the confusion gives way to understanding in the process of reading the textbook, working problems, and getting their questions answered. If they find a topic difficult, they work as many problems as necessary to grasp the subject. They don't wait for the understanding to come to them; they go out and get it by working lots of problems. In contrast, the students who lack confidence tend to give up when they become confused. Instead of working more problems, they sometimes stop working problems altogether, and that, of course, guarantees that they will remain confused.

If you are worried about this course because you lack confidence in your ability to understand trigonometry, and you want to change the way you feel about mathematics, then look forward to the first topic that causes you some confusion. As soon as that topic comes along, make it your goal to master it, in spite of your apprehension. You will see that each and every topic covered in this course is one you can eventually master, even if your initial introduction to it is accompanied by some confusion. As long as you have passed a college-level intermediate algebra course (or its equivalent), you are ready to take this course.

If you have decided to do well in trigonometry, the following list will be important to you:

HOW TO BE SUCCESSFUL IN TRIGONOMETRY

1. **Attend all class sessions on time.** You cannot know exactly what goes on in class unless you are there. Missing class and then expecting to find out what went on from someone else is not the same as being there yourself.
2. **Read the book.** It is best to read the section that will be covered in class beforehand. Reading in advance, even if you do not understand everything you read, is still better than going to class with no idea of what will be discussed.

3. **Work problems every day and check your answers.** The key to success in mathematics is working problems. The more problems you work, the better you will become at working them. The answers to the odd-numbered problems are given in the back of the book. When you have finished an assignment, be sure to compare your answers with those in the book. If you have made a mistake, find out what it is, and correct it.

4. **Do it on your own.** Don't be misled into thinking someone else's work is your own. Having someone else show you how to work a problem is not the same as working the same problem yourself. It is okay to get help when you are stuck. As a matter of fact, it is a good idea. Just be sure you do the work yourself.

5. **Review every day.** After you have finished the problems your instructor has assigned, take another 15 minutes and review a section you have already completed. The more you review, the longer you will retain the material you have learned.

6. **Don't expect to understand every new topic the first time you see it.** Sometimes you will understand everything you are doing, and sometimes you won't. That's just the way things are in mathematics. Expecting to understand each new topic the first time you see it can lead to disappointment and frustration. The process of understanding trigonometry takes time. It requires that you read the book, work problems, and get your questions answered.

7. **Spend as much time as it takes for you to master the material.** No set formula exists for the exact amount of time you need to spend on trigonometry to master it. You will find out as you go along what is or isn't enough time for you. If you end up spending 2 or more hours on each section in order to master the material there, then that's how much time it takes; trying to get by with less will not work.

8. **Relax.** It's probably not as difficult as you think.

THE SIX TRIGONOMETRIC FUNCTIONS

Without Thales there would not have been a Pythagoras—or such a Pythagoras; and without Pythagoras there would not have been a Plato—or such a Plato.

D. E. Smith

INTRODUCTION

The history of mathematics is a spiral of knowledge passed down from one generation to another. Each person in the history of mathematics is connected to the others along this spiral. In Italy, around 500 B.C., the Pythagoreans discovered a relationship between the sides of any right triangle. That discovery, known as the Pythagorean Theorem, is the foundation on which the Spiral of Roots shown in Figure 1 is built. The Spiral of Roots gives us a way to visualize square roots of positive integers.

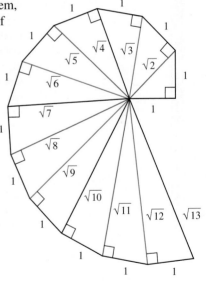

Figure 1

In Problem Set 1.1, you will have a chance to construct the Spiral of Roots yourself.

STUDY SKILLS FOR CHAPTER 1

At the beginning of the first few chapters of this book you will find a Study Skills section in which we list the skills that are necessary for success in trigonometry. If you have just completed an algebra class successfully, you have acquired most of these skills. If it has been some time since you have taken a math class, you must pay attention to the sections on study skills.

Here is a list of things you can do to develop effective study skills.

1. **Put Yourself on a Schedule**   The general rule is that you spend two hours on homework for every hour you are in class. Make a schedule for yourself, setting aside at least six hours a week to work on trigonometry. Once you make the schedule, stick to it. Don't just complete your assignments and then stop. Use all the time you have set aside. If you complete an assignment and have time left over, read the next section in the book, and work more problems. As the course progresses you may find that six hours a week is not enough time for you to master the material in this course. If it takes you longer than that to reach your goals for this course, then that's how much time it takes. Trying to get by with less will not work.

2. **Find Your Mistakes and Correct Them**   There is more to studying trigonometry than just working problems. You must always check your answers with those in the back of the book. When you have made a mistake, find out what it is and correct it. Making mistakes is part of the process of learning mathematics. The key to discovering what you do not understand can be found by correcting your mistakes.

3. **Imitate Success**   Your work should look like the work you see in this book and the work your instructor shows. The steps shown in solving problems in this book were written by someone who has been successful in mathematics. The same is true of your instructor. Your work should imitate the work of people who have been successful in mathematics.

4. **Memorize Definitions and Identities**   You may think that memorization is not necessary if you understand a topic you are studying. In trigonometry, memorization is especially important. In this first chapter, you will be presented with the definition of the six trigonometric functions that you will use throughout the rest of the course. We have seen many bright students struggle with trigonometry simply because they did not memorize the definitions and identities when they were first presented.

SECTION 1.1 | ANGLES, DEGREES, AND SPECIAL TRIANGLES

INTRODUCTION

Table 1 is taken from the trail map given to skiers at Northstar at Tahoe Ski Resort in Lake Tahoe, California. The table gives the length of each chair lift at Northstar, along with the change in elevation from the beginning of the lift to the end of the lift.

TABLE 1	FROM THE TRAIL MAP FOR NORTHSTAR AT TAHOE SKI RESORT	
Lift Information		
Lift	*Vertical Rise (ft)*	*Length (ft)*
Big Springs Gondola	480	4,100
Bear Paw Double	120	790
Echo Triple	710	4,890
Aspen Express Quad	900	5,100
Forest Double	1,170	5,750
Lookout Double	960	4,330
Comstock Express Quad	1,250	5,900
Rendezvous Triple	650	2,900
Schaffer Camp Triple	1,860	6,150
Chipmunk Tow Lift	28	280
Bear Cub Tow Lift	120	750

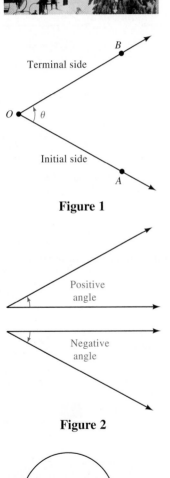

Figure 1

Figure 2

Right triangles are good mathematical models for chair lifts. In this section we review some important items from geometry, including right triangles. Let's begin by looking at some of the terminology associated with angles.

ANGLES IN GENERAL

An angle is formed by two rays with the same end point. The common end point is called the *vertex* of the angle, and the rays are called the *sides* of the angle.

In Figure 1 the vertex of angle θ (theta) is labeled O, and A and B are points on each side of θ. Angle θ can also be denoted by AOB, where the letter associated with the vertex is written between the letters associated with the points on each side.

We can think of θ as having been formed by rotating side OA about the vertex to side OB. In this case, we call side OA the *initial side* of θ and side OB the *terminal side* of θ.

When the rotation from the initial side to the terminal side takes place in a counterclockwise direction, the angle formed is considered a *positive angle*. If the rotation is in a clockwise direction, the angle formed is a *negative angle* (Figure 2).

DEGREE MEASURE

One way to measure the size of an angle is with degree measure. The angle formed by rotating a ray through one complete revolution has a measure of 360 degrees, written 360° (Figure 3).

One degree (1°), then, is 1/360 of a full rotation. Likewise, 180° is one-half of a full rotation, and 90° is half of that (or a quarter of a rotation). Angles that measure 90° are called *right angles,* while angles that measure 180° are called *straight angles.* Angles that measure between 0° and 90° are called *acute angles,* while angles that measure between 90° and 180° are called *obtuse angles.*

If two angles have a sum of 90°, then they are called *complementary angles,* and we say each is the *complement* of the other. Two angles with a sum of 180° are called *supplementary angles.*

One complete
revolution = 360°

Figure 3

NOTE   To be precise, we should say "two angles, the sum of the measures of which is 180°, are called supplementary angles" because there is a difference between an angle and its measure. However, in this book, we will not always draw the distinction between an angle and its measure. Many times we will refer to "angle θ" when we actually mean "the measure of angle θ."

NOTE   The little square by the vertex of the right angle in Figure 4 is used to indicate that the angle is a right angle. You will see this symbol often in the book.

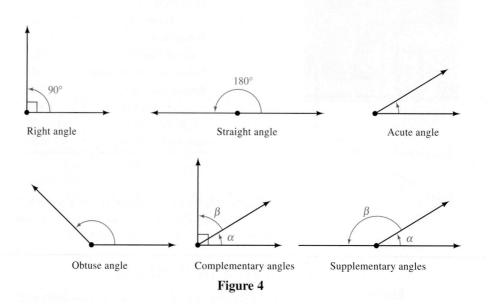

Right angle   Straight angle   Acute angle

Obtuse angle   Complementary angles   Supplementary angles

Figure 4

 **EXAMPLE 1**   Give the complement and the supplement of each angle.

a. 40°   **b.** 110°   **c.** θ

SOLUTION
a. The complement of 40° is 50° since 40° + 50° = 90°.
   The supplement of 40° is 140° since 40° + 140° = 180°.
b. The complement of 110° is −20° since 110° + (−20°) = 90°.
   The supplement of 110° is 70° since 110° + 70° = 180°.
c. The complement of θ is 90° − θ since θ + (90° − θ) = 90°.
   The supplement of θ is 180° − θ since θ + (180° − θ) = 180°.

SPECIAL TRIANGLES

A *right triangle* is a triangle in which one of the angles is a right angle. In every right triangle, the longest side is called the *hypotenuse,* and it is always opposite the right angle. The other two sides are called the *legs* of the right triangle. Since the sum of the angles in any triangle is 180°, the other two angles in a right triangle must be complementary, acute angles. The Pythagorean Theorem that we mentioned in the introduction to this chapter gives us the relationship that exists among the sides of a right triangle. First we state the theorem, then we will prove it.

PYTHAGOREAN THEOREM

In any right triangle, the square of the length of the longest side (called the hypotenuse) is equal to the sum of the squares of the lengths of the other two sides (called legs).

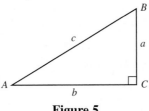

If $C = 90°$,
then $c^2 = a^2 + b^2$

Figure 5

We denote the lengths of the sides of triangle ABC in Figure 5 with lowercase letters and the angles or vertices with uppercase letters. It is standard practice in mathematics to label the sides and angles so that a is opposite A, b is opposite B, and c is opposite C.

Next we will prove the Pythagorean Theorem. Part of the proof involves finding the area of a triangle. In any triangle, the area is given by the formula

$$\text{Area} = \frac{1}{2}(\text{base})(\text{height})$$

For the right triangle shown in Figure 5, the base is b, and the height is a. Therefore the area is $A = \frac{1}{2}ab$.

A PROOF OF THE PYTHAGOREAN THEOREM

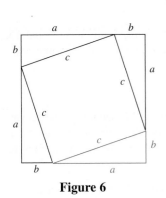

Figure 6

There are many ways to prove the Pythagorean Theorem. The Group Project at the end of this chapter introduces several of these ways. The method that we are offering here is based on the diagram shown in Figure 6 and the formula for the area of a triangle.

Figure 6 is constructed by taking the right triangle in the lower right corner and repeating it three times so that the final diagram is a square in which each side has length $a + b$.

To derive the relationship between a, b, and c, we simply notice that the area of the large square is equal to the sum of the areas of the four triangles and the inner square. In symbols we have

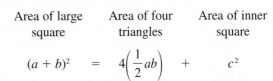

Area of large square	Area of four triangles	Area of inner square

$$(a + b)^2 = 4\left(\frac{1}{2}ab\right) + c^2$$

We expand the left side using the formula for the square of a binomial, from algebra. We simplify the right side by multiplying 4 with $\frac{1}{2}$.

$$a^2 + 2ab + b^2 = 2ab + c^2$$

Adding $-2ab$ to each side, we have the relationship we are after:

$$a^2 + b^2 = c^2$$

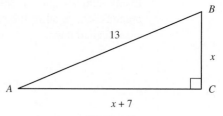

EXAMPLE 2 Solve for x in the right triangle in Figure 7.

Figure 7

SOLUTION Applying the Pythagorean Theorem gives us a quadratic equation to solve.

$$(x + 7)^2 + x^2 = 13^2$$

$$x^2 + 14x + 49 + x^2 = 169 \qquad \text{Expand } (x + 7)^2 \text{ and } 13^2$$

$$2x^2 + 14x + 49 = 169 \qquad \text{Combine similar terms}$$

$$2x^2 + 14x - 120 = 0 \qquad \text{Add } -169 \text{ to both sides}$$

$$x^2 + 7x - 60 = 0 \qquad \text{Divide both sides by 2}$$

$$(x - 5)(x + 12) = 0 \qquad \text{Factor the left side}$$

$$x - 5 = 0 \quad \text{or} \quad x + 12 = 0 \qquad \text{Set each factor to 0}$$

$$x = 5 \quad \text{or} \quad x = -12$$

Our only solution is $x = 5$. We cannot use $x = -12$ since x is the length of a side of triangle ABC and therefore cannot be negative.

NOTE The lengths of the sides of the triangle in Example 2 are 12, 13, and 15. Whenever the three sides in a right triangle are natural numbers, those three numbers are called a *Pythagorean triple*.

EXAMPLE 3 Table 1 in the introduction to this section gives the vertical rise of the Forest Double chair lift (Figure 8) as 1,170 feet and the length of the chair lift as 5,750 feet. To the nearest foot, find the horizontal distance covered by a person riding this lift.

SOLUTION Figure 9 is a model of the Forest Double chair lift. A rider gets on the lift at point A and exits at point B. The length of the lift is AB.

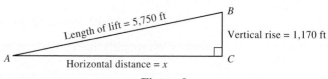

Figure 9

Figure 8

To find the horizontal distance covered by a person riding the chair lift we use the Pythagorean Theorem:

$$5{,}750^2 = x^2 + 1{,}170^2 \qquad \text{Pythagorean Theorem}$$

$$33{,}062{,}500 = x^2 + 1{,}368{,}900 \qquad \text{Simplify squares}$$

$$x^2 = 33{,}062{,}500 - 1{,}368{,}900 \qquad \text{Solve for } x^2$$

$$x^2 = 31{,}693{,}600 \qquad \text{Simplify the right side}$$

$$x = \sqrt{31{,}693{,}600}$$

$$x = 5{,}630 \text{ ft} \qquad \text{To the nearest foot}$$

A rider getting on the lift at point A and riding to point B will cover a horizontal distance of approximately 5,630 feet. ▪

Before leaving the Pythagorean Theorem we should mention something about Pythagoras and his followers, the Pythagoreans. They established themselves as a secret society around the year 540 B.C. The Pythagoreans kept no written record of their work; everything was handed down by spoken word. Their influence was not only in mathematics, but also in religion, science, medicine, and music. Among other things, they discovered the correlation between musical notes and the reciprocals of counting numbers, $\frac{1}{2}, \frac{1}{3}, \frac{1}{4}$, and so on. In their daily lives they followed strict dietary and moral rules to achieve a higher rank in future lives. The British philosopher Bertrand Russell has referred to Pythagoras as "intellectually one of the most important men that ever lived."

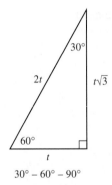

$30° - 60° - 90°$

Figure 10

THE 30°–60°–90° TRIANGLE

In any right triangle in which the two acute angles are 30° and 60°, the longest side (the hypotenuse) is always twice the shortest side (the side opposite the 30° angle), and the side of medium length (the side opposite the 60° angle) is always $\sqrt{3}$ times the shortest side (Figure 10).

NOTE   The shortest side t is opposite the smallest angle 30°. The longest side $2t$ is opposite the largest angle 90°.

To verify the relationship between the sides in this triangle, we draw an equilateral triangle (one in which all three sides are equal) and label half the base with t (Figure 11).

The altitude h (the colored line) bisects the base. We have two 30°–60°–90° triangles. The longest side in each is $2t$. We find that h is $t\sqrt{3}$ by applying the Pythagorean Theorem.

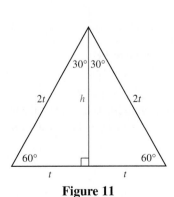

Figure 11

$$t^2 + h^2 = (2t)^2$$

$$h = \sqrt{4t^2 - t^2}$$

$$= \sqrt{3t^2}$$

$$= t\sqrt{3}$$

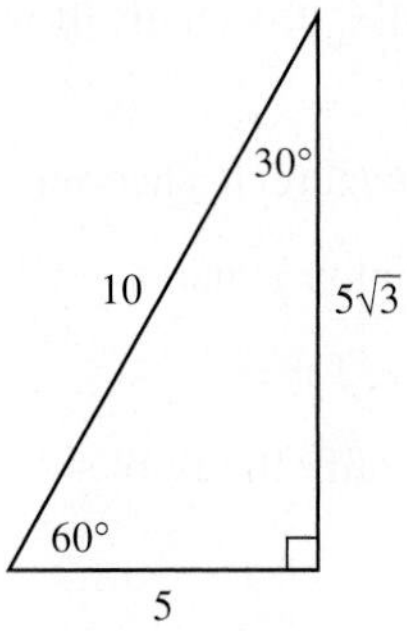

30°

10   $5\sqrt{3}$

60°

5

Figure 12

EXAMPLE 4   If the shortest side of a 30°–60°–90° triangle is 5, find the other two sides.

SOLUTION   The longest side is 10 (twice the shortest side), and the side opposite the 60° angle is $5\sqrt{3}$ (Figure 12).

EXAMPLE 5   A ladder is leaning against a wall. The top of the ladder is 4 feet above the ground and the bottom of the ladder makes an angle of 60° with the ground (Figure 13). How long is the ladder, and how far from the wall is the bottom of the ladder?

SOLUTION   The triangle formed by the ladder, the wall, and the ground is a 30°–60°–90° triangle. If we let x represent the distance from the bottom of the ladder to the wall, then the length of the ladder can be represented by $2x$. The distance from the top of the ladder to the ground is $x\sqrt{3}$, since it is opposite the 60° angle (Figure 14). It is also given as 4 feet. Therefore,

$$x\sqrt{3} = 4$$

$$x = \frac{4}{\sqrt{3}}$$

$$= \frac{4\sqrt{3}}{3} \qquad \begin{array}{l}\text{Rationalize the denominator} \\ \text{by multiplying the numerator} \\ \text{and denominator by } \sqrt{3}.\end{array}$$

Figure 13

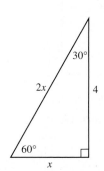

30°

$2x$ 4

60°

x

Figure 14

The distance from the bottom of the ladder to the wall, x, is $4\sqrt{3}/3$ feet, so the length of the ladder, $2x$, must be $8\sqrt{3}/3$ feet. Note that these lengths are given in exact values. If we want a decimal approximation for them, we can replace $\sqrt{3}$ with 1.732 to obtain

$$\frac{4\sqrt{3}}{3} \approx \frac{4(1.732)}{3} = 2.309 \text{ ft}$$

$$\frac{8\sqrt{3}}{3} \approx \frac{8(1.732)}{3} = 4.619 \text{ ft}$$

CALCULATOR NOTE   On a scientific calculator, this last calculation could be done as follows:

8 ⊗ 3 √ ÷ 3 =

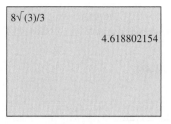

$8\sqrt{(3)}/3$

4.618802154

Figure 15

On a graphing calculator, the calculation is done like this:

$$8 \; \boxed{\times} \; \boxed{\sqrt{}} \; \boxed{(} \; \boxed{3} \; \boxed{)} \; \boxed{\div} \; 3 \; \boxed{\text{ENTER}}$$

Some graphing calculators use parentheses with certain functions, such as the square root function. For example, the TI-83 will automatically insert a left parenthesis, so TI-83 users should skip this key. Other models do not require them. For the sake of clarity, we will often include parentheses throughout this book. You may be able to omit one or both parentheses with your model.

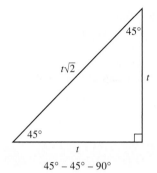

$45°$

$t\sqrt{2}$

t

$45°$

t

$45° - 45° - 90°$

Figure 16

THE 45°–45°–90° TRIANGLE

If the two acute angles in a right triangle are both $45°$, then the two shorter sides (the legs) are equal and the longest side (the hypotenuse) is $\sqrt{2}$ times as long as the shorter sides. That is, if the shorter sides are of length t, then the longest side has length $t\sqrt{2}$ (Figure 16).

To verify this relationship, we simply note that if the two acute angles are equal, then the sides opposite them are also equal. We apply the Pythagorean Theorem to find the length of the hypotenuse.

$$\text{hypotenuse} = \sqrt{t^2 + t^2}$$
$$= \sqrt{2t^2}$$
$$= t\sqrt{2}$$

EXAMPLE 6 A 10-foot rope connects the top of a tent pole to the ground. If the rope makes an angle of $45°$ with the ground, find the length of the tent pole (Figure 17).

SOLUTION Assuming that the tent pole forms an angle of $90°$ with the ground, the triangle formed by the rope, tent pole, and the ground is a $45°–45°–90°$ triangle (Figure 18).

If we let x represent the length of the tent pole, then the length of the rope, in terms of x, is $x\sqrt{2}$. It is also given as 10 feet. Therefore,

$$x\sqrt{2} = 10$$
$$x = \frac{10}{\sqrt{2}} = 5\sqrt{2}$$

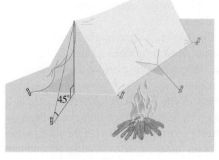

$45°$

Figure 17

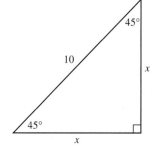

$45°$

10

x

$45°$

x

Figure 18

The length of the tent pole is $5\sqrt{2}$ feet. Again, $5\sqrt{2}$ is the exact value of the length of the tent pole. To find a decimal approximation, we replace $\sqrt{2}$ with 1.414 to obtain

$$5\sqrt{2} \approx 5(1.414) = 7.07 \text{ ft}$$

GETTING READY FOR CLASS

After reading through the preceding section, respond in your own words and in complete sentences.

a. What do we call the point where two rays come together to form an angle?

b. In your own words, define complementary angles.

c. In your own words, define supplementary angles.

d. Why is it important to recognize $30°$–$60°$–$90°$ and $45°$–$45°$–$90°$ triangles?

NOTE The icon next to a problem indicates that the complete solution to the problem is shown on the Digital Video Companion CD-ROM that accompanies the text. The ⬟ icon next to a problem indicates that the complete solution to the problem is shown on one of the videotapes in the videotape package.

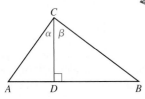

Figure 19

PROBLEM SET 1.1

Indicate which of the angles below are acute angles and which are obtuse angles. Then give the complement and the supplement of each angle.

1. $10°$ **2.** $50°$ **3.** $45°$ **4.** $90°$

5. $120°$ **6.** $160°$ **7.** x **8.** y

Problems 9 through 14 refer to Figure 19. (*Remember:* The sum of the three angles in any triangle is always $180°$.)

9. Find α if $A = 30°$.

10. Find B if $\beta = 45°$.

11. Find α if $A = \alpha$.

12. Find α if $A = 2\alpha$.

13. Find A if $B = 30°$ and $\alpha + \beta = 100°$.

14. Find B if $\alpha + \beta = 80°$ and $A = 80°$.

Figure 20 shows a walkway with a handrail. Angle α is the angle between the walkway and the horizontal, while angle β is the angle between the vertical posts of the handrail and the walkway. Use Figure 20 to work Problems 15 through 18. (Assume that the vertical posts are perpendicular to the horizontal.)

15. Are angles α and β complementary or supplementary angles?

16. If we did not know that the vertical posts were perpendicular to the horizontal, could we answer Problem 15?

17. Find β if $\alpha = 25°$.

18. Find α if $\beta = 52°$.

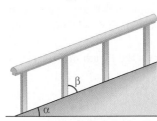

Figure 20

19. Rotating Light The red light on the top of a police car rotates through one complete revolution every 2 seconds. Through how many degrees does it rotate in 1 second?

20. Rotating Light A searchlight rotates through one complete revolution every 4 seconds. How long does it take the light to rotate through $90°$?

21. Clock Through how many degrees does the hour hand of a clock move in 4 hours?

22. Rotation of the Earth It takes the earth 24 hours to make one complete revolution on its axis. Through how many degrees does the earth turn in 12 hours?

23. Geometry An equilateral triangle is a triangle in which all three sides are equal. What is the measure of each angle in an equilateral triangle?

24. Geometry An isosceles triangle is a triangle in which two sides are equal in length. The angle between the two equal sides is called the vertex angle, while the other two angles are called the base angles. If the vertex angle is 40°, what is the measure of the base angles?

Problems 25 through 30 refer to right triangle ABC with $C = 90°$.

25. If $a = 4$ and $b = 3$, find c.   **26.** If $a = 6$ and $b = 8$, find c.

27. If $a = 8$ and $c = 17$, find b.   **28.** If $a = 2$ and $c = 6$, find b.

29. If $b = 12$ and $c = 13$, find a.   **30.** If $b = 10$ and $c = 26$, find a.

Solve for x in each of the following right triangles:

31.   **32.** 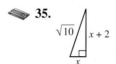   **33.**

34. 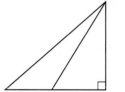   **35.**   **36.**

Problems 37 and 38 refer to Figure 21.

37. Find AB if $BC = 4$, $BD = 5$, and $AD = 2$.

38. Find BD if $BC = 5$, $AB = 13$, and $AD = 4$.

Problems 39 and 40 refer to Figure 22, which shows a circle with center at C and a radius of r, and right triangle ADC.

39. Find r if $AB = 4$ and $AD = 8$.

40. Find r if $AB = 8$ and $AD = 12$.

41. Pythagorean Theorem The roof of a house is to extend up 13.5 feet above the ceiling, which is 36 feet across (Figure 23). Find the length of one side of the roof.

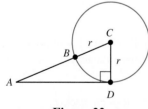

Figure 21

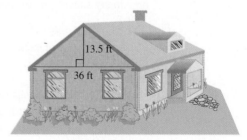

Figure 23

42. Surveying A surveyor is attempting to find the distance across a pond. From a point on one side of the pond he walks 25 yards to the end of the pond and then makes a 90° turn and walks another 60 yards before coming to a point directly across the pond from the point at which he started. What is the distance across the pond? (See Figure 24.)

Figure 22

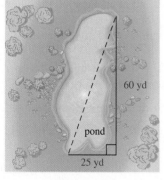

Figure 24

Find the remaining sides of a 30°–60°–90° triangle if

43. the shortest side is 1.

44. the shortest side is 3.

45. the longest side is 8.

46. the longest side is 5.

47. the side opposite 60° is 6.

48. the side opposite 60° is 4.

49. Escalator An escalator in a department store is to carry people a vertical distance of 20 feet between floors. How long is the escalator if it makes an angle of 30° with the ground?

50. Escalator What is the length of the escalator in Problem 49 if it makes an angle of 60° with the ground?

51. Tent Design A two-person tent is to be made so that the height at the center is 4 feet. If the sides of the tent are to meet the ground at an angle of 60°, and the tent is to be 6 feet in length, how many square feet of material will be needed to make the tent? (Figure 25; assume that the tent has a floor and is closed at both ends, and give your answer to the nearest tenth of a square foot.)

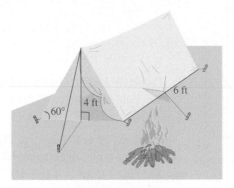

Figure 25

52. Tent Design If the height at the center of the tent in Problem 51 is to be 3 feet, how many square feet of material will be needed to make the tent?

Find the remaining sides of a 45°–45°–90° triangle if

53. the shorter sides are each $\frac{4}{5}$.

54. the shorter sides are each $\frac{1}{2}$.

55. the longest side is $8\sqrt{2}$.

56. the longest side is $5\sqrt{2}$.

57. the longest side is 4.

58. the longest side is 12.

59. Distance a Bullet Travels A bullet is fired into the air at an angle of 45°. How far does it travel before it is 1,000 feet above the ground? (Assume that the bullet travels in a straight line, neglect the forces of gravity, and give your answer to the nearest foot.)

60. Time a Bullet Travels If the bullet in Problem 59 is traveling at 2,828 feet per second, how long does it take for the bullet to reach a height of 1,000 feet?

Geometry: Characteristics of a Cube The object shown in Figure 26 is a cube (all edges are equal in length).

61. If the length of each edge of the cube shown in Figure 26 is 1 inch, find
  a. the length of diagonal *CH*.
  b. the length of diagonal *CF*.

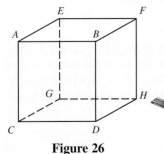

Figure 26

62. If the length of each edge of the cube shown in Figure 26 is 5 centimeters, find
   a. the length of diagonal *GD*.
   b. the length of diagonal *GB*.

63. If the length of each edge of the cube shown in Figure 26 is unknown, we can represent it with the variable *x*. Then we can write formulas for the lengths of any of the diagonals. Finish each of the following statements:
   a. If the length of each edge of a cube is *x*, then the length of the diagonal of any face of the cube will be _____.
   b. If the length of each edge of a cube is *x*, then the length of any diagonal that passes through the center of the cube will be _____.

64. What is the measure of ∠*GDH* in Figure 26?

65. Suppose the length of diagonal *CF* in Figure 26 is 3 feet. What is the length of edge *CD*?

66. Suppose the length of diagonal *CF* in Figure 26 is $5\sqrt{3}$ feet. What is the length of edge *CD*?

EXTENDING THE CONCEPTS

67. The Spiral of Roots The introduction to this chapter shows the Spiral of Roots. The following three figures (Figures 27, 28, and 29) show the first three stages in the construction of the Spiral of Roots. Using graph paper and a ruler, construct the Spiral of Roots, labeling each diagonal as you draw it, to the point where you can see a line segment with a length of $\sqrt{10}$.

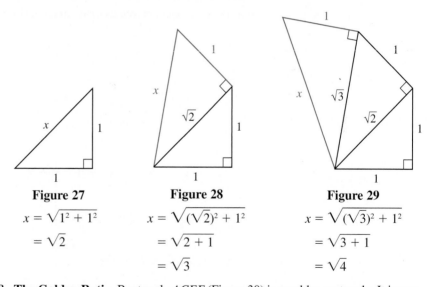

Figure 27	Figure 28	Figure 29
$x = \sqrt{1^2 + 1^2}$	$x = \sqrt{(\sqrt{2})^2 + 1^2}$	$x = \sqrt{(\sqrt{3})^2 + 1^2}$
$= \sqrt{2}$	$= \sqrt{2 + 1}$	$= \sqrt{3 + 1}$
	$= \sqrt{3}$	$= \sqrt{4}$

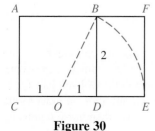

Figure 30

68. The Golden Ratio Rectangle *ACEF* (Figure 30) is a golden rectangle. It is constructed from square *ACDB* by holding line segment *OB* fixed at point *O* and then letting point *B* drop down until *OB* aligns with *CD*. The ratio of the length to the width in the golden rectangle is called the *golden ratio*. Find the lengths below to arrive at the golden ratio.
   a. Find the length of *OB*.
   b. Find the length of *OE*.
   c. Find the length of *CE*.
   d. Find the ratio $\dfrac{CE}{EF}$.

SECTION 1.2 | THE RECTANGULAR COORDINATE SYSTEM

René Descartes

The book *The Closing of the American Mind* by Allan Bloom was published in 1987 and spent many weeks on the bestseller list. In the book, Mr. Bloom recalls being in a restaurant in France and overhearing a waiter call another waiter a "Cartesian." He goes on to say that French people today define themselves in terms of the philosophy of either René Descartes (1595–1650) or Blaise Pascal (1623–1662). Followers of Descartes are sometimes referred to as *Cartesians*. As a philosopher, Descartes is responsible for the statement "I think, therefore I am." In mathematics, Descartes is credited with, among other things, the invention of the rectangular coordinate system, which we sometimes call the *Cartesian coordinate system*. Until Descartes invented his coordinate system in 1637, algebra and geometry were treated as separate subjects. The rectangular coordinate system allows us to connect algebra and geometry by associating geometric shapes with algebraic equations. For example, every nonvertical straight line (a geometric concept) can be paired with an equation of the form $y = mx + b$ (an algebraic concept), where m and b are real numbers, and x and y are variables that we associate with the axes of a coordinate system. In this section we will review some of the concepts developed around the rectangular coordinate system and graphing in two dimensions.

The rectangular (or Cartesian) coordinate system is constructed by drawing two number lines perpendicular to each other. The horizontal number line is called the *x-axis,* and the vertical number line is called the *y-axis*. Their point of intersection is called the *origin*. The axes divide the plane into four *quadrants* that are numbered I through IV in a counterclockwise direction (Figure 1).

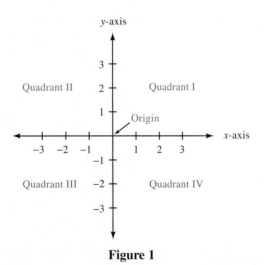

Figure 1

Among other things, the rectangular coordinate system is used to graph ordered pairs (a, b). For the ordered pair (a, b), a is called the *x-coordinate* (or *x*-component), and b is called the *y-coordinate* (or *y*-component). To graph the ordered pair (a, b) on a rectangular coordinate system, we start at the origin and move a units to the right or left (right if a is positive and left if a is negative). We then move b units up or down (up if b is positive and down if b is negative). The point where we end up after these two moves is the graph of the ordered pair (a, b).

EXAMPLE 1 Graph the ordered pairs $(1, 5)$, $(-2, 4)$, $(-3, -2)$, $(\frac{1}{2}, -4)$, $(3, 0)$, and $(0, -2)$.

SOLUTION To graph the ordered pair $(1, 5)$, we start at the origin and move 1 unit to the right and then 5 units up. We are now at the point whose coordinates are $(1, 5)$. The other ordered pairs are graphed in a similar manner, as in Figure 2.

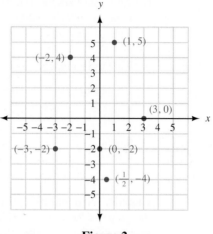

Figure 2

Looking at Figure 2, we see that any point in quadrant I will have both coordinates positive; that is, $(+, +)$. In quadrant II, the form is $(-, +)$. In quadrant III, the form is $(-, -)$, and in quadrant IV it is $(+, -)$. Also, any point on the x-axis will have a y-coordinate of 0 (it has no vertical displacement), and any point on the y-axis will have an x-coordinate of 0 (no horizontal displacement). A summary of this is given in Figure 3.

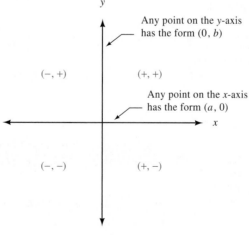

Figure 3

GRAPHING LINES

 **EXAMPLE 2** Graph the line $2x - 3y = 6$.

SOLUTION Since a line is determined by two points, we can graph the line $2x - 3y = 6$ by first finding two ordered pairs that satisfy its equation. To find ordered pairs that satisfy the equation $2x - 3y = 6$, we substitute any convenient number for either variable and then solve the equation that results for the corresponding value of the other variable.

If we let	$x = 0$	If	$y = 0$
the equation	$2x - 3y = 6$	the equation	$2x - 3y = 6$
becomes	$2(0) - 3y = 6$	becomes	$2x - 3(0) = 6$
	$-3y = 6$		$2x = 6$
	$y = -2$		$x = 3$

This gives us $(0, -2)$ as one solution to $2x - 3y = 6$.

This gives us $(3, 0)$ as a second solution.

Graphing the points $(0, -2)$ and $(3, 0)$ and then drawing a line through them, we have the graph of $2x - 3y = 6$ (Figure 4). The x-coordinate of the point where our graph crosses the x-axis is called the x-intercept; in this case, it is 3. The y-coordinate of the point where the graph crosses the y-axis is the y-intercept; in this case, -2.

NOTE Example 2 illustrates the connection between algebra and geometry that we mentioned in the introduction to this section. The rectangular coordinate system allows us to associate the equation $2x - 3y = 6$ (an algebraic concept) with a specific straight line (a geometric concept). The study of the relationship between equations in algebra and their associated geometric figures is called *analytic geometry* and is based on the coordinate system credited to Descartes.

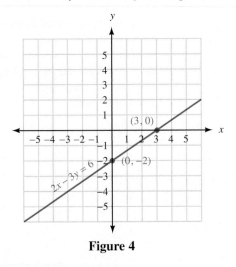

Figure 4

The 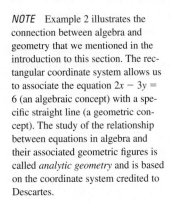 icon next to a Using Technology segment or example indicates that step-by-step calculator instructions are available on the Digital Video Companion CD-ROM.

USING TECHNOLOGY

To check the graph from Example 2 on a graphing calculator, we would first need to isolate the variable y:

$$y = \frac{2}{3}x - 2$$

Define this function as $Y_1 = (2/3)x - 2$. To match the graph shown in Figure 4, we can set the window variables so that $-5 \leq x \leq 5$ and $-5 \leq y \leq 5$. (By this, we mean that Xmin = -5, Xmax = 5, Ymin = -5, and Ymax = 5. We will assume that the scales for both axes, Xscl and Yscl, are set to 1 unless noted otherwise.)

Most graphing calculators have a command that will allow you to evaluate a function from the graph for a given value of x. We can use that command to verify the intercepts. Use the command first with $x = 3$ to check the x-intercept, and then again with $x = 0$ to check the y-intercept (Figure 5).

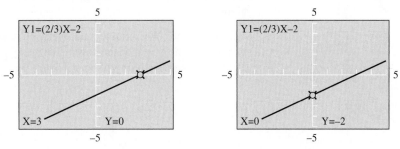

Figure 5

NOTE   There are many different models of graphing calculators, and each model has its own set of commands. For example, to evaluate a function on a TI-83 we would press 2nd CALC and select the **value** command. On a TI-86, it is the **EVAL** command found in the GRAPH menu. Because we have no way of knowing which model of calculator you are working with, we will generally avoid providing specific key icons or command names throughout the remainder of this book. Check your calculator manual to find the appropriate command for your particular model.

GRAPHING PARABOLAS

Recall from your algebra classes that any parabola that opens up or down can be described by an equation of the form

$$y = a(x - h)^2 + k$$

Likewise, any equation of this form will have a graph that is a parabola. The highest or lowest point on the parabola is called the vertex. The coordinates of the vertex are (h, k). Figure 6 illustrates this fact. The coefficient a stretches or shrinks the graph of the parabola. If $|a| < 1$, the graph is compressed toward the x-axis so that it is wider than the graph of $y = x^2$. If $|a| > 1$, the graph is stretched away from the x-axis and is narrower than the graph of $y = x^2$. Figure 7 illustrates this fact.

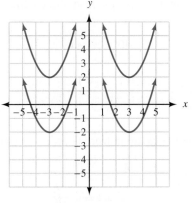

Figure 6

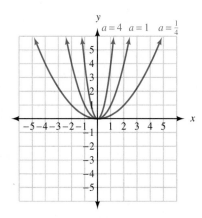

Figure 7

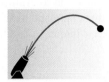

EXAMPLE 3 At the 1997 Washington County Fair in Oregon, David Smith, Jr., The Bullet, was shot from a cannon. As a human cannonball, he reached a height of 70 feet before landing in a net 160 feet from the cannon. Sketch the graph of his path, and then find the equation of the graph.

SOLUTION We assume that the path taken by the human cannonball is a parabola. If the origin of the coordinate system is at the opening of the cannon, then the net that catches him will be at 160 on the x-axis. Here is the graph:

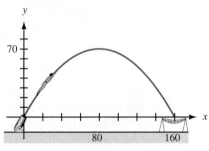

Figure 8

Since the curve is a parabola we know the equation will have the form

$$y = a(x - h)^2 + k$$

Since the vertex of the parabola is at (80, 70), we can fill in two of the three constants in our equation, giving us

$$y = a(x - 80)^2 + 70$$

To find a we note that the landing point will be (160, 0). Substituting the coordinates of this point into the equation, we solve for a.

$$0 = a(160 - 80)^2 + 70$$

$$0 = a(80)^2 + 70$$

$$0 = 6400a + 70$$

$$a = -\frac{70}{6400} = -\frac{7}{640}$$

The equation that describes the path of the human cannonball is

$$y = -\frac{7}{640}(x - 80)^2 + 70 \text{ for } 0 \le x \le 160$$

USING TECHNOLOGY

To verify that the equation from Example 3 is correct, we can graph the parabola and check the vertex and the x-intercepts. Graph the equation using the window settings shown below.

$$0 \le x \le 180, \text{ scale} = 20; 0 \le y \le 80, \text{ scale} = 10$$

Use the appropriate command on your calculator to find the maximum point on the graph (Figure 9), which is the vertex. Then evaluate the function at $x = 0$ and again at $x = 160$ to verify the x-intercepts (Figure 10).

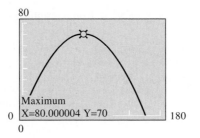

Figure 9   Figure 10

GRAPHING CIRCLES

The graph of an equation of the form

$$x^2 + y^2 = r^2$$

will be a circle with its center at the origin and a radius of r.

EXAMPLE 4 Graph each of the following circles:

a. $x^2 + y^2 = 16$ **b.** $x^2 + y^2 = 1$

SOLUTION   The graph of each equation will be a circle with its center at the origin. The first circle will have a radius of 4, and the second one will have a radius of 1. The graphs are shown in Figures 11 and 12.

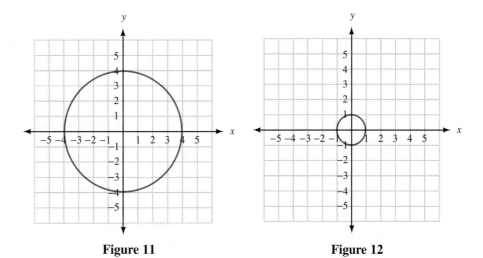

Figure 11   Figure 12

The circle shown in Figure 12 is called the *unit circle* because its radius is 1. As you will see, it will be an important part of one of the definitions we will give in Chapter 3.

🔅 — USING TECHNOLOGY →

```
Plot1 Plot2 Plot3
\Y1=√(1–X²)
\Y2=–√(1–X²)
\Y3=
\Y4=
\Y5=
\Y6=
\Y7=
```

Figure 13

To graph the circle $x^2 + y^2 = 1$ on a graphing calculator, we must first isolate y:

$$y^2 = 1 - x^2$$
$$y = \pm\sqrt{1 - x^2}$$

Because of the $\pm$ sign, we see that a circle is really the union of two separate functions. The positive square root represents the top half of the circle and the negative square root represents the bottom half. Define these functions separately (Figure 13).

When you graph both functions, you will probably see a circle that is not quite round and has gaps near the x-axis (Figure 14). Getting rid of the gaps can be tricky, and it requires a good understanding of how your calculator works. However, we can easily make the circle appear round using the zoom-square command (Figure 15).

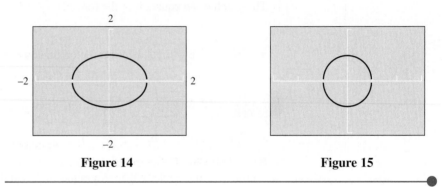

Figure 14   **Figure 15**

THE DISTANCE FORMULA

Our next definition gives us a formula for finding the distance between any two points on the coordinate system.

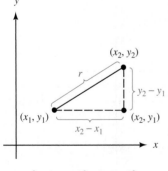

$$r^2 = (x_2 - x_1)^2 + (y_2 - y_1)^2$$

$$r = \sqrt{(x_2 - x_1)^2 + (y_2 - y_1)^2}$$

Figure 16

THE DISTANCE FORMULA

The distance between any two points (x_1, y_1) and (x_2, y_2) in a rectangular coordinate system is given by the formula

$$r = \sqrt{(x_2 - x_1)^2 + (y_2 - y_1)^2}$$

The distance formula can be derived by applying the Pythagorean Theorem to the right triangle in Figure 16.

EXAMPLE 5 Find the distance between the points (2, 1) and (−1, 5).

SOLUTION It makes no difference which of the points we call (x_1, y_1) and which we call (x_2, y_2) because this distance will be the same between the two points regardless (Figure 17).

$$r = \sqrt{(2 + 1)^2 + (1 - 5)^2}$$
$$= \sqrt{3^2 + (-4)^2}$$
$$= \sqrt{9 + 16}$$
$$= \sqrt{25}$$
$$= 5$$

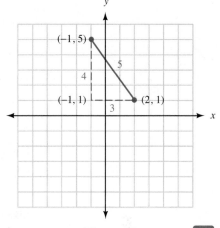

Figure 17

EXAMPLE 6 Find the distance from the origin to the point (x, y).

SOLUTION The coordinates of the origin are (0, 0). As shown in Figure 18, applying the distance formula, we have

$$r = \sqrt{(x - 0)^2 + (y - 0)^2}$$
$$= \sqrt{x^2 + y^2}$$

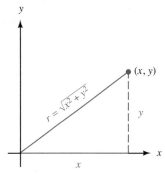

Figure 18

ANGLES IN STANDARD POSITION

DEFINITION

An angle is said to be in *standard position* if its initial side is along the positive *x*-axis and its vertex is at the origin.

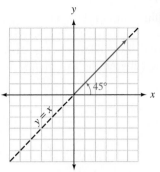

Figure 19

EXAMPLE 7 Draw an angle of 45° in standard position and find a point on the terminal side.

SOLUTION If we draw 45° in standard position, we see that the terminal side is along the line $y = x$ in quadrant I (Figure 19). Since the terminal side of 45° lies along the line $y = x$ in the first quadrant, any point on the terminal side will have positive coordinates that satisfy the equation $y = x$. Here are some of the points that do just that.

$$(1, 1) \qquad (2, 2) \qquad (3, 3) \qquad (\sqrt{2}, \sqrt{2}) \qquad \left(\frac{1}{2}, \frac{1}{2}\right) \qquad \left(\frac{7}{8}, \frac{7}{8}\right)$$

VOCABULARY

If angle θ is in standard position and the terminal side of θ lies in quadrant I, then we say θ lies in quadrant I and we abbreviate it like this:

$$\theta \in \text{QI}$$

Likewise, $\theta \in \text{QII}$ means θ is in standard position with its terminal side in quadrant II.

If the terminal side of an angle in standard position lies along one of the axes, then that angle is called a *quadrantal angle*. For example, an angle of 90° drawn in standard position would be a quadrantal angle, since the terminal side would lie along the positive y-axis. Likewise, 270° in standard position is a quadrantal angle because the terminal side would lie along the negative y-axis (Figure 20).

Two angles in standard position with the same terminal side are called *coterminal angles*. Figure 21 shows that 60° and −300° are coterminal angles when they are in standard position. Notice that these two angles differ by 360°. That is, 60° − (−300°) = 360°. Coterminal angles always differ from each other by some multiple of 360°.

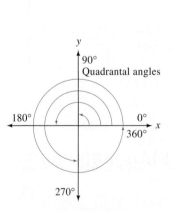

Figure 20

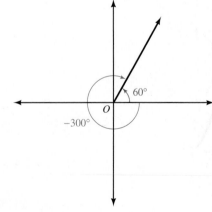

Figure 21

EXAMPLE 8 Draw $-90°$ in standard position, name some of the points on the terminal side, and find an angle between $0°$ and $360°$ that is coterminal with $-90°$.

SOLUTION Figure 22 shows $-90°$ in standard position. Since the terminal side is along the negative y-axis, the points $(0, -1)$, $(0, -2)$, $(0, -\frac{5}{2})$, and, in general, $(0, b)$ where b is a negative number, are all on the terminal side. The angle between $0°$ and $360°$ that is coterminal with $-90°$ is $270°$. ■

The diagram in Figure 23 shows some of the more common positive angles. We will see these angles often. Each angle is in standard position. Next to the terminal side of each angle is the degree measure of the angle. To simplify the diagram, we have made each of the terminal sides 1 unit long by drawing them out to the unit circle only.

Figure 22

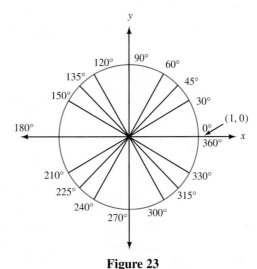

Figure 23

GETTING READY FOR CLASS

After reading through the preceding section, respond in your own words and in complete sentences.

a. Explain how you would construct a rectangular coordinate system from two real number lines.

b. Explain in words how you would graph the ordered pair $(3, -4)$.

c. How can you tell, by looking at the graph, if an ordered pair is a solution to an equation?

d. Explain how the distance formula and the Pythagorean Theorem are related.

PROBLEM SET 1.2

Graph each of the following ordered pairs on a rectangular coordinate system:

1. $(2, 4)$ **2.** $(2, -4)$ **3.** $(-2, 4)$ **4.** $(-2, -4)$

5. $(-3, -4)$ **6.** $(-4, -2)$ **7.** $(5, 0)$ **8.** $(-3, 0)$

9. $(0, -3)$ **10.** $(0, 2)$ **11.** $\left(-5, \dfrac{1}{2}\right)$ **12.** $\left(5, \dfrac{1}{2}\right)$

Graph each of the following lines:

13. $3x + 2y = 6$ **14.** $y = 2x - 1$ **15.** $y = \dfrac{1}{2}x$ **16.** $x = 3$

Graph each of the following parabolas:

17. $y = x^2 - 4$ **18.** $y = (x - 2)^2$

19. $y = (x + 2)^2 + 4$ **20.** $y = \dfrac{1}{4}(x + 2)^2 + 4$

21. Use your graphing calculator to graph $y = ax^2$ for $a = \frac{1}{10}, \frac{1}{5}, 1, 5,$ and 10. Copy all five graphs onto a single coordinate system and label each one. What happens to the shape of the parabola as the value of a gets close to zero? What happens to the shape of the parabola when the value of a gets large?

22. Use your graphing calculator to graph $y = ax^2$ for $a = \frac{1}{5}, 1,$ and 5, then again for $a = -\frac{1}{5}, -1,$ and -5. Copy all six graphs onto a single coordinate system and label each one. Explain how a negative value of a affects the parabola.

23. Use your graphing calculator to graph $y = (x - h)^2$ for $h = -3, 0,$ and 3. Copy all three graphs onto a single coordinate system and label each one. What happens to the position of the parabola when $h < 0$? What if $h > 0$?

24. Use your graphing calculator to graph $y = x^2 + k$ for $k = -3, 0,$ and 3. Copy all three graphs onto a single coordinate system and label each one. What happens to the position of the parabola when $k < 0$? What if $k > 0$?

Graph each of the following circles:

25. $x^2 + y^2 = 25$ **26.** $x^2 + y^2 = 36$

27. $x^2 + y^2 = 5$ **28.** $x^2 + y^2 = 6$

Graph the circle $x^2 + y^2 = 1$ with your graphing calculator. Use the feature on your calculator that allows you to evaluate a function from the graph to find the coordinates of all points on the circle that have the given x-coordinate. Write your answers as ordered pairs and round to four places past the decimal point when necessary.

29. $x = \dfrac{1}{2}$ **30.** $x = -\dfrac{1}{2}$ **31.** $x = \dfrac{\sqrt{2}}{2}$

32. $x = -\dfrac{\sqrt{2}}{2}$ **33.** $x = -\dfrac{\sqrt{3}}{2}$ **34.** $x = \dfrac{\sqrt{3}}{2}$

35. Use the graph of Problem 25 to name the points at which the line $x + y = 5$ will intersect the circle $x^2 + y^2 = 25$.

36. Use the graph of Problem 26 to name the points at which the line $x - y = 6$ will intersect the circle $x^2 + y^2 = 36$.

37. At what points will the line $y = x$ intersect the unit circle $x^2 + y^2 = 1$?

38. At what points will the line $y = -x$ intersect the unit circle $x^2 + y^2 = 1$?

Find the distance between the following points:

39. $(3, 7), (6, 3)$ **40.** $(4, 7), (8, 1)$ **41.** $(0, 12), (5, 0)$

42. $(-3, 0), (0, 4)$ **43.** $(-1, -2), (-10, 5)$ **44.** $(-3, 8), (-1, 6)$

45. Find the distance from the origin out to the point $(3, -4)$.

46. Find the distance from the origin out to the point $(12, -5)$.

47. Find x so the distance between $(x, 2)$ and $(1, 5)$ is $\sqrt{13}$.

48. Find y so the distance between $(7, y)$ and $(3, 3)$ is 5.

49. Pythagorean Theorem An airplane is approaching Los Angeles International Airport at an altitude of 2,640 feet. If the horizontal distance from the plane to the runway is 1.2 miles, use the Pythagorean Theorem to find the diagonal distance from the plane to the runway (Figure 24). (5,280 feet equals 1 mile.)

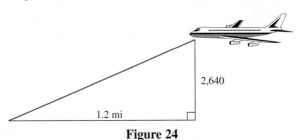

2,640

1.2 mi

Figure 24

50. Softball Diamond In softball, the distance from home plate to first base is 60 feet, as is the distance from first base to second base. If the lines joining home plate to first base and first base to second base form a right angle, how far does a catcher standing on home plate have to throw the ball so that it reaches the shortstop standing on second base (Figure 25)?

51. Softball and Rectangular Coordinates If a coordinate system is superimposed on the softball diamond in Problem 50 with the x-axis along the line from home plate to first base and the y-axis on the line from home plate to third base, what would be the coordinates of home plate, first base, second base, and third base?

52. Softball and Rectangular Coordinates If a coordinate system is superimposed on the softball diamond in Problem 50 with the origin on home plate and the positive x-axis along the line joining home plate to second base, what would be the coordinates of first base and third base?

53. In what two quadrants do all the points have negative x-coordinates?

54. In what two quadrants do all the points have negative y-coordinates?

55. For points (x, y) in quadrant I, the ratio x/y is always positive because x and y are always positive. In what other quadrant is the ratio x/y always positive?

56. For points (x, y) in quadrant II, the ratio x/y is always negative because x is negative and y is positive in quadrant II. In what other quadrant is the ratio x/y always negative?

57. Human Cannonball A human cannonball is shot from a cannon at the county fair. He reaches a height of 60 feet before landing in a net 160 feet from the cannon. Sketch the graph of his path, and then find the equation of the graph. Verify that your equation is correct using your graphing calculator.

58. Human Cannonball Referring to Problem 57, find the height reached by the human cannonball after he has traveled 30 feet horizontally, and after he has traveled 150 feet horizontally. Verify that your answers are correct using your graphing calculator.

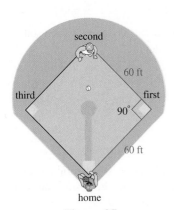

second

60 ft

third

first

90°

60 ft

home

Figure 25

Use the diagram in Figure 23 to help find the complement of each of the following angles.

59. 45°   **60.** 30°   **61.** 60°   **62.** 0°

Use the diagram in Figure 23 to help find the supplement of each of the following angles.

63. 120°   **64.** 150°   **65.** 90°   **66.** 45°

Use the diagram in Figure 23 to help you name an angle between 0° and 360° that is coterminal with each of the following angles.

67. −60°   **68.** −135°   **69.** −210°   **70.** −300°

Draw each of the following angles in standard position and then do the following:
a. Name a point on the terminal side of the angle.
b. Find the distance from the origin to that point.
c. Name another angle that is coterminal with the angle you have drawn.

71. 135°   **72.** 45°   **73.** 225°   **74.** 315°
75. 90°   **76.** 360°   **77.** −45°   **78.** −90°

79. Draw an angle of 30° in standard position. Then find a if the point $(a, 1)$ is on the terminal side of 30°.

80. Draw 60° in standard position. Then find b if the point $(2, b)$ is on the terminal side of 60°.

81. Draw an angle in standard position whose terminal side contains the point $(3, -2)$. Find the distance from the origin to this point.

82. Draw an angle in standard position whose terminal side contains the point $(2, -3)$. Find the distance from the origin to this point.

For problems 83 and 84, use the converse of the Pythagorean Theorem, which states that if $c^2 = a^2 + b^2$, then the triangle must be a right triangle.

83. Plot the points $(0, 0)$, $(5, 0)$, and $(5, 12)$ and show that, when connected, they are the vertices of a right triangle.

84. Plot the points $(0, 2)$, $(-3, 2)$, and $(-3, -2)$ and show that they form the vertices of a right triangle.

EXTENDING THE CONCEPTS

85. Descartes and Pascal In the introduction to this section we mentioned two French philosophers, Descartes and Pascal. Many people see the philosophies of the two men as being opposites. Why is this?

86. Pascal's Triangle Pascal has a triangular array of numbers named after him, Pascal's triangle. What part does Pascal's triangle play in the expansion of $(a + b)^n$, where n is a positive integer?

SECTION 1.3 | DEFINITION I: TRIGONOMETRIC FUNCTIONS

In this section we begin our work with trigonometry. The formal study of trigonometry dates back to the Greeks, when it was used mainly in the design of clocks and calendars and in navigation. The trigonometry of that period was spherical in nature, as it was based on measurement of arcs and chords associated with spheres. Unlike the trigonometry of the Greeks, our introduction to trigonometry takes place on a rectangular coordinate system. It concerns itself with angles, line segments, and points in the plane.

The definition of the trigonometric functions that begins this section is one of three definitions we will use. For us, it is the most important definition in the book. What should you do with it? Memorize it. Remember, in mathematics, definitions are simply accepted. That is, unlike theorems, there is no proof associated with a definition; we simply accept them exactly as they are written, memorize them, and then use them. When you are finished with this section, be sure you have memorized this first definition. It is the most valuable thing you can do for yourself at this point in your study of trigonometry.

DEFINITION I

If θ is an angle in standard position, and the point (x, y) is any point on the terminal side of θ other than the origin, then the six trigonometric functions of angle θ are defined as follows:

Function		Abbreviation		Definition
The sine of θ	$=$	$\sin\theta$	$=$	$\dfrac{y}{r}$
The cosine of θ	$=$	$\cos\theta$	$=$	$\dfrac{x}{r}$
The tangent of θ	$=$	$\tan\theta$	$=$	$\dfrac{y}{x}$
The cotangent of θ	$=$	$\cot\theta$	$=$	$\dfrac{x}{y}$
The secant of θ	$=$	$\sec\theta$	$=$	$\dfrac{r}{x}$
The cosecant of θ	$=$	$\csc\theta$	$=$	$\dfrac{r}{y}$

where $x^2 + y^2 = r^2$, or $r = \sqrt{x^2 + y^2}$. That is, r is the distance from the origin to (x, y).

As you can see, the six trigonometric functions are simply names given to the six possible ratios that can be made from the numbers x, y, and r as shown in Figure 1.

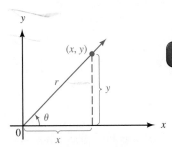

Figure 1

EXAMPLE 1 Find the six trigonometric functions of θ if θ is in standard position and the point $(-2, 3)$ is on the terminal side of θ.

SOLUTION We begin by making a diagram showing θ, $(-2, 3)$, and the distance r from the origin to $(-2, 3)$, as shown in Figure 2.

Applying the definition for the six trigonometric functions using the values $x = -2$, $y = 3$, and $r = \sqrt{13}$, we have

$$\sin\theta = \frac{y}{r} = \frac{3}{\sqrt{13}} \qquad \csc\theta = \frac{r}{y} = \frac{\sqrt{13}}{3}$$

$$\cos\theta = \frac{x}{r} = -\frac{2}{\sqrt{13}} \qquad \sec\theta = \frac{r}{x} = -\frac{\sqrt{13}}{2}$$

$$\tan\theta = \frac{y}{x} = -\frac{3}{2} \qquad \cot\theta = \frac{x}{y} = -\frac{2}{3}$$

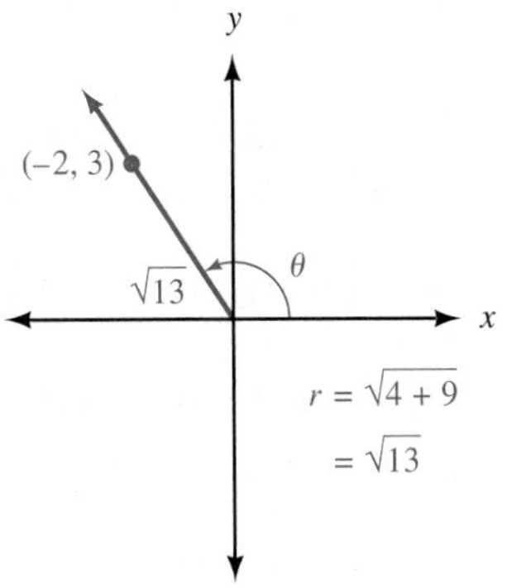

Figure 2

$(-2, 3)$

$\sqrt{13}$

θ

$r = \sqrt{4 + 9}$

$= \sqrt{13}$

NOTE In algebra, when we encounter expressions like $3/\sqrt{13}$ that contain a radical in the denominator, we usually rationalize the denominator; in this case, by multiplying the numerator and the denominator by $\sqrt{13}$.

$$\frac{3}{\sqrt{13}} = \frac{3}{\sqrt{13}} \cdot \frac{\sqrt{13}}{\sqrt{13}} = \frac{3\sqrt{13}}{13}$$

In trigonometry, it is sometimes convenient to use $3\sqrt{13}/13$, and at other times it is easier to use $3/\sqrt{13}$. For now, let's agree not to rationalize any denominators unless we are told to do so.

EXAMPLE 2 Find the sine and cosine of 45°.

SOLUTION According to the definition given earlier, we can find sin 45° and cos 45° if we know a point (x, y) on the terminal side of 45°, when 45° is in standard position. Figure 3 is a diagram of 45° in standard position. It shows that a convenient point on the terminal side of 45° is the point $(1, 1)$. (We say it is a convenient point because the coordinates are easy to work with.)

Since $x = 1$ and $y = 1$ and $r = \sqrt{x^2 + y^2}$, we have

$$r = \sqrt{1^2 + 1^2}$$
$$= \sqrt{2}$$

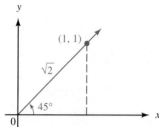

Figure 3

Substituting these values for x, y, and r into our definition for sine and cosine, we have

$$\sin 45° = \frac{y}{r} = \frac{1}{\sqrt{2}} \quad \text{and} \quad \cos 45° = \frac{x}{r} = \frac{1}{\sqrt{2}}$$

EXAMPLE 3 Find the six trigonometric functions of 270°.

SOLUTION Again, we need to find a point on the terminal side of 270°. From Figure 4, we see that the terminal side of 270° lies along the negative y-axis.

A convenient point on the terminal side of 270° is $(0, -1)$. Therefore,

$$r = \sqrt{x^2 + y^2}$$
$$= \sqrt{0^2 + (-1)^2}$$
$$= \sqrt{1}$$
$$= 1$$

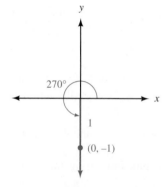

Figure 4

We have $x = 0$, $y = -1$, and $r = 1$. Here are the six trigonometric ratios for $\theta = 270°$.

$$\sin 270° = \frac{y}{r} = \frac{-1}{1} = -1 \qquad \csc 270° = \frac{r}{y} = \frac{1}{-1} = -1$$

$$\cos 270° = \frac{x}{r} = \frac{0}{1} = 0 \qquad \sec 270° = \frac{r}{x} = \frac{1}{0} = \text{undefined}$$

$$\tan 270° = \frac{y}{x} = \frac{-1}{0} = \text{undefined} \qquad \cot 270° = \frac{x}{y} = \frac{0}{-1} = 0$$

Note that tan 270° and sec 270° are undefined since division by 0 is undefined.

ALGEBRAIC SIGNS OF TRIGONOMETRIC FUNCTIONS

The algebraic sign, $+$ or $-$, of each of the six trigonometric functions will depend on the quadrant in which θ terminates. For example, in quadrant I all six trigonometric functions are positive since x, y, and r are all positive. In quadrant II, only $\sin \theta$ and $\csc \theta$ are positive since y and r are positive and x is negative. Table 1 shows the signs of all the ratios in each of the four quadrants.

TABLE 1

For θ in	QI	QII	QIII	QIV
$\sin \theta = \dfrac{y}{r}$ and $\csc \theta = \dfrac{r}{y}$	$+$	$+$	$-$	$-$
$\cos \theta = \dfrac{x}{r}$ and $\sec \theta = \dfrac{r}{x}$	$+$	$-$	$-$	$+$
$\tan \theta = \dfrac{y}{x}$ and $\cot \theta = \dfrac{x}{y}$	$+$	$-$	$+$	$-$

 EXAMPLE 4 If $\sin \theta = -5/13$, and θ terminates in quadrant III, find $\cos \theta$ and $\tan \theta$.

SOLUTION   Since $\sin \theta = -5/13$, we know the ratio of y to r, or y/r, is $-5/13$. We can let y be -5 and r be 13 and use these values of y and r to find x.

NOTE   We are not saying that if $y/r = -5/13$, then y *must* be -5 and r *must* be 13. We know from algebra that there are many pairs of numbers whose ratio is $-5/13$, not just -5 and 13. Our definition for sine and cosine, however, indicates we can choose *any* point on the terminal side of θ to find $\sin \theta$ and $\cos \theta$.

To find x, we use the fact that $x^2 + y^2 = r^2$.

$$x^2 + y^2 = r^2$$
$$x^2 + (-5)^2 = 13^2$$
$$x^2 + 25 = 169$$
$$x^2 = 144$$
$$x = \pm 12$$

Is x the number 12 or -12?

Since θ terminates in quadrant III, we know any point on its terminal side will have a negative x-coordinate; therefore,

$$x = -12$$

Using $x = -12$, $y = -5$, and $r = 13$ in our original definition, we have

$$\cos \theta = \frac{x}{r} = \frac{-12}{13} = -\frac{12}{13}$$

and

$$\tan \theta = \frac{y}{x} = \frac{-5}{-12} = \frac{5}{12}$$

As a final note, we should emphasize that the trigonometric functions of an angle are independent of the choice of the point (x, y) on the terminal side of the angle. Figure 5 shows an angle θ in standard position.

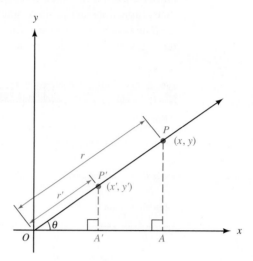

Figure 5

Points $P(x, y)$ and $P'(x', y')$ are both points on the terminal side of θ. Since triangles $P'OA'$ and POA are similar triangles, their corresponding sides are proportional. That is,

$$\sin \theta = \frac{y'}{r'} = \frac{y}{r} \qquad \cos \theta = \frac{x'}{r'} = \frac{x}{r} \qquad \tan \theta = \frac{y'}{x'} = \frac{y}{x}$$

GETTING READY FOR CLASS

After reading through the preceding section, respond in your own words and in complete sentences.

a. Find the six trigonometric functions of θ, if θ is an angle in standard position and the point (x, y) is a point on the terminal side of θ.

b. If r is the distance from the origin to the point (x, y), state the six ratios, or definitions, corresponding to the six trigonometric functions above.

c. Find the sine and cosine of $45°$.

d. Find the sine, cosine, and tangent of $270°$.

PROBLEM SET 1.3

Find all six trigonometric functions of θ if the given point is on the terminal side of θ. (In Problems 19 and 20, assume that a is a positive number.)

1. $(3, 4)$	**2.** $(-3, -4)$	**3.** $(-3, 4)$	**4.** $(3, -4)$
5. $(-5, 12)$	**6.** $(-12, 5)$	**7.** $(-1, -2)$	**8.** $(1, 2)$
9. (a, b)	**10.** (m, n)	**11.** $(-3, 0)$	**12.** $(0, -5)$
13. $(\sqrt{3}, -1)$	**14.** $(\sqrt{2}, \sqrt{2})$	**15.** $(-\sqrt{5}, 2)$	**16.** $(-3, \sqrt{7})$
17. $(60, 80)$	**18.** $(-80, 60)$	**19.** $(5a, -12a)$	**20.** $(-9a, 12a)$

In the diagrams below, angle θ is in standard position. In each case, find $\sin \theta$, $\cos \theta$, and $\tan \theta$.

21.

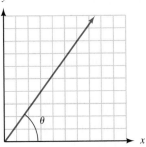

22.

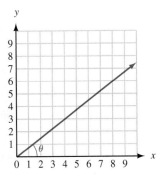

23.

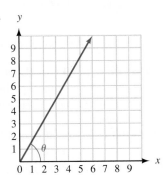

24.

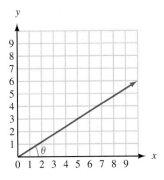

25. Use your calculator to find $\sin \theta$ and $\cos \theta$ if the point (9.36, 7.02) is on the terminal side of θ.

26. Use your calculator to find $\sin \theta$ and $\cos \theta$ if the point (6.36, 2.65) is on the terminal side of θ.

Draw each of the following angles in standard position, find a point on the terminal side, and then find the sine, cosine, and tangent of each angle:

27. $135°$ **28.** $225°$ **29.** $90°$ **30.** $180°$

31. $-45°$ **32.** $-90°$ **33.** $0°$ **34.** $-135°$

Indicate the quadrants in which the terminal side of θ must lie in order that

35. $\cos \theta$ is positive   **36.** $\cos \theta$ is negative

37. $\sin \theta$ is negative   **38.** $\tan \theta$ is negative

39. $\sin \theta$ is negative and $\tan \theta$ is positive

40. $\sin \theta$ is positive and $\cos \theta$ is negative

41. $\sin \theta$ and $\tan \theta$ have the same sign

42. $\cos \theta$ and $\cot \theta$ have the same sign

Find the remaining trigonometric functions of θ if

43. $\sin \theta = 12/13$ and θ terminates in QI

44. $\sin \theta = 12/13$ and θ terminates in QII

45. $\cos \theta = 24/25$ and θ terminates in QIV

46. $\cos \theta = 7/25$ and θ terminates in QIV

47. $\tan \theta = 3/4$ and θ terminates in QI

48. $\tan \theta = 12/5$ and θ terminates in QI

49. $\sin \theta = -20/29$ and θ terminates in QIII

50. $\cos \theta = -20/29$ and θ terminates in QII

51. $\sec \theta = 13/5$ and $\sin \theta < 0$

52. $\csc \theta = 13/5$ and $\cos \theta < 0$

53. $\cot \theta = 1/2$ and $\cos \theta > 0$

54. $\tan \theta = -1/2$ and $\sin \theta > 0$

55. $\cos \theta = \dfrac{\sqrt{3}}{2}$ and θ terminates in QIV

56. $\cos \theta = \dfrac{\sqrt{2}}{2}$ and θ terminates in QI

57. $\cot \theta = -2$ and θ terminates in QII

58. $\tan \theta = 2$ and θ terminates in QI

59. $\tan \theta = \dfrac{a}{b}$ where a and b are both positive

60. $\cot \theta = \dfrac{m}{n}$ where m and n are both positive

61. Find an angle θ for which $\sin \theta = 1$. (Look for an angle between $0°$ and $180°$.)

62. Find an angle θ in the first quadrant for which $\tan \theta = 1$.

63. Find an angle θ in the third quadrant for which $\tan \theta = 1$.

64. Find an angle θ between $0°$ and $360°$ for which $\cos \theta = -1$.

65. Find $\sin \theta$ and $\cos \theta$ if the terminal side of θ lies along the line $y = 2x$ in quadrant I.

66. Find $\sin \theta$ and $\cos \theta$ if the terminal side of θ lies along the line $y = 2x$ in quadrant III.

67. Find $\sin \theta$ and $\tan \theta$ if the terminal side of θ lies along the line $y = -3x$ in quadrant II.

68. Find $\sin \theta$ and $\tan \theta$ if the terminal side of θ lies along the line $y = -3x$ in quadrant IV.

69. Draw $45°$ and $-45°$ in standard position and then show that $\cos(-45°) = \cos 45°$.

70. Draw $45°$ and $-45°$ in standard position and then show that $\sin(-45°) = -\sin 45°$.

71. Find x if the point $(x, -3)$ is on the terminal side of θ and $\sin \theta = -3/5$.

72. Find y if the point $(5, y)$ is on the terminal side of θ and $\cos \theta = 5/13$.

SECTION 1.4 INTRODUCTION TO IDENTITIES

You may recall from the work you have done in algebra that an expression such as $\sqrt{x^2 + 9}$ cannot be simplified further because the square root of a sum is not equal to the sum of the square roots. (In other words, it would be a mistake to write $\sqrt{x^2 + 9}$ as $x + 3$.) However, expressions such as $\sqrt{x^2 + 9}$ occur frequently enough in mathematics that we would like to find expressions equivalent to them that do not contain square roots. As it turns out, the relationships that we develop in this section and the next are the key to rewriting expressions such as $\sqrt{x^2 + 9}$ in a more convenient form, which we will show in Section 1.5.

Before we begin our introduction to identities, we need to review some concepts from arithmetic and algebra.

In algebra, statements such as $2x = x + x$, $x^3 = x \cdot x \cdot x$, and $\dfrac{x}{4x} = \dfrac{1}{4}$ are called identities. They are identities because they are true for all replacements of the variable for which they are defined.

NOTE $\frac{x}{4x}$ is not equal to $\frac{1}{4}$ when x is 0. The statement $\frac{x}{4x} = \frac{1}{4}$ is still an identity, however, since it is true for all values of x for which $\frac{x}{4x}$ is defined.

The eight basic trigonometric identities we will work with in this section are all derived from our definition of the trigonometric functions. Since many trigonometric identities have more than one form, we will list the basic identity first and then give the most common equivalent forms of that identity.

RECIPROCAL IDENTITIES

Our definition for the sine and cosecant functions indicate that they are reciprocals; that is,

$$\csc \theta = \frac{1}{\sin \theta} \qquad \text{because} \qquad \frac{1}{\sin \theta} = \frac{1}{y/r} = \frac{r}{y} = \csc \theta$$

NOTE   We can also write this same relationship between $\sin \theta$ and $\csc \theta$ in another form as

$$\sin \theta = \frac{1}{\csc \theta} \qquad \text{because} \qquad \frac{1}{\csc \theta} = \frac{1}{r/y} = \frac{y}{r} = \sin \theta$$

The first identity we wrote, $\csc \theta = 1/\sin \theta$, is the basic identity. The second one, $\sin \theta = 1/\csc \theta$, is an equivalent form of the first.

From the discussion above and from the definition of $\cos \theta$, $\sec \theta$, $\tan \theta$, and $\cot \theta$, it is apparent that $\sec \theta$ is the reciprocal of $\cos \theta$, and $\cot \theta$ is the reciprocal of $\tan \theta$. Table 1 lists three basic reciprocal identities and their common equivalent forms.

TABLE 1 (MEMORIZE)	
Reciprocal identities	**Equivalent forms**
$\csc \theta = \dfrac{1}{\sin \theta}$	$\sin \theta = \dfrac{1}{\csc \theta}$
$\sec \theta = \dfrac{1}{\cos \theta}$	$\cos \theta = \dfrac{1}{\sec \theta}$
$\cot \theta = \dfrac{1}{\tan \theta}$	$\tan \theta = \dfrac{1}{\cot \theta}$

The examples that follow show some of the ways in which we use these reciprocal identities.

EXAMPLES

1. If $\sin \theta = \dfrac{3}{5}$, then $\csc \theta = \dfrac{5}{3}$, because

$$\csc \theta = \frac{1}{\sin \theta} = \frac{1}{3/5} = \frac{5}{3}$$

2. If $\cos \theta = -\dfrac{\sqrt{3}}{2}$, then $\sec \theta = -\dfrac{2}{\sqrt{3}}$.

(*Remember:* Reciprocals always have the same algebraic sign.)

3. If $\tan \theta = 2$, then $\cot \theta = \dfrac{1}{2}$.

4. If $\csc \theta = a$ then $\sin \theta = \dfrac{1}{a}$.

5. If $\sec \theta = 1$, then $\cos \theta = 1$ (1 is its own reciprocal).

6. If $\cot \theta = -1$, then $\tan \theta = -1$.

RATIO IDENTITIES

There are two ratio identities, one for $\tan \theta$ and one for $\cot \theta$.

TABLE 2 (MEMORIZE)		
Ratio identities		
$\tan \theta = \dfrac{\sin \theta}{\cos \theta}$	because	$\dfrac{\sin \theta}{\cos \theta} = \dfrac{y/r}{x/r} = \dfrac{y}{x} = \tan \theta$
$\cot \theta = \dfrac{\cos \theta}{\sin \theta}$	because	$\dfrac{\cos \theta}{\sin \theta} = \dfrac{x/r}{y/r} = \dfrac{x}{y} = \cot \theta$

EXAMPLE 7   If $\sin \theta = -3/5$ and $\cos \theta = 4/5$, find $\tan \theta$ and $\cot \theta$.

SOLUTION   Using the ratio identities, we have

$$\tan \theta = \frac{\sin \theta}{\cos \theta} = \frac{-3/5}{4/5} = -\frac{3}{4}$$

$$\cot \theta = \frac{\cos \theta}{\sin \theta} = \frac{4/5}{-3/5} = -\frac{4}{3}$$

NOTE   Once we found $\tan \theta$, we could have used a reciprocal identity to find $\cot \theta$.

$$\cot \theta = \frac{1}{\tan \theta} = \frac{1}{-3/4} = -\frac{4}{3}$$

NOTATION

The notation $\sin^2 \theta$ is a shorthand notation for $(\sin \theta)^2$. It indicates we are to square the number that is the sine of θ.

EXAMPLES

8. If $\sin \theta = \dfrac{3}{5}$, then $\sin^2 \theta = \left(\dfrac{3}{5}\right)^2 = \dfrac{9}{25}$.

9. If $\cos \theta = -\dfrac{1}{2}$, then $\cos^3 \theta = \left(-\dfrac{1}{2}\right)^3 = -\dfrac{1}{8}$.

PYTHAGOREAN IDENTITIES

To derive our first Pythagorean identity, we start with the relationship between x, y, and r as given in the definition of $\sin \theta$ and $\cos \theta$.

$$x^2 + y^2 = r^2$$

$$\frac{x^2}{r^2} + \frac{y^2}{r^2} = 1 \qquad \text{Divide through by } r^2$$

$$\left(\frac{x}{r}\right)^2 + \left(\frac{y}{r}\right)^2 = 1 \qquad \text{Property of exponents}$$

$$(\cos \theta)^2 + (\sin \theta)^2 = 1 \qquad \text{Definition of } \sin \theta \text{ and } \cos \theta$$

$$\cos^2\theta + \sin^2 \theta = 1 \qquad \text{Notation}$$

This last line is our first Pythagorean identity. We will use it many times throughout the book. It states that, for any angle θ, the sum of the squares of $\sin \theta$ and $\cos \theta$ is *always* 1.

There are two very useful equivalent forms of the first Pythagorean identity. One form occurs when we solve $\cos^2 \theta + \sin^2 \theta = 1$ for $\cos \theta$, and the other form is the result of solving for $\sin \theta$.

Solving for $\cos \theta$ we have

$$\cos^2 \theta + \sin^2 \theta = 1$$

$$\cos^2 \theta = 1 - \sin^2 \theta \qquad \text{Add } -\sin^2 \theta \text{ to both sides}$$

$$\cos \theta = \pm\sqrt{1 - \sin^2 \theta} \qquad \text{Take the square root of both sides}$$

Similarly, solving for $\sin \theta$ gives us

$$\sin^2 \theta = 1 - \cos^2 \theta$$

$$\sin \theta = \pm\sqrt{1 - \cos^2 \theta}$$

Our next Pythagorean identity is derived from the first Pythagorean identity, $\cos^2 \theta + \sin^2 \theta = 1$, by dividing both sides by $\cos^2 \theta$. Here is that derivation:

$$\cos^2 \theta + \sin^2 \theta = 1 \qquad \text{First Pythagorean identity}$$

$$\frac{\cos^2 \theta + \sin^2 \theta}{\cos^2 \theta} = \frac{1}{\cos^2 \theta} \qquad \text{Divide each side by } \cos^2 \theta$$

$$\frac{\cos^2 \theta}{\cos^2 \theta} + \frac{\sin^2 \theta}{\cos^2 \theta} = \frac{1}{\cos^2 \theta} \qquad \text{Write the left side as two fractions}$$

$$\left(\frac{\cos \theta}{\cos \theta}\right)^2 + \left(\frac{\sin \theta}{\cos \theta}\right)^2 = \left(\frac{1}{\cos \theta}\right)^2 \qquad \text{Property of exponents}$$

$$1 + \tan^2 \theta = \sec^2 \theta \qquad \text{Ratio and reciprocal identities}$$

This last expression, $1 + \tan^2 \theta = \sec^2 \theta$, is our second Pythagorean identity. To arrive at our third, and last, Pythagorean identity, we proceed as we have above, but instead of dividing each side by $\cos^2 \theta$, we divide by $\sin^2 \theta$. Without showing the work involved in doing so, the result is

$$1 + \cot^2 \theta = \csc^2 \theta$$

We summarize our derivations in Table 3.

TABLE 3 (MEMORIZE)	
Pythagorean identities	**Equivalent forms**
$\cos^2 \theta + \sin^2 \theta = 1$	$\cos \theta = \pm\sqrt{1 - \sin^2 \theta}$
	$\sin \theta = \pm\sqrt{1 - \cos^2 \theta}$
$1 + \tan^2 \theta = \sec^2 \theta$	
$1 + \cot^2 \theta = \csc^2 \theta$	

Notice the $\pm$ sign in the equivalent forms. It occurs as part of the process of taking the square root of both sides of the preceding equation. (*Remember:* We would obtain a similar result in algebra if we solved the equation $x^2 = 9$ to get $x = \pm 3$.) In Example 10 we will see how to deal with the $\pm$ sign.

 **EXAMPLE 10** If $\sin \theta = 3/5$ and θ terminates in quadrant II, find $\cos \theta$ and $\tan \theta$.

SOLUTION We begin by using the identity $\cos \theta = \pm\sqrt{1 - \sin^2 \theta}$.

$$\text{If} \qquad \sin \theta = \frac{3}{5}$$

$$\text{the identity} \qquad \cos \theta = \pm\sqrt{1 - \sin^2 \theta}$$

$$\text{becomes} \qquad \cos \theta = \pm\sqrt{1 - \left(\frac{3}{5}\right)^2}$$

$$= \pm\sqrt{1 - \frac{9}{25}}$$

$$= \pm\sqrt{\frac{16}{25}}$$

$$= \pm\frac{4}{5}$$

Now we know that $\cos \theta$ is either 4/5 or $-4/5$. Looking back to the original statement of the problem, however, we see that θ terminates in quadrant II; therefore, $\cos \theta$ must be negative.

$$\cos \theta = -\frac{4}{5}$$

To find $\tan \theta$, we use a ratio identity.

$$\tan \theta = \frac{\sin \theta}{\cos \theta}$$

$$= \frac{3/5}{-4/5}$$

$$= -\frac{3}{4} \quad \blacksquare$$

EXAMPLE 11 If $\cos \theta = 1/2$ and θ terminates in quadrant IV, find the remaining trigonometric ratios for θ.

SOLUTION   The first, and easiest, ratio to find is $\sec \theta$ because it is the reciprocal of $\cos \theta$.

$$\sec \theta = \frac{1}{\cos \theta} = \frac{1}{1/2} = 2$$

Next we find $\sin \theta$. Since θ terminates in QIV, $\sin \theta$ will be negative. Using one of the equivalent forms of our Pythagorean identity, we have

$$\sin \theta = -\sqrt{1 - \cos^2 \theta} \qquad \text{Negative sign because } \theta \in \text{QIV}$$

$$= -\sqrt{1 - \left(\frac{1}{2}\right)^2} \qquad \text{Substitute } \frac{1}{2} \text{ for } \cos \theta$$

$$= -\sqrt{1 - \frac{1}{4}} \qquad \text{Square } \frac{1}{2} \text{ to get } \frac{1}{4}$$

$$= -\sqrt{\frac{3}{4}} \qquad \text{Subtract}$$

$$= -\frac{\sqrt{3}}{2} \qquad \text{Take the square root of the numerator and denominator separately}$$

Now that we have $\sin \theta$ and $\cos \theta$, we can find $\tan \theta$ by using a ratio identity.

$$\tan \theta = \frac{\sin \theta}{\cos \theta} = \frac{-\sqrt{3}/2}{1/2} = -\sqrt{3}$$

Cot θ and csc θ are the reciprocals of $\tan \theta$ and $\sin \theta$, respectively. Therefore,

$$\cot \theta = \frac{1}{\tan \theta} = -\frac{1}{\sqrt{3}}$$

$$\csc \theta = \frac{1}{\sin \theta} = -\frac{2}{\sqrt{3}}$$

Here are all six ratios together:

$$\sin \theta = -\frac{\sqrt{3}}{2} \qquad \csc \theta = -\frac{2}{\sqrt{3}}$$

$$\cos \theta = \frac{1}{2} \qquad \sec \theta = 2$$

$$\tan \theta = -\sqrt{3} \qquad \cot \theta = -\frac{1}{\sqrt{3}}$$

As a final note, we should mention that the eight basic identities we have derived here, along with their equivalent forms, are very important in the study of trigonometry. It is essential that you memorize them. It may be a good idea to practice writing them from memory until you can write each of the eight, and their equivalent forms, perfectly. As time goes by, we will increase our list of identities, so you will want to keep up with them as we go along.

GETTING READY FOR CLASS

After reading through the preceding section, respond in your own words and in complete sentences.

a. State the reciprocal identities for csc θ, sec θ, and cot θ.

b. State the equivalent forms of the reciprocal identities for sin θ, cos θ, and tan θ.

c. State the ratio identities for tan θ and cot θ.

d. State the three Pythagorean identities.

PROBLEM SET 1.4

Give the reciprocal of each number.

1. 7   **2.** 4   **3.** $-2/3$ **4.** $-5/13$

5. $-1/\sqrt{2}$ **6.** $-\sqrt{3}/2$ **7.** x **8.** $1/a$

Use the reciprocal identities for the following problems.

9. If sin θ = 4/5, find csc θ.   **10.** If cos θ = $\sqrt{3}/2$, find sec θ.

11. If sec θ = -2, find cos θ.   **12.** If csc θ = $-13/12$, find sin θ.

13. If tan θ = a ($a \neq 0$), find cot θ.   **14.** If cot θ = $-b$ ($b \neq 0$), find tan θ.

Use a ratio identity to find tan θ if

15. $\sin \theta = \dfrac{3}{5}$ and $\cos \theta = -\dfrac{4}{5}$ **16.** $\sin \theta = \dfrac{2}{\sqrt{5}}$ and $\cos \theta = \dfrac{1}{\sqrt{5}}$

Use a ratio identity to find cot θ if

17. $\sin \theta = -\dfrac{5}{13}$ and $\cos \theta = -\dfrac{12}{13}$ **18.** $\sin \theta = \dfrac{2}{\sqrt{13}}$ and $\cos \theta = \dfrac{3}{\sqrt{13}}$

For Problems 19 through 22, recall that $\sin^2 \theta$ means $(\sin \theta)^2$.

19. If sin θ = $1/\sqrt{2}$, find $\sin^2 \theta$.   **20.** If cos θ = 1/3, find $\cos^2 \theta$.

21. If tan θ = 2, find $\tan^3 \theta$.   **22.** If sec θ = -3, find $\sec^3 \theta$.

For Problems 23 through 26, let sin θ = $-12/13$, and cos θ = $-5/13$, and find

23. tan θ **24.** cot θ **25.** sec θ **26.** csc θ

Use the equivalent forms of the first Pythagorean identity on Problems 27 through 36.

27. Find sin θ if $\cos \theta = \dfrac{3}{5}$ and θ terminates in QI.

28. Find sin θ if $\cos \theta = \dfrac{5}{13}$ and θ terminates in QI.

29. Find cos θ if $\sin \theta = \dfrac{1}{3}$ and θ terminates in QII.

30. Find cos θ if $\sin \theta = \dfrac{\sqrt{3}}{2}$ and θ terminates in QII.

31. If sin θ = $-4/5$ and θ terminates in QIII, find cos θ.

32. If sin θ = $-4/5$ and θ terminates in QIV, find cos θ.

33. If $\cos \theta = \sqrt{3}/2$ and θ terminates in QI, find $\sin \theta$.
34. If $\cos \theta = -1/2$ and θ terminates in QII, find $\sin \theta$.

35. If $\sin \theta = \dfrac{1}{\sqrt{5}}$ and $\theta \in$ QII, find $\cos \theta$.

36. If $\cos \theta = -\dfrac{1}{\sqrt{10}}$ and $\theta \in$ QIII, find $\sin \theta$.

37. Find $\tan \theta$ if $\sin \theta = 1/3$ and θ terminates in QI.
38. Find $\cot \theta$ if $\sin \theta = 2/3$ and θ terminates in QII.
39. Find $\sec \theta$ if $\tan \theta = 8/15$ and θ terminates in QIII.
40. Find $\csc \theta$ if $\cot \theta = -24/7$ and θ terminates in QIV.
41. Find $\csc \theta$ if $\cot \theta = -21/20$ and $\sin \theta > 0$.
42. Find $\sec \theta$ if $\tan \theta = 7/24$ and $\cos \theta < 0$.

Find the remaining trigonometric ratios of θ if

43. $\cos \theta = 12/13$ and θ terminates in QI
44. $\sin \theta = 12/13$ and θ terminates in QI
45. $\sin \theta = -1/2$ and θ is not in QIII
46. $\cos \theta = -1/3$ and θ is not in QII
47. $\sec \theta = 2$ and $\sin \theta$ is positive
48. $\csc \theta = 2$ and $\cos \theta$ is negative

49. $\cos \theta = \dfrac{2}{\sqrt{13}}$ and $\theta \in$ QIV

50. $\sin \theta = \dfrac{3}{\sqrt{10}}$ and $\theta \in$ QII

51. $\sec \theta = -3$ and $\theta \in$ QIII
52. $\sec \theta = -4$ and $\theta \in$ QII
53. $\csc \theta = a$ and θ terminates in QI
54. $\sec \theta = b$ and θ terminates in QI

Using your calculator and rounding your answers to the nearest hundredth, find the remaining trigonometric ratios of θ if

55. $\sin \theta = 0.23$ and $\theta \in$ QI
56. $\cos \theta = 0.51$ and $\theta \in$ QI
57. $\sec \theta = -1.24$ and $\theta \in$ QII
58. $\csc \theta = -2.45$ and $\theta \in$ QIII

Recall from algebra that the slope of the line through (x_1, y_1) and (x_2, y_2) is

$$m = \frac{y_2 - y_1}{x_2 - x_1}$$

It is the change in the y-coordinates divided by the change in the x-coordinates.
59. The line $y = 3x$ passes through the points $(0, 0)$ and $(1, 3)$. Find its slope.
60. Suppose the angle formed by the line $y = 3x$ and the positive x-axis is θ. Find the tangent of θ (Figure 1).
61. Find the slope of the line $y = mx$. [It passes through the origin and the point $(1, m)$.]
62. Find $\tan \theta$ if θ is the angle formed by the line $y = mx$ and the positive x-axis (Figure 2).

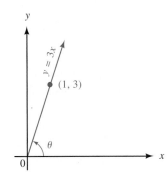

Figure 1

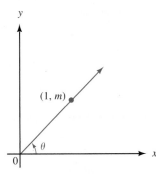

Figure 2

SECTION 1.5 | MORE ON IDENTITIES

The topics we will cover in this section are an extension of the work we did with identities in Section 1.4. We can use the eight basic identities we introduced in the previous section to rewrite or simplify certain expressions. It will be helpful at this point if you have already begun to memorize these identities.

The first topic involves writing any of the six trigonometric functions in terms of any of the others. Let's look at an example.

 EXAMPLE 1   Write $\tan \theta$ in terms of $\sin \theta$.

SOLUTION   When we say we want $\tan \theta$ written in terms of $\sin \theta$, we mean that we want to write an expression that is equivalent to $\tan \theta$ but involves no trigonometric function other than $\sin \theta$. Let's begin by using a ratio identity to write $\tan \theta$ in terms of $\sin \theta$ and $\cos \theta$.

$$\tan \theta = \frac{\sin \theta}{\cos \theta}$$

Now we need to replace $\cos \theta$ with an expression involving only $\sin \theta$. Since $\cos \theta = \pm\sqrt{1 - \sin^2 \theta}$

$$\tan \theta = \frac{\sin \theta}{\cos \theta}$$

$$= \frac{\sin \theta}{\pm\sqrt{1 - \sin^2 \theta}}$$

$$= \pm\frac{\sin \theta}{\sqrt{1 - \sin^2 \theta}}$$

This last expression is equivalent to $\tan \theta$ and is written in terms of $\sin \theta$ only. (In a problem like this, it is okay to include numbers and algebraic symbols with $\sin \theta$.) ■

Here is another example.

 EXAMPLE 2   Write $\sec \theta \tan \theta$ in terms of $\sin \theta$ and $\cos \theta$, and then simplify.

SOLUTION   Since $\sec \theta = 1/\cos \theta$ and $\tan \theta = \sin \theta/\cos \theta$, we have

$$\sec \theta \tan \theta = \frac{1}{\cos \theta} \cdot \frac{\sin \theta}{\cos \theta}$$

$$= \frac{\sin \theta}{\cos^2 \theta}$$ ■

The next examples show how we manipulate trigonometric expressions using algebraic techniques.

EXAMPLE 3 Add $\dfrac{1}{\sin \theta} + \dfrac{1}{\cos \theta}$.

SOLUTION   We can add these two expressions in the same way we would add $\frac{1}{3}$ and $\frac{1}{4}$—by first finding a least common denominator (LCD), and then writing each expression again with the LCD for its denominator.

$$\frac{1}{\sin \theta} + \frac{1}{\cos \theta} = \frac{1}{\sin \theta} \cdot \frac{\cos \theta}{\cos \theta} + \frac{1}{\cos \theta} \cdot \frac{\sin \theta}{\sin \theta} \qquad \text{The LCD is } \sin \theta \cos \theta$$

$$= \frac{\cos \theta}{\sin \theta \cos \theta} + \frac{\sin \theta}{\cos \theta \sin \theta}$$

$$= \frac{\cos \theta + \sin \theta}{\sin \theta \cos \theta} \quad \blacksquare$$

 **EXAMPLE 4** Multiply $(\sin \theta + 2)(\sin \theta - 5)$.

SOLUTION   We multiply these two expressions in the same way we would multiply $(x + 2)(x - 5)$. (In some algebra books this kind of multiplication is accomplished using the FOIL method.)

$$(\sin \theta + 2)(\sin \theta - 5) = \sin \theta \sin \theta - 5 \sin \theta + 2 \sin \theta - 10$$

$$= \sin^2 \theta - 3 \sin \theta - 10 \quad \blacksquare$$

In the introduction to Section 1.4, we mentioned that trigonometric identities are the key to writing the expression $\sqrt{x^2 + 9}$ without the square root symbol. Our next example shows how we do this using a trigonometric substitution.

EXAMPLE 5 Simplify the expression $\sqrt{x^2 + 9}$ as much as possible after substituting $3 \tan \theta$ for x.

SOLUTION   Our goal is to write the expression $\sqrt{x^2 + 9}$ without a square root by first making the substitution $x = 3 \tan \theta$.

If $\qquad\qquad\qquad x = 3 \tan \theta$

then the expression $\qquad \sqrt{x^2 + 9}$

becomes $\qquad\qquad \sqrt{(3 \tan \theta)^2 + 9} = \sqrt{9 \tan^2 \theta + 9}$

$$= \sqrt{9(\tan^2 \theta + 1)}$$

$$= \sqrt{9 \sec^2 \theta}$$

$$= 3 \left| \sec \theta \right| \quad \blacksquare$$

NOTE 1   We must use the absolute value symbol unless we know that $\sec \theta$ is positive. Remember, in algebra, $\sqrt{a^2} = a$ only when a is positive or zero. If it is possible that a is negative, then $\sqrt{a^2} = |a|$. In Section 4.6, we will see how to simplify our answer even further by removing the absolute value symbol.

NOTE 2   After reading through Example 5, you may be wondering if it is mathematically correct to make the substitution $x = 3 \tan \theta$. After all, x can be any real number because $x^2 + 9$ will always be positive. (*Remember:* We want to avoid taking the square root of a negative number.) How do we know that any real number can be written as $3 \tan \theta$? We will take care of this in Section 4.1 by showing that for any real number x, there is a value of θ between $-90°$ and $90°$ for which $x = 3 \tan \theta$.

In the examples that follow, we want to use the eight basic identities we developed in Section 1.4, along with some techniques from algebra, to show that some more complicated identities are true.

 **EXAMPLE 6**   Show that the following statement is true by transforming the left side into the right side.

$$\cos \theta \tan \theta = \sin \theta$$

SOLUTION   We begin by writing the left side in terms of $\sin \theta$ and $\cos \theta$.

$$\cos \theta \tan \theta = \cos \theta \cdot \frac{\sin \theta}{\cos \theta}$$

$$= \frac{\cos \theta \sin \theta}{\cos \theta}$$

$$= \sin \theta \qquad \text{Divide out the } \cos \theta \text{ common to the numerator and denominator}$$

Since we have succeeded in transforming the left side into the right side, we have shown that the statement $\cos \theta \tan \theta = \sin \theta$ is an identity. ■

 **EXAMPLE 7**   Prove the identity $(\sin \theta + \cos \theta)^2 = 1 + 2 \sin \theta \cos \theta$.

SOLUTION   Let's agree to prove the identities in this section, and the problem set that follows, by transforming the left side into the right side. In this case, we begin by expanding $(\sin \theta + \cos \theta)^2$. (Remember from algebra, $(a + b)^2 = a^2 + 2ab + b^2$.)

$$(\sin \theta + \cos \theta)^2 = \sin^2 \theta + 2 \sin \theta \cos \theta + \cos^2 \theta$$

Now we can rearrange the terms on the right side to get $\sin^2 \theta$ and $\cos^2 \theta$ together.

$$= (\sin^2 \theta + \cos^2 \theta) + 2 \sin \theta \cos \theta$$

$$= 1 + 2 \sin \theta \cos \theta \quad ■$$

We should mention that the ability to prove identities in trigonometry is not always obtained immediately. It usually requires a lot of practice. The more you work at it, the better you will become at it. In the meantime, if you are having trouble, check first to see that you have memorized the eight basic identities—reciprocal, ratio, Pythagorean, and their equivalent forms—as given in Section 1.4.

 GETTING READY FOR CLASS

After reading through the preceding section, respond in your own words and in complete sentences.

a. What do we mean when we want $\tan \theta$ written in terms of $\sin \theta$?

b. Write $\tan \theta$ in terms of $\sin \theta$.

c. What is the least common denominator for the expression $\dfrac{1}{\sin \theta} + \dfrac{1}{\cos \theta}$?

d. How would you prove the identity $\cos \theta \tan \theta = \sin \theta$?

PROBLEM SET 1.5

Write each of the following in terms of $\sin \theta$ only:

1. $\cos \theta$ **2.** $\csc \theta$ **3.** $\cot \theta$ **4.** $\sec \theta$

Write each of the following in terms of $\cos \theta$ only:

5. $\sec \theta$ **6.** $\sin \theta$ **7.** $\tan \theta$ **8.** $\csc \theta$

Write each of the following in terms of $\sin \theta$ and $\cos \theta$, and then simplify if possible:

9. $\csc \theta \cot \theta$ **10.** $\sec \theta \cot \theta$ **11.** $\csc \theta \tan \theta$ **12.** $\sec \theta \tan \theta \csc \theta$

13. $\dfrac{\sec \theta}{\csc \theta}$ **14.** $\dfrac{\csc \theta}{\sec \theta}$ **15.** $\dfrac{\sec \theta}{\tan \theta}$ **16.** $\dfrac{\csc \theta}{\cot \theta}$

17. $\dfrac{\tan \theta}{\cot \theta}$ **18.** $\dfrac{\cot \theta}{\tan \theta}$ **19.** $\dfrac{\sin \theta}{\csc \theta}$ **20.** $\dfrac{\cos \theta}{\sec \theta}$

21. $\tan \theta + \sec \theta$ **22.** $\cot \theta - \csc \theta$
23. $\sin \theta \cot \theta + \cos \theta$ **24.** $\cos \theta \tan \theta + \sin \theta$
25. $\sec \theta - \tan \theta \sin \theta$ **26.** $\csc \theta - \cot \theta \cos \theta$

Add and subtract as indicated. Then simplify your answers if possible. Leave all answers in terms of $\sin \theta$ and/or $\cos \theta$.

27. $\dfrac{\sin \theta}{\cos \theta} + \dfrac{1}{\sin \theta}$ **28.** $\dfrac{\cos \theta}{\sin \theta} + \dfrac{\sin \theta}{\cos \theta}$

29. $\dfrac{1}{\sin \theta} - \dfrac{1}{\cos \theta}$ **30.** $\dfrac{1}{\cos \theta} - \dfrac{1}{\sin \theta}$

31. $\sin \theta + \dfrac{1}{\cos \theta}$ **32.** $\cos \theta + \dfrac{1}{\sin \theta}$

33. $\dfrac{1}{\sin \theta} - \sin \theta$ **34.** $\dfrac{1}{\cos \theta} - \cos \theta$

Multiply.

35. $(\sin \theta + 4)(\sin \theta + 3)$ **36.** $(\cos \theta + 2)(\cos \theta - 5)$
37. $(2 \cos \theta + 3)(4 \cos \theta - 5)$ **38.** $(3 \sin \theta - 2)(5 \cos \theta - 4)$
39. $(1 - \sin \theta)(1 + \sin \theta)$ **40.** $(1 - \cos \theta)(1 + \cos \theta)$
41. $(1 - \tan \theta)(1 + \tan \theta)$ **42.** $(1 - \cot \theta)(1 + \cot \theta)$
43. $(\sin \theta - \cos \theta)^2$ **44.** $(\cos \theta + \sin \theta)^2$
45. $(\sin \theta - 4)^2$ **46.** $(\cos \theta - 2)^2$
47. Simplify the expression $\sqrt{x^2 + 4}$ as much as possible after substituting $2 \tan \theta$ for x.
48. Simplify the expression $\sqrt{x^2 + 1}$ as much as possible after substituting $\tan \theta$ for x.
49. Simplify the expression $\sqrt{9 - x^2}$ as much as possible after substituting $3 \sin \theta$ for x.
50. Simplify the expression $\sqrt{25 - x^2}$ as much as possible after substituting $5 \sin \theta$ for x.
51. Simplify the expression $\sqrt{x^2 - 36}$ as much as possible after substituting $6 \sec \theta$ for x.
52. Simplify the expression $\sqrt{x^2 - 64}$ as much as possible after substituting $8 \sec \theta$ for x.

53. Simplify the expression $\sqrt{4x^2 + 16}$ as much as possible after substituting $2 \tan \theta$ for x.

54. Simplify the expression $\sqrt{4x^2 + 100}$ as much as possible after substituting $5 \tan \theta$ for x.

55. Simplify the expression $\sqrt{36 - 9x^2}$ as much as possible after substituting $2 \sin \theta$ for x.

56. Simplify the expression $\sqrt{64 - 4x^2}$ as much as possible after substituting $4 \sin \theta$ for x.

57. Simplify the expression $\sqrt{9x^2 - 81}$ as much as possible after substituting $3 \sec \theta$ for x.

58. Simplify the expression $\sqrt{4x^2 - 144}$ as much as possible after substituting $6 \sec \theta$ for x.

Show that each of the following statements is an identity by transforming the left side of each one into the right side.

59. $\cos \theta \tan \theta = \sin \theta$

60. $\sin \theta \cot \theta = \cos \theta$

61. $\sin \theta \sec \theta \cot \theta = 1$

62. $\cos \theta \csc \theta \tan \theta = 1$

63. $\dfrac{\sin \theta}{\csc \theta} = \sin^2 \theta$

64. $\dfrac{\cos \theta}{\sec \theta} = \cos^2 \theta$

65. $\dfrac{\csc \theta}{\cot \theta} = \sec \theta$

66. $\dfrac{\sec \theta}{\tan \theta} = \csc \theta$

67. $\dfrac{\sec \theta}{\csc \theta} = \tan \theta$

68. $\dfrac{\csc \theta}{\sec \theta} = \cot \theta$

69. $\dfrac{\sec \theta \cot \theta}{\csc \theta} = 1$

70. $\dfrac{\csc \theta \tan \theta}{\sec \theta} = 1$

71. $\sin \theta \tan \theta + \cos \theta = \sec \theta$

72. $\cos \theta \cot \theta + \sin \theta = \csc \theta$

73. $\tan \theta + \cot \theta = \sec \theta \csc \theta$

74. $\tan^2 \theta + 1 = \sec^2 \theta$

75. $\csc \theta - \sin \theta = \dfrac{\cos^2 \theta}{\sin \theta}$

76. $\sec \theta - \cos \theta = \dfrac{\sin^2 \theta}{\cos \theta}$

77. $\csc \theta \tan \theta - \cos \theta = \dfrac{\sin^2 \theta}{\cos \theta}$

78. $\sec \theta \cot \theta - \sin \theta = \dfrac{\cos^2 \theta}{\sin \theta}$

79. $(1 - \cos \theta)(1 + \cos \theta) = \sin^2 \theta$

80. $(1 + \sin \theta)(1 - \sin \theta) = \cos^2 \theta$

81. $(\sin \theta + 1)(\sin \theta - 1) = -\cos^2 \theta$

82. $(\cos \theta + 1)(\cos \theta - 1) = -\sin^2 \theta$

83. $\dfrac{\cos \theta}{\sec \theta} + \dfrac{\sin \theta}{\csc \theta} = 1$

84. $1 - \dfrac{\sin \theta}{\csc \theta} = \cos^2 \theta$

85. $(\sin \theta - \cos \theta)^2 - 1 = -2 \sin \theta \cos \theta$

86. $(\cos \theta + \sin \theta)^2 - 1 = 2 \sin \theta \cos \theta$

87. $\sin \theta(\sec \theta + \csc \theta) = \tan \theta + 1$

88. $\sec \theta(\sin \theta + \cos \theta) = \tan \theta + 1$

89. $\sin \theta(\sec \theta + \cot \theta) = \tan \theta + \cos \theta$

90. $\cos \theta(\csc \theta + \tan \theta) = \cot \theta + \sin \theta$

91. $\sin \theta(\csc \theta - \sin \theta) = \cos^2 \theta$

92. $\cos \theta(\sec \theta - \cos \theta) = \sin^2 \theta$

CHAPTER 1	**SUMMARY**

EXAMPLES
We will use the margin to give examples of the topics being reviewed, whenever it is appropriate.

1.

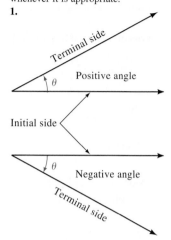

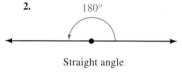

The number in brackets next to each heading indicates the section in which that topic is discussed.

Angles [1.1]

An angle is formed by two rays with a common end point. The common end point is called the *vertex,* and the rays are called *sides,* of the angle. If we think of an angle as being formed by rotating the initial side about the vertex to the terminal side, then a counterclockwise rotation gives a *positive angle,* and a clockwise rotation gives a *negative angle.*

2.   180°

Straight angle

90°

Right angle

Degree Measure [1.1]

There are 360° in a full rotation. This means that 1° is $\frac{1}{360}$ of a full rotation.

An angle that measures 90° is a *right angle.* An angle that measures 180° is a *straight angle.* Angles that measure between 0° and 90° are called *acute angles,* and angles that measure between 90° and 180° are called *obtuse angles.* *Complementary angles* have a sum of 90°, and *supplementary angles* have a sum of 180°.

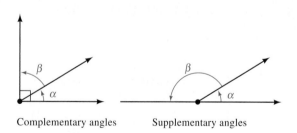

Complementary angles   Supplementary angles

3.   If *ABC* is a right triangle with $C = 90°$, and if $a = 4$ and $c = 5$, then

$$4^2 + b^2 = 5^2$$
$$16 + b^2 = 25$$
$$b^2 = 9$$
$$b = 3$$

Pythagorean Theorem [1.1]

In any right triangle, the square of the length of the longest side (the *hypotenuse*) is equal to the sum of the squares of the lengths of the other two sides (*legs*).

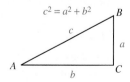

$$c^2 = a^2 + b^2$$

4. The distance between $(2, 7)$ and $(-1, 3)$ is

$$
\begin{aligned}
r &= \sqrt{(2 + 1)^2 + (7 - 3)^2} \\
&= \sqrt{9 + 16} \\
&= \sqrt{25} \\
&= 5
\end{aligned}
$$

Distance Formula [1.2]

The distance r between the points (x_1, y_1) and (x_2, y_2) is given by the formula

$$r = \sqrt{(x_2 - x_1)^2 + (y_2 - y_1)^2}$$

5. $135°$ in standard position is

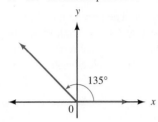

Standard Position for Angles [1.2]

An angle is said to be in standard position if its vertex is at the origin and its initial side is along the positive x-axis.

6. $60°$ and $-300°$ are coterminal angles.

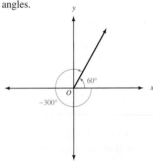

Coterminal Angles [1.2]

Two angles in standard position with the same terminal side are called *coterminal angles*. Coterminal angles always differ from each other by some multiple of $360°$.

7. If $(-3, 4)$ is on the terminal side of θ, then

$$r = \sqrt{9 + 16} = 5$$

and

$$
\sin \theta = \frac{4}{5} \qquad \csc \theta = \frac{5}{4}
$$

$$
\cos \theta = -\frac{3}{5} \qquad \sec \theta = -\frac{5}{3}
$$

$$
\tan \theta = -\frac{4}{3} \qquad \cot \theta = -\frac{3}{4}
$$

Trigonometric Functions [1.3]

If θ is an angle in standard position and (x, y) is any point on the terminal side of θ (other than the origin), then

$$
\sin \theta = \frac{y}{r} \qquad \csc \theta = \frac{r}{y}
$$

$$
\cos \theta = \frac{x}{r} \qquad \sec \theta = \frac{r}{x}
$$

$$
\tan \theta = \frac{y}{x} \qquad \cot \theta = \frac{x}{y}
$$

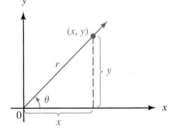

where $x^2 + y^2 = r^2$, or $r = \sqrt{x^2 + y^2}$. That is, r is the distance from the origin to (x, y).

8. If $\sin \theta > 0$, and $\cos \theta > 0$, then θ must terminate in QI.

If $\sin \theta > 0$, and $\cos \theta < 0$, then θ must terminate in QII.

If $\sin \theta < 0$, and $\cos \theta < 0$, then θ must terminate in QIII.

If $\sin \theta < 0$, and $\cos \theta > 0$, then θ must terminate in QIV.

Signs of the Trigonometric Functions [1.3]

The algebraic signs, $+$ or $-$, of the six trigonometric functions depend on the quadrant in which θ terminates.

For θ in	QI	QII	QIII	QIV
$\sin \theta$ and $\csc \theta$	+	+	−	−
$\cos \theta$ and $\sec \theta$	+	−	−	+
$\tan \theta$ and $\cot \theta$	+	−	+	−

9. If $\sin \theta = 1/2$ with θ in QI, then

$\cos \theta = \sqrt{1 - \sin^2 \theta}$

$\quad = \sqrt{1 - 1/4} = \dfrac{\sqrt{3}}{2}$

$\tan \theta = \dfrac{\sin \theta}{\cos \theta} = \dfrac{1/2}{\sqrt{3}/2} = \dfrac{1}{\sqrt{3}}$

$\cot \theta = \dfrac{1}{\tan \theta} = \sqrt{3}$

$\sec \theta = \dfrac{1}{\cos \theta} = \dfrac{2}{\sqrt{3}}$

$\csc \theta = \dfrac{1}{\sin \theta} = 2$

Basic Identities [1.4]

Reciprocal

$$\csc \theta = \frac{1}{\sin \theta} \qquad \sec \theta = \frac{1}{\cos \theta} \qquad \cot \theta = \frac{1}{\tan \theta}$$

Ratio

$$\tan \theta = \frac{\sin \theta}{\cos \theta} \qquad \cot \theta = \frac{\cos \theta}{\sin \theta}$$

Pythagorean

$$\cos^2 \theta + \sin^2 \theta = 1$$

$$1 + \tan^2 \theta = \sec^2 \theta$$

$$1 + \cot^2 \theta = \csc^2 \theta$$

CHAPTER 1 TEST

1. Find the complement and the supplement of 70°.

Solve for x in each of the following right triangles:

2.

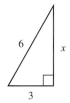

3.

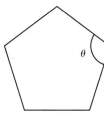

Figure 1

Problems 4 and 5 refer to Figure 1. (*Note: s* is the distance from A to D, and y is the distance from D to B.)

4. In order, find h, r, y, and x if $s = 5\sqrt{3}$.

5. In order, find x, h, s, and r if $y = 3$.

6. Figure 2 shows two right triangles drawn at 90° to one another. Find the length of DB if $DA = 6$, $AC = 5$, and $BC = 3$.

7. Through how many degrees does the hour hand of a clock move in 3 hours?

8. Find the remaining sides of a 30°–60°–90° triangle if the longest side is 5.

9. Escalator An escalator in a department store is to carry people a vertical distance of 15 feet between floors. How long is the escalator if it makes an angle of 45° with the ground?

10. Geometry Find the measure of one of the interior angles of a regular pentagon (Figure 3). (Try slicing the pentagon, like a pizza, into five congruent triangles.)

11. Geometry Find the measure of one of the interior angles of a regular hexagon (Figure 4).

Figure 2

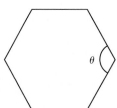

Figure 3

Figure 4

12. Graph the line $2x + 3y = 6$.
13. Find the distance between the points $(4, -2)$ and $(-1, 10)$.
14. Find the distance from the origin to the point (a, b).
15. Find x so that the distance between $(-2, 3)$ and $(x, 1)$ is $\sqrt{13}$.
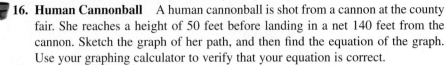 16. **Human Cannonball**   A human cannonball is shot from a cannon at the county fair. She reaches a height of 50 feet before landing in a net 140 feet from the cannon. Sketch the graph of her path, and then find the equation of the graph. Use your graphing calculator to verify that your equation is correct.

Find $\sin \theta$, $\cos \theta$, and $\tan \theta$ for each of the following values of θ:

17. $90°$
18. $-45°$

In which quadrant will θ lie if

19. $\sin \theta < 0$ and $\cos \theta > 0$
20. $\csc \theta > 0$ and $\cos \theta < 0$

Find all six trigonometric functions for θ if the given point lies on the terminal side of θ in standard position.

21. $(-6, 8)$
22. $(-3, -1)$

Find the remaining trigonometric functions of θ if

23. $\sin \theta = 1/2$ and θ terminates in QII
24. $\tan \theta = 12/5$ and θ terminates in QIII
25. Find $\sin \theta$ and $\cos \theta$ if the terminal side of θ lies along the line $y = -2x$ in quadrant IV.
26. If $\sin \theta = -3/4$, find $\csc \theta$.
27. If $\sec \theta = -2$, find $\cos \theta$.
28. If $\sin \theta = 1/3$, find $\sin^3 \theta$.
29. If $\sec \theta = 3$ with θ in QIV, find $\cos \theta$, $\sin \theta$, and $\tan \theta$.

30. If $\sin \theta = \dfrac{1}{a}$ with θ in QI, find $\cos \theta$, $\csc \theta$, and $\cot \theta$.

31. Multiply $(\sin \theta + 3)(\sin \theta - 7)$.
32. Expand and simplify $(\cos \theta - \sin \theta)^2$.

33. Subtract $\dfrac{1}{\sin \theta} - \sin \theta$.

34. Simplify the expression $\sqrt{4 - x^2}$ as much as possible after substituting $2 \sin \theta$ for x.

Show that each of the following statements is an identity by transforming the left side of each one into the right side.

35. $\dfrac{\cot \theta}{\csc \theta} = \cos \theta$

36. $\cot \theta + \tan \theta = \csc \theta \sec \theta$
37. $(1 - \sin \theta)(1 + \sin \theta) = \cos^2 \theta$
38. $\sin \theta(\csc \theta + \cot \theta) = 1 + \cos \theta$

CHAPTER **1** GROUP PROJECT

THE PYTHAGOREAN THEOREM

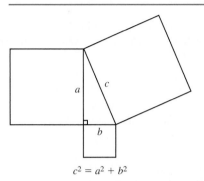

In a right angled triangle, the area of the square on the hypotenuse is the sum of the areas of the squares on the other two sides.

Theorem of Pythagoras

Objective: To prove the Pythagorean Theorem using several different geometric approaches.

$$c^2 = a^2 + b^2$$

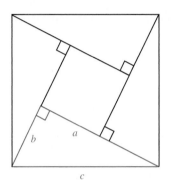

Figure 1

Although the Babylonians appear to have had an understanding of this relationship a thousand years before Pythagoras, his name is associated with the theorem because he was the first to provide a proof. In this project you will provide the details for several proofs of the theorem.

1. The diagram shown in Figure 1 was used by the Hindu mathematician Bhaskara to prove the theorem in the 12th century. His proof consisted only of the diagram and the word "Behold!" Use this figure to derive the Pythagorean Theorem. The right triangle with sides a, b, and c has been repeated three times. The area of the large square is equal to the sum of the areas of the four triangles and the area of the small square in the center. The key to this derivation is in finding the length of a side of the small square. Once you have the equation, simplify the right side to obtain the theorem.

2. The second proof, credited to former president James Garfield, is based on the diagram shown in Figure 2. Use this figure to derive the Pythagorean Theorem. The right triangle with sides a, b, and c has been repeated and a third triangle created by placing a line segment between points P and Q. The area of the entire figure (a trapezoid) is equal to the sum of the areas of the three triangles. Once you have the equation, simplify both sides and then isolate c^2 to obtain the theorem.

3. For the third "proof," consider the two squares shown in Figure 3. Both squares are of the same size, so their areas must be equal. Write a paragraph explaining how this diagram serves as a "visual proof" of the Pythagorean Theorem. The key to this argument is to make use of what the two squares have in common.

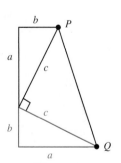

Figure 2

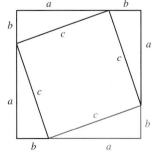
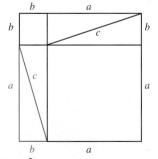

Figure 3

CHAPTER **1** RESEARCH PROJECT

PYTHAGORAS AND SHAKESPEARE

Pythagoras

Although Pythagoras preceded William Shakespeare by 2,000 years, the philosophy of the Pythagoreans is mentioned in Shakespeare's *The Merchant of Venice.* Here is a quote from that play:

> *Thou almost mak'st me waver in my faith,*
> *To hold opinion with Pythagoras,*
> *That souls of animals infuse themselves*
> *Into the trunks of men.*

Research the Pythagoreans. What were the main beliefs held by their society? What part of the philosophy of the Pythagoreans was Shakespeare referring to specifically with this quote? What present-day religions share a similar belief? Write a paragraph or two about your findings.

RIGHT TRIANGLE TRIGONOMETRY

*We are what we repeatedly do. Excellence, then,
is not an act, but a habit.*

Aristotle

INTRODUCTION

Most serious hikers are familiar with maps like the one shown in Figure 1. It is a *topographic map.* This particular map is of Bishop's Peak in San Luis Obispo, California.

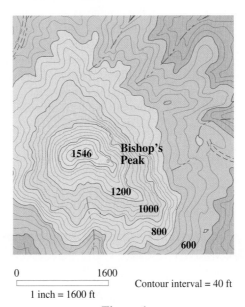

Contour interval = 40 ft
1 inch = 1600 ft

Figure 1

The curved lines on the map are called *contour lines;* they are used to show the changes in elevation of the land shown on the map. On this map, the change in elevation between any two contour lines is 40 feet, meaning that, if you were standing on one contour line and you were to hike to a position two contour lines away from your starting point, your elevation would change by 80 feet. In general, the closer the contour lines are together, the faster the land rises or falls. In this chapter, we use right triangle trigonometry to solve a variety of problems, some of which will involve topographic maps.

STUDY SKILLS FOR CHAPTER 2

The quote at the beginning of this chapter reinforces the idea that success in mathematics comes from practice. We practice trigonometry by working problems. Working problems is probably the single most important thing you can do to achieve excellence in trigonometry.

If you have successfully completed Chapter 1, then you have made a good start at developing the study skills necessary to succeed in all math classes. Here is the list of study skills for this chapter. Some are a continuation of the skills from Chapter 1, while others are new to this chapter.

1. **Continue to Set and Keep a Schedule**   Sometimes we find students do well in Chapter 1 and then become over-confident. They will spend less time with their homework. Don't do it. Keep to the same schedule.
2. **Continue to Memorize Definitions and Important Facts**   The important definitions in this chapter are those for trigonometric ratios in right triangles and exact values. We will point these out as we progress through the chapter.
3. **List Difficult Problems**   Begin to make lists of problems that give you the most difficulty. These are the problems that you are repeatedly making mistakes with. Try to spend a few minutes each day reworking some of the problems from your list. Once you are able to work these problems successfully, you will have much more confidence going into the next exam.
4. **Begin to Develop Confidence with Word Problems**   It seems that the main difference between people who are good at working word problems and those who are not, is confidence. People with confidence know that no matter how long it takes them, they will eventually be able to solve the problem they are working on. Those without confidence begin by saying to themselves, "I'll never be able to work this problem." If you are in the second category, then instead of telling yourself that you can't do word problems, that you don't like them, or that they're not good for anything anyway, decide to do whatever it takes to master them. As Aristotle said, it is practice that produces excellence. For us, practice means working problems, lots of problems.

SECTION 2.1 | DEFINITION II: RIGHT TRIANGLE TRIGONOMETRY

The word *trigonometry* is derived from two Greek words: *tri'gonon,* which translates as *triangle,* and *met'ron,* which means *measure.* Trigonometry, then, is triangle measure. In this section, we will give a second definition for the trigonometric functions that is, in fact, based on "triangle measure." We will define the trigonometric functions as the ratios of sides in right triangles. As you will see, this new definition does not conflict with the definition from Chapter 1 for the trigonometric functions.

DEFINITION II

If triangle ABC is a right triangle with $C = 90°$ (Figure 1), then the six trigonometric functions for A are defined as follows:

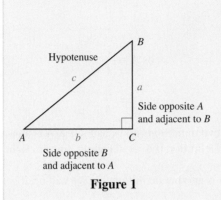

$$\sin A = \frac{\text{side opposite } A}{\text{hypotenuse}} = \frac{a}{c}$$

$$\cos A = \frac{\text{side adjacent } A}{\text{hypotenuse}} = \frac{b}{c}$$

$$\tan A = \frac{\text{side opposite } A}{\text{side adjacent } A} = \frac{a}{b}$$

$$\cot A = \frac{\text{side adjacent } A}{\text{side opposite } A} = \frac{b}{a}$$

$$\sec A = \frac{\text{hypotenuse}}{\text{side adjacent } A} = \frac{c}{b}$$

$$\csc A = \frac{\text{hypotenuse}}{\text{side opposite } A} = \frac{c}{a}$$

Figure 1

EXAMPLE 1 Triangle ABC is a right triangle with $C = 90°$. If $a = 6$ and $c = 10$, find the six trigonometric functions of A.

SOLUTION We begin by making a diagram of ABC, and then use the given information and the Pythagorean Theorem to solve for b (Figure 2).

$$b = \sqrt{c^2 - a^2}$$
$$= \sqrt{100 - 36}$$
$$= \sqrt{64}$$
$$= 8$$

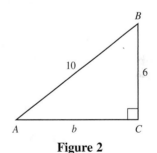

Figure 2

Now we write the six trigonometric functions of A using $a = 6$, $b = 8$, and $c = 10$.

$$\sin A = \frac{a}{c} = \frac{6}{10} = \frac{3}{5} \qquad \csc A = \frac{c}{a} = \frac{5}{3}$$

$$\cos A = \frac{b}{c} = \frac{8}{10} = \frac{4}{5} \qquad \sec A = \frac{c}{b} = \frac{5}{4}$$

$$\tan A = \frac{a}{b} = \frac{6}{8} = \frac{3}{4} \qquad \cot A = \frac{b}{a} = \frac{4}{3}$$

Now that we have done an example using our new definition, let's see how our new definition compares with Definition I from the previous chapter. We can place

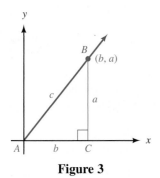

Figure 3

right triangle ABC on a rectangular coordinate system so that A is in standard position (Figure 3). We then note that a point on the terminal side of A is (b, a).

From Definition I in Chapter 1, we have

$$\sin A = \frac{a}{c}$$

$$\cos A = \frac{b}{c}$$

$$\tan A = \frac{a}{b}$$

From Definition II in this chapter, we have

$$\sin A = \frac{a}{c}$$

$$\cos A = \frac{b}{c}$$

$$\tan A = \frac{a}{b}$$

The two definitions agree as long as A is an acute angle. If A is not an acute angle, then Definition II does not apply, since in right triangle ABC, A must be an acute angle.

Here is another definition that we will need before we can take Definition II any further.

DEFINITION

Sine and *co*sine are *co*functions, as are tangent and *co*tangent, and secant and *co*secant. We say sine is the cofunction of cosine, and cosine is the cofunction of sine.

Now let's see what happens when we apply Definition II to B in right triangle ABC.

$$\sin B = \frac{\text{side opposite } B}{\text{hypotenuse}} = \frac{b}{c} = \cos A$$

$$\cos B = \frac{\text{side adjacent } B}{\text{hypotenuse}} = \frac{a}{c} = \sin A$$

$$\tan B = \frac{\text{side opposite } B}{\text{side adjacent } B} = \frac{b}{a} = \cot A$$

$$\cot B = \frac{\text{side adjacent } B}{\text{side opposite } B} = \frac{a}{b} = \tan A$$

$$\sec B = \frac{\text{hypotenuse}}{\text{side adjacent } B} = \frac{c}{a} = \csc A$$

$$\csc B = \frac{\text{hypotenuse}}{\text{side opposite } B} = \frac{c}{b} = \sec A$$

Figure 4

The prefix *co-* in *co*sine, *co*secant, and *co*tangent is a reference to the complement. Around the year 1463, Regiomontanus used the term *sinus rectus complementi,* presumably referring to the cosine as the sine of the complementary angle. In 1620, Edmund Gunter shortened this to *co.sinus,* which was further abbreviated as *cosinus* by John Newton in 1658.

As you can see in Figure 4, every trigonometric function of A is equal to the cofunction of B. That is, $\sin A = \cos B$, $\sec A = \csc B$, and $\tan A = \cot B$, to name a few. Since A and B are the acute angles in a right triangle, they are always complementary angles; that is, their sum is always $90°$. What we actually have here is another property of trigonometric functions: The sine of an angle is the cosine of its complement, the secant of an angle is the cosecant of its complement, and the tangent of an angle is the cotangent of its complement. Or, in symbols,

$$\text{if } A + B = 90°, \text{ then } \begin{cases} \sin A = \cos B \\ \sec A = \csc B \\ \tan A = \cot B \end{cases}$$

and so on.

We generalize this discussion with the following theorem.

COFUNCTION THEOREM

A trigonometric function of an angle is always equal to the cofunction of the complement of the angle.

To clarify this further, if two angles are complementary, such as 40° and 50°, then a trigonometric function of one is equal to the cofunction of the other. That is, $\sin 40° = \cos 50°$, $\sec 40° = \csc 50°$, and $\tan 40° = \cot 50°$.

EXAMPLE 2 Fill in the blanks so that each expression becomes a true statement.

a. $\sin \underline{\hspace{1cm}} = \cos 30°$ **b.** $\tan y = \cot \underline{\hspace{1cm}}$ **c.** $\sec 75° = \csc \underline{\hspace{1cm}}$

SOLUTION Using the theorem on cofunctions of complementary angles, we fill in the blanks as follows:

a. $\sin \underline{\hspace{0.3cm}60°\hspace{0.3cm}} = \cos 30°$ since sine and cosine are cofunctions and $60° + 30° = 90°$

b. $\tan y = \cot \underline{(90° - y)}$ since tangent and cotangent are cofunctions and $y + (90° - y) = 90°$

c. $\sec 75° = \csc \underline{\hspace{0.3cm}15°\hspace{0.3cm}}$ since secant and cosecant are cofunctions and $75° + 15° = 90°$

For our next application of Definition II, we need to recall the two special triangles we introduced in Chapter 1. They are the 30°–60°–90° triangle and the 45°–45°–90° triangle (Figure 5).

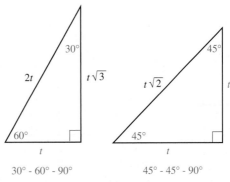

30° - 60° - 90°   45° - 45° - 90°

Figure 5

We can use these two special triangles to find the trigonometric functions of 30°, 45°, and 60°. For example,

$$\sin 60° = \frac{\text{side opposite } 60°}{\text{hypotenuse}} = \frac{t\sqrt{3}}{2t} = \frac{\sqrt{3}}{2}$$ Dividing out the common factor t

$$\tan 30° = \frac{\text{side opposite } 30°}{\text{side adjacent } 30°} = \frac{t}{t\sqrt{3}} = \frac{1}{\sqrt{3}}$$

We could go on in this manner—using Definition II and the two special triangles—to find the six trigonometric ratios for 30°, 45°, and 60°. Instead, let us vary things a little and use the information just obtained for sin 60° and tan 30° and the theorem on cofunctions of complementary angles to find cos 30° and cot 60°.

$$\cos 30° = \sin 60° = \frac{\sqrt{3}}{2}$$ Cofunction Theorem

$$\cot 60° = \tan 30° = \frac{1}{\sqrt{3}}$$

To vary things even more, we can use some reciprocal identities to find csc 60° and cot 30°.

$$\csc 60° = \frac{1}{\sin 60°} = \frac{2}{\sqrt{3}}$$

$$= \frac{2\sqrt{3}}{3}$$ If the denominator is rationalized

$$\cot 30° = \frac{1}{\tan 30°} = \sqrt{3}$$

The idea behind using the different methods listed here is not to make things confusing. We have a number of tools at hand, and it does not hurt to show the different ways they can be used.

If we were to continue finding the sine, cosine, and tangent for these special angles, we would obtain the results summarized in Table 1 and illustrated in Figure 6.

Table 1 is called a table of exact values to distinguish it from a table of approximate values. Later in this chapter we will work with a calculator to obtain tables of approximate values.

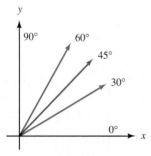

Figure 6

TABLE 1 EXACT VALUES *(MEMORIZE)*			
θ	$\sin \theta$	$\cos \theta$	$\tan \theta$
30°	$\dfrac{1}{2}$	$\dfrac{\sqrt{3}}{2}$	$\dfrac{1}{\sqrt{3}} = \dfrac{\sqrt{3}}{3}$
45°	$\dfrac{1}{\sqrt{2}} = \dfrac{\sqrt{2}}{2}$	$\dfrac{1}{\sqrt{2}} = \dfrac{\sqrt{2}}{2}$	1
60°	$\dfrac{\sqrt{3}}{2}$	$\dfrac{1}{2}$	$\sqrt{3}$

EXAMPLE 3 Use the exact values from Table 1 to show that the following are true.

a. $\cos^2 30° + \sin^2 30° = 1$ **b.** $\cos^2 45° + \sin^2 45° = 1$

SOLUTION

a. $\cos^2 30° + \sin^2 30° = \left(\dfrac{\sqrt{3}}{2}\right)^2 + \left(\dfrac{1}{2}\right)^2 = \dfrac{3}{4} + \dfrac{1}{4} = 1$

b. $\cos^2 45° + \sin^2 45° = \left(\dfrac{1}{\sqrt{2}}\right)^2 + \left(\dfrac{1}{\sqrt{2}}\right)^2 = \dfrac{1}{2} + \dfrac{1}{2} = 1$ ■

 EXAMPLE 4 Let $x = 30°$ and $y = 45°$ in each of the expressions that follow, and then simplify each expression as much as possible.

a. $2 \sin x$ **b.** $\sin 2y$ **c.** $4 \sin (3x - 90°)$

SOLUTION

a. $2 \sin x = 2 \sin 30° = 2\left(\dfrac{1}{2}\right) = 1$

b. $\sin 2y = \sin 2 (45°) = \sin 90° = 1$
c. $4 \sin (3x - 90°) = 4 \sin [3(30°) - 90°] = 4 \sin 0° = 4(0) = 0$ ■

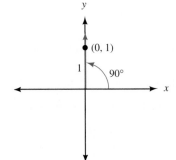

Figure 7

As you will see as we progress through the book, there are times when it is appropriate to use Definition II for the trigonometric functions and times when it is more appropriate to use Definition I. To illustrate, suppose we wanted to find $\sin \theta$, $\cos \theta$, and $\tan \theta$ for $\theta = 90°$. We would not be able to use Definition II because we can't draw a right triangle in which one of the *acute* angles is 90°. Instead, we would draw 90° in standard position, locate a point on the terminal side, and use Definition I to find $\sin 90°$, $\cos 90°$, and $\tan 90°$. Figure 7 illustrates this.

A point on the terminal side of 90° is (0, 1). The distance from the origin to (0, 1) is 1. Therefore, $x = 0$, $y = 1$, and $r = 1$. From Definition I we have:

$$\sin 90° = \frac{y}{r} = \frac{1}{1} = 1$$

$$\cos 90° = \frac{x}{r} = \frac{0}{1} = 0$$

$$\tan 90° = \frac{y}{x} = \frac{1}{0}, \text{ which is undefined}$$

To show how the properties, definitions, and theorems you are learning can be used together, let's use the above information, along with the Cofunction Theorem and a ratio identity, to find $\sin \theta$, $\cos \theta$, and $\tan \theta$ for $\theta = 0°$.

First, since 0° and 90° are complementary angles, we can use the Cofunction Theorem to write

A trigonometric function of an angle		is equal to the cofunction of its complement
$\sin 0°$	$=$	$\cos 90° = 0$
$\cos 0°$	$=$	$\sin 90° = 1$

Now that we have sin 0° and cos 0°, we can find tan 0° by using a ratio identity.

$$\tan 0° = \frac{\sin 0°}{\cos 0°} = \frac{0}{1} = 0$$

To conclude this section, we take the information just obtained for 0° and 90°, along with the exact values in Table 1, and summarize them in Table 2. To make the information in Table 2 a little easier to memorize, we have written some of the exact values differently than we usually do. For example, in Table 2 we have written 2 as $\sqrt{4}$, 0 as $\sqrt{0}$, and 1 as $\sqrt{1}$.

TABLE 2

θ	0°	30°	45°	60°	90°
$\sin \theta$	$\dfrac{\sqrt{0}}{2}$	$\dfrac{\sqrt{1}}{2}$	$\dfrac{\sqrt{2}}{2}$	$\dfrac{\sqrt{3}}{2}$	$\dfrac{\sqrt{4}}{2}$
$\cos \theta$	$\dfrac{\sqrt{4}}{2}$	$\dfrac{\sqrt{3}}{2}$	$\dfrac{\sqrt{2}}{2}$	$\dfrac{\sqrt{1}}{2}$	$\dfrac{\sqrt{0}}{2}$
$\tan \theta$	0	$\dfrac{\sqrt{3}}{3}$	1	$\sqrt{3}$	undefined

GETTING READY FOR CLASS

After reading through the preceding section, respond in your own words and in complete sentences.

a. In a right triangle, which side is opposite the right angle?

b. If A is an acute angle in a right triangle, how do you define sin A, cos A, and tan A?

c. State the Cofunction Theorem.

d. How are sin 30° and cos 60° related?

PROBLEM SET 2.1

Problems 1 through 6 refer to right triangle ABC with $C = 90°$. In each case, use the given information to find the six trigonometric functions of A.

1. $b = 3, c = 5$ **2.** $b = 5, c = 13$
3. $a = 2, b = 1$ **4.** $a = 3, b = 2$
5. $a = 2, b = \sqrt{5}$ **6.** $a = 3, b = \sqrt{7}$

In each right triangle below, find sin A, cos A, tan A, and sin B, cos B, tan B.

7.

8.

9.

10.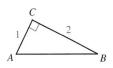

11.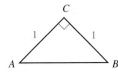

12.

13.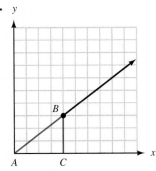

14.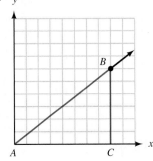

In each diagram below, angle A is in standard position. In each case, find the coordinates of point B and then find $\sin A$, $\cos A$, and $\tan A$.

15. y

16. y

Use the Cofunction Theorem to fill in the blanks so that each expression becomes a true statement.

17. $\sin 10° = \cos$ _____

18. $\cos 40° = \sin$ _____

19. $\tan 8° = \cot$ _____

20. $\cot 12° = \tan$ _____

21. $\sin x = \cos$ _____

22. $\sin y = \cos$ _____

23. $\tan (90° - x) = \cot$ _____

24. $\tan (90° - y) = \cot$ _____

Complete the following tables using exact values. Do not rationalize any denominators.

25.

x	$\sin x$	$\csc x$
0°	0	
30°	$\dfrac{1}{2}$	
45°	$\dfrac{1}{\sqrt{2}}$	
60°	$\dfrac{\sqrt{3}}{2}$	
90°	1	

26.

x	$\cos x$	$\sec x$
0°	1	
30°	$\dfrac{\sqrt{3}}{2}$	
45°	$\dfrac{1}{\sqrt{2}}$	
60°	$\dfrac{1}{2}$	
90°	0	

Simplify each expression by first substituting values from the table of exact values and then simplifying the resulting expression.

27. $4 \sin 30°$ **28.** $5 \sin^2 30°$
29. $(2 \cos 30°)^2$ **30.** $\sin^3 30°$
31. $(\sin 60° + \cos 60°)^2$ **32.** $\sin^2 60° + \cos^2 60°$
33. $\sin^2 45° - 2 \sin 45° \cos 45° + \cos^2 45°$
34. $(\sin 45° - \cos 45°)^2$
35. $(\tan 45° + \tan 60°)^2$ **36.** $\tan^2 45° + \tan^2 60°$

For each expression that follows, replace x with 30°, y with 45°, and z with 60°, and then simplify as much as possible.

37. $2 \sin x$ **38.** $4 \cos y$
39. $4 \cos (z - 30°)$ **40.** $-2 \sin (y + 45°)$
41. $-3 \sin 2x$ **42.** $3 \sin 2y$
43. $2 \cos (3x - 45°)$ **44.** $2 \sin (90° - z)$

Find exact values for each of the following:

45. $\sec 30°$ **46.** $\csc 30°$ **47.** $\csc 60°$ **48.** $\sec 60°$
49. $\cot 45°$ **50.** $\cot 30°$ **51.** $\sec 45°$ **52.** $\csc 45°$

Problems 53 through 56 refer to right triangle ABC with $C = 90°$. In each case, use a calculator to find $\sin A$, $\cos A$, $\sin B$, and $\cos B$. Round your answers to the nearest hundredth.

53. $a = 3.42$, $c = 5.70$ **54.** $b = 8.88$, $c = 9.62$
55. $a = 19.44$, $b = 5.67$ **56.** $a = 11.28$, $b = 8.46$

57. Suppose each edge of the cube shown in Figure 8 is 5 inches long. Find the sine and cosine of the angle formed by diagonals CF and CH.

58. Suppose each edge of the cube shown in Figure 8 is 3 inches long. Find the sine and cosine of the angle formed by diagonals DE and DG.

59. Suppose each edge of the cube shown in Figure 8 is x inches long. Find the sine and cosine of the angle formed by diagonals CF and CH.

60. Suppose each edge of the cube shown in Figure 8 is x inches long. Find the sine and cosine of the angle formed by diagonals DE and DG.

Figure 8

REVIEW PROBLEMS

From here on, each Problem Set will end with a series of review problems. In mathematics, it is very important to review. The more you review, the better you will understand the topics we cover and the longer you will remember them. Also, there will be times when material that seemed confusing earlier will be less confusing the second time around.

The problems that follow review material we covered in Section 1.2.

Find the distance between each pair of points.

61. $(5, 1)$, $(2, 5)$ **62.** $(3, -2)$, $(-1, -4)$
63. Find x so that the distance between $(x, 2)$ and $(1, 5)$ is $\sqrt{13}$.
64. Graph the line $2x - 3y = 6$.   **65.** Graph the line $y = 2x - 1$.

Draw each angle in standard position and name a point on the terminal side.

66. $135°$ **67.** $45°$

For each angle below, name a coterminal angle between 0° and 360°.

68. $-90°$ **69.** $-135°$ **70.** $-210°$ **71.** $-300°$

SECTION 2.2 | CALCULATORS AND TRIGONOMETRIC FUNCTIONS OF AN ACUTE ANGLE

In this section, we will see how calculators can be used to find approximations for trigonometric functions of angles between 0° and 90°. Before we begin our work with calculators, we need to look at degree measure in more detail.

We previously defined 1 degree (1°) to be $\frac{1}{360}$ of a full rotation. A degree itself can be broken down further. If we divide 1° into 60 equal parts, each one of the parts is called 1 minute, denoted 1′. One minute is $\frac{1}{60}$ of a degree; in other words, there are 60 minutes in every degree. The next smaller unit of angle measure is a second. One second, 1″, is $\frac{1}{60}$ of a minute. There are 60 seconds in every minute.

$$1° = 60' \qquad \text{or} \qquad 1' = \left(\frac{1}{60}\right)°$$

$$1' = 60'' \qquad \text{or} \qquad 1'' = \left(\frac{1}{60}\right)'$$

Table 1 shows how to read angles written in degree measure.

TABLE 1

The Expression	Is Read
52° 10′	52 degrees, 10 minutes
5° 27′ 30″	5 degrees, 27 minutes, 30 seconds
13° 24′ 15″	13 degrees, 24 minutes, 15 seconds

  Add 48° 49′ and 72° 26′.

SOLUTION   We can add in columns with degrees in the first column and minutes in the second column.

$$\begin{array}{r} 48° \ 49' \\ + \ 72° \ 26' \\ \hline 120° \ 75' \end{array}$$

Since 60 minutes is equal to 1 degree, we can carry 1 degree from the minutes column to the degrees column.

$$120° \ 75' = 121° \ 15' \quad \blacksquare$$

  Subtract 24° 14′ from 90°.

SOLUTION   To subtract 24° 14′ from 90°, we will have to "borrow" 1° and write that 1° as 60′.

$$\begin{array}{rcl} 90° & = & 89° \ 60' \ \text{(Still 90°)} \\ - \ 24° \ 14' & & - \ 24° \ 14' \\ \hline & & 65° \ 46' \end{array} \quad \blacksquare$$

DECIMAL DEGREES

An alternative to using minutes and seconds to break down degrees into smaller units is decimal degrees. For example, 30.5°, 101.75°, and 62.831° are measures of angles written in decimal degrees.

To convert from decimal degrees to degrees and minutes, we simply multiply the fractional part of the angle (the part to the right of the decimal point) by 60 to convert it to minutes.

 **EXAMPLE 3** Change 27.25° to degrees and minutes.

SOLUTION   Multiplying 0.25 by 60, we have the number of minutes equivalent to 0.25°.

$$27.25° = 27° + 0.25°$$
$$= 27° + 0.25(60')$$
$$= 27° + 15'$$
$$= 27° 15'$$

Of course in actual practice, we would not show all of these steps. They are shown here simply to indicate why we multiply only the decimal part of the decimal degree by 60 to change to degrees and minutes. ▪

```
48°49'+72°26'►DMS
                    121°15'0"
90−24°14'►DMS
                    65°46'0"
27.25►DMS
                    27°15'0"
```

Figure 1

CALCULATOR NOTE   Most scientific and graphing calculators have keys that let you enter angles in degrees, minutes, and seconds format (DMS) or in decimal degrees (DD). You may also have keys or commands that let you convert from one format to the other. Figure 1 shows how Examples 1 through 3 might look when solved on a TI-83 graphing calculator. Consult your calculator's manual to see how this is done with your particular model.

 **EXAMPLE 4** Change 10° 45' to decimal degrees.

SOLUTION   We have to reverse the process we used in Example 3. To change 45' to a decimal, we must divide by 60.

$$10° 45' = 10° + 45'$$
$$= 10° + \left(\frac{45}{60}\right)°$$
$$= 10° + 0.75°$$
$$= 10.75° ▪$$

TABLE 2	
Decimal Degree	**Minutes**
0.1°	6'
0.2°	12'
0.3°	18'
0.4°	24'
0.5°	30'
0.6°	36'
0.7°	42'
0.8°	48'
0.9°	54'
1.0°	60'

The process of converting back and forth between decimal degrees and degrees and minutes can become more complicated when we use decimal numbers with more digits or when we convert to degrees, minutes, and seconds. In this book, most of the angles written in decimal degrees will be written to the nearest tenth or, at most, the nearest hundredth. The angles written in degrees, minutes, and seconds will rarely go beyond the minutes column.

Table 2 lists the most common conversions between decimal degrees and minutes.

TRIGONOMETRIC FUNCTIONS AND ACUTE ANGLES

Until now, we have been able to determine trigonometric functions only for angles for which we could find a point on the terminal side or angles that were part of special triangles. We can find decimal approximations for trigonometric functions of any acute angle by using a calculator with keys for sine, cosine, and tangent.

First, there are a couple of things you should know. Just as distance can be measured in feet and also in meters, angles can be measured in degrees and also in radians. We will cover radian measure in Chapter 3. For now, you simply need to be sure that your calculator is set to work in degrees. We will refer to this setting as being in *degree mode.* The most common mistake students make when finding values of trigonometric functions on a calculator is working in the wrong mode. As a rule, you should *always* check the mode settings on your calculator before evaluating a trigonometric function.

 EXAMPLE 5   Use a calculator to find cos 37.8°.

SOLUTION   First, be sure your calculator is set to degree mode. Then, depending on the type of calculator you have, press the indicated keys.

Scientific Calculator	**Graphing Calculator**
37.8 $\boxed{\cos}$	$\boxed{\cos}$ $\boxed{(}$ 37.8 $\boxed{)}$ $\boxed{\text{ENTER}}$

Your calculator will display a number that rounds to 0.7902. The number 0.7902 is just an approximation of cos 37.8°, which is actually an irrational number, as are the trigonometric functions of most angles.

NOTE   We will give answers accurate to four places past the decimal point. You can set your calculator to four-place fixed-point mode, and it will show you the same results without your having to round your answers mentally.

CALCULATOR NOTE   As we mentioned in Section 1.1, some graphing calculators use parentheses with certain functions. For example, the TI-83 will automatically insert a left parenthesis when the $\boxed{\cos}$ key is pressed, so TI-83 users can skip the $\boxed{(}$ key. Other models do not require them at all. For the sake of clarity, we will often include parentheses throughout this book. You may be able to omit one or both parentheses with your model. Just be sure that you are able to obtain the same results shown for each example.

 EXAMPLE 6   Find tan 58.75°.

SOLUTION   This time, we use the $\boxed{\tan}$ key:

Scientific Calculator	**Graphing Calculator**
58.75 $\boxed{\tan}$	$\boxed{\tan}$ $\boxed{(}$ 58.75 $\boxed{)}$ $\boxed{\text{ENTER}}$

Rounding to four places past the decimal point, we have

$$\tan 58.75° = 1.6479$$

 **EXAMPLE 7** Find $\sin^2 14°$.

SOLUTION Since $\sin^2 14° = (\sin 14°)^2$, the calculator sequence is:

Scientific Calculator   **Graphing Calculator**

Rounding to four digits past the decimal point, we have

$$\sin^2 14° = 0.0585 \quad \blacksquare$$

Most calculators do not have additional keys for the secant, cosecant, or cotangent functions. To find a value for one of these functions, we will use the appropriate reciprocal identity.

EXAMPLE 8 Find $\sec 78°$.

SOLUTION Since $\sec 78° = \dfrac{1}{\cos 78°}$, the calculator sequence is:

Scientific Calculator   **Graphing Calculator**

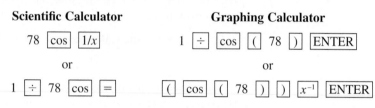

Rounding to four digits past the decimal point, we have

$$\sec 78° = 4.8097 \quad \blacksquare$$

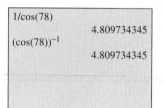

Figure 2

EXAMPLE 9 To further justify the Cofunction Theorem introduced in the previous section, use a calculator to find $\sin 24.3°$ and $\cos 65.7°$.

SOLUTION Note that the sum of $24.3°$ and $65.7°$ is $24.3° + 65.7° = 90°$; the two angles are complementary. Using a calculator, and rounding our answers as we have above, we find that the sine of $24.3°$ is the cosine of its complement $65.7°$.

$$\sin 24.3° = 0.4115 \qquad \text{and} \qquad \cos 65.7° = 0.4115 \quad \blacksquare$$

USING TECHNOLOGY

WORKING WITH TABLES

Most graphing calculators have the ability to display several values of one or more functions simultaneously in a table format. We can use the table feature to display approximate values for the sine and cosine functions. Then we can easily compare the different trigonometric function values for a given angle, or compare values of a single trigonometric function for different angles.

Create a table for the two functions $Y_1 = \sin x$ and $Y_2 = \cos x$. Be sure your calculator is set to degree mode. Set up your table so that you can input the values

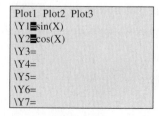

Figure 3

```
TABLE SETUP
 TblStart=0
 ΔTbl=1
Indpnt: Auto Ask
Depend: Auto Ask
```

Figure 4

of x yourself. On some calculators, this is done by setting the independent variable to Ask (Figure 4). Display the table, and input the following values for x:

$$x = 0, 30, 45, 60, 90$$

Once you have entered all five x-values, your table should look something like the one shown in Figure 5. Compare the approximate values from your calculator with the exact values we found in Section 2.1, shown here in Table 3.

X	Y1	Y2
0	0	1
30	.5	.86603
45	.70711	.70711
60	.86603	.5
90	1	0
▬		

X=

Figure 5

TABLE 3 EXACT VALUES		
X	**sin X**	**cos X**
0°	0	1
30°	$\dfrac{1}{2}$	$\dfrac{\sqrt{3}}{2}$
45°	$\dfrac{1}{\sqrt{2}}$	$\dfrac{1}{\sqrt{2}}$
60°	$\dfrac{\sqrt{3}}{2}$	$\dfrac{1}{2}$
90°	1	0

The $\boxed{\sin}$, $\boxed{\cos}$, and $\boxed{\tan}$ keys allow us to find the value of a trigonometric function when we know the measure of the angle. (*Remember:* We can think of this value as being the ratio of the lengths of two sides of a right triangle.) There are some problems, however, where we may be in the opposite situation and need to do the reverse. That is, we may know the value of the trigonometric function and need to find the angle. The calculator has another set of keys for this purpose. They are the $\boxed{\sin^{-1}}$, $\boxed{\cos^{-1}}$, and $\boxed{\tan^{-1}}$ keys. At first glance, the notation on these keys may lead you to believe that they will give us the reciprocals of the trigonometric functions. Instead, this notation is used to denote an inverse function, something we will cover in more detail in Chapter 4. For the time being, just remember that these keys are used to find the angle given the value of one of the trigonometric functions of the angle.

CALCULATOR NOTE   Some calculators do not have a key labeled as $\boxed{\sin^{-1}}$. You may need to press a combination of keys, such as $\boxed{\text{INV}}$ $\boxed{\sin}$, $\boxed{\text{ARC}}$ $\boxed{\sin}$, or $\boxed{\text{2nd}}$ $\boxed{\sin}$. If you are not sure which key to press, look in the index of your calculator manual under inverse trigonometric function.

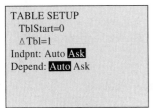 **EXAMPLE 10**   Find the acute angle θ for which $\tan \theta = 3.152$. Round your answer to the nearest tenth of a degree.

SOLUTION   We are looking for the angle whose tangent is 3.152. We must use the $\boxed{\tan^{-1}}$ key. First, be sure your calculator is set to degree mode. Then, press the indicated keys.

Scientific Calculator	**Graphing Calculator**
3.152 $\boxed{\tan^{-1}}$	$\boxed{\tan^{-1}}$ $\boxed{(}$ 3.152 $\boxed{)}$ $\boxed{\text{ENTER}}$

To the nearest tenth of a degree the answer is 72.4°. That is, if $\tan \theta = 3.152$, then $\theta = 72.4°$. ▬

 **EXAMPLE 11** Find the acute angle A for which $\sin A = 0.3733$. Round your answer to the nearest tenth of a degree.

SOLUTION The sequences are:

Scientific Calculator   **Graphing Calculator**

0.3733 $\boxed{\sin^{-1}}$ $\boxed{\sin^{-1}}$ $\boxed{(}$ 0.3733 $\boxed{)}$ $\boxed{\text{ENTER}}$

The result is $A = 21.9°$.

 **EXAMPLE 12** To the nearest hundredth of a degree, find the acute angle B for which $\sec B = 1.0768$.

SOLUTION Since we do not have a secant key on the calculator, we must first use a reciprocal to convert this problem into a problem involving $\cos B$ (as in Example 8).

If $\sec B = 1.0768$

then $\dfrac{1}{\sec B} = \dfrac{1}{1.0768}$ Take the reciprocal of each side

$\cos B = \dfrac{1}{1.0768}$ Since the cosine is the reciprocal of the secant

From this last line we see that the keys to press are:

Scientific Calculator   **Graphing Calculator**

1.0768 $\boxed{1/x}$ $\boxed{\cos^{-1}}$ $\boxed{\cos^{-1}}$ $\boxed{(}$ 1.0768 $\boxed{x^{-1}}$ $\boxed{)}$ $\boxed{\text{ENTER}}$

See Figure 6.

To the nearest hundredth of a degree our answer is $B = 21.77°$.

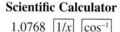

$\cos^{-1}(1.0768^{-1})$

21.77039922

Figure 6

 **EXAMPLE 13** Find the acute angle C for which $\cot C = 0.0975$. Round to the nearest degree.

SOLUTION First, we rewrite the problem in terms of $\tan C$.

If $\cot C = 0.0975$

then $\dfrac{1}{\cot C} = \dfrac{1}{0.0975}$ Take the reciprocal of each side

$\tan C = \dfrac{1}{0.0975}$ Since the tangent is the reciprocal of the cotangent

From this last line we see that the keys to press are:

Scientific Calculator   **Graphing Calculator**

0.0975 $\boxed{1/x}$ $\boxed{\tan^{-1}}$ $\boxed{\tan^{-1}}$ $\boxed{(}$ 0.0975 $\boxed{x^{-1}}$ $\boxed{)}$ $\boxed{\text{ENTER}}$

To the nearest degree our answer is $C = 84°$.

GETTING READY FOR CLASS

After reading through the preceding section, respond in your own words and in complete sentences.

a. What is 1 minute of angle measure?

b. How do you convert from decimal degrees to degrees and minutes?

c. How do you find sin 58.75° on your calculator?

d. If tan θ = 3.152, how do you use a calculator to find θ?

PROBLEM SET 2.2

Add or subtract as indicated.

1. (37° 45′) + (26° 24′)   **2.** (41° 20′) + (32° 16′)
3. (51° 55′) + (37° 45′)   **4.** (63° 38′) + (24° 52′)
5. (61° 33′) + (45° 16′)   **6.** (77° 21′) + (23° 16′)
7. 90° − (34° 12′)   **8.** 90° − (62° 25′)
9. 180° − (120° 17′)   **10.** 180° − (112° 19′)
11. (76° 24′) − (22° 34′)   **12.** (89° 38′) − (28° 58′)
13. (70° 40′) − (30° 50′)   **14.** (80° 50′) − (50° 56′)

Convert each of the following to degrees and minutes. Round to the nearest minute.

15. 35.4°   **16.** 63.2°   **17.** 16.25°   **18.** 18.75°
19. 92.55°   **20.** 34.45°   **21.** 19.9°   **22.** 18.8°

Change each of the following to decimal degrees. If rounding is necessary, round to the nearest hundredth of a degree.

23. 45° 12′   **24.** 74° 18′   **25.** 62° 36′   **26.** 21° 48′
27. 17° 20′   **28.** 29° 40′   **29.** 48° 27′   **30.** 78° 21′

Use a calculator to find each of the following. Round all answers to four places past the decimal point.

31. sin 27.2°   **32.** cos 82.9°   **33.** cos 18°   **34.** sin 42°
35. tan 87.32°   **36.** tan 81.43°   **37.** cot 31°   **38.** cot 24°
39. sec 48.2°   **40.** sec 71.8°   **41.** csc 14.15°   **42.** csc 12.21°

Use a calculator to find each of the following. Round all answers to four places past the decimal point.

43. cos 24° 30′   **44.** sin 35° 10′   **45.** tan 42° 15′ **46.** tan 19° 45′
47. sin 56° 40′   **48.** cos 66° 40′   **49.** sec 45° 54′ **50.** sec 84° 48′

Use a calculator to complete the following tables. (Be sure your calculator is in degree mode.) Round all answers to four digits past the decimal point. If you have a graphing calculator with table-building capabilities, use it to construct the tables.

51.

X	sin X	csc X
0°		
30°		
45°		
60°		
90°		

52.

X	cos X	sec X
0°		
30°		
45°		
60°		
90°		

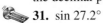

Use a calculator to complete the following tables. (Be sure your calculator is in degree mode.) Round all answers to four digits past the decimal point. If you have a graphing calculator with table-building capabilities, use it to construct the tables.

53.

X	sin X	cos X	tan X
0°			
15°			
30°			
45°			
60°			
75°			
90°			

54.

X	csc X	sec X	cot X
0°			
15°			
30°			
45°			
60°			
75°			
90°			

Find θ if θ is between 0° and 90°. Round your answers to the nearest tenth of a degree.

55. $\cos \theta = 0.9770$ **56.** $\sin \theta = 0.3971$

57. $\tan \theta = 0.6873$ **58.** $\cos \theta = 0.5490$

59. $\sin \theta = 0.9813$ **60.** $\tan \theta = 0.6273$

61. $\sec \theta = 1.0191$ **62.** $\sec \theta = 1.0801$

63. $\csc \theta = 1.8214$ **64.** $\csc \theta = 1.4293$

65. $\cot \theta = 0.6873$ **66.** $\cot \theta = 0.4327$

Use a calculator to find a value of θ between 0° and 90° that satisfies each statement below. Write your answer in degrees and minutes rounded to the nearest minute.

67. $\sin \theta = 0.7038$ **68.** $\cos \theta = 0.9153$

69. $\cos \theta = 0.4112$ **70.** $\sin \theta = 0.9954$

71. $\cot \theta = 5.5764$ **72.** $\cot \theta = 4.6252$

73. $\sec \theta = 1.0129$ **74.** $\csc \theta = 7.0683$

To further justify the Cofunction Theorem, use your calculator to find a value for each pair of trigonometric functions below. In each case, the trigonometric functions are cofunctions of one another, and the angles are complementary angles. Round your answers to four places past the decimal point.

75. $\sin 23°$, $\cos 67°$ **76.** $\sin 33°$, $\cos 57°$

77. $\sec 34.5°$, $\csc 55.5°$ **78.** $\sec 56.7°$, $\csc 33.3°$

79. $\tan 4° 30'$, $\cot 85° 30'$ **80.** $\tan 10° 30'$, $\cot 79° 30'$

Work each of the following problems on your calculator. Do not write down or round off any intermediate answers. Each answer should be 1.

81. $\cos^2 37° + \sin^2 37°$ **82.** $\cos^2 85° + \sin^2 85°$

83. $\sin^2 10° + \cos^2 10°$ **84.** $\sin^2 8° + \cos^2 8°$

85. What happens when you try to find A for $\sin A = 1.234$ on your calculator? Why does it happen?

86. What happens when you try to find B for $\sin B = 4.321$ on your calculator? Why does this happen?

87. What happens when you try to find $\tan 90°$ on your calculator? Why does this happen?

88. What happens when you try to find $\cot 0°$ on your calculator? Why does this happen?

Complete each of the following tables. Round all answers to the nearest tenth.

89. a.

X	tan X
87°	
87.5°	
88°	
88.5°	
89°	
89.5°	
90°	

b.

X	tan X
89.4°	
89.5°	
89.6°	
89.7°	
89.8°	
89.9°	
90°	

90. a.

X	cot X
3°	
2.5°	
2°	
1.5°	
1°	
0.5°	
0°	

b.

X	cot X
0.6°	
0.5°	
0.4°	
0.3°	
0.2°	
0.1°	
0°	

REVIEW PROBLEMS

The problems that follow review material we covered in Section 1.3. Find $\sin \theta$, $\cos \theta$, and $\tan \theta$ if the given point is on the terminal side of θ.

91. $(3, -2)$ **92.** $(-\sqrt{3}, 1)$

Find $\sin \theta$, $\cos \theta$, and $\tan \theta$ for each value of θ. (Do not use calculators.)

93. $90°$ **94.** $135°$

Find the remaining trigonometric functions of θ if

95. $\cos \theta = -5/13$ and θ terminates in QIII
96. $\tan \theta = -3/4$ and θ terminates in QII

In which quadrant must the terminal side of θ lie if

97. $\sin \theta > 0$ and $\cos \theta < 0$ **98.** $\tan \theta > 0$ and $\sec \theta < 0$

SECTION 2.3 | SOLVING RIGHT TRIANGLES

The first Ferris wheel was designed and built by American engineer George W. G. Ferris in 1893. The diameter of this wheel was 250 feet. It had 36 cars, each of which held 40 passengers. The top of the wheel was 264 feet above the ground. It took 20 minutes to complete one revolution. As you will see as we progress through the book, trigonometric functions can be used to model the motion of a rider on a Ferris wheel. The model can be used to give information about the position of the rider at any time during a ride. For instance, in the last example in this section, we will use Definition II for the trigonometric functions to find the height a rider is above the ground at certain positions on a Ferris wheel.

In this section, we will use Definition II for trigonometric functions of an acute angle, along with our calculators, to find the missing parts to some right triangles. Before we begin, however, we need to talk about significant digits.

> **DEFINITION**
>
> The number of *significant digits* (or figures) in a number is found by counting all the digits from left to right beginning with the first nonzero digit on the left.

According to this definition,

0.042 has two significant digits

0.005 has one significant digit

20.5 has three significant digits

6.000 has four significant digits

9,200 has four significant digits

700 has three significant digits

NOTE   In actual practice it is not always possible to tell how many significant digits an integer like 700 has. For instance, if exactly 700 people signed up to take history at your school, then 700 has three significant digits. On the other hand, if 700 is the result of a calculation in which the answer, 700, has been rounded to the nearest ten, then it has two significant digits. There are ways to write integers like 700 so that the number of significant digits can be determined exactly. One way is with scientific notation. However, to simplify things, in this book we will assume that integers have the greatest possible number of significant digits. In the case of 700, that number is three.

The relationship between the accuracy of the sides of a triangle and the accuracy of the angles in the same triangle is shown in Table 1.

TABLE 1

Accuracy of sides	Accuracy of angles
Two significant digits	Nearest degree
Three significant digits	Nearest 10 minutes or tenth of a degree
Four significant digits	Nearest minute or hundredth of a degree

We are now ready to use Definition II to solve right triangles. We solve a right triangle by using the information given about it to find all of the missing sides and angles. In all of the examples and in the Problem Set that follows, we will assume C is the right angle in all of our right triangles, unless otherwise noted.

Unless stated otherwise, we round our answers so that the number of significant digits in our answers matches the number of significant digits in the least significant number given in the original problem. Also, we round our answers only and not any of the numbers in the intermediate steps. Finally, we are showing the values of the trigonometric functions to four significant digits simply to avoid cluttering the page with long decimal numbers. This does not mean that you should stop halfway through a problem and round the values of trigonometric functions to four significant digits before continuing.

 **EXAMPLE 1** In right triangle ABC, $A = 40°$ and $c = 12$ centimeters. Find a, b, and B.

SOLUTION We begin by making a diagram of the situation (Figure 1). The diagram is very important because it lets us visualize the relationship between the given information and the information we are asked to find.

To find B, we use the fact that the sum of the two acute angles in any right triangle is 90°.

$$B = 90° - A$$
$$= 90° - 40°$$
$$B = 50°$$

To find a, we can use the formula for $\sin A$.

$$\sin A = \frac{a}{c}$$

Multiplying both sides of this formula by c and then substituting in our given values of A and c we have

$$a = c \sin A$$
$$= 12 \sin 40°$$
$$= 12(0.6428) \qquad \sin 40° = 0.6428$$
$$a = 7.7 \text{ cm} \qquad \text{Answer rounded to two significant digits}$$

There is more than one way to find b.

Using $\cos A = \dfrac{b}{c}$, we have   Using the Pythagorean Theorem, we have

$$b = c \cos A \qquad\qquad\qquad c^2 = a^2 + b^2$$
$$= 12 \cos 40° \qquad\qquad\quad b = \sqrt{c^2 - a^2}$$
$$= 12(0.7660) \qquad\qquad\quad = \sqrt{12^2 - (7.7)^2}$$
$$b = 9.2 \text{ cm} \qquad\qquad\qquad = \sqrt{144 - 59.29}$$
$$\qquad\qquad\qquad\qquad\qquad = \sqrt{84.71}$$
$$\qquad\qquad\qquad\qquad\qquad b = 9.2 \quad \blacksquare$$

In Example 2, we are given two sides and asked to find the remaining parts of a right triangle.

EXAMPLE 2 In right triangle ABC, $a = 2.73$ and $b = 3.41$. Find the remaining side and angles.

SOLUTION Figure 2 is a diagram of the triangle.
We can find A by using the formula for $\tan A$.

$$\tan A = \frac{a}{b}$$
$$= \frac{2.73}{3.41}$$
$$\tan A = 0.8006$$

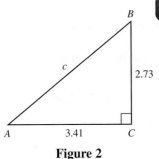

Figure 1

Figure 2

Now, to find A, we use a calculator.

$$A = \tan^{-1}(0.8006) = 38.7°$$

Next we find B.

$$B = 90.0° - A$$

$$= 90.0° - 38.7°$$

$$B = 51.3°$$

Notice we are rounding each angle to the nearest tenth of a degree since the sides we were originally given have three significant digits.

We can find c using the Pythagorean Theorem or one of our trigonometric functions. Let's start with a trigonometric function.

If $\sin A = \dfrac{a}{c}$

then $c = \dfrac{a}{\sin A}$ Multiply each side by c, then divide each side by $\sin A$

$$= \dfrac{2.73}{\sin 38.7°}$$

$$= \dfrac{2.73}{0.6252}$$

$$= 4.37$$ To three significant digits

Using the Pythagorean Theorem, we obtain the same result.

If $c^2 = a^2 + b^2$

then $c = \sqrt{a^2 + b^2}$

$$= \sqrt{(2.73)^2 + (3.41)^2}$$

$$= \sqrt{19.081}$$

$$= 4.37 \quad \blacksquare$$

EXAMPLE 3   The circle in Figure 3 has its center at C and a radius of 18 inches. If triangle ADC is a right triangle and A is 35°, find x, the distance from A to B.

SOLUTION   In triangle ADC, the side opposite A is 18 and the hypotenuse is $x + 18$. We can use $\sin A$ to write an equation that will allow us to solve for x.

$$\sin 35° = \dfrac{18}{x + 18}$$

$$(x + 18)\sin 35° = 18 \qquad \text{Multiply each side by } x + 18$$

$$x + 18 = \dfrac{18}{\sin 35°} \qquad \text{Divide each side by } \sin 35°$$

$$x = \dfrac{18}{\sin 35°} - 18 \qquad \text{Subtract 18 from each side}$$

$$= \dfrac{18}{0.5736} - 18$$

$$= 13 \text{ inches} \qquad \text{To two significant digits} \quad \blacksquare$$

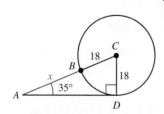

Figure 3

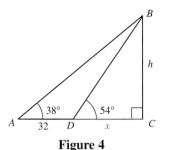

EXAMPLE 4 In Figure 4, the distance from A to D is 32 feet. Use the information in Figure 4 to solve for x, the distance between D and C.

SOLUTION To find x we write two equations, each of which contains the variables x and h. Then we solve each equation for h and set the two expressions for h equal to each other.

Two equations involving both x and h $\begin{cases} \tan 54° = \dfrac{h}{x} \quad\Rightarrow h = x \tan 54° \\[2mm] \tan 38° = \dfrac{h}{x+32} \Rightarrow h = (x+32)\tan 38° \end{cases}$ Solve each equation for h

Figure 4

Setting the two expressions for h equal to each other gives us an equation that involves only x.

Since $h = h$

we have $x \tan 54° = (x + 32)\tan 38°$

$$x \tan 54° = x \tan 38° + 32 \tan 38° \quad \text{Distributive property}$$

$$x \tan 54° - x \tan 38° = 32 \tan 38° \quad \begin{matrix}\text{Subtract } x \tan 38° \\ \text{from each side}\end{matrix}$$

$$x(\tan 54° - \tan 38°) = 32 \tan 38° \quad \begin{matrix}\text{Factor } x \text{ from each} \\ \text{term on the left side}\end{matrix}$$

$$x = \frac{32 \tan 38°}{\tan 54° - \tan 38°} \quad \begin{matrix}\text{Divide each side by} \\ \text{the coefficient of } x\end{matrix}$$

$$= \frac{32(0.7813)}{1.3764 - 0.7813}$$

$$= 42 \text{ ft} \quad \text{To two significant digits}$$

EXAMPLE 5 In the introduction to this section, we gave some of the facts associated with the first Ferris wheel. Figure 5 is a simplified model of that Ferris wheel. If θ is the central angle formed as a rider moves from position P_0 to position P_1, find the rider's height above the ground h when θ is 45.0°.

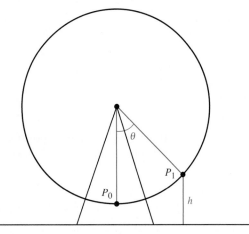

Figure 5

SOLUTION   We know from the introduction to this section that the diameter of the first Ferris wheel was 250 feet, which means the radius was 125 feet. Since the top of the wheel was 264 feet above the ground, the distance from the ground to the bottom of the wheel was 14 feet (the distance to the top minus the diameter of the wheel). To form a right triangle, we draw a horizontal line from P_1 to the vertical line connecting the center of the wheel with P_0. This information is shown in Figure 6.

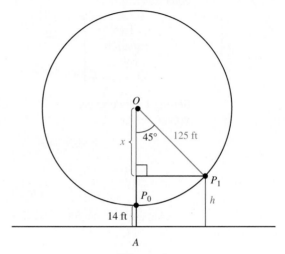

Figure 6

The key to solving this problem is recognizing that x is the difference between OA (the distance from the center of the wheel to the ground) and h. Since OA is 139 feet (the radius of the wheel plus the distance between the bottom of the wheel and the ground: $125 + 14 = 139$), we have

$$x = 139 - h$$

We use a cosine ratio to write an equation that contains h.

$$\cos 45.0° = \frac{x}{125}$$

$$= \frac{139 - h}{125}$$

Solving for h we have

$$125 \cos 45.0° = 139 - h$$

$$h = 139 - 125 \cos 45.0°$$

$$= 139 - 125(0.7071)$$

$$= 139 - 88.4$$

$$= 50.6 \text{ ft}$$

If $\theta = 45.0°$, a rider at position P_1 is $\frac{1}{8}$ of the way around the wheel. At that point, the rider is approximately 50.6 feet above the ground.

GETTING READY FOR CLASS

After reading through the preceding section, respond in your own words and in complete sentences.

a. How do you determine the number of significant digits in a number?

b. Explain the relationship between the accuracy of the sides and the accuracy of the angles in a triangle.

c. In right triangle *ABC*, angle *A* is 40°. How do you find the measure of angle *B*?

d. In right triangle *ABC*, angle *A* is 40° and side *c* is 12 centimeters. How do you find the length of side *a*?

PROBLEM SET 2.3

Problems 1 through 14 refer to right triangle *ABC* with *C* = 90°. Begin each problem by drawing a picture of the triangle with both the given and asked for information labeled appropriately. Also, write your answers for angles in decimal degrees.

1. If *A* = 42° and *c* = 15 ft, find *a*.

2. If *A* = 42° and *c* = 89 cm, find *b*.

3. If *A* = 34° and *a* = 22 m, find *c*.

4. If *A* = 34° and *b* = 55 m, find *c*.

5. If *B* = 24.5° and *c* = 2.34 ft, find *a*.

6. If *B* = 16.9° and *c* = 7.55 cm, find *b*.

7. If *B* = 55.33° and *b* = 12.34 yd, find *a*.

8. If *B* = 77.66° and *a* = 43.21 inches, find *b*.

9. If *a* = 16 cm and *b* = 26 cm, find *A*.

10. If *a* = 42.3 inches and *b* = 32.4 inches, find *B*.

11. If *b* = 6.7 m and *c* = 7.7 m, find *A*.

12. If *b* = 9.8 mm and *c* = 12 mm, find *B*.

13. If *c* = 45.54 ft and *a* = 23.32 ft, find *B*.

14. If *c* = 5.678 ft and *a* = 4.567 ft, find *A*.

Problems 15 through 32 refer to right triangle *ABC* with *C* = 90°. In each case, solve for all the missing parts using the given information. (In Problems 27 through 32, write your angles in decimal degrees.)

15. *A* = 25°, *c* = 24 m

16. *A* = 41°, *c* = 36 m

17. *A* = 32.6°, *a* = 43.4 inches

18. *A* = 48.3°, *a* = 3.48 inches

19. *A* = 10° 42′, *b* = 5.932 cm

20. *A* = 66° 54′, *b* = 28.28 cm

21. *B* = 76°, *c* = 5.8 ft

22. *B* = 21°, *c* = 4.2 ft

23. *B* = 26° 30′, *b* = 324 mm

24. *B* = 53° 30′, *b* = 725 mm

25. *B* = 23.45°, *a* = 5.432 mi

26. *B* = 44.44°, *a* = 5.555 mi

27. *a* = 37 ft, *b* = 87 ft

28. *a* = 91 ft, *b* = 85 ft

29. *a* = 2.75 cm, *c* = 4.05 cm

30. *a* = 62.3 cm, *c* = 73.6 cm

31. *b* = 12.21 inches, *c* = 25.52 inches

32. *b* = 377.3 inches, *c* = 588.5 inches

In Problems 33 and 34, use the information given in the diagram to find A to the nearest degree.

33.

34.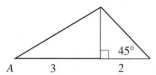

The circle in Figure 7 has a radius of r and center at C. The distance from A to B is x. For Problems 35 through 38, redraw Figure 7, label it as indicated in each problem, and then solve the problem.

35. If $A = 31°$ and $r = 12$, find x.

36. If $C = 26°$ and $r = 20$, find x.

37. If $A = 45°$ and $x = 15$, find r.

38. If $C = 65°$ and $x = 22$, find r.

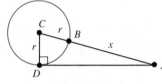

Figure 7

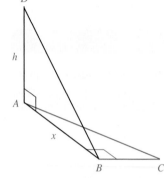

Figure 8

Figure 8 shows two right triangles drawn at $90°$ to each other. For Problem 39 through 42, redraw Figure 8, label it as the problem indicates, and then solve the problem.

39. If $\angle ABD = 27°$, $C = 62°$, and $BC = 42$, find x and then find h.

40. If $\angle ABD = 53°$, $C = 48°$, and $BC = 42$, find x and then find h.

41. If $AC = 32$, $h = 19$, and $C = 41°$, find $\angle ABD$.

42. If $AC = 19$, $h = 32$, and $C = 49°$, find $\angle ABD$.

In Figure 9, the distance from A to D is y, the distance from D to C is x, and the distance from C to B is h. Use Figure 9 to solve Problems 43 through 48.

43. If $A = 41°$, $\angle BDC = 58°$, $AB = 18$, and $DB = 14$, find x, then y.

44. If $A = 32°$, $\angle BDC = 48°$, $AB = 17$, and $DB = 12$, find x, then y.

45. If $A = 41°$, $\angle BDC = 58°$, and $AB = 28$, find h, then x.

46. If $A = 32°$, $\angle BDC = 48°$, and $AB = 56$, find h, then x.

47. If $A = 43°$, $\angle BDC = 57°$, and $y = 10$, find x.

48. If $A = 32°$, $\angle BDC = 41°$, and $y = 14$, find x.

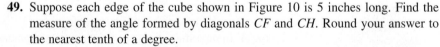

Figure 9

49. Suppose each edge of the cube shown in Figure 10 is 5 inches long. Find the measure of the angle formed by diagonals CF and CH. Round your answer to the nearest tenth of a degree.

50. Suppose each edge of the cube shown in Figure 10 is 3 inches long. Find the measure of the angle formed by diagonals DE and DG. Round your answer to the nearest tenth of a degree.

51. Suppose each edge of the cube shown is x inches long. Find the measure of the angle formed by diagonals CF and CH in Figure 10. Round your answer to the nearest tenth of a degree.

52. Suppose each edge of the cube shown is y inches long. Find the measure of the angle formed by diagonals DE and DG in Figure 10. Round your answer to the nearest tenth of a degree.

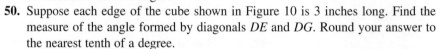

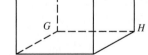

Figure 10

Repeat Example 5 from this section for the following values of θ.

53. $\theta = 30°$

54. $\theta = 60°$

55. $\theta = 120°$

56. $\theta = 135°$

57. Ferris Wheel In 1897, a Ferris wheel was built in Vienna that still stands today. It is named the Riesenrad, which translates to the *Great Wheel*. The diameter of the Riesenrad is 197 feet. The top of the wheel stands 209 feet above the ground. Figure 11 is a model of the Riesenrad with angle θ the central angle that is formed as a rider moves from the initial position P_0 to position P_1. The rider is h feet above the ground at position P_1. (Round to the nearest tenth.)

  a. Find h if θ is 120°.
  b. Find h if θ is 210°.
  c. Find h if θ is 315°.

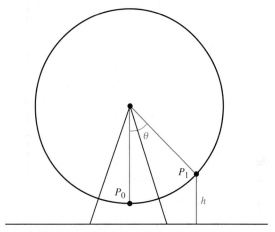

Figure 11

58. Ferris Wheel A Ferris wheel with a diameter of 165 feet was built in St. Louis in 1986. It is called Colossus. The top of the wheel stands 174 feet above the ground. Use the diagram in Figure 11 as a model of Colossus. (Round to the nearest tenth.)

  a. Find h if θ is 150°.
  b. Find h if θ is 240°.
  c. Find h if θ is 315°.

REVIEW PROBLEMS

The following problems review material we covered in Section 1.4.

59. If $\sec B = 2$, find $\cos^2 B$.
60. If $\csc B = 3$, find $\sin^2 B$.
61. If $\sin \theta = 1/3$ and θ terminates in QI, find $\cos \theta$.
62. If $\cos \theta = -2/3$ and θ terminates in QIII, find $\sin \theta$.
63. If $\cos A = 2/5$ with A in QIV, find $\sin A$.
64. If $\sin A = 1/4$ with A in QII, find $\cos A$.

Find the remaining trigonometric ratios for θ, if

65. $\sin \theta = \sqrt{3}/2$ with θ in QII
66. $\cos \theta = 1/\sqrt{5}$ with θ in QIV
67. $\sec \theta = -2$ with θ in QIII
68. $\csc \theta = -2$ with θ in QIII

EXTENDING THE CONCEPTS

69. Human Cannonball In Example 3 of Section 1.2, we found the equation of the path of the human cannonball. At the 1997 Washington County Fair in Oregon, David Smith, Jr., The Bullet, was shot from a cannon. As a human cannonball, he reached a height of 70 feet before landing in a net 160 feet from the cannon. In that example we found the equation that describes his path is

$$y = -\frac{7}{640}(x - 80)^2 + 70 \text{ for } 0 \le x \le 160$$

Graph this equation using the window

$$0 \le x \le 180, \text{ scale} = 20; 0 \le y \le 80, \text{ scale} = 10$$

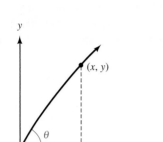

Figure 12

Then zoom in on the curve near the origin until the graph has the appearance of a straight line. Use ⬚TRACE⬚ to find the coordinates of any point on the graph. This point defines an approximate right triangle (Figure 12). Use the triangle to find the angle between the cannon and the horizontal.

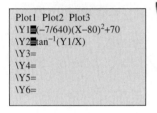

Plot1 Plot2 Plot3
\Y1⬛(-7/640)(X-80)²+70
\Y2⬛tan⁻¹(Y1/X)
\Y3=
\Y4=
\Y5=
\Y6=

Figure 13

70. Human Cannonball To get a better estimate of the angle described in Problem 69, we can use the table feature of your calculator. From Figure 12, we see that for any point (x, y) on the curve, $\theta \approx \tan^{-1}(y/x)$. Define functions Y1 and Y2 as shown in Figure 13. Then set up your table so that you can input the following values of x:

$$x = 10, 5, 1, 0.5, 0.1, 0.01$$

Based on the results, what is the angle between the cannon and the horizontal?

SECTION 2.4 | APPLICATIONS

As mentioned in the introduction to this chapter, we can use right triangle trigonometry to solve a variety of problems, such as problems involving topographic maps. In this section we will see how this is done by looking at a number of applications of right triangle trigonometry.

 EXAMPLE 1 The two equal sides of an isosceles triangle are each 24 centimeters. If each of the two equal angles measures 52°, find the length of the base and the altitude.

SOLUTION An isosceles triangle is any triangle with two equal sides. The angles opposite the two equal sides are called the base angles, and they are always equal. To the left is a picture of our isosceles triangle.

We have labeled the altitude x. We can solve for x using a sine ratio.

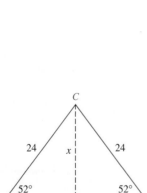

Figure 1

$$\text{If} \quad \sin 52° = \frac{x}{24}$$

$$\text{then} \quad x = 24 \sin 52°$$

$$= 24(0.7880)$$

$$= 19 \text{ cm} \quad \text{Rounded to 2 significant digits}$$

We have labeled half the base with y. To solve for y, we can use a cosine ratio.

$$\text{If} \quad \cos 52° = \frac{y}{24}$$

$$\text{then} \quad y = 24 \cos 52°$$

$$= 24(0.6157)$$

$$= 14.8 \text{ cm}$$

The base is $2y = 2(14.8) = 30$ cm to the nearest centimeter.

For our next applications, we need the following definition.

DEFINITION

An angle measured from the horizontal up is called an *angle of elevation*. An angle measured from the horizontal down is called an *angle of depression* (Figure 2).

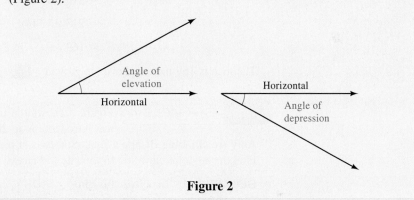

Figure 2

These angles of elevation and depression are always considered positive angles. Also, if an observer positioned at the vertex of the angle views an object in the direction of the nonhorizontal side of the angle, then this side is sometimes called the *line of sight* of the observer.

Figure 3

EXAMPLE 2 If a 75.0-foot flagpole casts a shadow 43.0 feet long, to the nearest 10 minutes what is the angle of elevation of the sun from the tip of the shadow?

SOLUTION We begin by making a diagram of the situation (Figure 3).
If we let θ = the angle of elevation of the sun, then

$$\tan \theta = \frac{75.0}{43.0}$$

$$\tan \theta = 1.7442$$

which means $\theta = \tan^{-1}(1.7442) = 60° \, 10'$ to the nearest 10 minutes.

EXAMPLE 3 A man climbs 213 meters up the side of a pyramid and finds that the angle of depression to his starting point is 52.6°. How high off the ground is he?

SOLUTION Again, we begin by making a diagram of the situation (Figure 4).

Figure 4

If x is the height above the ground, we can solve for x using a sine ratio.

$$\text{If} \quad \sin 52.6° = \frac{x}{213}$$

$$\text{then} \quad x = 213 \sin 52.6°$$

$$= 213(0.7944)$$

$$= 169 \text{ m} \qquad \text{To 3 significant digits}$$

The man is 169 meters above the ground. ■

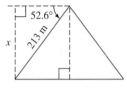 **EXAMPLE 4** Figure 5 shows the topographic map we mentioned in the introduction to this chapter. Suppose Stacey and Amy are climbing Bishop's Peak. Stacey is at position S, and Amy is at position A. Find the angle of elevation from Amy to Stacey.

SOLUTION To solve this problem, we have to use two pieces of information from the legend on the map. First, we need to find the horizontal distance between the two people. The legend indicates that 1 inch on the map corresponds to an actual horizontal distance of 1,600 feet. If we measure the distance from Amy to Stacey with a ruler, we find it is $\frac{3}{8}$ inch. Multiplying this by 1,600, we have

$$\frac{3}{8} \cdot 1,600 = 600 \text{ ft}$$

which is the actual horizontal distance from Amy to Stacey.

Next, we need the vertical distance Stacey is above Amy. We find this by counting the number of contour intervals between them. There are three. From the legend on the map we know that the elevation changes by 40 feet between any two contour lines. Therefore, Stacey is 120 feet above Amy. Figure 6 shows a triangle that models the information we have so far.

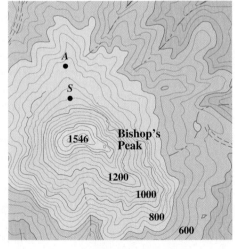

1546 Bishop's Peak

1200

1000

800

600

0   1600

1 inch = 1600 ft

Contour interval = 40 ft

Figure 5

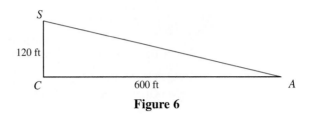

Figure 6

The angle of elevation from Amy to Stacey is angle A. To find A, we use the tangent ratio.

$$\tan A = \frac{120}{600} = 0.2$$

$$A = \tan^{-1}(0.2) = 11.3° \qquad \text{To the nearest tenth of a degree}$$

Amy must look up at $11.3°$ from straight ahead to see Stacey.

Our next applications are concerned with what is called the *bearing of a line*. It is used in navigation and surveying.

> **DEFINITION**
>
> The *bearing of a line l* is the acute angle formed by the north-south line and the line l. The notation used to designate the bearing of a line begins with N or S (for north or south), followed by the number of degrees in the angle, and ends with E or W (for east or west).

Figure 7 shows some examples.

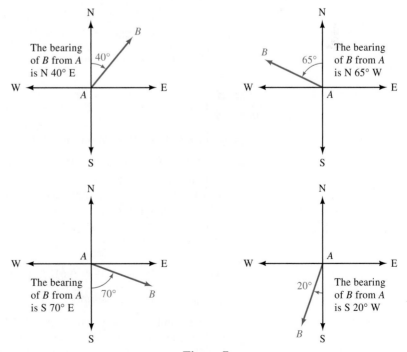

Figure 7

EXAMPLE 5 San Luis Obispo, California, is 12 miles due north of Grover Beach. If Arroyo Grande is 4.6 miles due east of Grover Beach, what is the bearing of San Luis Obispo from Arroyo Grande?

SOLUTION Since we are looking for the bearing of San Luis Obispo *from* Arroyo Grande, we will put our N-S-E-W system on Arroyo Grande (Figure 8).

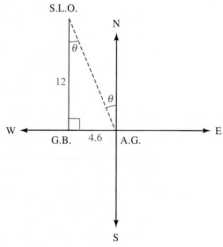

Figure 8

We solve for θ using the tangent ratio.

$$\tan \theta = \frac{4.6}{12}$$

$$\tan \theta = 0.3833$$

$$\theta = \tan^{-1}(0.3833) = 21° \qquad \text{To the nearest degree}$$

The bearing of San Luis Obispo from Arroyo Grande is N 21° W.

EXAMPLE 6 A boat travels on a course of bearing N 52° 40′ E for a distance of 238 miles. How many miles north and how many miles east has the boat traveled?

SOLUTION In the diagram of the situation we put our N-S-E-W system at the boat's starting point (Figure 9).

Solving for *x* with a sine ratio and *y* with a cosine ratio and rounding our answers to three significant digits, we have

If $\quad \sin 52° 40' = \dfrac{x}{238}$ $\qquad\qquad$ If $\quad \cos 52° 40' = \dfrac{y}{238}$

then $\qquad\quad x = 238(0.7951)$ $\qquad$ then $\qquad\qquad y = 238(0.6065)$

$\qquad\qquad\quad = 189 \text{ mi}$ $\qquad\qquad\qquad\qquad\quad = 144 \text{ mi}$

Traveling 238 miles on a line with bearing N 52° 40′ E will get you to the same place as traveling 144 miles north and then 189 miles east.

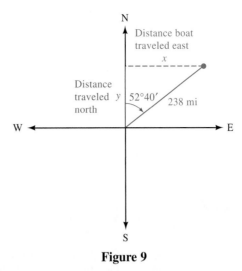

Figure 9

> **EXAMPLE 7**   Figure 10 is a diagram that shows how Diane estimates the height of a flagpole. She can't measure the distance between herself and the flagpole directly because there is a fence in the way. So she stands at point A facing the pole and finds the angle of elevation from point A to the top of the pole to be 61.7°. Then she turns 90° and walks 25.0 feet to point B, where she measures the angle between her path and a line from B to the base of the pole. She finds that angle is 54.5°. Use this information to find the height of the pole.

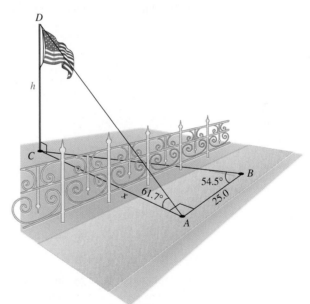

Figure 10

SOLUTION   First we find x in right triangle ABC with a tangent ratio.

$$\tan 54.5° = \frac{x}{25.0}$$

$$x = 25.0 \tan 54.5°$$

$$= 25.0(1.4019)$$

$$= 35.0487 \text{ ft}$$

Without rounding x, we use it to find h in right triangle ACD using another tangent ratio.

$$\tan 61.7° = \frac{h}{35.0487}$$

$$h = 35.0487(1.8572)$$

$$= 65.1 \text{ ft} \qquad \text{To 3 significant digits}$$

Note that if it weren't for the fence, she could measure x directly and use just one triangle to find the height of the flagpole.

EXAMPLE 8   A helicopter is hovering over the desert when it develops mechanical problems and is forced to land. After landing, the pilot radios his position to a pair of radar stations located 25 miles apart along a straight road running north and south. The bearing of the helicopter from one station is N 13° E, and from the other it is S 19° E. After doing a few trigonometric calculations, one of the stations instructs the pilot to walk due west for 3.5 miles to reach the road. Is this information correct?

SOLUTION   Figure 11 is a three-dimensional diagram of the situation. The helicopter is hovering at point D and lands at point C. The radar stations are at A and B, respectively. Since the road runs north and south, the shortest distance from C to the road is due west of C toward point F. To see if the pilot has the correct information, we must find y, the distance from C to F.

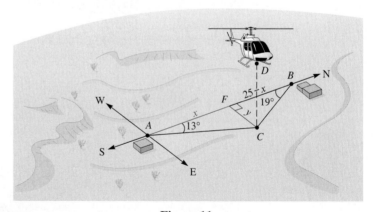

Figure 11

Since the radar stations are 25 miles apart, $AB = 25$. If we let $AF = x$, then $FB = 25 - x$. If we use cotangent ratios in triangles AFC and BFC, we will save ourselves some work.

In triangle AFC $\qquad \cot 13° = \dfrac{x}{y}$

so $\qquad\qquad\qquad\qquad x = y \cot 13°$

In triangle BFC $\qquad \cot 19° = \dfrac{25 - x}{y}$

so $\qquad\qquad\qquad\qquad 25 - x = y \cot 19°$

Solving this equation for x we have

$$-x = -25 + y \cot 19°$$ Add -25 to each side

$$x = 25 - y \cot 19°$$ Multiply each side by -1

Next, we set our two values of x equal to each other.

$$x = x$$

$$y \cot 13° = 25 - y \cot 19°$$

$$y \cot 13° + y \cot 19° = 25$$ Add $y \cot 19°$ to each side

$$y(\cot 13° + \cot 19°) = 25$$ Factor y from each term

$$y = \frac{25}{\cot 13° + \cot 19°}$$ Divide by the coefficient of y

$$= \frac{25}{4.3315 + 2.9042}$$

$$= \frac{25}{7.2357}$$

$$= 3.5 \text{ mi}$$ To 2 significant digits

The information given to the pilot is correct.

GETTING READY FOR CLASS

After reading through the preceding section, respond in your own words and in complete sentences.

a. What is an angle of elevation?

b. What is an angle of depression?

c. How do you define the bearing of line l?

d. Draw a diagram that shows that point B is N 40° E from point A.

PROBLEM SET 2.4

Solve each of the following problems. In each case, be sure to make a diagram of the situation with all the given information labeled.

1. Geometry The two equal sides of an isosceles triangle are each 42 centimeters. If the base measures 30 centimeters, find the height and the measure of the two equal angles.

2. Geometry An equilateral triangle (one with all sides the same length) has an altitude of 4.3 inches. Find the length of the sides.

 3. Geometry The height of a right circular cone is 25.3 centimeters. If the diameter of the base is 10.4 centimeters, what angle does the side of the cone make with the base (Figure 12)?

4. Geometry The diagonal of a rectangle is 348 millimeters, while the longer side is 278 millimeters. Find the shorter side of the rectangle and the angles the diagonal makes with the sides.

25.3 cm

10.4 cm

Figure 12

5. **Length of an Escalator** How long should an escalator be if it is to make an angle of 33° with the floor and carry people a vertical distance of 21 feet between floors?

6. **Height of a Hill** A road up a hill makes an angle of 5.1° with the horizontal. If the road from the bottom of the hill to the top of the hill is 2.5 miles long, how high is the hill?

7. **Length of a Rope** A 72.5-foot rope from the top of a circus tent pole is anchored to the ground 43.2 feet from the bottom of the pole. What angle does the rope make with the pole? (Assume the pole is perpendicular to the ground.)

8. **Angle of a Ladder** A ladder is leaning against the top of a 7.0-foot wall. If the bottom of the ladder is 4.5 feet from the wall, what is the angle between the ladder and the wall?

9. **Angle of Elevation** If a 73.0-foot flagpole casts a shadow 51.0 feet long, what is the angle of elevation of the sun (to the nearest tenth of a degree)?

10. **Angle of Elevation** If the angle of elevation of the sun is 63.4° when a building casts a shadow of 37.5 feet, what is the height of the building?

11. **Angle of Depression** A person standing 150 centimeters from a mirror notices that the angle of depression from his eyes to the bottom of the mirror is 12°, while the angle of elevation to the top of the mirror is 11°. Find the vertical dimension of the mirror (Figure 13).

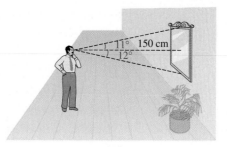

Figure 13

12. **Width of a Sand Pile** A person standing on top of a 15-foot high sand pile wishes to estimate the width of the pile. He visually locates two rocks on the ground below at the base of the sand pile. The rocks are on opposite sides of the sand pile, and he and the two rocks are in the same vertical plane. If the angles of depression from the top of the sand pile to each of the rocks are 27° and 19°, how far apart are the rocks?

Figure 14 shows the topographic map we used in Example 4 of this section. Recall that Stacey is at position *S* and Amy is at position *A*. In Figure 14, Travis, a third hiker, is at position *T*.

13. **Topographic Map Reading** If the distance between *A* and *T* on the map in Figure 14 is 0.5 inch, find each of the following:
    a. the horizontal distance between Amy and Travis
    b. the difference in elevation between Amy and Travis
    c. the angle of elevation from Travis to Amy

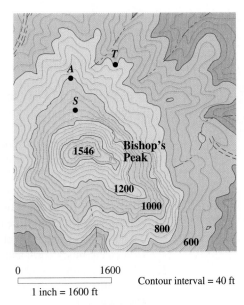

Figure 14

14. **Topographic Map Reading** If the distance between S and T on the map in Figure 14 is $\frac{5}{8}$ inch, find each of the following:
    a. the horizontal distance between Stacey and Travis
    b. the difference in elevation between Stacey and Travis
    c. the angle of elevation from Travis to Stacey

15. **Height of a Door** From a point on the floor the angle of elevation to the top of a door is $47°$, while the angle of elevation to the ceiling above the door is $59°$. If the ceiling is 10 feet above the floor, what is the vertical dimension of the door (Figure 15)?

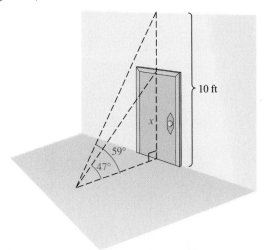

Figure 15

16. **Height of a Building** A man standing on the roof of a building 60.0 feet high looks down to the building next door. He finds the angle of depression to the roof of that building from the roof of his building to be $34.5°$, while the angle of depression from the roof of his building to the bottom of the building next door is $63.2°$. How tall is the building next door?

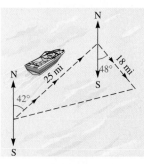

Figure 16

17. Distance and Bearing A boat leaves the harbor entrance and travels 25 miles in the direction N 42° E. The captain then turns the boat 90° and travels another 18 miles in the direction S 48° E. At that time, how far is the boat from the harbor entrance, and what is the bearing of the boat from the harbor entrance (Figure 16)?

18. Distance and Bearing A man wandering in the desert walks 2.3 miles in the direction S 31° W. He then turns 90° and walks 3.5 miles in the direction N 59° W. At that time, how far is he from his starting point, and what is his bearing from his starting point?

19. Distance and Bearing Lompoc, California, is 18 miles due south of Nipomo. Buellton, California, is due east of Lompoc and S 65° E from Nipomo. How far is Lompoc from Buellton?

20. Distance and Bearing A tree on one side of a river is due west of a rock on the other side of the river. From a stake 21 yards north of the rock, the bearing of the tree is S 18.2° W. How far is it from the rock to the tree?

21. Distance A boat travels on a course of bearing N 37° 10′ W for 79.5 miles. How many miles north and how many miles west has the boat traveled?

22. Distance A boat travels on a course of bearing S 63° 50′ E for 100 miles. How many miles south and how many miles east has the boat traveled?

23. Distance In Figure 17, a person standing at point A notices that the angle of elevation to the top of the antenna is 47° 30′. A second person standing 33.0 feet farther from the antenna than the person at A finds the angle of elevation to the top of the antenna to be 42° 10′. How far is the person at A from the base of the antenna?

24. Height of a Tree Two people decide to find the height of a tree. They position themselves 25 feet apart in line with, and on the same side of, the tree. If they find that the angles of elevation from the ground where they are standing to the top of the tree are 65° and 44°, how tall is the tree?

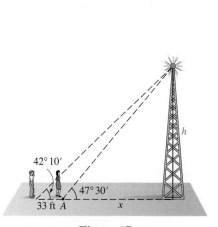

Figure 17

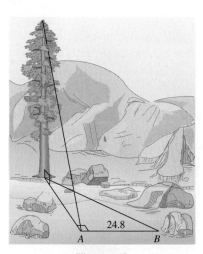

Figure 18

25. Height of a Tree An ecologist wishes to find the height of a redwood tree that is on the other side of a creek, as shown in Figure 18. From point A he finds that the angle of elevation to the top of the tree is 10.7°. He then walks 24.8 feet at a right angle from point A to point B. There he finds that the angle between AB and a line extending from B to the tree is 86.6°. What is the height of the tree?

26. **Rescue** A helicopter makes a forced landing at sea. The last radio signal received at station C gives the bearing of the helicopter from C as N 57.5° E at an altitude of 426 feet. An observer at C sights the helicopter and gives $\angle DCB$ as 12.3°. How far will a rescue boat at A have to travel to reach any survivors at B (Figure 19)?

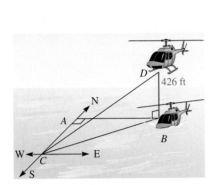

Figure 19

Figure 20

27. **Height of a Flagpole** Two people decide to estimate the height of a flagpole. One person positions himself due north of the pole and the other person stands due east of the pole. If the two people are the same distance from the pole and 25 feet from each other, find the height of the pole if the angle of elevation from the ground to the top of the pole at each person's position is 56° (Figure 20).

28. **Height of a Tree** To estimate the height of a tree, one person positions himself due south of the tree, while another person stands due east of the tree. If the two people are the same distance from the tree and 35 feet from each other, what is the height of the tree if the angle of elevation from the ground at each person's position to the top of the tree is 48°?

29. **Radius of the Earth** A satellite is circling 112 miles above the earth, as shown in Figure 21. When the satellite is directly above point B, angle A is found to be 76.6°. Use this information to find the radius of the earth.

30. **Distance** Suppose Figure 21 is an exaggerated diagram of a plane flying above the earth. If the plane is 4.55 miles above the earth and the radius of the earth is 4,000 miles, how far is it from the plane to the horizon? What is the measure of angle A?

31. **Distance** A ship is anchored off a long straight shoreline that runs north and south. From two observation points 15 miles apart on shore, the bearings of the ship are N 31° E and S 53° E. What is the shortest distance from the ship to the shore?

32. **Distance** Pat and Tim position themselves 2.5 miles apart to watch a missile launch from Vandenberg Air Force Base. When the missile is launched, Pat estimates its bearing from him to be S 75° W, while Tim estimates the bearing of the missile from his position to be N 65° W. If Tim is due south of Pat, how far is Tim from the missile when it is launched?

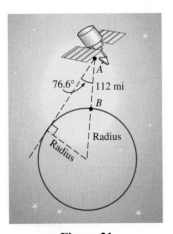

Figure 21

Spiral of Roots Figure 22 shows the Spiral of Roots we mentioned in the previous chapter. Notice that we have labeled the angles at the center of the spiral with θ_1, θ_2, θ_3, and so on.

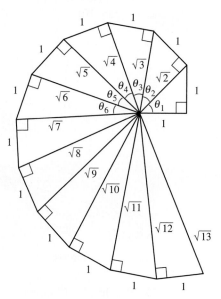

Figure 22

33. Find the values of θ_1, θ_2, and θ_3, accurate to the nearest hundredth of a degree.

34. If θ_n stands for the nth angle formed at the center of the Spiral of Roots, find a formula for $\sin \theta_n$.

REVIEW PROBLEMS

The following problems review material we covered in Section 1.5.

35. Expand and simplify: $(\sin \theta - \cos \theta)^2$

36. Subtract: $\dfrac{1}{\cos \theta} - \cos \theta$

Show that each of the following statements is true by transforming the left side of each one into the right side.

37. $\sin \theta \cot \theta = \cos \theta$

38. $\cos \theta \csc \theta \tan \theta = 1$

39. $\dfrac{\sec \theta}{\tan \theta} = \csc \theta$

40. $(1 - \cos \theta)(1 + \cos \theta) = \sin^2 \theta$

41. $\sec \theta - \cos \theta = \dfrac{\sin^2 \theta}{\cos \theta}$

42. $1 - \dfrac{\cos \theta}{\sec \theta} = \sin^2 \theta$

EXTENDING THE CONCEPTS

43. One of the items we discussed in this section was topographic maps. The process of making one of these maps is an interesting one. It involves aerial photography and different colored projections of the resulting photographs. Research the process used to draw the contour lines on a topographic map, and then give a detailed explanation of that process.

Figure 23

44. Albert lives in New Orleans. At noon on a summer day, the angle of elevation of the sun is 83.5°. The window in Albert's room is 4.0 feet high and 6.5 feet wide. (See Figure 23.)

   a. Calculate the area of the floor surface in Albert's room that is illuminated by the sun when the angle of elevation of the sun is 83.5°.

   b. One winter day, the angle of elevation of the sun outside Albert's window is 36.5°. Will the illuminated area of the floor in Albert's room be greater on the summer day, or on the winter day?

SECTION 2.5 | VECTORS: A GEOMETRIC APPROACH

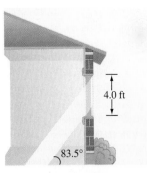

William Rowan Hamilton

By most accounts, the study of vectors is based on the work of Irish mathematician Sir William Hamilton (1805–1865). Hamilton was actually studying complex numbers (a topic we will cover in Chapter 8) when he made the discoveries that led to what we now call vectors.

Today, vectors are treated both algebraically and geometrically. In this section, we will focus our attention on the geometric representation of vectors. We will cover the algebraic approach later in the book in Section 7.5. We begin with a discussion of vector quantities and scalar quantities.

Many of the quantities that describe the world around us have both magnitude and direction, while others have only magnitude. Quantities that have magnitude and direction are called *vector quantities,* while quantities with magnitude only are called *scalars.* Some examples of vector quantities are force, velocity, and acceleration. For example, a car traveling 50 miles per hour due south has a different velocity from another car traveling due north at 50 miles per hour, while a third car traveling at 25 miles per hour due north has a velocity that is different from both of the first two.

One way to represent vector quantities geometrically is with arrows. The direction of the arrow represents the direction of the vector quantity, and the length of the arrow corresponds to the magnitude. For example, the velocities of the three cars we mentioned above could be represented as in Figure 1:

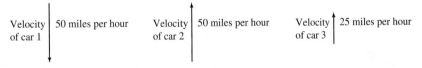

Figure 1

NOTATION

To distinguish between vectors and scalars, we will write the letters used to represent vectors with boldface type, such as **U** or **V**. (When you write them on paper, put an arrow above them like this: $\vec{U}$ or $\vec{V}$.) The magnitude of a vector is represented with absolute value symbols. For example, the magnitude of **V** is written $|\mathbf{V}|$. Table 1 illustrates further.

TABLE 1	
Notation	**The quantity is**
$\mathbf{V}$	a vector
$\vec{V}$	a vector
$\overrightarrow{AB}$	a vector
x	a scalar
$|\mathbf{V}|$	the magnitude of vector $\mathbf{V}$, a scalar

EQUALITY FOR VECTORS

The position of a vector in space is unimportant. Two vectors are equivalent if they have the same magnitude and direction.

In Figure 2, $\mathbf{V}_1 = \mathbf{V}_2 \neq \mathbf{V}_3$. The vectors $\mathbf{V}_1$ and $\mathbf{V}_2$ are equivalent because they have the same magnitude and the same direction. Notice also that $\mathbf{V}_3$ and $\mathbf{V}_4$ have the same magnitude but opposite directions. This means that $\mathbf{V}_4$ is the opposite of $\mathbf{V}_3$, or $\mathbf{V}_4 = -\mathbf{V}_3$.

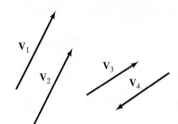

Figure 2

ADDITION AND SUBTRACTION OF VECTORS

The sum of the vectors $\mathbf{U}$ and $\mathbf{V}$, written $\mathbf{U} + \mathbf{V}$, is called the *resultant vector*. It is the vector that extends from the tail of $\mathbf{U}$ to the tip of $\mathbf{V}$ when the tail of $\mathbf{V}$ is placed at the tip of $\mathbf{U}$, as illustrated in Figure 3. Note that this diagram shows the resultant vector to be a diagonal in the parallelogram that has $\mathbf{U}$ and $\mathbf{V}$ as adjacent sides. This being the case, we could also add the vectors by putting the tails of $\mathbf{U}$ and $\mathbf{V}$ together to form adjacent sides of that same parallelogram, as shown in Figure 4. In either case, the resultant vector is the diagonal that starts at the tail of $\mathbf{U}$.

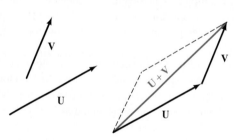

Figure 3

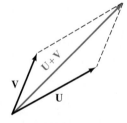

Figure 4

To subtract one vector from another, we can add its opposite. That is,

$$\mathbf{U} - \mathbf{V} = \mathbf{U} + (-\mathbf{V})$$

If $\mathbf{U}$ and $\mathbf{V}$ are the vectors shown in Figure 3, then their difference, $\mathbf{U} - \mathbf{V}$, is shown in Figure 5.

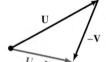

Figure 5

 **EXAMPLE 1** A boat is crossing a river that runs due north. The boat is pointed due east and is moving through the water at 12 miles per hour. If the current of the river is a constant 5.1 miles per hour, find the true course of the boat through the water to two significant digits.

SOLUTION Problems like this are a little difficult to read the first time they are encountered. Even though the boat is "headed" due east, as it travels through the water the water itself is moving northward, so it is actually on a course that will take it east and a little north.

It may help to imagine the first hour of the journey as a sequence of separate events. First, the boat travels 12 miles directly across the river while the river remains still. We can represent this part of the trip by a vector pointing due east with magnitude 12. Then the boat turns off its engine, and we allow the river to flow while the boat remains still (although the boat will be carried by the river). We use a vector pointing due north with magnitude 5.1 to represent this second part of the journey. The end result is that each hour the boat travels both east and north (Figure 6).

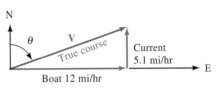

Figure 6

We find θ using a tangent ratio.

$$\tan \theta = \frac{12}{5.1}$$

$$\tan \theta = 2.3529$$

$$\theta = \tan^{-1}(2.3529) = 67° \qquad \text{To the nearest degree}$$

If we let **V** represent the true course of the boat, then we can find the magnitude of **V** using the Pythagorean Theorem or a trigonometric ratio. Using the Pythagorean Theorem, we have

$$|\mathbf{V}| = \sqrt{12^2 + 5.1^2}$$

$$= 13 \qquad \text{To 2 significant digits}$$

The true course of the boat is 13 miles per hour at N 67° E. That is, the vector **V**, which represents the motion of the boat with respect to the banks of the river, has a magnitude of 13 miles per hour and a direction of N 67° E. ▪

HORIZONTAL AND VERTICAL VECTOR COMPONENTS

In Example 1, we saw that **V** was the sum of a vector pointing east (in a horizontal direction) and a second vector pointing north (in a vertical direction). Many times it is convenient to express a vector as the sum of a horizontal and vertical vector. To do so, we first superimpose a coordinate system on the vector in question so that the tail of the vector is at the origin. We call this *standard position* for a vector. Figure 7 shows a vector with magnitude 10 making an angle of 52° with the horizontal.

Two horizontal and vertical vectors whose sum is **V** are shown in Figure 8. Note that in Figure 8 we labeled the horizontal vector as $\mathbf{V}_x$ and the vertical as $\mathbf{V}_y$. We call $\mathbf{V}_x$ the *horizontal vector component* of **V** and $\mathbf{V}_y$ the *vertical vector component* of **V**. We can find the magnitudes of these vectors by using sine and cosine ratios.

$$|\mathbf{V}_x| = |\mathbf{V}| \cos 52° = 10(0.6157) = 6.2 \text{ to the nearest tenth}$$

$$|\mathbf{V}_y| = |\mathbf{V}| \sin 52° = 10(0.7880) = 7.9 \text{ to the nearest tenth}$$

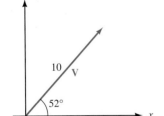

Figure 7

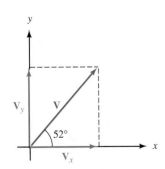

Figure 8

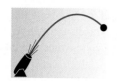

 EXAMPLE 2 The human cannonball is shot from a cannon with an initial velocity of 50 miles per hour at an angle of 60° from the horizontal. Find the magnitudes of the horizontal and vertical vector components of the velocity vector.

SOLUTION Figure 9 is a diagram of the situation.

The magnitudes of $\mathbf{V}_x$ and $\mathbf{V}_y$ from Figure 9 to two significant digits are as follows:

$$\left| \mathbf{V}_x \right| = 50 \cos 60° = 25 \text{ mi/hr}$$

$$\left| \mathbf{V}_y \right| = 50 \sin 60° = 43 \text{ mi/hr}$$

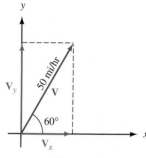

Figure 9

The human cannonball has a horizontal velocity of 25 miles per hour and an initial vertical velocity of 43 miles per hour.

The magnitude of a vector can be written in terms of the magnitude of its horizontal and vertical vector components (Figure 10).

By the Pythagorean Theorem we have

$$\left| \mathbf{V} \right| = \sqrt{\left| \mathbf{V}_x \right|^2 + \left| \mathbf{V}_y \right|^2}$$

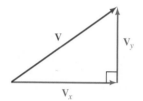

Figure 10

 EXAMPLE 3 An arrow is shot into the air so that its horizontal velocity is 25 feet per second and its vertical velocity is 15 feet per second. Find the velocity of the arrow.

SOLUTION Figure 11 shows the velocity vector along with the angle of elevation of the velocity vector.

The magnitude of the velocity is given by

$$\left| \mathbf{V} \right| = \sqrt{25^2 + 15^2}$$

$$= 29 \text{ ft/sec} \qquad \text{To the nearest whole number}$$

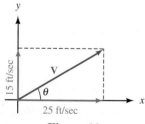

Figure 11

We can find the angle of elevation using a tangent ratio.

$$\tan \theta = \frac{\left| \mathbf{V}_y \right|}{\left| \mathbf{V}_x \right|}$$

$$= \frac{15}{25}$$

$$\tan \theta = 0.6$$

$$\theta = \tan^{-1}(0.6) = 31° \qquad \text{To the nearest degree}$$

The arrow was shot into the air at 29 feet per second at an angle of elevation of 31°.

EXAMPLE 4 A boat travels 72 miles on a course of bearing N 27° E and then changes its course to travel 37 miles at N 55° E. How far north and how far east has the boat traveled on this 109-mile trip?

SOLUTION   We can solve this problem by representing each part of the trip with a vector and then writing each vector in terms of its horizontal and vertical vector components. Figure 12 shows the vectors that represent the two parts of the trip. As Figure 12 indicates, the total distance traveled east is given by the sum of the horizontal components, while the total distance traveled north is given by the sum of the vertical components.

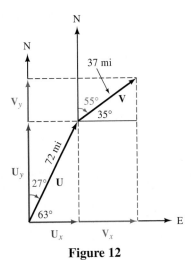

Figure 12

$$\text{Total distance traveled east} = \left| \mathbf{U}_x \right| + \left| \mathbf{V}_x \right|$$
$$= 72 \cos 63° + 37 \cos 35°$$
$$= 63 \text{ mi}$$

$$\text{Total distance traveled north} = \left| \mathbf{U}_y \right| + \left| \mathbf{V}_y \right|$$
$$= 72 \sin 63° + 37 \sin 35°$$
$$= 85 \text{ mi}$$

FORCE

Another important vector quantity is *force*. We can loosely define force as a push or a pull. The most intuitive force in our lives is the force of gravity that pulls us towards the center of the earth. The magnitude of this force is our weight; the direction of this force is always straight down toward the center of the earth.

Imagine a 10-pound bronze sculpture sitting on a coffee table. The force of gravity pulls the sculpture downward with a force of magnitude 10 pounds. At the same time, the table pushes the sculpture upward with a force of magnitude 10 pounds. The net result is that the sculpture remains motionless; the two forces, represented by vectors, add to zero (Figure 13).

Although there may be many forces acting on an object at the same time, if the object is stationary, the sum of the forces must be 0. This leads us to our next definition.

Figure 13

DEFINITION STATIC EQUILIBRIUM

When an object is stationary (at rest) we say it is in a state of *static equilibrium*. When an object is in this state, the sum of the forces acting on the object must be 0.

EXAMPLE 5 Danny is 5 years old and weighs 42.0 pounds. He is sitting on a swing when his sister Stacey pulls him and the swing back horizontally through an angle of 30.0° and then stops. Find the tension in the ropes of the swing and the magnitude of the force exerted by Stacey. (Figure 14 is a diagram of the situation.)

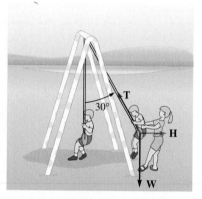

Figure 14

SOLUTION As you can see from Figure 14, there are three forces acting on Danny (and the swing), which we have labeled **W**, **H**, and **T**. The vector **W** is due to the force of gravity, pulling him toward the center of the earth. Its magnitude is $|\mathbf{W}| = 42.0$ pounds, and its direction is straight down. The vector **H** represents the force with which Stacey is pulling Danny horizontally, and **T** is the force acting on Danny in the direction of the ropes. We call this force the *tension* in the ropes.

If we rearrange the vectors in the diagram from Figure 14, we can get a better picture of the situation. Since Stacey is holding Danny in the position shown in Figure 14, he is in a state of static equilibrium. Therefore, the sum of the forces acting on him is 0. We add the three vectors **W**, **H**, and **T** using the tip-to-tail rule as described on page 92. The resultant vector, extending from the tail of **W** to the tip of **T**, must have a length of 0. This means that the tip of **T** has to coincide with the tail of **W**, forming a right triangle, as shown in Figure 15.

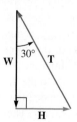

Figure 15

The lengths of the sides of the right triangle shown in Figure 15 are given by the magnitudes of the vectors. We use right triangle trigonometry to find the magnitude of **T**.

$$\cos 30.0° = \frac{|\mathbf{W}|}{|\mathbf{T}|} \qquad \text{Definition of cosine}$$

$$|\mathbf{T}| = \frac{|\mathbf{W}|}{\cos 30.0°} \qquad \text{Solve for } |\mathbf{T}|$$

$$= \frac{42.0}{0.8660} \qquad \text{The magnitude of } \mathbf{W} \text{ is } 42.0$$

$$= 48.5 \text{ lb} \qquad \text{To the nearest tenth of a pound}$$

Next, let's find the magnitude of the force with which Stacey pulls on Danny to keep him in static equilibrium.

$$\tan 30.0° = \frac{|\mathbf{H}|}{|\mathbf{W}|} \qquad \text{Definition of tangent}$$

$$|\mathbf{H}| = |\mathbf{W}| \tan 30.0° \qquad \text{Solve for } |\mathbf{H}|$$

$$= 42.0(0.5774)$$

$$= 24.2 \text{ lb} \qquad \text{To the nearest tenth of a pound}$$

Stacey must pull horizontally with a force of magnitude 24.2 pounds to hold Danny at an angle of 30.0° from vertical. ▊

WORK

One application of vectors that is related to the concept of force is *work*. Intuitively, work is a measure of the "effort" expended when moving an object by applying a force to it. For example, if you have ever had to push a stalled automobile or lift a heavy object, then you have experienced work. Work is calculated as the product of the magnitude of the force and the distance the object is moved. A common unit of measure for work is the foot-pound (ft-lb).

> **DEFINITION   WORK**
>
> If a constant force **F** is applied to an object and moves the object in a straight line a distance d in the direction of the force, then the *work* W performed by the force is $W = |\mathbf{F}| \cdot d$.

It is important to understand that the definition of work requires that the object move in the direction of the force. If this is not the case, then only part of the force may count when calculating the work. Our next example illustrates this situation.

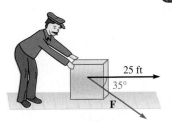

EXAMPLE 6   A shipping clerk pushes a heavy package across the floor. He applies a force of 60 pounds in a downward direction, making an angle of 35° with the horizontal. If the package is moved 25 feet, how much work is done by the clerk?

SOLUTION   A diagram of the problem is shown in Figure 16.

Since the package moves in a horizontal direction, and not in the direction of the force, we must first find the amount of force that is directed horizontally. This will be the magnitude of the horizontal vector component of **F**.

25 ft
35°
F

Figure 16

$$|\mathbf{F}_x| = |\mathbf{F}| \cos 35° = 60 \cos 35° \text{ lb}$$

To find the work, we multiply this value by the distance the package moves.

$$\text{Work} = (60 \cos 35°)(25)$$

$$= 1{,}200 \text{ ft-lb} \qquad \text{To 2 significant digits}$$

1,200 foot-pounds of work are performed by the shipping clerk in moving the package. ▨

GETTING READY FOR CLASS

After reading through the preceding section, respond in your own words and in complete sentences.

a. What is a vector?

b. How are a vector and a scalar different?

c. How do you add two vectors geometrically?

d. Explain what is meant by static equilibrium.

PROBLEM SET 2.5

Draw vectors representing the following velocities:

1. 30 mi/hr due north

2. 30 mi/hr due south

3. 30 mi/hr due east

4. 30 mi/hr due west

 5. 50 cm/sec N 30° W

6. 50 cm/sec N 30° E

7. 20 ft/min S 60° E

8. 20 ft/min S 60° W

9. Bearing and Distance A person is riding in a hot air balloon. For the first hour and a half the wind current is a constant 22.0 miles per hour in the direction N 37.5° E. Then the wind current changes to 18.5 miles per hour and heads the balloon in the direction S 52.5° E. If this continues for another 2 hours, how far is the balloon from its starting point? What is the bearing of the balloon from its starting point (Figure 17)?

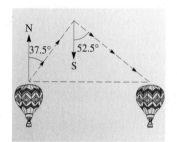

Figure 17

10. Bearing and Distance Two planes take off at the same time from an airport. The first plane is flying at 255 miles per hour on a bearing of S 45° E. The second plane is flying in the direction S 45° W at 275 miles per hour. If there are no wind currents blowing, how far apart are they after 2 hours? What is the bearing of the second plane from the first after 2 hours?

11. Navigation If a navigation error puts a plane 3.0° off course, how far off course is the plane after flying 130 miles?

12. Navigation A ship is 2.8° off course. If the ship is traveling at 14 miles per hour, how far off course will it be after 2 hours?

Each problem below refers to a vector $\mathbf{V}$ with magnitude $|\mathbf{V}|$ that forms an angle θ with the positive x-axis. In each case, give the magnitudes of the horizontal and vertical vector components of $\mathbf{V}$, namely $\mathbf{V}_x$ and $\mathbf{V}_y$, respectively.

13. $|\mathbf{V}| = 13.8, \theta = 24.2°$

14. $|\mathbf{V}| = 17.6, \theta = 67.2°$

15. $|\mathbf{V}| = 420, \theta = 36° \, 10'$

16. $|\mathbf{V}| = 380, \theta = 16° \, 40'$

17. $|\mathbf{V}| = 64, \theta = 0°$

18. $|\mathbf{V}| = 48, \theta = 90°$

For each problem below, the magnitudes of the horizontal and vertical vector components, $\mathbf{V}_x$ and $\mathbf{V}_y$, of vector $\mathbf{V}$ are given. In each case find the magnitude of $\mathbf{V}$.

19. $|\mathbf{V}_x| = 35.0, |\mathbf{V}_y| = 26.0$

20. $|\mathbf{V}_x| = 45.0, |\mathbf{V}_y| = 15.0$

21. $|\mathbf{V}_x| = 4.5, |\mathbf{V}_y| = 3.8$

22. $|\mathbf{V}_x| = 2.2, |\mathbf{V}_y| = 5.8$

23. **Velocity of a Bullet** A bullet is fired into the air with an initial velocity of 1,200 feet per second at an angle of 45° from the horizontal. Find the magnitude of the horizontal and vertical vector components of the velocity vector.

24. **Velocity of a Bullet** A bullet is fired into the air with an initial velocity of 1,800 feet per second at an angle of 60° from the horizontal. Find the magnitudes of the horizontal and vertical vector components of the velocity vector.

25. **Distance Traveled by a Bullet** Use the results of Problem 23 to find the horizontal distance traveled by the bullet in 3 seconds. (Neglect the resistance of air on the bullet.)

26. **Distance Traveled by a Bullet** Use the results of Problem 24 to find the horizontal distance traveled by the bullet in 2 seconds.

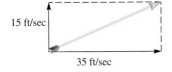

15 ft/sec

35 ft/sec

Figure 18

27. **Velocity of an Arrow** An arrow is shot into the air so that its horizontal velocity is 35.0 feet per second and its vertical velocity is 15.0 feet per second (Figure 18). Find the velocity of the arrow.

28. **Velocity of an Arrow** The horizontal and vertical components of the velocity of an arrow shot into the air are 15.0 feet per second and 25.0 feet per second, respectively. Find the velocity of the arrow.

29. **Distance** A ship travels 135 kilometers on a bearing of S 42° E. How far east and how far south has it traveled?

30. **Distance** A plane flies for 3 hours at 230 kilometers per hour in the direction S 35° W. How far west and how far south does it travel in the 3 hours?

31. **Distance** A plane travels 175 miles on a bearing of N 18° E and then changes its course to N 49° E and travels another 120 miles. Find the total distance traveled north and the total distance traveled east.

32. **Distance** A ship travels in the direction S 12° E for 68 miles and then changes its course to S 60° E and travels another 112 miles. Find the total distance south and the total distance east that the ship traveled.

33. **Static Equilibrium** Repeat the swing problem shown in Example 5 if Stacey pulls Danny through an angle of 45.0° and then holds him at static equilibrium. Find the magnitudes of both **H** and **T**.

34. **Static Equilibrium** The diagram in Figure 15 of this section would change if Stacey were to push Danny forward through an angle of 30° and then hold him in that position. Draw the diagram that corresponds to this new situation.

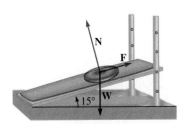

Figure 19

35. **Force** A 10-pound weight is lying on a sit-up bench at the gym. If the bench is inclined at an angle of 15°, there are three forces acting on the weight, as shown in Figure 19. **N** is called the normal force and it acts in the direction perpendicular to the bench. **F** is the force due to friction that holds the weight on the bench. If the weight does not move, then the sum of these three forces is 0. Find the magnitude of **N** and the magnitude of **F**.

36. **Force** Repeat Problem 35 for a 25.0-pound weight and a bench inclined at 10.0°.

37. **Force** Danny and Stacey have gone from the swing (Example 5) to the slide at the park. The slide is inclined at an angle of 52.0°. Danny weighs 42.0 pounds. He is sitting in a cardboard box with a piece of wax paper on the bottom. Stacey is at the top of the slide holding on to the cardboard box (Figure 20). Find the magnitude of the force Stacey must pull with, in order to keep Danny from sliding down the slide. (We are assuming that the wax paper makes the slide into a frictionless surface, so that the only force keeping Danny from sliding is the force with which Stacey pulls.)

38. **Force** Repeat Problem 37 for a slide inclined at an angle of 65.0°.

39. **Work** A package is pushed across a floor a distance of 75 feet by exerting a force of 40 pounds downward at an angle of 20° with the horizontal. How much work is done?

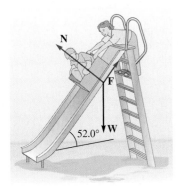

Figure 20

Figure 21

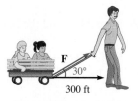

Figure 22

40. Work A package is pushed across a floor a distance of 50 feet by exerting a force of 15 pounds downward at an angle of 25° with the horizontal. How much work is done?

41. Work An automobile is pushed down a level street by exerting a force of 85 pounds at an angle of 15° with the horizontal (Figure 21). How much work is done in pushing the car 100 feet?

42. Work Mark pulls Allison and Mattie in a wagon by exerting a force of 25 pounds on the handle at an angle of 30° with the horizontal (Figure 22). How much work is done by Mark in pulling the wagon 300 feet?

REVIEW PROBLEMS

The problems that follow review material we covered in Section 1.3.

43. Draw 135° in standard position, locate a convenient point on the terminal side, and then find sin 135°, cos 135°, and tan 135°.

44. Draw −270° in standard position, locate a convenient point on the terminal side, and then find sine, cosine, and tangent of −270°.

45. Find $\sin \theta$ and $\cos \theta$ if the terminal side of θ lies along the line $y = 2x$ in quadrant I.

46. Find $\sin \theta$ and $\cos \theta$ if the terminal side of θ lies along the line $y = -x$ in quadrant II.

47. Find x if the point $(x, -8)$ is on the terminal side of θ and $\sin \theta = -\frac{4}{5}$.

48. Find y if the point $(-6, y)$ is on the terminal side of θ and $\cos \theta = -\frac{3}{5}$.

> ## CHAPTER 2 SUMMARY

EXAMPLES

1.

Definition II for Trigonometric Functions [2.1]

If triangle ABC is a right triangle with $C = 90°$, then the six trigonometric functions for angle A are

$$\sin A = \frac{\text{side opposite } A}{\text{hypotenuse}} = \frac{a}{c}$$

$$\cos A = \frac{\text{side adjacent } A}{\text{hypotenuse}} = \frac{b}{c}$$

$$\tan A = \frac{\text{side opposite } A}{\text{side adjacent } A} = \frac{a}{b}$$

$$\cot A = \frac{\text{side adjacent } A}{\text{side opposite } A} = \frac{b}{a}$$

$$\sec A = \frac{\text{hypotenuse}}{\text{side adjacent } A} = \frac{c}{b}$$

$$\csc A = \frac{\text{hypotenuse}}{\text{side opposite } A} = \frac{c}{a}$$

$\sin A = \dfrac{3}{5} = \cos B$

$\cos A = \dfrac{4}{5} = \sin B$

$\tan A = \dfrac{3}{4} = \cot B$

$\cot A = \dfrac{4}{3} = \tan B$

$\sec A = \dfrac{5}{4} = \csc B$

$\csc A = \dfrac{5}{3} = \sec B$

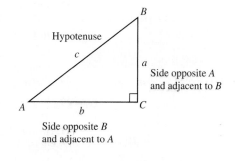

2. $\sin 3° = \cos 87°$
$\cos 10° = \sin 80°$
$\tan 15° = \cot 75°$
$\cot A = \tan (90° - A)$
$\sec 30° = \csc 60°$
$\csc 45° = \sec 45°$

Cofunction Theorem [2.1]

A trigonometric function of an angle is always equal to the cofunction of its complement. In symbols, since the complement of x is $90° - x$, we have

$$\sin x = \cos (90° - x)$$

$$\sec x = \csc (90° - x)$$

$$\tan x = \cot (90° - x)$$

3. The values given in the table are called *exact values* because they are not decimal approximations as you would find on a calculator.

Trigonometric Functions of Special Angles [2.1]

θ	0°	30°	45°	60°	90°
$\sin \theta$	0	$\dfrac{1}{2}$	$\dfrac{1}{\sqrt{2}}$ or $\dfrac{\sqrt{2}}{2}$	$\dfrac{\sqrt{3}}{2}$	1
$\cos \theta$	1	$\dfrac{\sqrt{3}}{2}$	$\dfrac{1}{\sqrt{2}}$ or $\dfrac{\sqrt{2}}{2}$	$\dfrac{1}{2}$	0
$\tan \theta$	0	$\dfrac{1}{\sqrt{3}}$ or $\dfrac{\sqrt{3}}{3}$	1	$\sqrt{3}$	undefined

4. $\quad 47° \ 30'$
$\underline{+ \ 23° \ 50'}$
$70° \ 80' = 71° \ 20'$

Degrees, Minutes, and Seconds [2.2]

There are 360° (degrees) in one complete rotation, 60′ (minutes) in one degree, and 60″ (seconds) in one minute. This is equivalent to saying 1 minute is 1/60 of a degree, and 1 second is 1/60 of a minute.

5. $74.3° = 74° + 0.3°$
$\quad\quad = 74° + 0.3(60')$
$\quad\quad = 74° + 18'$
$\quad\quad = 74° \ 18'$

$42° \ 48' = 42° + \left(\dfrac{48}{60}\right)°$
$\quad\quad\quad = 42° + 0.8°$
$\quad\quad\quad = 42.8°$

Converting to and from Decimal Degrees [2.2]

To convert from decimal degrees to degrees and minutes, multiply the fractional part of the angle (that which follows the decimal point) by 60 to get minutes.

To convert from degrees and minutes to decimal degrees, divide minutes by 60 to get the fractional part of the angle.

6. These angles and sides correspond in accuracy.

$a = 24$	$A = 39°$
$a = 5.8$	$A = 45°$
$a = 62.3$	$A = 31.3°$
$a = 0.498$	$A = 42.9°$
$a = 2.77$	$A = 37° \ 10'$
$a = 49.87$	$A = 43° \ 18'$
$a = 6.932$	$A = 24.81°$

Significant Digits [2.3]

The number of *significant digits* (or figures) in a number is found by counting the number of digits from left to right, beginning with the first nonzero digit on the left and disregarding the decimal point.

The relationship between the accuracy of the sides in a triangle and the accuracy of the angles in the same triangle is given below.

Accuracy of Sides	**Accuracy of Angles**
Two significant digits	Nearest degree
Three significant digits	Nearest 10 minutes or tenth of a degree
Four significant digits	Nearest minute or hundredth of a degree

7.

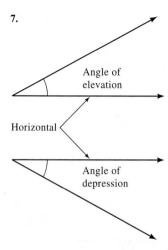

Angle of Elevation and Angle of Depression [2.4]

An angle measured from the horizontal up is called an *angle of elevation*. An angle measured from the horizontal down is called an *angle of depression*.

8.

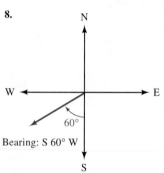

Bearing: S 60° W

Direction [2.4]

One way to specify the direction of a line or vector is called *bearing*. The notation used to designate bearing begins with N or S, followed by the number of degrees in the angle, and ends with E or W, as in S 60° W.

9. If a car is traveling at 50 miles per hour due south, then its velocity can be represented with a vector.

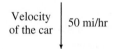

Vectors [2.5]

Quantities that have both magnitude and direction are called *vector quantities,* while quantities that have only magnitude are called *scalar quantities.* We represent vectors graphically by using arrows. The length of the arrow corresponds to the magnitude of the vector, and the direction of the arrow corresponds to the direction of the vector. In symbols, we denote the magnitude of vector **V** with $|\mathbf{V}|$.

10.

Addition of Vectors [2.5]

The sum of the vectors **U** and **V**, written **U** + **V**, is the vector that extends from the tail of **U** to the tip of **V** when the tail of **V** coincides with the tip of **U**. The sum of **U** and **V** is called the *resultant vector.*

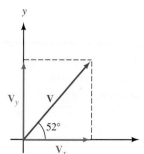

Horizontal and Vertical Vector Components [2.5]

The horizontal and vertical vector components of vector **V** are the horizontal and vertical vectors whose sum is **V**. The horizontal vector component is denoted by $\mathbf{V}_x$, and the vertical vector component is denoted by $\mathbf{V}_y$.

CHAPTER 2 TEST

Find sin A, cos A, tan A, and sin B, cos B, and tan B in right triangle ABC, with $C = 90°$, if

1. $a = 1$ and $b = 2$

2. $b = 3$ and $c = 6$

3. $a = 3$ and $c = 5$

4. $a = 5$ and $b = 12$

Fill in the blanks to make each statement true.

5. $\sin 14° = \cos$ _____

6. $\sec$ _____ $= \csc 73°$

Simplify each expression as much as possible.

7. $\sin^2 45° + \cos^2 30°$

8. $\tan 45° + \cot 45°$

9. $\sin^2 60° - \cos^2 30°$

10. $\dfrac{1}{\sec 30°}$

11. Add $48° \, 18'$ and $24° \, 52'$.

12. Subtract $15° \, 32'$ from $25° \, 15'$.

Convert to degrees and minutes.

13. $73.2°$

14. $16.45°$

Convert to decimal degrees.

15. $2° \, 48'$

16. $79° \, 30'$

Use a calculator to find the following:

17. $\sin 24° \, 20'$

18. $\cos 37.8°$

19. $\tan 63° \, 50'$

20. $\cot 71° \, 20'$

Use a calculator to find θ to the nearest tenth of a degree if θ is an acute angle and

21. $\tan \theta = 0.0816$

22. $\sec \theta = 1.923$

23. $\sin \theta = 0.9465$

24. $\cos \theta = 0.9730$

Give the number of significant digits in each number.

25. 49.35

26. 0.0028

The following problems refer to right triangle ABC with $C = 90°$. In each case, find all the missing parts.

27. $a = 68.0$ and $b = 104$

28. $a = 24.3$ and $c = 48.1$

29. $b = 305$ and $B = 24.9°$

30. $c = 0.462$ and $A = 35° \, 30'$

31. Geometry If the altitude of an isosceles triangle is 25 centimeters and each of the two equal angles measures $17°$, how long are the two equal sides?

32. Angle of Elevation If the angle of elevation of the sun is $75° \, 30'$, how tall is a post that casts a shadow 1.5 feet long?

33. Distance Two guy wires from the top of a 35-foot tent pole are anchored to the ground below by two stakes so that the two stakes and the tent pole lie along the same line. If the angles of depression from the top of the pole to each of the stakes are $47°$ and $43°$, how far apart are the stakes? (Assume the tent pole is perpendicular to the ground and between the two stakes.)

34. Vector Magnitude If vector $\mathbf{V}$ has magnitude 5.0 and makes an angle of $30°$ with the positive x-axis, find the magnitudes of the horizontal and vertical vector components of $\mathbf{V}$.

35. Vector Angle Vector $\mathbf{V}$ has a horizontal vector component with magnitude 11 and a vertical vector component with magnitude 31. What is the acute angle formed by $\mathbf{V}$ and the positive x-axis?

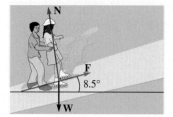

Figure 1

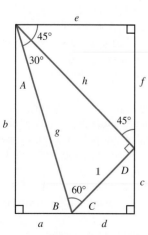

Figure 2

36. Velocity A bullet is fired into the air with an initial velocity of 800 feet per second at an angle of 62° from the horizontal. Find the magnitudes of the horizontal and vertical vector components of the velocity vector.

37. Distance and Bearing A ship travels 120 miles on a bearing of S 60° E. How far east and how far south has the ship traveled?

38. Force Tyler and his cousin Kelly have attached a rope to the branch of a tree and tied a board to the other end to form a swing. Tyler sits on the board while his cousin pushes him through an angle of 25.5° and holds him there. If Tyler weighs 95.5 pounds, find the magnitude of the force Kelly must push with horizontally to keep Tyler in static equilibrium. See Figure 1.

39. Force After they are done swinging, Tyler and Kelly decide to rollerskate. They come to a hill that is inclined at 8.5°. Tyler pushes Kelly halfway up the hill and then holds her there (Figure 2). If Kelly weighs 58.0 pounds, find the magnitude of the force Tyler must push with to keep Kelly from rolling down the hill. (We are assuming that the rollerskates make the hill into a frictionless surface so that the only force keeping Kelly from rolling backwards down the hill is the force Tyler is pushing with.)

40. Work William pulls his cousins Hannah and Serah on a sled by exerting a force of 35 pounds on a rope at an angle of elevation of 40° (Figure 3). How much work is done by William in pulling the sled 80 feet?

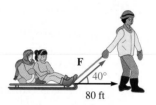

Figure 3

CHAPTER 2 GROUP PROJECT

THE 15°–75°–90° TRIANGLE

Objective: To find exact values of the six trigonometric functions for 15° and 75° angles.

In Section 2.1 we saw how to obtain exact values for the trigonometric functions using two special triangles, the 30°–60°–90° triangle and the 45°–45°–90° triangle. In this project you will use both of these triangles to create a 15°–75°–90° triangle. Once your triangle is complete, you will be able to use it to find additional exact values of the trigonometric functions.

The diagram shown in Figure 1 is called an Ailles rectangle. It is named after a high school teacher, Doug Ailles, who first discovered it. Notice that the rectangle is constructed from four triangles. The triangle in the middle is a 30°–60°–90° triangle, and the triangle above it in the upper right corner is a 45°–45°–90° triangle.

1. Find the measures of the remaining four angles labeled *A*, *B*, *C*, and *D*.
2. Use your knowledge of a 30°–60°–90° triangle to find sides *g* and *h*.
3. Now use your knowledge of a 45°–45°–90° triangle to find sides *e* and *f*.
4. Find the remaining sides *a*, *b*, *c*, and *d*.

Figure 1

5. Now that the diagram is complete, you should have a 15°–75°–90° triangle with all three sides labeled. Use this triangle to fill in the table below with exact values for each trigonometric function.

θ	**15°**	**75°**
$\sin \theta$		
$\cos \theta$		
$\tan \theta$		
$\csc \theta$		
$\sec \theta$		
$\cot \theta$		

6. Use your calculator to check each of the values in your table. For example, find sin 15° on your calculator and make sure it agrees with the value in your table. (You will need to approximate the value in your table with your calculator also.)

CHAPTER **2** RESEARCH PROJECT

SHADOWY ORIGINS

The origins of the sine function are found in the tables of chords for a circle constructed by the Greek astronomers/mathematicians Hipparchus and Ptolemy. However, the origins of the tangent and cotangent functions lie primarily with Arabic and Islamic astronomers. Called the *umbra recta* and *umbra versa,* their connection was not to the chord of a circle but to the gnomon of a sundial.

Research the origins of the tangent and cotangent functions. What was the connection to sundials? What other contributions did Arabic astronomers make to trigonometry? Write a paragraph or two about your findings.

RADIAN MEASURE

The great book of nature can be read only by those who know the language in which it is written. And this language is mathematics.

Galileo

INTRODUCTION

The first cable car was built in San Francisco, California, by Andrew Smith Hallidie. It was completed on August 2, 1873. Today, three cable car lines remain in operation in San Francisco. The cars are driven by steel cables that vary in length from 17,600 feet to 21,500 feet. Inside the powerhouse, these cables are driven by large 14-foot drive wheels, called *sheaves,* that turn at a rate of 19 revolutions per minute (Figure 1).

With radian measure, we can solve problems that involve angular speeds such as revolutions per minute. For example, we will be able to use this information to determine the linear speed at which the cable car travels.

Figure 1

STUDY SKILLS FOR CHAPTER 3

The study skills for this chapter focus on getting ready to take an exam.

1. **Getting Ready to Take an Exam** Try to arrange your daily study habits so that you have very little studying to do the night before your next exam. The next two goals will help you achieve goal number 1.
2. **Review with the Exam in Mind** Review material that will be covered on the next exam every day. Your review should consist of working problems. Preferably, the problems you work should be problems from the list of difficult problems you have been recording.
3. **Continue to List Difficult Problems** This study skill was started in the previous chapter. You should continue to list and rework the problems that give you the most difficulty. Use this list to study for the next exam. Your goal is to go into the next exam knowing that you can successfully work any problem from your list of hard problems.

4. Pay Attention to Instructions Taking a test is different than doing homework. When you take a test, the problems will be mixed up. When you do your homework, you usually work a number of similar problems. Sometimes students do very well on their homework, but become confused when they see the same problems on a test because they have not paid attention to the instructions on their homework.

SECTION 3.1 | REFERENCE ANGLE

In the previous chapter we found exact values for trigonometric functions of certain angles between 0° and 90°. By using what are called *reference angles,* we can find exact values for trigonometric functions of angles outside the interval 0° to 90°.

> **DEFINITION**
>
> The *reference angle* (sometimes called *related angle*) for any angle θ in standard position is the positive acute angle between the terminal side of θ and the x-axis. In this book, we will denote the reference angle for θ by $\hat{\theta}$.

Note that, for this definition, $\hat{\theta}$ is always positive and always between 0° and 90°. That is, a reference angle is always an acute angle.

EXAMPLE 1 Name the reference angle for each of the following angles.

a. 30° **b.** 135° **c.** 240° **d.** 330°

SOLUTION We draw each angle in standard position. The reference angle is the positive acute angle formed by the terminal side of the angle in question and the x-axis (Figure 1).

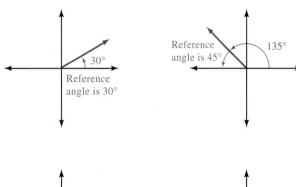

Figure 1

We can generalize the results of Example 1 as follows: If θ is a positive angle between 0° and 360°, then

$$\text{If } \theta \in \text{QI, then } \hat{\theta} = \theta$$

$$\text{If } \theta \in \text{QII, then } \hat{\theta} = 180° - \theta$$

$$\text{If } \theta \in \text{QIII, then } \hat{\theta} = \theta - 180°$$

$$\text{If } \theta \in \text{QIV, then } \hat{\theta} = 360° - \theta$$

We can use our information on reference angles and the signs of the trigonometric functions to write the following theorem.

REFERENCE ANGLE THEOREM

A trigonometric function of an angle and its reference angle differ at most in sign.

We will not give a detailed proof of this theorem, but rather, justify it by example. Let's look at the sines of all the angles between 0° and 360° that have a reference angle of 30°. These angles are 30°, 150°, 210°, and 330° (Figure 2).

$$\sin 150° = \sin 30° = \frac{1}{2}$$

They differ in sign only.

$$\sin 210° = \sin 330° = -\frac{1}{2}$$

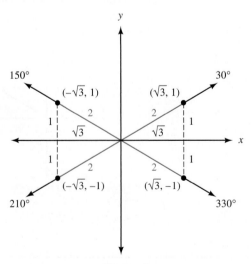

Figure 2

As you can see, any angle with a reference angle of 30° will have a sine of $\frac{1}{2}$ or $-\frac{1}{2}$. The sign, + or −, will depend on the quadrant in which the angle terminates. Using this discussion as justification, we write the following steps to find trigonometric functions of angles between 0° and 360°.

Step 1. Find $\hat{\theta}$, the reference angle.

Step 2. Determine the sign of the trigonometric function based on the quadrant in which θ terminates.

Step 3. Write the original trigonometric function of θ in terms of the same trigonometric function of $\hat{\theta}$.

Step 4. Find the trigonometric function of $\hat{\theta}$.

EXAMPLE 2 Find the exact value of sin 240°.

SOLUTION For this first example, we will list the steps just given as we use them. Figure 3 is a diagram of the situation.

Step 1. We find $\hat{\theta}$ by subtracting 180° from θ.

$$\hat{\theta} = 240° - 180° = 60°$$

Step 2. Since θ terminates in quadrant III, and the sine function is negative in quadrant III, our answer will be negative. That is, in this case, $\sin\theta = -\sin\hat{\theta}$.

Step 3. Using the results of Steps 1 and 2, we write

$$\sin 240° = -\sin 60°$$

Step 4. We finish by finding sin 60°.

$$\sin 240° = -\sin 60° = -\frac{\sqrt{3}}{2}$$

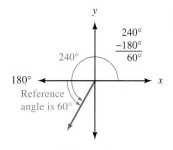

$$\begin{array}{r} 240° \\ -180° \\ \hline 60° \end{array}$$

Figure 3

EXAMPLE 3 Find the exact value of tan 315°.

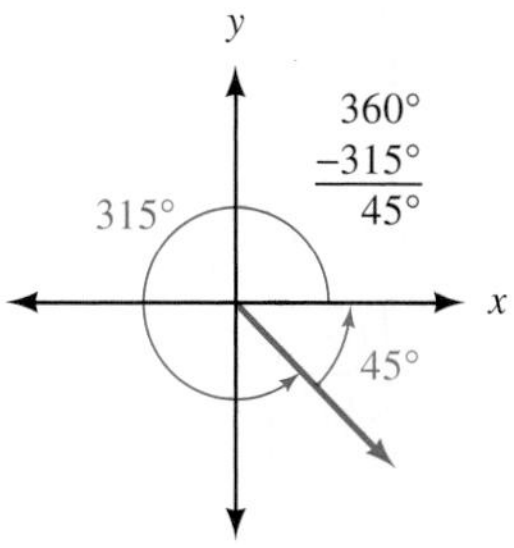

SOLUTION The reference angle is 360° − 315° = 45°. Since 315° terminates in quadrant IV, its tangent will be negative (Figure 4).

$$\tan 315° = -\tan 45° \qquad \text{Because tangent is negative in QIV}$$

$$= -1$$

$$\begin{array}{r} 360° \\ -315° \\ \hline 45° \end{array}$$

Figure 4

EXAMPLE 4 Find the exact value of csc 300°.

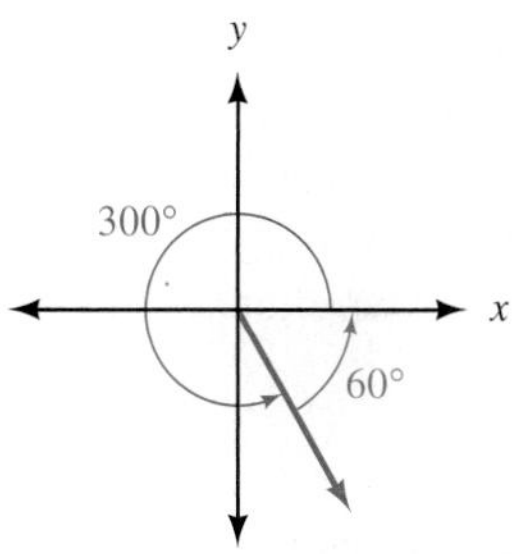

SOLUTION The reference angle is 360° − 300° = 60°. To find the exact value of csc 60°, we use the fact that cosecant and sine are reciprocals (Figure 5).

$$\csc 300° = -\csc 60° \qquad \text{Because cosecant is negative in QIV}$$

$$= -\frac{1}{\sin 60°}$$

$$= -\frac{1}{\sqrt{3}/2}$$

$$= -\frac{2}{\sqrt{3}}$$

Figure 5

Recall from Section 1.2 that coterminal angles always differ from each other by multiples of 360°. For example, −45° and 315° are coterminal, as are 10° and 370° (Figure 6).

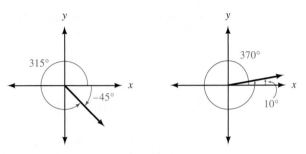

Figure 6

The trigonometric functions of an angle and any angle coterminal to it are always equal. For sine and cosine, we can write this in symbols as follows:

for any integer k,

$$\sin (\theta + 360°k) = \sin \theta \qquad \text{and} \qquad \cos (\theta + 360°k) = \cos \theta$$

To find values of trigonometric functions for an angle larger than 360° or smaller than 0°, we simply find an angle between 0° and 360° that is coterminal to it and then use the steps outlined in Examples 2 through 4.

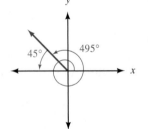

Figure 7

EXAMPLE 5 Find the exact value of cos 495°.

SOLUTION By subtracting 360° from 495°, we obtain 135°, which is coterminal to 495°. The reference angle for 135° is 45°. Since 495° terminates in quadrant II, its cosine is negative (Figure 7).

$$\cos 495° = \cos 135° \qquad 495° \text{ and } 135° \text{ are coterminal}$$
$$= -\cos 45° \qquad \text{In QII } \cos \theta = -\cos \hat{\theta}$$
$$= -\frac{1}{\sqrt{2}} \qquad \text{Exact value} \quad \blacksquare$$

APPROXIMATIONS

To find trigonometric functions of angles that do not lend themselves to exact values, we use a calculator. To find an approximation for sin θ, cos θ, or tan θ, we simply enter the angle and press the appropriate key on the calculator. Check to see that you can obtain the following values for sine, cosine, and tangent of 250° and −160° on your calculator. (These answers are rounded to the nearest ten-thousandth.) Make sure your calculator is set to degree mode.

$$\sin 250° = -0.9397 \qquad \sin (-160°) = -0.3420$$
$$\cos 250° = -0.3420 \qquad \cos (-160°) = -0.9397$$
$$\tan 250° = 2.7475 \qquad \tan (-160°) = 0.3640$$

To find csc 250°, sec 250°, and cot 250°, we must use the reciprocals of sin 250°, cos 250°, and tan 250°.

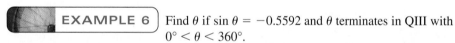

	Scientific Calculator	**Graphing Calculator**

$\csc 250° = \dfrac{1}{\sin 250°}$ 250 [sin] [1/x]   1 [÷] [sin] [(] 250 [)] [ENTER]

$= -1.0642$

$\sec 250° = \dfrac{1}{\cos 250°}$ 250 [cos] [1/x]   1 [÷] [cos] [(] 250 [)] [ENTER]

$= -2.9238$

$\cot 250° = \dfrac{1}{\tan 250°}$ 250 [tan] [1/x]   1 [÷] [tan] [(] 250 [)] [ENTER]

$= 0.3640$

Next we use a calculator to find an approximation for θ, given one of the trigonometric functions of θ and the quadrant in which θ terminates. To do this we will need to use the [sin⁻¹], [cos⁻¹], and [tan⁻¹] keys, which we introduced earlier in Section 2.2. In using one of these keys to find a reference angle, always enter a *positive* value for the trigonometric function. (*Remember:* A reference angle is between 0° and 90°, so all six trigonometric functions of that angle will be positive.)

CALCULATOR NOTE   Some calculators do not have a key labeled as [sin⁻¹]. You may need to press a combination of keys, such as [inv] [sin], [arc] [sin], or [2nd] [sin].

EXAMPLE 6   Find θ if $\sin \theta = -0.5592$ and θ terminates in QIII with $0° < \theta < 360°$.

SOLUTION   First we find the reference angle using the [sin⁻¹] key with the positive value 0.5592. From this, we get $\hat{\theta} = 34°$. The angle in QIII whose reference angle is 34° is (Figure 8)

$$\theta = 180° + \hat{\theta}$$
$$= 180° + 34°$$
$$= 214°$$

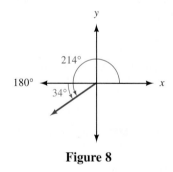

Figure 8

If we wanted to list *all* the angles that terminate in quadrant III and have a sine of -0.5592, we would write

$$\theta = 214° + 360°k \qquad \text{where } k \text{ is an integer}$$

This gives us all angles coterminal with 214°.

CALCULATOR NOTE   There is a big difference between the [sin] key and the [sin⁻¹] key. The [sin] key can be used to find the value of the sine function for *any* angle. However, the reverse is not true with the [sin⁻¹] key. For example, if you were to try Example 6 by entering -0.5592 and using the [sin⁻¹] key, you would not obtain the angle 214°. To see why this happens you will have to wait until we cover inverse trigonometric functions in Chapter 4. In the meantime, to use a calculator on this kind of problem, use it to find the reference angle and then proceed as we did in Example 6.

EXAMPLE 7 Find θ to the nearest tenth of a degree if $\tan \theta = -0.8541$ and θ terminates in QIV with $0° < \theta < 360°$.

SOLUTION Using the $\boxed{\tan^{-1}}$ key with the positive value 0.8541 gives the reference angle $\hat{\theta} = 40.5°$. The angle in QIV with a reference angle of 40.5° is

$$\theta = 360° - 40.5° = 319.5°$$

Again, if we wanted to list *all* angles in quadrant IV with a tangent of -0.8541, we would write

$$\theta = 319.5° + 360°k \qquad k \text{ is an integer}$$

to include not only 319.5° but all angles coterminal with it.

 EXAMPLE 8 Find θ if $\sin \theta = -\frac{1}{2}$ and θ terminates in QIII with $0° < \theta < 360°$.

SOLUTION Using our calculator, we find the reference angle to be 30°. The angle in QIII with a reference angle of 30° is $180° + 30° = 210°$.

EXAMPLE 9 Find θ to the nearest degree if $\sec \theta = 3.8637$ and θ terminates in QIV with $0° < \theta < 360°$.

SOLUTION To find the reference angle on a calculator, we must use the fact that $\sec \theta$ is the reciprocal of $\cos \theta$. That is,

$$\text{If } \sec \theta = 3.8637, \quad \text{then } \cos \theta = \frac{1}{3.8637}$$

From this last line we see that the keys to press are:

Scientific Calculator

3.8637 $\boxed{1/x}$ $\boxed{\cos^{-1}}$

Graphing Calculator

$\boxed{\cos^{-1}}$ $\boxed{(}$ 3.8637 $\boxed{x^{-1}}$ $\boxed{)}$ $\boxed{\text{ENTER}}$

To the nearest degree, the reference angle is $\hat{\theta} = 75°$. Since we want θ to terminate in QIV, we subtract 75° from 360° to get

$$\theta = 360° - 75° = 285°$$

We can check our result on a calculator by entering 285°, finding its cosine, and then finding the reciprocal of the result.

Scientific Calculator

285 $\boxed{\cos}$ $\boxed{1/x}$

Graphing Calculator

1 $\boxed{\div}$ $\boxed{\cos}$ $\boxed{(}$ 285 $\boxed{)}$ $\boxed{\text{ENTER}}$

The calculator gives a result of 3.8637.

EXAMPLE 10 Find θ to the nearest degree if $\cot \theta = -1.6003$ and θ terminates in QII, with $0° < \theta < 360°$.

SOLUTION To find the reference angle on a calculator, we ignore the negative sign in -1.6003 and use the fact that $\cot \theta$ is the reciprocal of $\tan \theta$.

$$\text{If } \cot \theta = 1.6003, \quad \text{then } \tan \theta = \frac{1}{1.6003}$$

From this last line we see that the keys to press are:

Scientific Calculator   **Graphing Calculator**

1.6003 $\boxed{1/x}$ $\boxed{\tan^{-1}}$ $\boxed{\tan^{-1}}$ $\boxed{(}$ 1.6003 $\boxed{x^{-1}}$ $\boxed{)}$ $\boxed{\text{ENTER}}$

To the nearest degree, the reference angle is $\hat{\theta} = 32°$. Since we want θ to terminate in QII, we subtract 32° from 180° to get $\theta = 148°$.

Again, we can check our result on a calculator by entering 148°, finding its tangent, and then finding the reciprocal of the result.

Scientific Calculator   **Graphing Calculator**

148 $\boxed{\tan}$ $\boxed{1/x}$ 1 $\boxed{÷}$ $\boxed{\tan}$ $\boxed{(}$ 148 $\boxed{)}$ $\boxed{\text{ENTER}}$

The calculator gives a result of -1.6003. ■

GETTING READY FOR CLASS

After reading through the preceding section, respond in your own words and in complete sentences.

a. Define reference angle.

b. State the reference angle theorem.

c. What is the first step in finding the exact value of cos 495°?

d. Explain how to find θ to the nearest tenth of a degree, if $\tan \theta = -0.8541$ and θ terminates in QIV with $0° < \theta < 360°$.

PROBLEM SET 3.1

Draw each of the following angles in standard position and then name the reference angle.

1. 210°   **2.** 150°   **3.** 143.4°   **4.** 253.8°
5. 311.7°   **6.** 93.2°   **7.** 195° 10′   **8.** 171° 40′
9. −300°   **10.** −330°   **11.** −120°   **12.** −150°

Find the exact value of each of the following.

13. cos 225°   **14.** cos 135°   **15.** sin 120°   **16.** sin 210°
17. tan 135°   **18.** tan 315°   **19.** cos 240°   **20.** cos 150°
21. csc 330°   **22.** sec 330°   **23.** sec 300°   **24.** csc 300°
25. sin 390°   **26.** cos 420°   **27.** cot 480°   **28.** cot 510°

Use a calculator to find the following.

29. cos 347°   **30.** cos 238°   **31.** sec 101.8°   **32.** csc 166.7°
33. tan 143.4°   **34.** tan 253.8°   **35.** sec 311.7°   **36.** csc 93.2°
37. cot 390°   **38.** cot 420°   **39.** csc 575.4°   **40.** sec 590.9°
41. cos (−315°)   **42.** sin (−225°)   **43.** tan 195° 10′   **44.** tan 171° 40′
45. sec 314° 40′   **46.** csc 670° 20′   **47.** sin (−120°)   **48.** cos (−150°)

Use a calculator to find θ to the nearest tenth of a degree, if $0° < \theta < 360°$ and

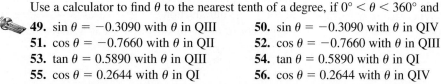

49. sin θ = −0.3090 with θ in QIII   **50.** sin θ = −0.3090 with θ in QIV
51. cos θ = −0.7660 with θ in QII   **52.** cos θ = −0.7660 with θ in QIII
53. tan θ = 0.5890 with θ in QIII   **54.** tan θ = 0.5890 with θ in QI
55. cos θ = 0.2644 with θ in QI   **56.** cos θ = 0.2644 with θ in QIV

57. $\sin \theta = 0.9652$ with θ in QII

58. $\sin \theta = 0.9652$ with θ in QI

59. $\sec \theta = 1.4325$ with θ in QIV

60. $\csc \theta = 1.4325$ with θ in QII

61. $\csc \theta = 2.4957$ with θ in QII

62. $\sec \theta = -3.4159$ with θ in QII

63. $\cot \theta = -0.7366$ with θ in QII

64. $\cot \theta = -0.1234$ with θ in QIV

65. $\sec \theta = -1.7876$ with θ in QIII

66. $\csc \theta = -1.7876$ with θ in QIII

Find θ, if $0° < \theta < 360°$ and

67. $\sin \theta = -\dfrac{\sqrt{3}}{2}$ and θ in QIII

68. $\sin \theta = -\dfrac{1}{\sqrt{2}}$ and θ in QIII

69. $\cos \theta = -\dfrac{1}{\sqrt{2}}$ and θ in QII

70. $\cos \theta = -\dfrac{\sqrt{3}}{2}$ and θ in QIII

71. $\sin \theta = -\dfrac{\sqrt{3}}{2}$ and θ in QIV

72. $\sin \theta = \dfrac{1}{\sqrt{2}}$ and θ in QII

73. $\tan \theta = \sqrt{3}$ and θ in QIII

74. $\tan \theta = \dfrac{1}{\sqrt{3}}$ and θ in QIII

75. $\sec \theta = -2$ with θ in QII

76. $\csc \theta = 2$ with θ in QII

77. $\csc \theta = \sqrt{2}$ with θ in QII

78. $\sec \theta = \sqrt{2}$ with θ in QIV

79. $\cot \theta = -1$ with θ in QIV

80. $\cot \theta = \sqrt{3}$ with θ in QIII

REVIEW PROBLEMS

The problems that follow review material we covered in Sections 1.1 and 2.1. Give the complement and supplement of each angle.

81. $70°$ **82.** $120°$ **83.** x **84.** $90° - y$

85. If the longest side in a $30°$–$60°$–$90°$ triangle is 10, find the length of the other two sides.

86. If the two shorter sides of a $45°$–$45°$–$90°$ triangle are both $\dfrac{3}{4}$, find the length of the hypotenuse.

Simplify each expression by substituting values from the table of exact values and then simplifying the resulting expression.

87. $\sin 30° \cos 60°$

88. $4 \sin 60° - 2 \cos 30°$

89. $\sin^2 45° + \cos^2 45°$

90. $(\sin 45° + \cos 45°)^2$

SECTION 3.2 | RADIANS AND DEGREES

Although James Thomson is credited with the first printed use of the term *radian,* it is believed that the concept of radian measure was originally proposed by Roger Cotes (1682–1716), who was also the first to calculate 1 radian in degrees.

If you think back to the work you have done with functions of the form $y = f(x)$ in your algebra class, you will see that the variables x and y were always real numbers. The trigonometric functions we have worked with so far have had the form $y = f(\theta)$, where θ is measured in degrees. In order to apply the knowledge we have about functions from algebra to our trigonometric functions, we need to write our angles as real numbers, not degrees. The key to doing this is called *radian measure.* Radian measure is a relatively new concept in the history of mathematics. The term *radian* was first introduced by physicist James T. Thomson in examination questions in 1873. The introduction of radian measure will allow us to do a number of useful things. For instance, in Chapter 4 we will graph the function $y = \sin x$ on a rectangular coordinate system, where the units on the x and y axes are given by real numbers, just as they would be if we were graphing $y = 2x + 3$ or $y = x^2$. To understand the definition for radian measure, we have to recall from geometry that a central angle in a circle is an angle with its vertex at the center of the circle. Here is the definition for an angle with a measure of 1 radian.

DEFINITION

In a circle, a central angle that cuts off an arc equal in length to the radius of the circle has a measure of 1 *radian* (rad). Figure 1 illustrates this.

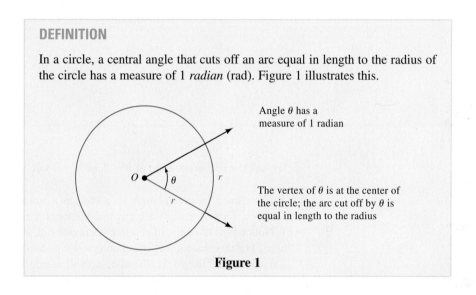

Angle θ has a measure of 1 radian

The vertex of θ is at the center of the circle; the arc cut off by θ is equal in length to the radius

Figure 1

To find the radian measure of *any* central angle, we must find how many radii are in the arc it cuts off. To do so, we divide the arc length by the radius. If the radius is 2 centimeters and the arc cut off by central angle θ is 6 centimeters, then the radian measure of θ is $\frac{6}{2} = 3$ rad. Here is the formal definition:

DEFINITION   RADIAN MEASURE

If a central angle θ, in a circle of radius r, cuts off an arc of length s, then the measure of θ, in radians, is given by s/r (Figure 2).

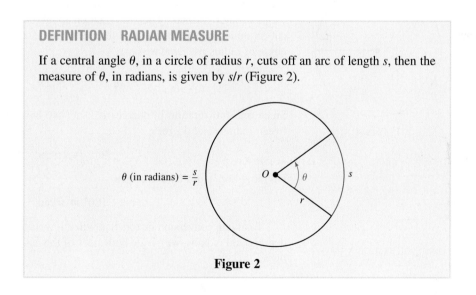

θ (in radians) $= \dfrac{s}{r}$

Figure 2

As you will see later in this section, one radian is equal to approximately $57.3°$.

EXAMPLE 1 A central angle θ in a circle of radius 3 centimeters cuts off an arc of length 6 centimeters What is the radian measure of θ?

SOLUTION We have $r = 3$ cm and $s = 6$ cm (Figure 3); therefore,

$$\theta \text{ (in radians)} = \frac{s}{r}$$

$$= \frac{6 \text{ cm}}{3 \text{ cm}}$$

$$= 2$$

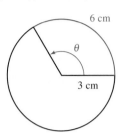

Figure 3

We say the radian measure of θ is 2, or $\theta = 2$ rad.

NOTE Since radian measure is defined as a ratio, s/r, technically it is a unitless measure. That is, radians are just real numbers. To see why, look again at Example 1. Notice how the units of centimeters divide out, leaving the number 2 without any units. For this reason, it is common practice to omit the word *radian* when using radian measure. (To avoid confusion, we will sometimes use the term *rad* in this book as if it were a unit.) If no units are showing, an angle is understood to be measured in radians; with degree measure, the degree symbol ° must be written.

$$\theta = 2 \quad \text{means the measure of } \theta \text{ is 2 radians}$$

$$\theta = 2° \quad \text{means the measure of } \theta \text{ is 2 degrees}$$

To see the relationship between degrees and radians, we can compare the number of degrees and the number of radians in one full rotation (Figure 4).

The angle formed by one full rotation about the center of a circle of radius r will cut off an arc equal to the circumference of the circle. Since the circumference of a circle of radius r is $2\pi r$, we have

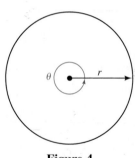

Figure 4

θ measures one full rotation $\qquad \theta = \dfrac{2\pi r}{r} = 2\pi \qquad$ The measure of θ in radians is 2π

Since one full rotation in degrees is 360°, we have the following relationship between radians and degrees.

$$360° = 2\pi \text{ rad}$$

Dividing both sides by 2 we have

$$180° = \pi \text{ rad}$$

To obtain conversion factors that will allow us to change back and forth between degrees and radians, we divide both sides of this last equation alternately by 180 and by π.

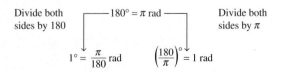

Divide both sides by 180 $\qquad$ ——— $180° = \pi$ rad ——— $\qquad$ Divide both sides by π

$$1° = \frac{\pi}{180} \text{ rad} \qquad \left(\frac{180}{\pi}\right)° = 1 \text{ rad}$$

To gain some insight into the relationship between degrees and radians, we can approximate π with 3.14 to obtain the approximate number of degrees in 1 radian.

$$1 \text{ rad} = 1\left(\frac{180}{\pi}\right)^\circ$$

$$\approx 1\left(\frac{180}{3.14}\right)^\circ$$

$$= 57.3^\circ \qquad \text{To the nearest tenth}$$

We see that 1 radian is approximately 57°. A radian is much larger than a degree. Figure 5 illustrates the relationship between 20° and 20 radians.

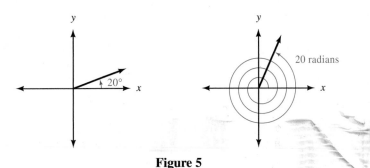

Figure 5

Here are some further conversions between degrees and radians.

CONVERTING FROM DEGREES TO RADIANS

 Convert 45° to radians.

SOLUTION   Since $1^\circ = \dfrac{\pi}{180}$ radians, and 45° is the same as 45(1°), we have

$$45^\circ = 45\left(\frac{\pi}{180}\right) \text{rad} = \frac{\pi}{4} \text{ rad}$$

as illustrated in Figure 6.

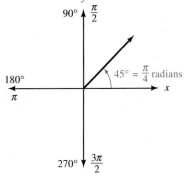

Figure 6

When we have our answer in terms of π, as in $\pi/4$, we are writing an exact value. If we wanted a decimal approximation, we would substitute 3.14 for π.

$$\text{Exact value} \qquad \frac{\pi}{4} \approx \frac{3.14}{4} = 0.785 \qquad \text{Approximate value}$$

Note that if we wanted the radian equivalent of 90°, we could simply multiply $\pi/4$ by 2, since 90° = 2 × 45°.

$$90° = 2 \times 45° = 2 \times \frac{\pi}{4} = \frac{\pi}{2}$$

EXAMPLE 3 Convert 450° to radians.

SOLUTION As illustrated in Figure 7, multiplying by $\pi/180$ we have

$$450° = 450\left(\frac{\pi}{180}\right) = \frac{5\pi}{2} \text{ rad}$$

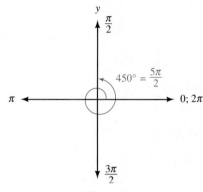

Figure 7

Again, $5\pi/2$ is the exact value. If we wanted a decimal approximation, we would substitute 3.14 for π.

Exact value $\qquad \dfrac{5\pi}{2} \approx \dfrac{5(3.14)}{2} = 7.85 \qquad$ Approximate value

CONVERTING FROM RADIANS TO DEGREES

EXAMPLE 4 Convert $\pi/6$ to degrees.

SOLUTION To convert from radians to degrees, we multiply by $180/\pi$.

$$\frac{\pi}{6} \text{ (rad)} = \frac{\pi}{6}\left(\frac{180}{\pi}\right)°$$
$$= 30°$$

Note that 60° is twice 30°, so $2(\pi/6) = \pi/3$ must be the radian equivalent of 60°. Figure 8 illustrates.

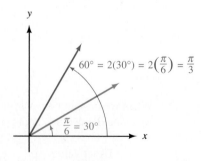

Figure 8

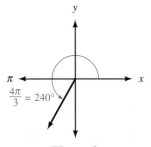

Figure 9

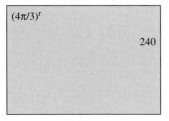

Figure 10

EXAMPLE 5) Convert $4\pi/3$ to degrees.

SOLUTION   Multiplying by $180/\pi$ we have

$$\frac{4\pi}{3} \text{ (rad)} = \frac{4\pi}{3} \left(\frac{180}{\pi}\right)^{\circ} = 240^{\circ}$$

Note that the reference angle for the angle shown in Figure 9 can be given in either degrees or radians.

$$\text{In degrees: } \hat{\theta} = 240^{\circ} - 180^{\circ} = 60^{\circ}$$

$$\text{In radians: } \hat{\theta} = \frac{4\pi}{3} - \pi = \frac{4\pi}{3} - \frac{3\pi}{3} = \frac{\pi}{3}$$

CALCULATOR NOTE   Some calculators have the ability to convert angles from degree measure to radian measure, and vice versa. For example, Figure 10 shows how Example 5 could be done on a graphing calculator that is set to degree mode. Consult your calculator manual to see if your model is able to perform angle conversions.

As is apparent from the preceding examples, changing from degrees to radians and radians to degrees is simply a matter of multiplying by the appropriate conversion factors.

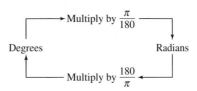

Table 1 displays the conversions between degrees and radians for the special angles, and also summarizes the exact values of the sine, cosine, and tangent of these angles for your convenience.

TABLE 1

θ		Value of Trigonometric Function		
Degrees	*Radians*	*sin θ*	*cos θ*	*tan θ*
0°	0	0	1	0
30°	$\dfrac{\pi}{6}$	$\dfrac{1}{2}$	$\dfrac{\sqrt{3}}{2}$	$\dfrac{1}{\sqrt{3}}$
45°	$\dfrac{\pi}{4}$	$\dfrac{1}{\sqrt{2}}$	$\dfrac{1}{\sqrt{2}}$	1
60°	$\dfrac{\pi}{3}$	$\dfrac{\sqrt{3}}{2}$	$\dfrac{1}{2}$	$\sqrt{3}$
90°	$\dfrac{\pi}{2}$	1	0	Undefined
180°	π	0	−1	0
270°	$\dfrac{3\pi}{2}$	−1	0	Undefined
360°	2π	0	1	0

EXAMPLE 6 Find $\sin \dfrac{\pi}{6}$.

SOLUTION Since $\pi/6$ and $30°$ are equivalent, so are their sines.

$$\sin \frac{\pi}{6} = \sin 30° = \frac{1}{2}$$

CALCULATOR NOTE To work this problem on a calculator, we must first set the calculator to radian mode. (Consult the manual that came with your calculator to see how to do this.) If your calculator does not have a key labeled π, use 3.1416. Here is the sequence to key in your calculator to work the problem given in Example 6.

Scientific Calculator   **Graphing Calculator**

3.1416 $\boxed{\div}$ 6 $\boxed{=}$ $\boxed{\sin}$ $\boxed{\sin}$ $\boxed{(}$ $\boxed{\pi}$ $\boxed{\div}$ 6 $\boxed{)}$ $\boxed{\text{ENTER}}$

EXAMPLE 7 Find $4 \sin \dfrac{7\pi}{6}$.

SOLUTION Since $7\pi/6$ terminates in QIII, its sine will be negative. The reference angle is $7\pi/6 - \pi = \pi/6$ (Figure 11).

$$4 \sin \frac{7\pi}{6} = 4\left(-\sin \frac{\pi}{6}\right)$$

$$= 4\left(-\frac{1}{2}\right)$$

$$= -2$$

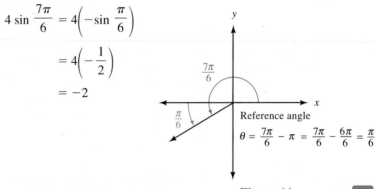

Reference angle
$$\theta = \frac{7\pi}{6} - \pi = \frac{7\pi}{6} - \frac{6\pi}{6} = \frac{\pi}{6}$$

Figure 11

EXAMPLE 8 Evaluate $4 \sin (2x + \pi)$ when $x = \pi/6$.

SOLUTION Substituting $\pi/6$ for x and simplifying, we have

$$4 \sin \left(2 \cdot \frac{\pi}{6} + \pi\right) = 4 \sin \left(\frac{\pi}{3} + \pi\right)$$

$$= 4 \sin \frac{4\pi}{3}$$

$$= 4 \left(-\sin \frac{\pi}{3}\right)$$

$$= 4 \left(-\frac{\sqrt{3}}{2}\right)$$

$$= -2\sqrt{3}$$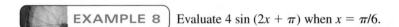

GETTING READY FOR CLASS

After reading through the preceding section, respond in your own words and in complete sentences.

a. What is a radian?

b. What is radian measure?

c. Explain how to convert from radian measure to degree measure.

d. Explain how to convert from degree measure to radian measure.

PROBLEM SET 3.2

Find the radian measure of angle θ, if θ is a central angle in a circle of radius r, and θ cuts off an arc of length s.

 1. $r = 3$ cm, $s = 9$ cm   **2.** $r = 6$ cm, $s = 3$ cm

3. $r = 10$ inches, $s = 5$ inches   **4.** $r = 5$ inches, $s = 10$ inches

5. $r = 4$ inches, $s = 12\pi$ inches   **6.** $r = 3$ inches, $s = 12$ inches

7. $r = \dfrac{1}{4}$ cm, $s = \dfrac{1}{2}$ cm   **8.** $r = \dfrac{1}{4}$ cm, $s = \dfrac{1}{8}$ cm

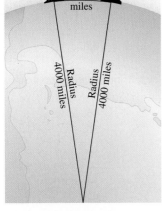

Los Angeles   San Francisco

450 miles

Radius 4000 miles

Radius 4000 miles

Figure 12

9. Angle Between Cities Los Angeles and San Francisco are approximately 450 miles apart on the surface of the earth. Assuming that the radius of the earth is 4,000 miles, find the radian measure of the central angle with its vertex at the center of the earth that has Los Angeles on one side and San Francisco on the other side (Figure 12).

10. Angle Between Cities Los Angeles and New York City are approximately 2,500 miles apart on the surface of the earth. Assuming that the radius of the earth is 4,000 miles, find the radian measure of the central angle with its vertex at the center of the earth that has Los Angeles on one side and New York City on the other side.

For each angle below:

a. Draw the angle in standard position.

b. Convert to radian measure using exact values.

c. Name the reference angle in both degrees and radians.

 11. $30°$ **12.** $60°$ **13.** $90°$ **14.** $270°$

15. $260°$ **16.** $340°$ **17.** $-150°$ **18.** $-210°$

19. $420°$ **20.** $390°$ **21.** $-135°$ **22.** $-120°$

For Problems 23–26, use 3.1416 for π unless your calculator has a key marked π.

23. Use a calculator to convert $120° \, 40'$ to radians. Round your answer to the nearest hundredth. (First convert to decimal degrees, then multiply by the appropriate conversion factor to convert to radians.)

24. Use a calculator to convert $256° \, 20'$ to radians to the nearest hundredth of a radian.

25. Use a calculator to convert $1'$ (1 min) to radians to three significant digits.

26. Use a calculator to convert $1°$ to radians to three significant digits.

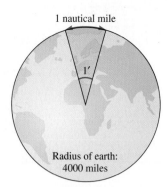

1 nautical mile

Radius of earth:
4000 miles

Figure 13

27. Nautical Miles If a central angle with its vertex at the center of the earth has a measure of $1'$, then the arc on the surface of the earth that is cut off by this angle has a measure of 1 nautical mile (Figure 13). Find the number of regular (statute) miles in 1 nautical mile to the nearest hundredth of a mile. (Use 4,000 miles for the radius of the earth.)

28. Nautical Miles If two ships are 20 nautical miles apart on the ocean, how many statute miles apart are they? (Use the result of Problem 27 to do the calculations.)

29. Clock Through how many radians does the minute hand of a clock turn during a 5-minute period?

30. Clock Through how many radians does the minute hand of a clock turn during a 25-minute period?

For each angle below:
a. Convert to degree measure.
b. Draw the angle in standard position.
c. Label the reference angle in both degrees and radians.

31. $\dfrac{\pi}{3}$ **32.** $\dfrac{\pi}{4}$ **33.** $\dfrac{2\pi}{3}$ **34.** $\dfrac{3\pi}{4}$

35. $\dfrac{-7\pi}{6}$ **36.** $\dfrac{-5\pi}{6}$ **37.** $\dfrac{5\pi}{3}$ **38.** $\dfrac{7\pi}{3}$

39. 4π **40.** 3π **41.** $\dfrac{\pi}{12}$ **42.** $\dfrac{5\pi}{12}$

Use a calculator to convert each of the following to degree measure to the nearest tenth of a degree.

43. 1 **44.** 2 **45.** 1.3 **46.** 2.4
47. 0.75 **48.** 0.25 **49.** 5 **50.** 6

Give the exact value of each of the following:

51. $\sin \dfrac{4\pi}{3}$ **52.** $\cos \dfrac{4\pi}{3}$ **53.** $\tan \dfrac{\pi}{6}$ **54.** $\cot \dfrac{\pi}{3}$

55. $\sec \dfrac{2\pi}{3}$ **56.** $\csc \dfrac{3\pi}{2}$ **57.** $\csc \dfrac{5\pi}{6}$ **58.** $\sec \dfrac{5\pi}{6}$

59. $4 \sin\left(-\dfrac{\pi}{4}\right)$ **60.** $4 \cos\left(-\dfrac{\pi}{4}\right)$

61. $-\sin \dfrac{\pi}{4}$ **62.** $-\cos \dfrac{\pi}{4}$ **63.** $2 \cos \dfrac{\pi}{6}$ **64.** $2 \sin \dfrac{\pi}{6}$

Evaluate each of the following expressions when x is $\pi/6$. In each case, use exact values.

65. $\sin 2x$ **66.** $\sin 3x$ **67.** $6 \cos 3x$ **68.** $6 \cos 2x$

69. $\sin\left(x + \dfrac{\pi}{2}\right)$ **70.** $\sin\left(x - \dfrac{\pi}{2}\right)$

71. $4 \cos\left(2x + \dfrac{\pi}{3}\right)$ **72.** $4 \cos\left(3x + \dfrac{\pi}{6}\right)$

For the following expressions, find the value of y that corresponds to each value of x, then write your results as ordered pairs (x, y).

73. $y = \sin x$ for $x = 0, \dfrac{\pi}{4}, \dfrac{\pi}{2}, \dfrac{3\pi}{4}, \pi$

74. $y = \cos x$ for $x = 0, \dfrac{\pi}{4}, \dfrac{\pi}{2}, \dfrac{3\pi}{4}, \pi$

75. $y = 2 \sin x$ for $x = 0, \dfrac{\pi}{2}, \pi, \dfrac{3\pi}{2}, 2\pi$

76. $y = \dfrac{1}{2} \cos x$ for $x = 0, \dfrac{\pi}{2}, \pi, \dfrac{3\pi}{2}, 2\pi$

77. $y = \sin 2x$ for $x = 0, \dfrac{\pi}{4}, \dfrac{\pi}{2}, \dfrac{3\pi}{4}, \pi$

78. $y = \cos 3x$ for $x = 0, \dfrac{\pi}{6}, \dfrac{\pi}{3}, \dfrac{\pi}{2}, \dfrac{2\pi}{3}$

79. $y = \sin \left(x - \dfrac{\pi}{2} \right)$ for $x = \dfrac{\pi}{2}, \pi, \dfrac{3\pi}{2}, 2\pi, \dfrac{5\pi}{2}$

80. $y = \cos \left(x - \dfrac{\pi}{6} \right)$ for $x = \dfrac{\pi}{6}, \dfrac{\pi}{3}, \dfrac{2\pi}{3}, \pi, \dfrac{7\pi}{6}$

81. $y = 3 \sin \left(2x + \dfrac{\pi}{2} \right)$ for $x = -\dfrac{\pi}{4}, 0, \dfrac{\pi}{4}, \dfrac{\pi}{2}, \dfrac{3\pi}{4}$

82. $y = 5 \cos \left(2x - \dfrac{\pi}{3} \right)$ for $x = \dfrac{\pi}{6}, \dfrac{\pi}{3}, \dfrac{2\pi}{3}, \pi, \dfrac{7\pi}{6}$

83. Cycling The Shimano WH-R540 aluminum wheel has 8 pairs of spokes evenly distributed around the rim of the wheel. What is the measure, in radians, of the central angle formed by adjacent pairs of spokes (Figure 14)?

Figure 14

84. Cycling The Mavic Ksyrium Elite wheel has 18 spokes evenly distributed around the rim of the wheel. What is the measure, in radians, of the central angle formed by adjacent spokes (Figure 15)?

REVIEW PROBLEMS

The problems that follow review material we covered in Section 1.3.

Find all six trigonometric functions of θ, if the given point is on the terminal side of θ.

85. $(1, -3)$ **86.** $(-1, 3)$ **87.** (m, n) **88.** (a, b)

89. Find the remaining trigonometric functions of θ, if $\sin \theta = \dfrac{1}{2}$ and θ terminates in QII.

90. Find the remaining trigonometric functions of θ, if $\cos \theta = -1/\sqrt{2}$ and θ terminates in QII.

91. Find all six trigonometric functions of θ, if the terminal side of θ lies along the line $y = 2x$ in QI.

92. Find the six trigonometric functions of θ, if the terminal side of θ lies along the line $y = 2x$ in QIII.

Figure 15

SECTION 3.3 | DEFINITION III: CIRCULAR FUNCTIONS

In this section, we will give a third and final definition for the trigonometric functions. Rather than give the new definition first and then show that it does not conflict with our previous definitions, we will do the reverse and show that either of our first two definitions leads to conclusions that imply a third definition.

To begin, recall that the unit circle (Figure 1) is the circle with center at the origin and radius 1. The equation of the unit circle is $x^2 + y^2 = 1$.

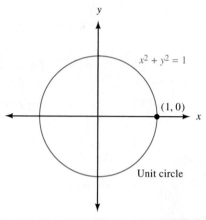

Figure 1

Suppose the terminal side of angle θ, in standard position, intersects the unit circle at point (x, y) as shown in Figure 2.

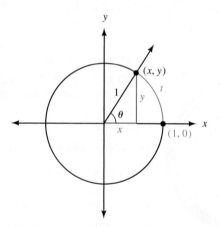

Figure 2

Since the radius of the unit circle is 1, the distance from the origin to the point (x, y) is 1. We can use our first definition for the trigonometric functions from Section 1.3 to write

$$\cos \theta = \frac{x}{r} = \frac{x}{1} = x \qquad \text{and} \qquad \sin \theta = \frac{y}{r} = \frac{y}{1} = y$$

In other words, the ordered pair (x, y) shown in Figure 2 can also be written as $(\cos \theta, \sin \theta)$. We can arrive at this same result by applying our second definition for the trigonometric functions (from Section 2.1). Since θ is an acute angle in a right triangle, by Definition II we have

$$\cos \theta = \frac{\text{side adjacent } \theta}{\text{hypotenuse}} = \frac{x}{1} = x$$

$$\sin \theta = \frac{\text{side opposite } \theta}{\text{hypotenuse}} = \frac{y}{1} = y$$

As you can see, both of our first two definitions lead to the conclusion that the point (x, y) shown in Figure 2 can be written as $(\cos \theta, \sin \theta)$.

Next, consider the length of the arc from $(1, 0)$ to (x, y), which we have labeled t in Figure 2. If θ is measured in radians, then by definition

$$\theta = \frac{t}{r} = \frac{t}{1} = t$$

The length of the arc from $(1, 0)$ to (x, y) is exactly the same as the radian measure of angle θ. Therefore, we can write

$$\cos \theta = \cos t = x \qquad \text{and} \qquad \sin \theta = \sin t = y$$

These results give rise to a third definition for the trigonometric functions.

DEFINITION III CIRCULAR FUNCTIONS

If (x, y) is any point on the unit circle, and t is the distance from $(1, 0)$ to (x, y) along the circumference of the unit circle, then, as Figure 3 illustrates,

$\cos t = x$

$\sin t = y$

$\tan t = \dfrac{y}{x}$

$\cot t = \dfrac{x}{y}$

$\csc t = \dfrac{1}{y}$

$\sec t = \dfrac{1}{x}$

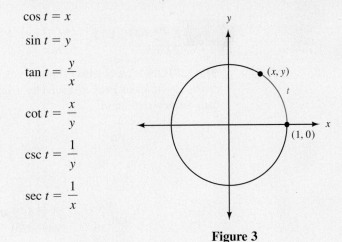

Figure 3

As we travel around the unit circle, starting at the point $(1, 0)$, the points we come across all have coordinates $(\cos t, \sin t)$, where t is the distance we have traveled. When we define the trigonometric functions this way, we call them circular functions because of their relationship to the unit circle.

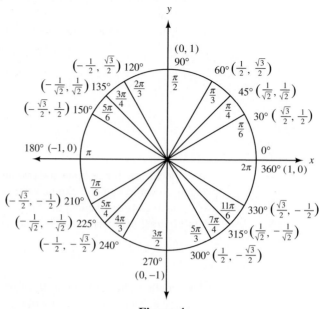

Figure 4

Figure 4 shows an enlarged version of the unit circle with multiples of $\pi/6$ and $\pi/4$ marked off. Each angle is given in both degrees and radians. The radian measure of each angle is the same as the distance from $(1, 0)$ to the point on the terminal side of the angle, as measured along the circumference of the circle. The x- and y-coordinate of each point shown are the cosine and sine, respectively, of the associated angle or distance.

 **EXAMPLE 1** Use Figure 4 to find the six trigonometric functions of $5\pi/6$.

SOLUTION We obtain cosine and sine directly from Figure 4. The other trigonometric functions of $5\pi/6$ are found by using the ratio and reciprocal identities, rather than the new definition.

$$\sin \frac{5\pi}{6} = y = \frac{1}{2}$$

$$\cos \frac{5\pi}{6} = x = -\frac{\sqrt{3}}{2}$$

$$\tan \frac{5\pi}{6} = \frac{\sin (5\pi/6)}{\cos (5\pi/6)} = \frac{1/2}{-\sqrt{3}/2} = -\frac{1}{\sqrt{3}}$$

$$\cot \frac{5\pi}{6} = \frac{1}{\tan (5\pi/6)} = \frac{1}{-1/\sqrt{3}} = -\sqrt{3}$$

$$\sec \frac{5\pi}{6} = \frac{1}{\cos (5\pi/6)} = \frac{1}{-\sqrt{3}/2} = -\frac{2}{\sqrt{3}}$$

$$\csc \frac{5\pi}{6} = \frac{1}{\sin (5\pi/6)} = \frac{1}{1/2} = 2$$ ◼

Figure 4 is helpful in visualizing the relationships among the angles shown and the trigonometric functions of those angles. You may want to make a larger copy of this diagram yourself. In the process of doing so you will become more familiar with the relationship between degrees and radians and the exact values of the angles in the diagram.

 **EXAMPLE 2**   Use the unit circle to find all values of t between 0 and 2π for which $\cos t = \frac{1}{2}$.

SOLUTION   We look for all ordered pairs on the unit circle with an x-coordinate of $\frac{1}{2}$. The angles, or distances, associated with these points are the angles for which $\cos t = \frac{1}{2}$. They are $t = \pi/3$ or $60°$ and $t = 5\pi/3$ or $300°$. ▪

USING TECHNOLOGY

GRAPHING THE UNIT CIRCLE

We can use Definition III to graph the unit circle with a set of parametric equations. We will cover parametric equations in more detail in Section 6.4. First, set your graphing calculator to parametric mode and to radian mode. Then define the following pair of functions (Figure 5).

$$X_1 = \cos t$$

$$Y_1 = \sin t$$

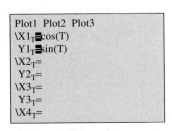

Plot1 Plot2 Plot3
\X1ₜ=cos(T)
 Y1ₜ=sin(T)
\X2ₜ=
 Y2ₜ=
\X3ₜ=
 Y3ₜ=
\X4ₜ=

Figure 5

Set the window variables so that

$$0 \le t \le 2\pi, \text{ step} = \pi/12; \ -1.5 \le x \le 1.5; \ -1.5 \le y \le 1.5$$

Graph the equations using the zoom-square command. This will make the circle appear round and not like an ellipse. Now you can trace the circle in increments of $\pi/12$. For example, to find the sine and cosine of $\pi/3$ radians, we would trace through four increments (Figure 6).

$$4\left(\frac{\pi}{12}\right) = \frac{4\pi}{12}$$

$$= \frac{\pi}{3}$$

$$\approx 1.0472$$

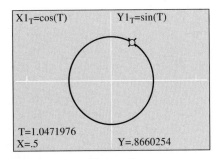

Figure 6

We see that $\cos(\pi/3) = 0.5$ and $\sin(\pi/3) = 0.8660$ to the nearest ten-thousandth.

Because it is easier for a calculator to display angles exactly in degrees rather than in radians, it is sometimes helpful to graph the unit circle using degrees. Set your graphing calculator to degree mode. Change the window settings so that $0 \le t \le 360$ with scale $= 5$. Graph the circle once again. Now when you trace the graph, each increment represents $5°$.

We can use this graph to verify our results in Example 2. Trace around the unit circle until the x-coordinate is 0.5. The calculator shows us there are two such points, at $t = 60°$ and at $t = 300°$ (Figure 7).

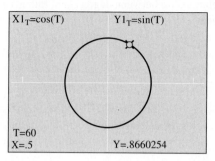 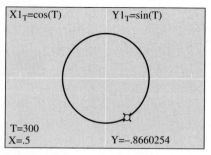

Figure 7

EVEN AND ODD FUNCTIONS

Recall from algebra the definitions of even and odd functions.

DEFINITION

An *even function* is a function for which

$$f(-x) = f(x) \text{ for all } x \text{ in the domain of } f$$

An even function is a function for which replacing x with $-x$ leaves the equation that defines the function unchanged. If a function is even, then every time the point (x, y) is on the graph, so is the point $(-x, y)$. The function $f(x) = x^2 + 3$ is an even function since

$$f(-x) = (-x)^2 + 3 = x^2 + 3 = f(x)$$

DEFINITION

An *odd function* is a function for which

$$f(-x) = -f(x) \text{ for all } x \text{ in the domain of } f$$

An odd function is a function for which replacing x with $-x$ changes the sign of the equation that defines the function. If a function is odd, then every time the point (x, y) is on the graph, so is the point $(-x, -y)$. The function $f(x) = x^3 - x$ is an odd function since

$$f(-x) = (-x)^3 - (-x) = -x^3 + x = -(x^3 - x) = -f(x)$$

From the unit circle it is apparent that sine is an odd function and cosine is an even function. To begin to see that this is true, we locate $\pi/6$ and $-\pi/6$ ($-\pi/6$ is coterminal with $11\pi/6$) on the unit circle and notice that

$$\cos\left(-\frac{\pi}{6}\right) = \frac{\sqrt{3}}{2} = \cos\frac{\pi}{6}$$

and

$$\sin\left(-\frac{\pi}{6}\right) = -\frac{1}{2} = -\sin\frac{\pi}{6}$$

We can generalize this result by drawing an angle θ and its opposite $-\theta$ in standard position and then labeling the points where their terminal sides intersect the unit circle with (x, y) and $(x, -y)$, respectively. (Can you see from Figure 8 why we label these two points in this way? That is, does it make sense that if (x, y) is on the terminal side of θ, then $(x, -y)$ must be on the terminal side of $-\theta$?)

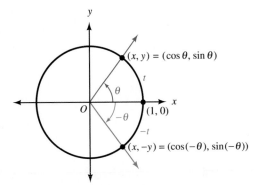

Figure 8

Since, on the unit circle, $\cos\theta = x$ and $\sin\theta = y$, we have

$$\cos(-\theta) = x = \cos\theta$$

indicating that cosine is an even function and

$$\sin(-\theta) = -y = -\sin\theta$$

indicating that sine is an odd function.

Now that we have established that sine is an odd function and cosine is an even function, we can use our ratio and reciprocal identities to find which of the other trigonometric functions are even and which are odd. Example 3 shows how this is done for the cosecant function.

 EXAMPLE 3   Show that cosecant is an odd function.

SOLUTION   We must prove that $\csc(-\theta) = -\csc\theta$. That is, we must turn $\csc(-\theta)$ into $-\csc\theta$. Here is how it goes:

$$\csc(-\theta) = \frac{1}{\sin(-\theta)} \qquad \text{Reciprocal identity}$$

$$= \frac{1}{-\sin\theta} \qquad \text{Sine is an odd function}$$

$$= -\frac{1}{\sin\theta} \qquad \text{Algebra}$$

$$= -\csc\theta \qquad \text{Reciprocal identity} \quad \blacksquare$$

EXAMPLE 4 Use the even and odd function relationships to find exact values for each of the following.

a. $\sin(-60°)$ **b.** $\cos\left(-\dfrac{2\pi}{3}\right)$ **c.** $\csc(-225°)$

SOLUTION

a. $\sin(-60°) = -\sin 60°$ Sine is an odd function

$\qquad\qquad = -\dfrac{\sqrt{3}}{2}$ Unit circle

b. $\cos\left(-\dfrac{2\pi}{3}\right) = \cos\left(\dfrac{2\pi}{3}\right)$ Cosine is an even function

$\qquad\qquad\quad = -\dfrac{1}{2}$ Unit circle

c. $\csc(-225°) = \dfrac{1}{\sin(-225°)}$ Reciprocal identity

$\qquad\qquad = \dfrac{1}{-\sin 225°}$ Sine is an odd function

$\qquad\qquad = \dfrac{1}{-(-1/\sqrt{2})}$ Unit circle

$\qquad\qquad = \sqrt{2}$

There is an interesting diagram we can draw using the ideas given in this section. Figure 9 shows a point $P(x, y)$ that is t units from the point $(1, 0)$ on the circumference of the unit circle. Therefore, $\cos t = x$ and $\sin t = y$. Can you see why QB is labeled $\tan t$? It has to do with the fact that $\tan t$ is the ratio of $\sin t$ to $\cos t$ along with the fact that triangles POA and QOB are similar.

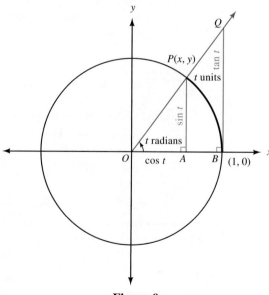

Figure 9

There are many concepts that can be visualized from Figure 9. One of the more important is the variations that occur in sin t, cos t, and tan t as P travels around the unit circle. To illustrate, imagine P traveling once around the unit circle starting at $(1, 0)$ and ending 2π units later at the same point. As P moves from $(1, 0)$ to $(0, 1)$, t increases from 0 to $\pi/2$. At the same time, sin t increases from 0 to 1, and cos t decreases from 1 down to 0. As we continue around the unit circle, sin t and cos t simply oscillate between -1 and 1; furthermore, everywhere sin t is 1 or -1, cos t is 0. We will look at these oscillations in more detail in Chapter 4.

GETTING READY FOR CLASS

After reading through the preceding section, respond in your own words and in complete sentences.

a. If (x, y) is any point on the unit circle, and t is the distance from $(1, 0)$ to (x, y) along the circumference of the unit circle, what are the six trigonometric functions of t?

b. What is the first step in finding all values of t between 0 and 2π on the unit circle for which cos $t = \frac{1}{2}$?

c. What is the definition of an even function?

d. What is the definition of an odd function?

PROBLEM SET 3.3

Use the unit circle to find the six trigonometric functions of each angle.

1. $150°$ **2.** $135°$ **3.** $11\pi/6$ **4.** $5\pi/3$
5. $180°$ **6.** $270°$ **7.** $3\pi/4$ **8.** $5\pi/4$

Use the unit circle and the fact that cosine is an even function to find each of the following:

9. $\cos(-60°)$ **10.** $\cos(-120°)$

11. $\cos\left(-\dfrac{5\pi}{6}\right)$ **12.** $\cos\left(-\dfrac{4\pi}{3}\right)$

Use the unit circle and the fact that sine is an odd function to find each of the following:

13. $\sin(-30°)$ **14.** $\sin(-90°)$

15. $\sin\left(-\dfrac{3\pi}{4}\right)$ **16.** $\sin\left(-\dfrac{7\pi}{4}\right)$

Use the unit circle to find all values of θ between 0 and 2π for which

17. $\sin\theta = 1/2$ **18.** $\sin\theta = -1/2$
19. $\cos\theta = -\sqrt{3}/2$ **20.** $\cos\theta = 0$
21. $\tan\theta = -\sqrt{3}$ **22.** $\cot\theta = \sqrt{3}$

▶ Graph the unit circle using parametric equations with your calculator set to radian mode. Use a scale of $\pi/12$. Trace the circle to find the sine and cosine of each angle to the nearest ten-thousandth.

23. $\dfrac{\pi}{4}$ **24.** $\dfrac{4\pi}{3}$ **25.** $\dfrac{7\pi}{6}$ **26.** $\dfrac{5\pi}{12}$

▶ Graph the unit circle using parametric equations with your calculator set to radian mode. Use a scale of $\pi/12$. Trace the circle to find all values of t between 0 and 2π satisfying each of the following statements. Round your answers to the nearest ten-thousandth.

27. $\sin t = \dfrac{1}{2}$ **28.** $\cos t = -\dfrac{1}{2}$

29. $\cos t = -1$ **30.** $\sin t = 1$

▶ Graph the unit circle using parametric equations with your calculator set to degree mode. Use a scale of 5. Trace the circle to find the sine and cosine of each angle to the nearest ten-thousandth.

31. $120°$ **32.** $225°$ **33.** $75°$ **34.** $310°$

▶ Graph the unit circle using parametric equations with your calculator set to degree mode. Use a scale of 5. Trace the circle to find all values of t between $0°$ and $360°$ satisfying each of the following statements.

35. $\sin t = -\dfrac{1}{2}$ **36.** $\cos t = 0$

37. $\sin t = \cos t$ **38.** $\sin t = -\cos t$

39. If angle θ is in standard position and intersects the unit circle at the point $(1/\sqrt{5}, -2/\sqrt{5})$, find $\sin \theta$, $\cos \theta$, and $\tan \theta$.

40. If angle θ is in standard position and intersects the unit circle at the point $(-1/\sqrt{10}, -3/\sqrt{10})$, find $\sin \theta$, $\cos \theta$, and $\tan \theta$.

41. If $\sin \theta = -1/3$, find $\sin(-\theta)$.

42. If $\cos \theta = -1/3$, find $\cos(-\theta)$.

If we start at the point $(1, 0)$ and travel once around the unit circle, we travel a distance of 2π units and arrive back where we started at the point $(1, 0)$. If we continue around the unit circle a second time, we will repeat all the values of x and y that occurred during our first trip around. Use this discussion to evaluate the following expressions:

43. $\sin\left(2\pi + \dfrac{\pi}{2}\right)$ **44.** $\cos\left(2\pi + \dfrac{\pi}{2}\right)$

45. $\sin\left(2\pi + \dfrac{\pi}{6}\right)$ **46.** $\cos\left(2\pi + \dfrac{\pi}{3}\right)$

47. $\sin \dfrac{5\pi}{2}$ **48.** $\cos \dfrac{5\pi}{2}$

49. $\sin \dfrac{13\pi}{6}$ **50.** $\cos \dfrac{7\pi}{3}$

Make a diagram of the unit circle with an angle θ in quadrant I and its supplement $180° - \theta$ in quadrant II. Label the point on the terminal side of θ and the unit circle with (x, y) and the point on the terminal side of $180° - \theta$ and the unit circle with $(-x, y)$. Use the diagram to show that

51. $\sin (180° - \theta) = \sin \theta$

52. $\cos (180° - \theta) = -\cos \theta$

 53. Show that tangent is an odd function.

54. Show that cotangent is an odd function.

Prove each identity.

55. $\sin (-\theta) \cot (-\theta) = \cos \theta$

56. $\cos (-\theta) \tan \theta = \sin \theta$

57. $\sin (-\theta) \sec (-\theta) \cot (-\theta) = 1$

58. $\cos (-\theta) \csc (-\theta) \tan (-\theta) = 1$

59. $\csc \theta + \sin (-\theta) = \dfrac{\cos^2 \theta}{\sin \theta}$

60. $\sec \theta - \cos (-\theta) = \dfrac{\sin^2 \theta}{\cos \theta}$

61. Geometry Redraw the diagram in Figure 9 from this section and label the line segment that corresponds to sec t.

62. Geometry Make a diagram similar to the diagram in Figure 9 from this section, but instead of labeling the point $(1, 0)$ with B, label the point $(0, 1)$ with B. Then place Q on the line OP and connect Q to B so that QB is perpendicular to the y-axis. Now, if $P(x, y)$ is t units from $(1, 0)$, label the line segments that correspond to sin t, cos t, cot t, and csc t.

REVIEW PROBLEMS

The problems that follow review material we covered in Section 2.3.

Problems 63–70 refer to right triangle ABC in which $C = 90°$. Solve each triangle.

63. $A = 42°, c = 36$

64. $A = 58°, c = 17$

65. $B = 22°, b = 320$

66. $B = 48°, b = 270$

67. $a = 20.5, b = 31.4$

68. $a = 16.3, b = 20.8$

69. $a = 4.37, c = 6.21$

70. $a = 7.12, c = 8.44$

SECTION 3.4 | ARC LENGTH AND AREA OF A SECTOR

In Chapter 2, we discussed some of the aspects of the first Ferris wheel, which was built by George Ferris in 1893. The first topic we will cover in this section is *arc length*. Our study of arc length will allow us to find the distance traveled by a rider on a Ferris wheel at any point during the ride.

In Section 3.2, we found that if a central angle θ, measured in radians, in a circle of radius r cuts off an arc of length s, then the relationship between s, r, and θ can be written as $\theta = \dfrac{s}{r}$. Figure 1 illustrates this.

If we multiply both sides of this equation by r, we will obtain the equation that gives arc length s in terms of r and θ.

$$\theta = \frac{s}{r} \qquad \text{Definition of radian measure}$$

$$r \cdot \theta = r \cdot \frac{s}{r} \qquad \text{Multiply both sides by } r$$

$$r\theta = s$$

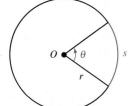

Figure 1

ARC LENGTH

If θ (in radians) is a central angle in a circle with radius r, then the length of the arc cut off by θ is given by

$$s = r\theta \qquad (\theta \text{ in radians})$$

 EXAMPLE 1   Give the length of the arc cut off by a central angle of 2 radians in a circle of radius 4.3 inches.

SOLUTION   We have $\theta = 2$ and $r = 4.3$ inches. Applying the formula $s = r\theta$ gives us

$$s = r\theta$$
$$= 4.3(2)$$
$$= 8.6 \text{ inches}$$

Figure 2 illustrates this example.

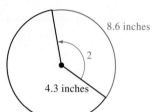

Figure 2

 EXAMPLE 2   Figure 3 is a model of George Ferris's Ferris wheel. Recall that the diameter of the wheel is 250 feet, and θ is the central angle formed as a rider travels from his or her initial position P_0 to position P_1. Find the distance traveled by the rider if $\theta = 45°$ and if $\theta = 105°$.

SOLUTION   The formula for arc length, $s = r\theta$, requires θ to be given in radians. Since θ is given in degrees, we must multiply it by $\pi/180$ to convert to radians. Also, since the diameter of the wheel is 250 feet, the radius is 125 feet.

For $\theta = 45°$:

$$s = r\theta$$
$$= 125(45)\left(\frac{\pi}{180}\right)$$
$$= \frac{125\pi}{4} \text{ ft}$$
$$\approx 98.2 \text{ ft}$$

For $\theta = 105°$:

$$s = r\theta$$
$$= 125(105)\left(\frac{\pi}{180}\right)$$
$$= \frac{875\pi}{12} \text{ ft}$$
$$\approx 229.1 \text{ ft}$$

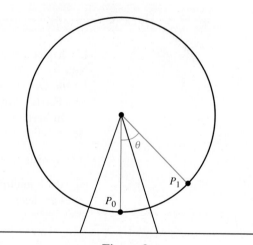

Figure 3

EXAMPLE 3   The minute hand of a clock is 1.2 centimeters long. To two significant digits, how far does the tip of the minute hand move in 20 minutes?

SOLUTION   We have $r = 1.2$ cm. Since we are looking for s, we need to find θ. We can use a proportion to find θ. Since one complete rotation is 60 minutes and 2π radians, we say θ is to 2π as 20 minutes is to 60 minutes, or

$$\text{If} \quad \frac{\theta}{2\pi} = \frac{20}{60} \quad \text{then} \quad \theta = \frac{2\pi}{3}$$

Now we can find s.

$$s = r\theta$$

$$= 1.2\left(\frac{2\pi}{3}\right)$$

$$= \frac{2.4\pi}{3}$$

$$\approx 2.5 \text{ cm}$$

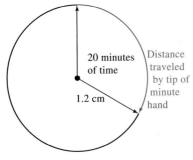

Figure 4

Figure 4 illustrates this example.

The tip of the minute hand will travel approximately 2.5 centimeters every 20 minutes.

If we are working with relatively small central angles in circles with large radii, we can use the length of the intercepted arc to approximate the length of the associated chord. For example, Figure 5 shows a central angle of 1° in a circle of radius 1,800 feet, along with the arc and chord cut off by 1°. (Figure 5 is not drawn to scale.)

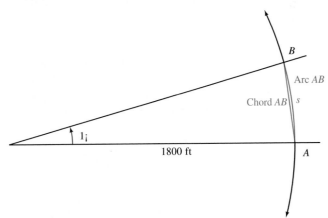

Figure 5

To find the length of arc AB, we convert θ to radians by multiplying by $\pi/180$. Then we apply the formula $s = r\theta$.

$$s = r\theta = 1{,}800(1)\left(\frac{\pi}{180}\right) = 10\pi \approx 31.4 \text{ ft}$$

If we had carried out the calculation of arc AB to six significant digits, we would have obtained $s = 31.4159$. The length of the chord AB is 31.4155 to six significant digits (found by using the law of sines, which we will cover in Chapter 7). As you can see, the first five digits in each number are the same. It seems reasonable then to approximate the length of chord AB with the length of arc AB.

As our next example illustrates, we can also use the procedure just outlined in the reverse order to find the radius of a circle by approximating arc length with the length of the associated chord.

EXAMPLE 4 A person standing on the earth notices that a 747 Jumbo Jet flying overhead subtends an angle of 0.45°. If the length of the jet is 230 feet, find its altitude to the nearest thousand feet.

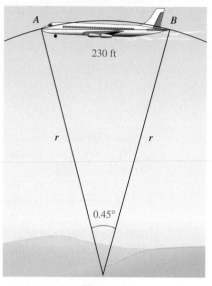

Figure 6

SOLUTION Figure 6 is a diagram of the situation. Since we are working with a relatively small angle in a circle with a large radius, we use the length of the airplane (chord AB in Figure 6) as an approximation of the length of the arc AB, and r as an approximation for the altitude of the plane.

Since $s = r\theta$, $r = \dfrac{s}{\theta}$

$$\text{so } r = \frac{230}{(0.45)(\pi/180)} \qquad \text{We multiply 0.45° by } \pi/180 \text{ to change to radian measure}$$

$$= \frac{230(180)}{(0.45)(\pi)}$$

$$= 29{,}000 \text{ ft} \qquad \text{To the nearest thousand feet}$$

AREA OF A SECTOR

Next we want to derive the formula for the area of the sector formed by a central angle θ (Figure 7).

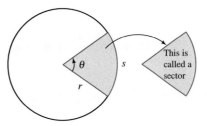

Figure 7

If we let A represent the area of the sector formed by central angle θ, we can find A by setting up a proportion as follows: We say the area A of the sector is to the area of the circle as θ is to one full rotation. That is,

Area of sector $\longrightarrow \dfrac{A}{\pi r^2} = \dfrac{\theta}{2\pi}$ $\longleftarrow$ Central angle θ
Area of circle $\longrightarrow$ $\qquad\qquad\longleftarrow$ One full rotation

We solve for A by multiplying both sides of the proportion by πr^2.

$$\pi r^2 \cdot \frac{A}{\pi r^2} = \frac{\theta}{2\pi} \cdot \pi r^2$$

$$A = \frac{1}{2} r^2 \theta$$

AREA OF A SECTOR

If θ (in radians) is a central angle in a circle with radius r, then the area of the sector formed by angle θ is given by

$$A = \frac{1}{2} r^2 \theta \qquad (\theta \text{ in radians})$$

 EXAMPLE 5   Find the area of the sector formed by a central angle of 1.4 radians in a circle of radius 2.1 meters.

SOLUTION   We have $r = 2.1$ m and $\theta = 1.4$. Applying the formula for A gives us

$$A = \frac{1}{2} r^2 \theta$$

$$= \frac{1}{2}(2.1)^2(1.4)$$

$$= 3.1 \text{ m}^2 \qquad \text{To the nearest tenth}$$

REMEMBER   Area is measured in square units. When $r = 2.1$ m, $r^2 = (2.1 \text{ m})^2 = 4.41 \text{ m}^2$.

EXAMPLE 6   If the sector formed by a central angle of 15° has an area of $\pi/3$ cm², find the radius of the circle.

SOLUTION   We first convert 15° to radians.

$$\theta = 15\left(\frac{\pi}{180}\right) = \frac{\pi}{12}$$

Then we substitute $\theta = \pi/12$ and $A = \pi/3$ into the formula for A and solve for r.

$$A = \frac{1}{2}r^2\theta$$

$$\frac{\pi}{3} = \frac{1}{2}r^2\frac{\pi}{12}$$

$$\frac{\pi}{3} = \frac{\pi}{24}r^2$$

$$r^2 = \frac{\pi}{3} \cdot \frac{24}{\pi}$$

$$r^2 = 8$$

$$r = 2\sqrt{2} \text{ cm}$$

Note that we need only use the positive square root of 8, since we know our radius must be measured with positive units.   ▪

 EXAMPLE 7   A lawn sprinkler located at the corner of a yard is set to rotate through 90° and project water out 30 feet. To three significant digits, what area of lawn is watered by the sprinkler?

SOLUTION   We have $\theta = 90° = \frac{\pi}{2} \approx 1.57$ rad and $r = 30$ ft. Figure 8 illustrates this example.

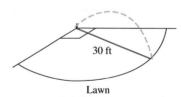

30 ft

Lawn

Figure 8

$$A = \frac{1}{2}r^2\theta$$

$$\approx \frac{1}{2}(30)^2(1.57)$$

$$= 707 \text{ ft}^2 \quad ▪$$

GETTING READY FOR CLASS

After reading through the preceding section, respond in your own words and in complete sentences.

a. What is the definition and formula for arc length?

b. Which is longer, the chord AB or the corresponding arc AB?

c. What is a sector?

d. What is the definition and formula for the area of a sector?

PROBLEM SET 3.4

Unless otherwise stated, all answers in this Problem Set that need to be rounded should be rounded to three significant digits.

For each problem below, θ is a central angle in a circle of radius r. In each case, find the length of arc s cut off by θ.

1. $\theta = 2, r = 3$ inches

2. $\theta = 3, r = 2$ inches

3. $\theta = 1.5, r = 1.5$ ft

4. $\theta = 2.4, r = 1.8$ ft

5. $\theta = \pi/6, r = 12$ cm

6. $\theta = \pi/3, r = 12$ cm

 7. $\theta = 60°, r = 4$ mm

8. $\theta = 30°, r = 4$ mm

9. $\theta = 240°, r = 10$ inches

10. $\theta = 315°, r = 5$ inches

11. Arc Length The minute hand of a clock is 2.4 centimeters long. How far does the tip of the minute hand travel in 20 minutes?

12. Arc Length The minute hand of a clock is 1.2 centimeters long. How far does the tip of the minute hand travel in 40 minutes?

13. Arc Length A space shuttle 200 miles above the earth is orbiting the earth once every 6 hours. How far does the shuttle travel in 1 hour? (Assume the radius of the earth is 4,000 miles.) Give your answer as both an exact value and an approximation to three significant digits (Figure 9).

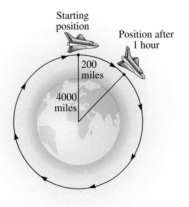

Figure 9

14. Arc Length How long, in hours, does it take the space shuttle in Problem 13 to travel 8,400 miles? Give both the exact value and an approximate value for your answer.

15. Arc Length The pendulum on a grandfather clock swings from side to side once every second. If the length of the pendulum is 4 feet and the angle through which it swings is 20°, how far does the tip of the pendulum travel in 1 second?

16. Arc Length Find the total distance traveled in 1 minute by the tip of the pendulum on the grandfather clock in Problem 15.

17. Cable Car Drive System The current San Francisco cable railway is driven by two large 14-foot-diameter drive wheels, called *sheaves*. Because of the figure-eight system used, the cable subtends a central angle of 270° on each sheave. Find the length of cable riding on one of the drive sheaves at any given time (Figure 10).

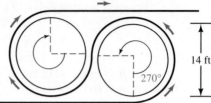

Figure 10

18. Cable Car Drive System The first cable railway to make use of the figure-eight drive system was the Sutter Street Railway in San Francisco in 1883 (Figure 11). Each drive sheave was 12 feet in diameter. Find the length of cable riding on one of the drive sheaves. (See Problem 17.)

Figure 11

19. Drum Brakes The Isuzu NPR 250 light truck with manual transmission has a circular brake drum with a diameter of 320 millimeters. Each brake pad, which presses against the drum, is 307 millimeters long. What central angle is subtended by one of the brake pads? Write your answer in both radians and degrees (Figure 12).

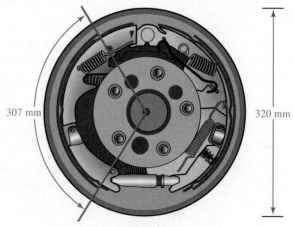

Figure 12

20. **Drum Brakes** The Isuzu NPR 250 truck with automatic transmission has a circular brake drum with a diameter of 320 millimeters. Each brake pad, which presses against the drum, is 335 millimeters long. What central angle is subtended by one of the brake pads? Write your answer in both radians and degrees.

21. **Diameter of the Moon** From the earth, the moon subtends an angle of approximately 0.5°. If the distance to the moon is approximately 240,000 miles, find an approximation for the diameter of the moon accurate to the nearest hundred miles. (See Example 4 and the discussion that precedes it.)

22. **Diameter of the Sun** If the distance to the sun is approximately 93 million miles, and, from the earth, the sun subtends an angle of approximately 0.5°, estimate the diameter of the sun to the nearest 10,000 miles.

Repeat Example 2 from this section for the following values of θ.

23. $\theta = 30°$ 24. $\theta = 60°$

25. $\theta = 220°$ 26. $\theta = 315°$

27. **Ferris Wheel** In Problem Set 2.3, we mentioned a Ferris wheel built in Vienna in 1897, known as the Great Wheel. The diameter of this wheel is 197 feet. Use Figure 3 from this section as a model of the Great Wheel, and find the distance traveled by a rider in going from initial position P_0 to position P_1 if

   a. θ is 60°

   b. θ is 210°

   c. θ is 285°

28. **Ferris Wheel** A Ferris Wheel called Colossus that we mentioned in Problem Set 2.3 has a diameter of 165 feet. Using Figure 3 from this section as a model, find the distance traveled by someone starting at initial position P_0 and moving to position P_1 if

   a. θ is 150°

   b. θ is 240°

   c. θ is 345°

In each problem below, θ is a central angle that cuts off an arc of length s. In each case, find the radius of the circle.

29. $\theta = 6$, $s = 3$ ft   30. $\theta = 1$, $s = 2$ ft

31. $\theta = 1.4$, $s = 4.2$ inches   32. $\theta = 5.1$, $s = 10.2$ inches

 33. $\theta = \pi/4$, $s = \pi$ cm   34. $\theta = 3\pi/4$, $s = \pi$ cm

35. $\theta = 90°$, $s = \pi/2$ m   36. $\theta = 180°$, $s = \pi/2$ m

37. $\theta = 225°$, $s = 4$ km   38. $\theta = 150°$, $s = 5$ km

Find the area of the sector formed by central angle θ in a circle of radius r if

39. $\theta = 2$, $r = 3$ cm   40. $\theta = 3$, $r = 2$ cm

41. $\theta = 2.4$, $r = 4$ inches   42. $\theta = 1.8$, $r = 2$ inches

43. $\theta = \pi/5$, $r = 3$ m   44. $\theta = 2\pi/5$, $r = 3$ m

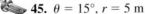 45. $\theta = 15°$, $r = 5$ m   46. $\theta = 15°$, $r = 10$ m

47. **Area of a Sector** A central angle of 2 radians cuts off an arc of length 4 inches. Find the area of the sector formed.

48. **Area of a Sector** An arc of length 3 feet is cut off by a central angle of $\pi/4$ radians. Find the area of the sector formed.

49. **Radius of a Circle** If the sector formed by a central angle of 30° has an area of $\pi/3$ square centimeters, find the radius of the circle.

50. **Arc Length** What is the length of the arc cut off by angle θ in Problem 49?

51. **Radius of a Circle** A sector of area $2\pi/3$ square inches is formed by a central angle of 45°. What is the radius of the circle?

52. **Radius of a Circle** A sector of area 25 square inches is formed by a central angle of 4 radians. Find the radius of the circle.

53. **Lawn Sprinkler** A lawn sprinkler is located at the corner of a yard. The sprinkler is set to rotate through 90° and project water out 60 feet. What is the area of the yard watered by the sprinkler?

54. **Windshield Wiper** An automobile windshield wiper 10 inches long rotates through an angle of 60°. If the rubber part of the blade covers only the last 9 inches of the wiper, find the area of the windshield cleaned by the windshield wiper.

55. **Cycling** The Shimano WH-R540 aluminum wheel, which has a diameter of 700 millimeters, has 8 pairs of spokes evenly distributed around the rim of the wheel. What is the length of the rim subtended by adjacent pairs of spokes (Figure 13)?

Figure 13

56. **Cycling** The Mavic Ksyrium Elite wheel, which has a diameter of 700 millimeters, has 18 spokes evenly distributed around the rim of the wheel. What is the length of the rim subtended by adjacent spokes (Figure 14)?

Figure 14

REVIEW PROBLEMS

The problems that follow review material we covered in Section 2.4.

57. Angle of Elevation If a 75-foot flagpole casts a shadow 43 feet long, what is the angle of elevation of the sun from the tip of the shadow?

58. Height of a Hill A road up a hill makes an angle of 5° with the horizontal. If the road from the bottom of the hill to the top of the hill is 2.5 miles long, how high is the hill?

59. Angle of Depression A person standing 5.2 feet from a mirror notices that the angle of depression from his eyes to the bottom of the mirror is 13°, while the angle of elevation to the top of the mirror is 12°. Find the vertical dimension of the mirror.

60. Distance and Bearing A boat travels on a course of bearing S 63°50′ E for 114 miles. How many miles south and how many miles east has the boat traveled?

61. Geometry The height of a right circular cone is 35.8 centimeters. If the diameter of the base is 20.5 centimeters, what angle does the side of the cone make with the base?

62. Distance and Bearing A ship leaves the harbor entrance and travels 35 miles in the direction N 42° E. The captain then turns the ship 90° and travels another 24 miles in the direction S 48° E. At that time, how far is the ship from the harbor entrance, and what is the bearing of the ship from the harbor entrance?

63. Angle of Depression A man standing on the roof of a building 86.0 feet above the ground looks down to the building next door. He finds the angle of depression to the roof of that building from the roof of his building to be 14.5°, while the angle of depression from the roof of his building to the bottom of the building next door is 43.2°. How tall is the building next door?

64. Height of a Tree Two people decide to find the height of a tree. They position themselves 35 feet apart in line with, and on the same side of, the tree. If they find the angles of elevation from the ground where they are standing to the top of the tree are 65° and 44°, how tall is the tree?

EXTENDING THE CONCEPTS

Apparent Diameter
As we mentioned in this section, for small central angles in circles with large radii, the intercepted arc and the chord are approximately the same length. Figure 15 shows a diagram of a person looking at the moon. The arc and the chord are essentially the same length and are both good approximations to the diameter.

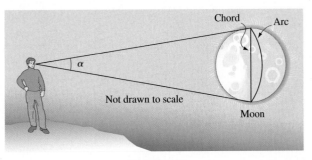

Figure 15

In astronomy, the angle α at which an object is seen by the eye is called the *apparent diameter* of the object. Apparent diameter depends on the size of the object and its distance from the observer. To three significant figures, the diameter of the moon is 3,470 kilometers and its average distance to the earth is 384,000 kilometers. The diameter of the sun is 1,390,000 kilometers and its mean distance to the earth is 150,000,000 kilometers.

65. Calculate the apparent diameter of the moon as seen from the earth. Give your answer in both radians and in degrees. Round each to the nearest thousandth.

66. Suppose you have a friend who is 6 feet tall. At what distance must your friend stand so that he or she has the same apparent diameter as the moon?

67. The moon and the sun are, at first sight, about equal in size. How can you explain this apparent similarity in size even though the diameter of the sun is about 400 times greater than the diameter of the moon?

68. The distance between earth and moon varies as the moon travels its orbit around the earth. Accordingly, the apparent diameter of the moon must also vary. The smallest distance between earth and moon is 356,340 kilometers, while the greatest distance is 406,630 kilometers. Find the largest and smallest apparent diameter of the moon.

SECTION 3.5 | VELOCITIES

The specifications for the first Ferris wheel indicate that one trip around the wheel took 20 minutes. How fast was a rider traveling around the wheel? There are a number of ways to answer this question. The most intuitive measure of the rate at which the rider is traveling around the wheel is what we call *linear velocity*. The units of linear velocity are miles per hour, feet per second, and so forth. Another way to specify how fast the rider is traveling around the wheel is with what we call *angular velocity*. Angular velocity is given as the amount of central angle through which the rider travels over a given amount of time. The central angle swept out by a rider traveling once around the wheel is 360°, or 2π radians. If one trip around the wheel takes 20 minutes, then the angular velocity of a rider is

$$\frac{2\pi \text{ rad}}{20 \text{ min}} = \frac{\pi}{10} \text{ rad/min}$$

In this section, we will learn more about angular velocity and linear velocity and the relationship between them. Let's start with the formal definition for the linear velocity of a point moving on the circumference of a circle.

DEFINITION

If P is a point on a circle of radius r, and P moves a distance s on the circumference of the circle in an amount of time t, then the *linear velocity, v,* of P is given by the formula

$$v = \frac{s}{t}$$

 **EXAMPLE 1** A point on a circle travels 5 centimeters in 2 seconds. Find the linear velocity of the point.

SOLUTION   Substituting $s = 5$ and $t = 2$ into the equation $v = s/t$ gives us

$$v = \frac{5 \text{ cm}}{2 \text{ sec}}$$

$$= 2.5 \text{ cm/sec}$$

NOTE   In all the examples and problems in this section, we are assuming that the point on the circle moves with uniform circular motion. That is, the velocity of the point is constant.

DEFINITION

If P is a point moving with uniform circular motion on a circle of radius r, and the line from the center of the circle through P sweeps out a central angle θ in an amount of time t, then the *angular velocity,* ω (omega), of P is given by the formula

$$\omega = \frac{\theta}{t} \qquad \text{where } \theta \text{ is measured in radians}$$

EXAMPLE 2 A point on a circle rotates through $3\pi/4$ radians in 3 seconds. Give the angular velocity of P.

SOLUTION   Substituting $\theta = 3\pi/4$ and $t = 3$ into the equation $\omega = \theta/t$ gives us

$$\omega = \frac{3\pi/4 \text{ rad}}{3 \text{ sec}}$$

$$= \frac{\pi}{4} \text{ rad/sec}$$

NOTE   There are a number of equivalent ways to express the units of velocity. For example, the answer to Example 2 can be expressed in each of the following ways; they are all equivalent.

$$\frac{\pi}{4} \text{ radians per second} = \frac{\pi}{4} \text{ rad/sec} = \frac{\pi \text{ radians}}{4 \text{ seconds}} = \frac{\pi \text{ rad}}{4 \text{ sec}}$$

Likewise, you can express the answer to Example 1 in any of the following ways:

$$2.5 \text{ centimeters per second} = 2.5 \text{ cm/sec} = \frac{2.5 \text{ cm}}{1 \text{ sec}}$$

EXAMPLE 3 A bicycle wheel with a radius of 13 inches turns with an angular velocity of 3 radians per second. Find the distance traveled by a point on the bicycle tire in 1 minute.

SOLUTION We have $\omega = 3$ rad/sec, $r = 13$ inches, and $t = 60$ sec. First we find θ using $\omega = \theta/t$.

$$\text{If} \qquad \omega = \frac{\theta}{t}$$

$$\text{then} \qquad \theta = \omega t$$

$$= (3 \text{ rad/sec})(60 \text{ sec})$$

$$= 180 \text{ rad}$$

To find the distance traveled by the point in 60 seconds, we use the formula $s = r\theta$ from Section 3.4, with $r = 13$ inches and $\theta = 180$.

$$s = 13(180)$$

$$= 2{,}340 \text{ inches}$$

If we want this result expressed in feet, we divide by 12.

$$s = \frac{2{,}340}{12}$$

$$= 195 \text{ ft}$$

A point on the tire of the bicycle will travel 195 feet in 1 minute. If the bicycle were being ridden under these conditions, the rider would travel 195 feet in 1 minute. ■

EXAMPLE 4 Figure 1 shows a fire truck parked on the shoulder of a freeway next to a long block wall. The red light on the top of the truck is 10 feet from the wall and rotates through one complete revolution every 2 seconds. Find the equations that give the lengths d and l in terms of time t.

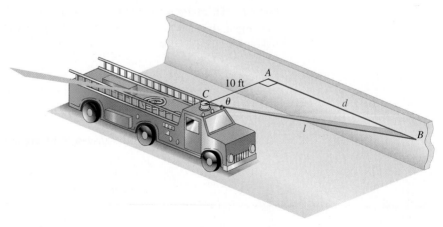

Figure 1

SOLUTION   The angular velocity of the rotating red light is

$$\omega = \frac{\theta}{t} = \frac{2\pi \text{ rad}}{2 \text{ sec}} = \pi \text{ rad/sec}$$

From right triangle ABC, we have the following relationships:

$$\tan \theta = \frac{d}{10} \qquad \text{and} \qquad \sec \theta = \frac{l}{10}$$

$$d = 10 \tan \theta \qquad \qquad l = 10 \sec \theta$$

Now, these equations give us d and l in terms of θ. To write d and l in terms of t, we solve $\omega = \theta/t$ for θ to obtain $\theta = \omega t = \pi t$. Substituting this for θ in each equation, we have d and l expressed in terms of t.

$$d = 10 \tan \pi t \quad l = 10 \sec \pi t \quad \blacksquare$$

THE RELATIONSHIP BETWEEN THE TWO VELOCITIES

To find the relationship between the two kinds of velocities we have developed so far, we can take the equation that relates arc length and central angle measure, $s = r\theta$, and divide both sides by time, t.

$$\text{If} \qquad s = r\theta$$

$$\text{then} \qquad \frac{s}{t} = \frac{r\theta}{t}$$

$$\frac{s}{t} = r\frac{\theta}{t}$$

$$v = r\omega$$

Linear velocity is the product of the radius and the angular velocity.

NOTE   When using this formula to relate linear velocity to angular velocity, keep in mind that the angular velocity ω must be expressed in radians per unit of time.

EXAMPLE 5   A phonograph record is turning at 45 revolutions per minute (rpm). If the distance from the center of the record to a point on the edge of the record is 3 inches, find the angular velocity and the linear velocity of the point in feet per minute.

SOLUTION   The quantity 45 revolutions per minute is another way of expressing the rate at which the point on the record is moving. We can obtain the angular velocity from it by remembering that one complete revolution is equivalent to 2π radians. Therefore,

$$\omega = 45 \text{ rev/min} = \frac{45 \text{ rev}}{1 \text{ min}} \cdot \frac{2\pi \text{ rad}}{1 \text{ rev}}$$

$$= 90\pi \text{ rad/min}$$

Since one revolution is equivalent to 2π radians, the fraction

$$\frac{2\pi \text{ rad}}{1 \text{ rev}}$$

is just another form of the number 1. We call this kind of fraction a *conversion factor.* Notice how the units of revolutions divide out in much the same way that common factors divide out when we reduce fractions to lowest terms. The conversion factor allows us to convert from revolutions to radians by dividing out revolutions.

To find the linear velocity, we multiply ω by the radius.

$$v = r\omega$$

$$= (3 \text{ inches})(90\pi \text{ rad/min})$$

$$= 270\pi \text{ inches/min} \qquad \text{Exact value}$$

$$\approx 270(3.14) \text{ inches/min} \qquad \text{Approximate value}$$

$$= 848 \text{ inches/min} \qquad \text{To 3 significant digits}$$

To convert 848 inches per minute to feet per minute, we use another conversion factor relating feet to inches. Here is our work:

$$848 \text{ inches/min} = \frac{848 \text{ inches}}{1 \text{ min}} \cdot \frac{1 \text{ ft}}{12 \text{ inches}} = \frac{848}{12} \text{ ft/min} \approx 70.7 \text{ ft/min} \quad \blacksquare$$

EXAMPLE 6 The Ferris wheel shown in Figure 2 is a model of the one we mentioned in the introduction to this section. If the diameter of the wheel is 250 feet and one complete revolution takes 20 minutes, find the linear velocity of a person riding on the wheel. Give the answer in miles per hour.

SOLUTION We found the angular velocity in the introduction to this section. It is

$$\omega = \frac{\pi}{10} \text{ rad/min}$$

Next, we use the formula $v = r\omega$ to find the linear velocity. That is, we multiply the angular velocity by the radius to find the linear velocity.

$$v = r\omega$$

$$= (125 \text{ ft})\left(\frac{\pi}{10} \text{ rad/min}\right)$$

$$= \frac{25\pi}{2} \text{ ft/min}$$

$$\approx 39.27 \text{ ft/min (intermediate answer)}$$

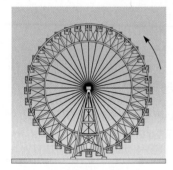

Figure 2

To convert to miles per hour, we use the facts that there are 60 minutes in 1 hour and 5,280 feet in 1 mile.

$$39.27 \text{ ft/min} = \frac{39.27 \text{ ft}}{1 \text{ min}} \cdot \frac{60 \text{ min}}{1 \text{ hr}} \cdot \frac{1 \text{ mi}}{5,280 \text{ ft}}$$

$$= \frac{(39.27)(60) \text{ mi}}{5,280 \text{ hr}}$$

$$\approx 0.45 \text{ mi/hr} \quad \blacksquare$$

NOTE In light of our previous discussion about units, you may be wondering how we went from (ft)(rad/min) to just ft/min in Example 6. Remember that radians are just real numbers and technically have no units. We sometimes write them as if they did for our own convenience, but one of the advantages of radian measure is that it does not introduce a new unit into calculations.

To gain a more intuitive understanding of the relationship between the radius of the circle and the linear velocity of a point on the circle, imagine that a bird is sitting on one of the wires that connects the center of the Ferris wheel in Example 6 to the wheel itself. Imagine further that the bird is sitting exactly halfway between the center of the wheel and a rider on the wheel. How does the linear velocity of the bird compare with the linear velocity of the rider? Since the angular velocity of the bird and the rider are equal (both sweep out the same amount of central angle in a given amount of time), and linear velocity is the product of the radius and the angular velocity, we can simply multiply the linear velocity of the rider by $\frac{1}{2}$ to obtain the linear velocity of the bird. Therefore, the bird is traveling at

$$\frac{1}{2} \cdot 0.45 \text{ mi/hr} = 0.225 \text{ mi/hr}$$

In one revolution around the wheel, the rider will travel a greater distance than the bird in the same amount of time, so the rider's velocity is greater than that of the bird. Twice as great, to be exact.

GETTING READY FOR CLASS

After reading through the preceding section, respond in your own words and in complete sentences.

a. Define linear velocity and give its formula.

b. Define angular velocity and give its formula.

c. What is the relationship between linear and angular velocity?

d. Explain the difference in linear velocities between a rider on a Ferris wheel and a bird positioned between the rider and the center of the wheel. Why is their angular velocity the same?

PROBLEM SET 3.5

In this Problem Set, round any answers that need rounding to three significant digits.

Find the linear velocity of a point moving with uniform circular motion, if the point covers a distance s in an amount of time t, where

1. $s = 3$ ft and $t = 2$ min   **2.** $s = 10$ ft and $t = 2$ min
3. $s = 12$ cm and $t = 4$ sec   **4.** $s = 12$ cm and $t = 2$ sec
5. $s = 30$ mi and $t = 2$ hr   **6.** $s = 100$ mi and $t = 4$ hr

Find the distance s covered by a point moving with linear velocity v for a time t if

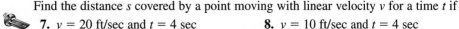

7. $v = 20$ ft/sec and $t = 4$ sec   **8.** $v = 10$ ft/sec and $t = 4$ sec
9. $v = 45$ mi/hr and $t = \frac{1}{2}$ hr   **10.** $v = 55$ mi/hr and $t = \frac{1}{2}$ hr
11. $v = 21$ mi/hr and $t = 20$ min   **12.** $v = 63$ mi/hr and $t = 10$ sec

Point P sweeps out central angle θ as it rotates on a circle of radius r as given below. In each case, find the angular velocity of point P.

13. $\theta = 2\pi/3$, $t = 5$ sec

14. $\theta = 3\pi/4$, $t = 5$ sec

15. $\theta = 12$, $t = 3$ min

16. $\theta = 24$, $t = 6$ min

17. $\theta = 8\pi$, $t = 3\pi$ sec

18. $\theta = 12\pi$, $t = 5\pi$ sec

19. $\theta = 45\pi$, $t = 1.2$ hr

20. $\theta = 24\pi$, $t = 1.8$ hr

21. Rotating Light Figure 3 shows a lighthouse that is 100 feet from a long straight wall on the beach. The light in the lighthouse rotates through one complete rotation once every 4 seconds. Find an equation that gives the distance d in terms of time t, then find d when t is $\frac{1}{2}$ second and $\frac{3}{2}$ seconds. What happens when you try $t = 1$ second in the equation? How do you interpret this?

22. Rotating Light Using the diagram in Figure 3, find an equation that expresses l in terms of time t. Find l when t is 0.5 second, 1.0 second, and 1.5 seconds. (Assume the light goes through one rotation every 4 seconds.)

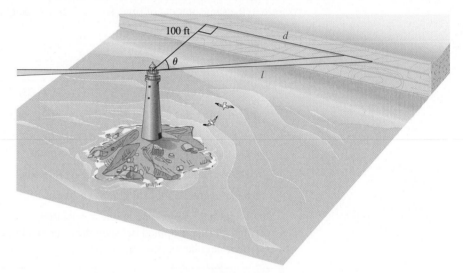

Figure 3

In the problems that follow, point P moves with angular velocity ω on a circle of radius r. In each case, find the distance s traveled by the point in time t.

 23. $\omega = 4$ rad/sec, $r = 2$ inches, $t = 5$ sec

24. $\omega = 2$ rad/sec, $r = 4$ inches, $t = 5$ sec

25. $\omega = 3\pi/2$ rad/sec, $r = 4$ m, $t = 30$ sec

26. $\omega = 4\pi/3$ rad/sec, $r = 8$ m, $t = 20$ sec

27. $\omega = 15$ rad/sec, $r = 5$ ft, $t = 1$ min

28. $\omega = 10$ rad/sec, $r = 6$ ft, $t = 2$ min

For each of the following problems, find the angular velocity, in radians per minute, associated with the given rpm.

 29. 10 rpm   **30.** 20 rpm   **31.** $33\frac{1}{3}$ rpm   **32.** $16\frac{2}{3}$ rpm

33. 5.8 rpm   **34.** 7.2 rpm

For each problem below, a point is rotating with uniform circular motion on a circle of radius *r*.

35. Find *v* if *r* = 2 inches and *ω* = 5 rad/sec.

36. Find *v* if *r* = 8 inches and *ω* = 4 rad/sec.

37. Find *ω* if *r* = 6 cm and *v* = 3 cm/sec.

38. Find *ω* if *r* = 3 cm and *v* = 8 cm/sec.

39. Find *v* if *r* = 4 ft and the point rotates at 10 rpm.

40. Find *v* if *r* = 1 ft and the point rotates at 20 rpm.

41. Velocity at the Equator The earth rotates through one complete revolution every 24 hours. Since the axis of rotation is perpendicular to the equator, you can think of a person standing on the equator as standing on the edge of a disc that is rotating through one complete revolution every 24 hours. Find the angular velocity of a person standing on the equator.

42. Velocity at the Equator Assuming the radius of the earth is 4,000 miles, use the information from Problem 41 to find the linear velocity of a person standing on the equator.

43. Velocity of a Mixer Blade A mixing blade on a food processor extends out 3 inches from its center. If the blade is turning at 600 revolutions per minute, what is the linear velocity of the tip of the blade in feet per minute?

44. Velocity of a Lawnmower Blade A gasoline-driven lawnmower has a blade that extends out 1 foot from its center. The tip of the blade is traveling at the speed of sound, which is 1,100 feet per second. Through how many revolutions per minute is the blade turning?

45. Cable Cars The San Francisco cable cars travel by clamping onto a steel cable that circulates in a channel beneath the streets. This cable is driven by a large 14-foot-diameter pulley, called a *sheave* (Figure 4). The sheave turns at a rate of 19 revolutions per minute. Find the speed of the cable car, in miles per hour, by determining the linear velocity of the cable. (1 mi = 5,280 ft)

Figure 4

46. Cable Cars The Los Angeles Cable Railway was driven by a 13-foot-diameter drum that turned at a rate of 18 revolutions per minute. Find the speed of the cable car, in miles per hour, by determining the linear velocity of the cable.

47. Cable Cars The old Sutter Street cable car line in San Francisco (Figure 5) used a 12-foot-diameter sheave to drive the cable. In order to keep the cable cars moving at a linear velocity of 10 miles per hour, how fast would the sheave need to turn (in revolutions per minute)?

Figure 5

48. Cable Cars The Cleveland City Cable Railway had a 14-foot-diameter pulley to drive the cable. In order to keep the cable cars moving at a linear velocity of 12 miles per hour, how fast would the pulley need to turn (in revolutions per minute)?

49. Ski Lift A ski lift operates by driving a wire rope, from which chairs are suspended, around a bullwheel (Figure 6). If the bullwheel is 12 feet in diameter and turns at a rate of 9 revolutions per minute, what is the linear velocity, in feet per second, of someone riding the lift?

Figure 6

50. Ski Lift An engineering firm is designing a ski lift. The wire rope needs to travel with a linear velocity of 2.0 meters per second, and the angular velocity of the bullwheel will be 10 revolutions per minute. What diameter bullwheel should be used to drive the wire rope?

51. Velocity of a Ferris Wheel Figure 7 is a model of the Ferris wheel known as the Riesenrad, or Great Wheel, that was built in Vienna in 1897. The diameter of the wheel is 197 feet, and one complete revolution takes 15 minutes. Find the linear velocity of a person riding on the wheel. Give your answer in miles per hour and round to the nearest hundredth.

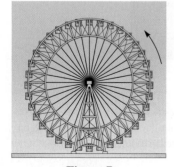

Figure 7

52. **Velocity of a Ferris Wheel** Use Figure 7 as a model of the Ferris wheel called Colossus that was built in St. Louis in 1986. The diameter of the wheel is 165 feet. A brochure that gives some statistics associated with Colossus indicates that it rotates at 1.5 revolutions per minute. The same brochure also indicates that a rider on the wheel is traveling at 10 miles per hour. Explain why these two numbers, 1.5 revolutions per minute and 10 miles per hour, cannot both be correct.

53. **Velocity of a Bike Wheel** A woman rides a bicycle for 1 hour and travels 16 kilometers (about 10 miles). Find the angular velocity of the wheel if the radius is 30 centimeters.

54. **Velocity of a Bike Wheel** Find the number of revolutions per minute for the wheel in Problem 53.

Figure 8

Cycling Lance Armstrong, five-time winner of the Tour de France, rides a Trek 5900 bicycle equipped with Dura-Ace components (Figure 8). When Lance pedals, he turns a gear, called a *chainring.* The angular velocity of the chainring will determine the linear speed at which the chain travels. The chain connects the chainring to a smaller gear, called a *sprocket,* which is attached to the rear wheel (Figure 9). The angular velocity of the sprocket depends upon the linear speed of the chain. The sprocket and rear wheel rotate at the same rate, and the diameter of the rear wheel is 700 millimeters. The speed at which Lance travels is determined by the angular velocity of his rear wheel. Use this information to answer Problems 55–60.

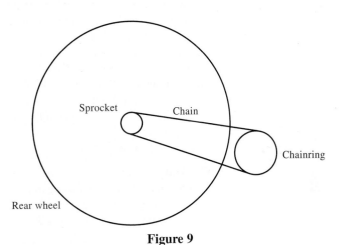

Figure 9

55. When Lance Armstrong blazed up Mount Ventoux in the 2002 Tour, he was equipped with a 150-millimeter-diameter chainring and a 95-millimeter-diameter sprocket. Lance is known for maintaining a very high *cadence,* or pedal rate. If he was pedaling at a rate of 90 revolutions per minute, find his speed in kilometers per hour. (1 km = 1,000,000 mm)

56. On level ground, Lance would use a larger chainring and a smaller sprocket. If he shifted to a 210-millimeter-diameter chainring and a 40-millimeter-diameter sprocket, how fast would he be traveling in kilometers per hour if he pedaled at a rate of 80 revolutions per minute?

57. If Lance was using his 210-millimeter-diameter chainring and pedaling at a rate of 85 revolutions per minute, what diameter sprocket would he need in order to maintain a speed of 45 kilometers per hour?

58. If Lance was using his 150-millimeter-diameter chainring and pedaling at a rate of 95 revolutions per minute, what diameter sprocket would he need in order to maintain a speed of 24 kilometers per hour?

59. Suppose Lance was using a 150-millimeter-diameter chainring and an 80-millimeter-diameter sprocket. How fast would he need to pedal, in revolutions per minute, in order to maintain a speed of 20 kilometers per hour?

60. Suppose Lance was using a 210-millimeter-diameter chainring and a 40-millimeter-diameter sprocket. How fast would he need to pedal, in revolutions per minute, in order to maintain a speed of 40 kilometers per hour?

REVIEW PROBLEMS

The problems that follow review material we covered in Section 2.5.

61. True Course A boat is crossing a river that runs due west. The direction of the boat is due north, and it is moving through the water at 9.50 miles per hour. If the current is running at a constant 4.25 miles per hour, find the true course of the boat.

62. Navigation A ship is 4° off course. How far off course is it after traveling for 75 miles?

63. Magnitude of a Vector Find the magnitudes of the horizontal and vertical vector components of a velocity vector of 68 feet per second with angle of elevation 37°.

64. Magnitude of a Vector The magnitude of the horizontal component of a vector is 75, while the magnitude of its vertical component is 45. What is the magnitude of the vector?

65. Distance and Bearing A ship sails for 85.5 miles on a bearing of S 57.3° W. How far west and how far south has the boat traveled?

66. Distance and Bearing A plane flying with a constant speed of 285.5 miles per hour flies for 2 hours on a course with bearing N 48.7° W. How far north and how far west does the plane fly?

EXTENDING THE CONCEPTS

67. Winternationals Jim Rizzoli owns, maintains, and races an alcohol dragster. On board the dragster is a computer that compiles data into a number of different categories during each of Jim's races. The table below shows some of the data from a race Jim was in during the 1993 Winternationals.

Time in Seconds	Speed in Miles/Hour	Front Axle RPM
0	0	0
1	72.7	1,107
2	129.9	1,978
3	162.8	2,486
4	192.2	2,919
5	212.4	3,233
6	228.1	3,473

The front wheels, at each end of the front axle, are 22.07 inches in diameter. Derive a formula that will convert rpm's from a 22.07-inch wheel into miles per hour. Test your formula on all six rows of the table. Explain any discrepancies between the table values and the values obtained from your formula.

CHAPTER 3 SUMMARY

EXAMPLES

1. 30° is the reference angle for 30°, 150°, 210°, and 330°.

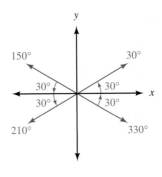

Reference Angle [3.1]

The *reference angle* $\hat{\theta}$ for any angle θ in standard position is the positive acute angle between the terminal side of θ and the *x*-axis.

A trigonometric function of an angle and its reference angle differ at most in sign.

We find trigonometric functions for angles between 0° and 360° by first finding the reference angle. We then find the value of the trigonometric function of the reference angle and use the quadrant in which the angle terminates to assign the correct sign.

$$\sin 150° = \sin 30° = \frac{1}{2}$$

$$\sin 210° = -\sin 30° = -\frac{1}{2}$$

$$\sin 330° = -\sin 30° = -\frac{1}{2}$$

2.

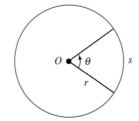

Radian Measure [3.2]

In a circle with radius *r*, if central angle θ cuts off an arc of length *s*, then the radian measure of θ is given by

$$\theta = \frac{s}{r}$$

3. Radians to degrees

$$\frac{4\pi}{3}\text{ rad} = \frac{4\pi}{3}\left(\frac{180}{\pi}\right)°$$

$$= 240°$$

Degrees to radians

$$450° = 450\left(\frac{\pi}{180}\right)$$

$$= \frac{5\pi}{2}\text{ rad}$$

Radians and Degrees [3.2]

Changing from degrees to radians and radians to degrees is simply a matter of multiplying by the appropriate conversion factor.

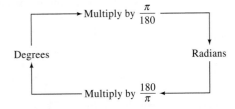

4.

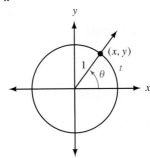

Circular Functions [3.3]

The unit circle is the circle with its center at the origin and a radius of 1. The equation of the unit circle is $x^2 + y^2 = 1$. Because the radius of the unit circle is 1, any point (x, y) on the circle can be written

$$(x, y) = (\cos \theta, \sin \theta) = (\cos t, \sin t)$$

5. Tan θ is an odd function.

$$\tan(-\theta) = \frac{\sin(-\theta)}{\cos(-\theta)}$$

$$= \frac{-\sin\theta}{\cos\theta}$$

$$= -\frac{\sin\theta}{\cos\theta}$$

$$= -\tan\theta$$

Even and Odd Functions [3.3]

An *even function* is a function for which

$$f(-x) = f(x) \text{ for all } x \text{ in the domain of } f$$

and an *odd function* is a function for which

$$f(-x) = -f(x) \text{ for all } x \text{ in the domain of } f$$

Cosine is an even function, and sine is an odd function. That is,

$$\cos(-\theta) = \cos\theta \qquad \text{Cosine is an even function}$$

and

$$\sin(-\theta) = -\sin\theta \qquad \text{Sine is an odd function}$$

6. The arc cut off by 2.5 radians in a circle of radius 4 inches is

$$s = 4(2.5) = 10 \text{ inches}$$

Arc Length [3.4]

If s is an arc cut off by a central angle θ, measured in radians, in a circle of radius r, then

$$s = r\theta$$

7. The area of the sector formed by a central angle of 2.5 radians in a circle of radius 4 inches is

$$A = \frac{1}{2}(4)^2(2.5) = 20 \text{ inches}^2$$

Area of a Sector [3.4]

The area of the sector formed by a central angle θ in a circle of radius r is

$$A = \frac{1}{2}r^2\theta$$

where θ is measured in radians.

8. If a point moving at a uniform speed on a circle travels 12 centimeters every 3 seconds, then the linear velocity of the point is

$$v = \frac{12 \text{ cm}}{3 \text{ sec}} = 4 \text{ cm/sec}$$

Linear Velocity [3.5]

If P is a point on a circle of radius r, and P moves a distance s on the circumference of the circle in an amount of time t, then the *linear velocity, v,* of P is given by the formula

$$v = \frac{s}{t}$$

9. If a point moving at uniform speed on a circle of radius 4 inches rotates through $3\pi/4$ radians every 3 seconds, then the angular velocity of the point is

$$\omega = \frac{3\pi/4 \text{ rad}}{3 \text{ sec}} = \frac{\pi}{4} \text{ rad/sec}$$

The linear velocity of the same point is given by

$$v = 4\left(\frac{\pi}{4}\right) = \pi \text{ inches/sec}$$

Angular Velocity [3.5]

If P is a point moving with uniform circular motion on a circle of radius r, and the line from the center of the circle through P sweeps out a central angle θ in an amount of time t, then the *angular velocity, ω,* of P is given by the formula

$$\omega = \frac{\theta}{t} \qquad \text{where } \theta \text{ is measured in radians}$$

The relationship between linear velocity and angular velocity is given by the formula

$$v = r\omega$$

CHAPTER 3 TEST

Draw each of the following angles in standard position and then name the reference angle:

1. 235°   **2.** 117.8°

3. 410° 20'   **4.** −225°

Use a calculator to find each of the following:

5. cot 320°   **6.** cot (−25°)

7. csc (−236.7°)   **8.** sec 322.3°

9. sec 140° 20'   **10.** csc 188° 50'

Use a calculator to find θ, to the nearest tenth of a degree, if θ is between 0° and 360° and

11. $\sin \theta = 0.1045$ with θ in QII   **12.** $\cos \theta = -0.4772$ with θ in QIII

13. $\cot \theta = 0.9659$ with θ in QIII   **14.** $\sec \theta = 1.545$ with θ in QIV

Give the exact value of each of the following:

15. sin 225°   **16.** cos 135°

17. tan 330°   **18.** sec 390°

Convert each of the following to radian measure. Write each answer as an exact value.

19. 250°   **20.** −390°

Convert each of the following to degree measure:

21. $4\pi/3$ **22.** $7\pi/12$

Give the exact value of each of the following:

23. $\sin \dfrac{2\pi}{3}$ **24.** $\cos \dfrac{2\pi}{3}$

25. $4 \cos \left(-\dfrac{3\pi}{4}\right)$ **26.** $2 \cos \left(-\dfrac{5\pi}{3}\right)$

27. $\sec \dfrac{5\pi}{6}$ **28.** $\csc \dfrac{5\pi}{6}$

29. Evaluate $2 \cos \left(3x - \dfrac{\pi}{2}\right)$ when x is $\dfrac{\pi}{3}$.

30. Evaluate $4 \sin \left(2x + \dfrac{\pi}{4}\right)$ when x is $\dfrac{\pi}{4}$.

31. Show that cotangent is an odd function.

32. Prove the identity $\sin (-\theta) \sec (-\theta) \cot (-\theta) = 1$.

For Problems 33 and 34, θ is a central angle in a circle of radius r. In each case, find the length of arc s cut off by θ to the nearest hundredth.

33. $\theta = \pi/6$, $r = 12$ m   **34.** $\theta = 60°$, $r = 6$ ft

In Problems 35 and 36, θ is a central angle that cuts off an arc of length s. In each case, find the radius of the circle.

35. $\theta = \pi/4$, $s = \pi$ cm   **36.** $\theta = 2\pi/3$, $s = \pi/4$ cm

Find the area of the sector formed by central angle θ in a circle of radius r if

37. $\theta = 90°$, $r = 4$ inches **38.** $\theta = 2.4$, $r = 3$ cm

39. Arc Length The minute hand of a clock is 2 centimeters long. How far does the tip of the minute hand travel in 30 minutes?

40. Distance A boy is twirling a model airplane on a string 5 feet long. If he twirls the plane at 0.5 revolutions per minute, how far does the plane travel in 2 minutes?

41. Area of a Sector A central angle of 4 radians cuts off an arc of length 8 inches. Find the area of the sector formed.

Find the distance s covered by a point moving with linear velocity v for a time t if

42. $v = 30$ ft/sec and $t = 3$ sec **43.** $v = 66$ ft/sec and $t = 1$ min

In the problems that follow, point P moves with angular velocity ω on a circle of radius r. In each case, find the distance s traveled by the point in time t.

44. $\omega = 4$ rad/sec, $r = 3$ inches, $t = 6$ sec
45. $\omega = 3\pi/4$ rad/sec, $r = 8$ ft, $t = 20$ sec

For Problems 46 and 47, find the angular velocity, in radians per minute, associated with the given revolutions per minute.

46. 6 rpm **47.** 2 rpm

For each problem below, a point is rotating with uniform circular motion on a circle of radius r.

48. Find ω if $r = 10$ cm and $v = 5$ cm/sec.
49. Find ω if $r = 3$ cm and $v = 5$ cm/sec.
50. Find v if $r = 2$ ft and the point rotates at 20 rpm.
51. Find v if $r = 1$ ft and the point rotates at 10 rpm.

52. Angular Velocity A belt connects a pulley of radius 8 centimeters to a pulley of radius 6 centimeters. Each point on the belt is traveling at 24 centimeters per second. Find the angular velocity of each pulley (Figure 1).

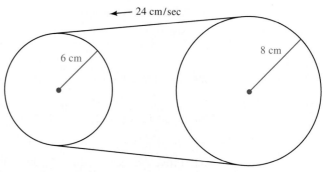

Figure 1

53. Linear Velocity A propeller with radius 1.50 feet is rotating at 900 revolutions per minute. Find the linear velocity of the tip of the propeller. Give the exact value and an approximation to three significant digits.

54. Velocity of a Diskette A $3\frac{1}{2}$-inch diskette, when placed in the disk drive of a computer, rotates at 300 revolutions per minute. Find the linear velocity of a point 1.5 inches from the center of the diskette.

55. Velocity of a Zip Disk A 100-MB zip disk, when placed in the internal disk drive of a computer, rotates at 2,941 revolutions per minute. Find the linear velocity, in miles per hour, of a point 1 inch from the center of the disk. Then find the linear velocity, in miles per hour, of a point 1.5 inches from the center of the disk. Round approximate answers to the nearest tenth.

56. Cable Cars If a 14-foot-diameter sheave is used to drive a cable car, at what angular velocity must the sheave turn in order for the cable car to travel 11 miles per hour? Give your answer in revolutions per minute. (1 mi = 5,280 ft)

57. Cycling Roberto Heras is riding in the Vuelta a España. He is using a 210-millimeter-diameter chainring and a 50-millimeter-diameter sprocket with a 700-millimeter-diameter rear wheel. Find his linear velocity if he is pedaling at a rate of 75 revolutions per minute (Figure 2). Give your answer in kilometers per hour. (1 km = 1,000,000 mm)

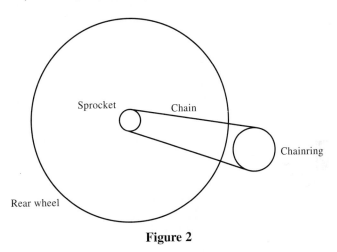

Figure 2

CHAPTER **3** GROUP PROJECT

MODELING A DOUBLE FERRIS WHEEL

Objective: To find a model for the height of a rider on a double Ferris wheel.

In 1939, John Courtney invented the double Ferris wheel, called a Sky Wheel, consisting of two smaller wheels spinning at the ends of a rotating arm.

For this project, we will model a double Ferris wheel with a 50-foot arm that is spinning at a rate of 3 revolutions per minute in a counterclockwise direction. The center of the arm is 44 feet above the ground. The diameter of each wheel is 32 feet, and the wheels turn at a rate of 5 revolutions per minute in a clockwise direction. A diagram of the situation is shown in Figure 1. *M* is the midpoint of the arm, and *O* is the center of the lower wheel. Assume the rider is initially at point *P* on the wheel.

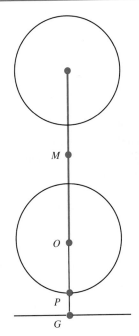

Figure 1

1. Find the lengths of *MO*, *OP*, and *MG*.

Figure 2 shows the location of the rider after a short amount of time has passed. The arm has rotated counterclockwise through an angle θ, while the wheel has rotated clockwise through an angle ϕ relative to the direction of the arm. Point P shows the current position of the rider, and the height of the rider is h.

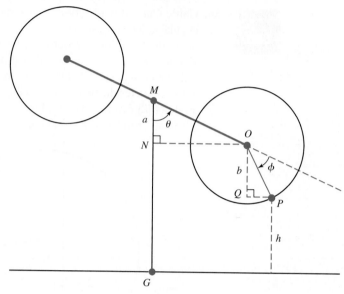

Figure 2

2. Find angle QOP in terms of θ and ϕ.

3. Use right triangle trigonometry to find lengths a and b, and then the height of the rider h, in terms of θ and ϕ.

We know that the angular velocity of the arm is 3 revolutions per minute, and the angular velocity of the wheel is 5 revolutions per minute. The last step is to find the radian measure of angles θ and ϕ.

4. Let t be the number of minutes that have passed since the ride began. Use the angular velocities of the arm and wheel to find θ and ϕ in radians in terms of t. Replace θ and ϕ in your answer from Step 3 to obtain the height h as a function of time t (in minutes).

5. Use this function to find the following:
  a. The height of the rider after 30 seconds have passed.
  b. The height of the rider after the arm has completed its first revolution.
  c. The height of the rider after the wheel has completed two revolutions.

6. Graph this function with your graphing calculator. Make sure your calculator is set to radian mode. Use the graph to find the following:
  a. The maximum height of the rider.
  b. The minimum number of minutes required for the rider to return to the original position (point P in Figure 1).

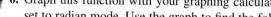

CHAPTER **3** RESEARCH PROJECT

THE THIRD MAN

The Riesenrad in Vienna

We mentioned in Chapter 2 that the Ferris wheel called the *Riesenrad,* built in Vienna in 1897, is still in operation today. A brochure that gives some statistics associated with the Riesenrad indicates that passengers riding it travel at 2 feet, 6 inches per second. The Orson Welles movie *The Third Man* contains a scene in which Orson Welles rides the Riesenrad through one complete revolution. Play *The Third Man* on a VCR, so you can view the Riesenrad in operation. Then devise a method of using the movie to estimate the angular velocity of the wheel. Give a detailed account of the procedure you use to arrive at your estimate. Finally, use your results either to prove or to disprove the claim that passengers travel at 2.5 feet per second on the Riesenrad.

GRAPHING AND INVERSE FUNCTIONS

A page of sheet music represents a piece of music; the music itself is what you get when the notes on the page are sung or performed on a musical instrument.

Keith Devlin

INTRODUCTION

Trigonometric functions are periodic functions because they repeat all their range values at regular intervals. The riders on the Ferris wheels we have been studying repeat their positions around the wheel periodically as well. That is why trigonometric functions are good models for the motion of a rider on a Ferris wheel. When we use a rectangular coordinate system to plot the distance between the ground and a rider during a ride, we find that the shape of the graph matches exactly the shape of the graph of one of the trigonometric functions.

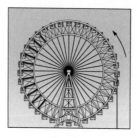 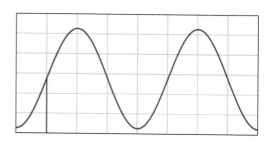

Once we have modeled the motion of the rider with a trigonometric function, we have a new set of mathematical tools with which to investigate the motion of the rider on the wheel.

STUDY SKILLS FOR CHAPTER 4

The study skills for this chapter are about attitude. They are points of view that point toward success.

1. **Be Focused, Not Distracted**   We have students who begin their assignments by asking themselves, "Why am I taking this class?" If you are asking yourself similar questions, you are distracting yourself from doing the things that will produce the results you want in this course. Don't dwell on questions and evaluations of the class that can be used as excuses for not doing well. If you want to succeed in this course, focus your energy and efforts toward success, rather than distracting yourself from your goals.

2. **Be Resilient**   Don't let setbacks keep you from your goals. You want to put yourself on the road to becoming a person who can succeed in this class, or any college class. Failing a test or quiz, or having a difficult time on some topics, is normal. No one goes through college without some setbacks. Don't let a temporary disappointment keep you from succeeding in this course. A low grade on a test or quiz is simply a signal that some reevaluation of your study habits needs to take place.

3. **Intend to Succeed**   We always have a few students who simply go through the motions of studying without intending on mastering the material. It is more important to them to look like they are studying than to actually study. You need to study with the intention of being successful in the course. Intend to master the material, no matter what it takes.

SECTION 4.1   BASIC GRAPHS AND AMPLITUDE

THE SINE GRAPH

To graph the equation $y = \sin x$, we begin by making a table of values of x and y that satisfy the equation (Table 1), and then use the information in the table to sketch the graph. To make it easy on ourselves, we will let x take on values that are multiples of $\pi/4$. As an aid in sketching the graphs, we will approximate $1/\sqrt{2}$ with 0.7.

Graphing each ordered pair and then connecting them with a smooth curve, we obtain the following graph:

TABLE 1

x	$y = \sin x$
0	$\sin 0 = 0$
$\dfrac{\pi}{4}$	$\sin \dfrac{\pi}{4} = \dfrac{1}{\sqrt{2}}$
$\dfrac{\pi}{2}$	$\sin \dfrac{\pi}{2} = 1$
$\dfrac{3\pi}{4}$	$\sin \dfrac{3\pi}{4} = \dfrac{1}{\sqrt{2}}$
π	$\sin \pi = 0$
$\dfrac{5\pi}{4}$	$\sin \dfrac{5\pi}{4} = -\dfrac{1}{\sqrt{2}}$
$\dfrac{3\pi}{2}$	$\sin \dfrac{3\pi}{2} = -1$
$\dfrac{7\pi}{4}$	$\sin \dfrac{7\pi}{4} = -\dfrac{1}{\sqrt{2}}$
2π	$\sin 2\pi = 0$

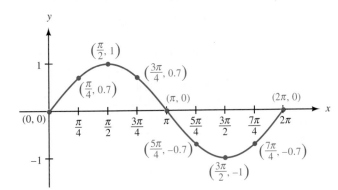

Figure 1

GRAPHING Y = SIN X USING THE UNIT CIRCLE

We can also obtain the graph of the sine function by using the unit circle definition we introduced in Section 3.3 (Definition III): If the point (x, y) is t units from $(1, 0)$ along the circumference of the unit circle, then $\sin t = y$. Therefore, if we start at the point $(1, 0)$ and travel once around the unit circle (a distance of 2π units), we can find the value of y in the equation $y = \sin t$ by simply keeping track of the y-coordinates of the points that are t units from $(1, 0)$.

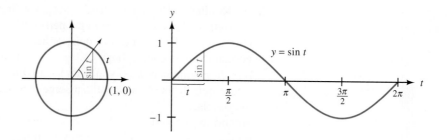

unit circle

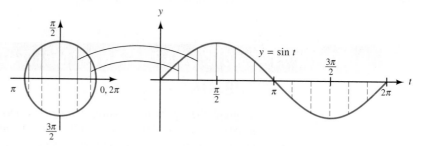

Figure 2

EXTENDING THE SINE GRAPH

Figures 1 and 2 each show one complete cycle of $y = \sin x$. We can extend the graph of $y = \sin x$ to the right of $x = 2\pi$ by realizing that, once we go past $x = 2\pi$, we will begin to name angles that are coterminal with the angles between 0 and 2π. Because of this, we will start to repeat the values of $\sin x$. Likewise, if we let x take on values to the left of $x = 0$, we will simply get the values of $\sin x$ between 0 and 2π in the reverse order. Figure 3 shows the graph of $y = \sin x$ extended beyond the interval from $x = 0$ to $x = 2\pi$.

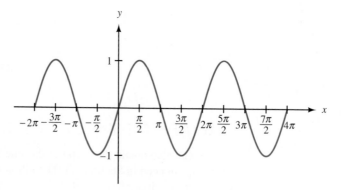

Figure 3

Our graph of $y = \sin x$ never goes above 1 or below -1, and it repeats itself every 2π units on the x-axis. This gives rise to the following two definitions.

DEFINITION PERIOD

For any function $y = f(x)$, the smallest positive number p for which

$$f(x + p) = f(x) \qquad \text{for all } x$$

is called the *period* of $f(x)$. In the case of $y = \sin x$, the period is $p = 2\pi$ since 2π is the smallest positive number for which

$$\sin(x + p) = \sin x \qquad \text{for all } x$$

DEFINITION AMPLITUDE

If the greatest value of y is M and the least value of y is m, then the *amplitude* of the graph of y is defined to be

$$A = \frac{1}{2}\left| M - m \right|$$

In the case of $y = \sin x$, the amplitude is 1 because $\frac{1}{2}\left| 1 - (-1) \right| = \frac{1}{2}(2) = 1$.

Along with the definitions for period and amplitude, we need to review the definitions for domain and range from algebra.

DEFINITION DOMAIN AND RANGE

The *domain* for the function $y = f(x)$ is the set of values that x can assume. For the function $y = \sin x$, the domain is all real numbers, since there is no value of x for which $\sin x$ is undefined.

The *range* for the function $y = f(x)$ is the set of values that y assumes. From the graph of $y = \sin x$ shown in Figure 3, we find that the range is the set

$$\{y \mid -1 \le y \le 1\}$$

THE COSINE GRAPH

The graph of $y = \cos x$ has the same general shape as the graph of $y = \sin x$.

EXAMPLE 1 Sketch the graph of $y = \cos x$.

SOLUTION We can arrive at the graph by making a table of convenient values of x and y.

TABLE 2

x	$y = \cos x$
0	$\cos 0 = 1$
$\dfrac{\pi}{4}$	$\cos \dfrac{\pi}{4} = \dfrac{1}{\sqrt{2}}$
$\dfrac{\pi}{2}$	$\cos \dfrac{\pi}{2} = 0$
$\dfrac{3\pi}{4}$	$\cos \dfrac{3\pi}{4} = -\dfrac{1}{\sqrt{2}}$
π	$\cos \pi = -1$
$\dfrac{5\pi}{4}$	$\cos \dfrac{5\pi}{4} = -\dfrac{1}{\sqrt{2}}$
$\dfrac{3\pi}{2}$	$\cos \dfrac{3\pi}{2} = 0$
$\dfrac{7\pi}{4}$	$\cos \dfrac{7\pi}{4} = \dfrac{1}{\sqrt{2}}$
2π	$\cos 2\pi = 1$

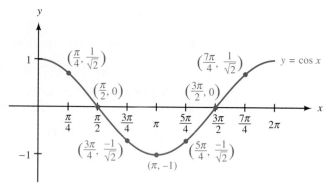

Figure 4

We can generate the graph of the cosine function using the unit circle just as we did for the sine function. By Definition III, if the point (x, y) is t units from $(1, 0)$ along the circumference of the unit circle, then $\cos t = x$. We start at the point $(1, 0)$ and travel once around the unit circle, keeping track of the x-coordinates of the points that are t units from $(1, 0)$. To help visualize how the x-coordinates generate the cosine graph, we have rotated the unit circle 90° counter-clockwise so that we may represent the x-coordinates as vertical line segments (Figure 5).

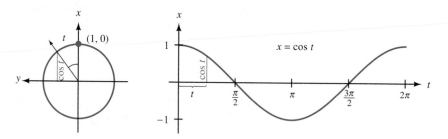

unit circle (rotated)

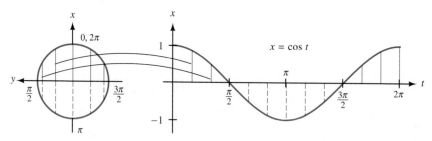

Figure 5

CALCULATOR NOTE To graph one cycle of the sine or cosine function using your graphing calculator in radian mode, define $Y_1 = \sin(x)$ or $Y_1 = \cos(x)$ and set your window variables so that

$$0 \le x \le 2\pi, \text{ scale } = \pi/2;\ -1.5 \le y \le 1.5$$

To graph either function in degree mode, set your window variables to

$$0 \le x \le 360, \text{ scale } = 90;\ -1.5 \le y \le 1.5$$

Figure 6 shows the graph of $y = \cos x$ in degree mode with the trace feature being used to observe ordered pairs along the graph.

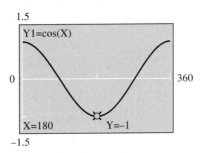

Figure 6

THE TANGENT GRAPH

Table 3 lists some solutions to $y = \tan x$ between $x = 0$ and $x = \pi$. Note that, at $\pi/2$, tangent is undefined. Since $\tan x = (\sin x)/(\cos x)$, it must be undefined at $\pi/2$ since $\cos x$ is 0 at multiples of $\pi/2$ and division by 0 is undefined. If we continue Table 3 beyond $x = \pi$, then the values of $\tan x = (\sin x)/(\cos x)$ will begin to repeat since these ratios are the same in quadrant III as they are in quadrant I. (The division of two negatives is a positive.) Figure 7 shows the graph based on the information from Table 3.

TABLE 3

x	$\tan x$
0	0
$\dfrac{\pi}{4}$	1
$\dfrac{\pi}{3}$	$\sqrt{3} \approx 1.7$
$\dfrac{\pi}{2}$	undefined
$\dfrac{2\pi}{3}$	$-\sqrt{3} \approx -1.7$
$\dfrac{3\pi}{4}$	-1
π	0

Figure 7

In Table 3, $y = \tan x$ is undefined because there is no corresponding value of y when $x = \pi/2$. That is, there will be no point on the graph with an x-coordinate of $\pi/2$. To help us remember this, we have drawn a dotted vertical line through $x = \pi/2$. This vertical line is called an *asymptote*. Our graph will never cross or touch this line. If we were to calculate values of $\tan x$ when x is very close to $\pi/2$ (or very close to 90° in degree mode), we would find that $\tan x$ would become very large for values of x just to the left of the asymptote and very small for values of x just to the right of the asymptote, as shown in Table 4.

TABLE 4

x	$\tan x$
85°	11.4
89°	57.3
89.9°	573.0
89.99°	5729.6
95°	-11.4
91°	-57.3
90.1°	-573.0
90.01°	-5729.6

Extending this graph to the right of π and to the left of 0 we obtain the graph shown in Figure 8. As this figure indicates, the period of $y = \tan x$ is π. The tangent function has no amplitude since there is no largest or smallest value of y on the graph of $y = \tan x$.

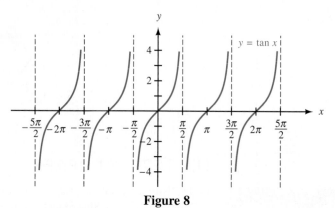

Figure 8

THE COSECANT GRAPH

EXAMPLE 2   Sketch the graph of $y = \csc x$.

SOLUTION   To graph $y = \csc x$, we can use the fact that $\csc x = 1/\sin x$ is the reciprocal of $\sin x$. In Table 5, we use the values of $\sin x$ from Table 1 and take reciprocals. Filling in with some additional points, we obtain the graph shown in Figure 9.

TABLE 5

x	$\sin x$	$\csc x = 1/\sin x$
0	0	undefined
$\dfrac{\pi}{4}$	$\dfrac{1}{\sqrt{2}}$	$\sqrt{2} \approx 1.4$
$\dfrac{\pi}{2}$	1	1
$\dfrac{3\pi}{4}$	$\dfrac{1}{\sqrt{2}}$	$\sqrt{2} \approx 1.4$
π	0	undefined
$\dfrac{5\pi}{4}$	$\dfrac{-1}{\sqrt{2}}$	$-\sqrt{2} \approx -1.4$
$\dfrac{3\pi}{2}$	-1	-1
$\dfrac{7\pi}{4}$	$\dfrac{-1}{\sqrt{2}}$	$-\sqrt{2} \approx -1.4$
2π	0	undefined

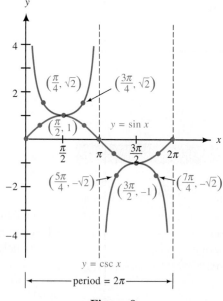

Figure 9

Because $y = \sin x$ repeats every 2π, so do the reciprocals of $\sin x$, so the period of $y = \csc x$ is 2π. As was the case with $y = \tan x$, there is no amplitude.

USING TECHNOLOGY

GRAPHING WITH ASYMPTOTES

When graphing a function with vertical asymptotes, such as $y = \csc x$, we must be careful how we interpret what the graphing calculator shows us. For example, having defined $Y_1 = 1/\sin(x)$, Figure 10 shows the graph of one cycle of this function in radian mode with the window variables set so that

$$0 \leq x \leq 2\pi, \text{ scale } = \pi/2; \ -4 \leq y \leq 4$$

and Figure 11 shows the same graph in degree mode with window settings

$$0 \leq x \leq 360, \text{ scale } = 90; \ -4 \leq y \leq 4$$

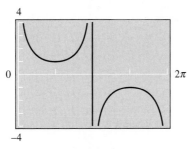

Figure 10

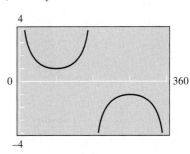

Figure 11

Because the calculator graphs a function by plotting points and connecting them, the vertical line seen in Figure 10 will sometimes appear where an asymptote exists. This line is not part of the graph of the cosecant function, as Figure 11 indicates.

THE COTANGENT AND SECANT GRAPHS

In Problem Set 4.1, you will be asked to graph $y = \cot x$ and $y = \sec x$. These graphs are shown in Figures 12 and 13 for reference.

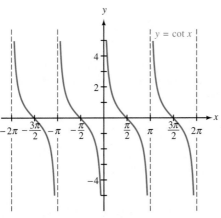

Figure 12

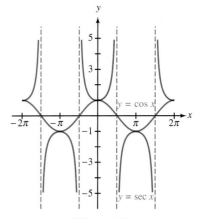

Figure 13

Here is a summary of the important facts associated with the graphs of our trigonometric functions.

Functions	One Cycle of the Graph	Domain	Range	Amplitude	Period
$y = \sin x$		all real numbers	$-1 \le y \le 1$	1	2π
$y = \cos x$		all real numbers	$-1 \le y \le 1$	1	2π
$y = \tan x$		all real numbers except $x = \dfrac{\pi}{2} + k\pi$ where k is an integer	all real numbers	none	π
$y = \cot x$		all real numbers except $x = k\pi$ where k is an integer	all real numbers	none	π
$y = \sec x$		all real numbers except $x = \dfrac{\pi}{2} + k\pi$ where k is an integer	$y \le -1$ or $y \ge 1$	none	2π
$y = \csc x$		all real numbers except $x = k\pi$ where k is an integer	$y \le -1$ or $y \ge 1$	none	2π

AMPLITUDE

Now we will consider the effect on the graph of multiplying a trigonometric function by a numerical factor.

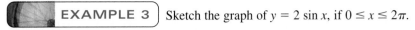

EXAMPLE 3   Sketch the graph of $y = 2 \sin x$, if $0 \leq x \leq 2\pi$.

SOLUTION   The coefficient 2 on the right side of the equation will simply multiply each value of $\sin x$ by a factor of 2. Therefore, the values of y in $y = 2 \sin x$ should all be twice the corresponding values of y in $y = \sin x$. Table 6 contains some values for $y = 2 \sin x$. Figure 14 shows the graphs of $y = \sin x$ and $y = 2 \sin x$. (We are including the graph of $y = \sin x$ simply for reference and comparison. With both graphs to look at, it is easier to see what change is brought about by the coefficient 2.)

TABLE 6

x	$y = 2 \sin x$	(x, y)
0	$y = 2 \sin 0 = 2(0) = 0$	$(0, 0)$
$\dfrac{\pi}{2}$	$y = 2 \sin \dfrac{\pi}{2} = 2(1) = 2$	$\left(\dfrac{\pi}{2}, 2\right)$
π	$y = 2 \sin \pi = 2(0) = 0$	$(\pi, 0)$
$\dfrac{3\pi}{2}$	$y = 2 \sin \dfrac{3\pi}{2} = 2(-1) = -2$	$\left(\dfrac{3\pi}{2}, -2\right)$
2π	$y = 2 \sin 2\pi = 2(0) = 0$	$(2\pi, 0)$

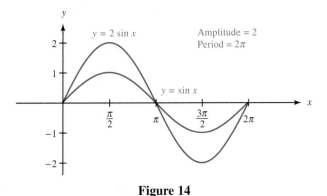

Figure 14

The coefficient 2 in $y = 2 \sin x$ changes the amplitude from 1 to 2 but does not affect the period. That is, we can think of the graph of $y = 2 \sin x$ as if it were the graph of $y = \sin x$ with the amplitude extended to 2 instead of 1. ■

EXAMPLE 4 Sketch one complete cycle of the graph of $y = \frac{1}{2} \cos x$.

SOLUTION Table 7 gives us some points on the curve $y = \frac{1}{2} \cos x$. Figure 15 shows the graphs of both $y = \frac{1}{2} \cos x$ and $y = \cos x$ on the same set of axes, from $x = 0$ to $x = 2\pi$.

TABLE 7

x	$y = \dfrac{1}{2} \cos x$	(x, y)
0	$y = \dfrac{1}{2} \cos 0 = \dfrac{1}{2}(1) = \dfrac{1}{2}$	$\left(0, \dfrac{1}{2}\right)$
$\dfrac{\pi}{2}$	$y = \dfrac{1}{2} \cos \dfrac{\pi}{2} = \dfrac{1}{2}(0) = 0$	$\left(\dfrac{\pi}{2}, 0\right)$
π	$y = \dfrac{1}{2} \cos \pi = \dfrac{1}{2}(-1) = -\dfrac{1}{2}$	$\left(\pi, -\dfrac{1}{2}\right)$
$\dfrac{3\pi}{2}$	$y = \dfrac{1}{2} \cos \dfrac{3\pi}{2} = \dfrac{1}{2}(0) = 0$	$\left(\dfrac{3\pi}{2}, 0\right)$
2π	$y = \dfrac{1}{2} \cos 2\pi = \dfrac{1}{2}(1) = \dfrac{1}{2}$	$\left(2\pi, \dfrac{1}{2}\right)$

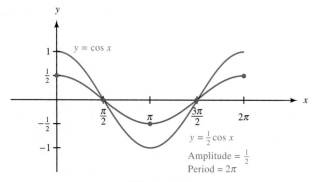

Figure 15

The coefficient $\frac{1}{2}$ in $y = \frac{1}{2} \cos x$ determines the amplitude of the graph.

SUMMARY

Generalizing the results of these last two examples, we can say that if A is a positive number, then the graphs of $y = A \sin x$ and $y = A \cos x$ will have amplitude A.

Unlike the sine and cosine functions, the other four trigonometric functions do not have a greatest value of y or a least value of y. For this reason they do not have a defined amplitude. However, the coefficient will have an effect on their graphs that is similar, as the next example shows.

EXAMPLE 5 Graph $y = \frac{1}{4} \csc x$ for $0 \le x \le 2\pi$.

SOLUTION   Although the cosecant function does not have a defined amplitude, the numerical factor of $\frac{1}{4}$ will have a similar effect on the graph as was found with the sine and cosine functions. That is, for the same x, the value of y in $y = \frac{1}{4} \csc x$ will be one-fourth the corresponding value of y in $y = \csc x$. The graph is shown in Figure 17.

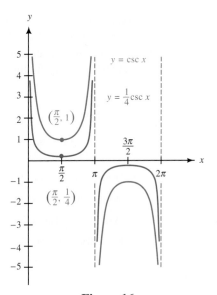

Figure 16

As you can see from Figure 16, the points on the graph of $y = \frac{1}{4} \csc x$ are all closer to the x-axis by a factor of $\frac{1}{4}$.

In Section 1.5, we found that the expression $\sqrt{x^2 + 9}$ could be rewritten without a square root by making the substitution $x = 3 \tan \theta$. Then we noted that the substitution itself was questionable because we did not know at that time if every real number x could be written as $3 \tan \theta$, for some value of θ. We can clear up this point by looking at the graph of $y = 3 \tan x$.

EXAMPLE 6 Graph $y = 3 \tan x$ for $-\dfrac{\pi}{2} < x < \dfrac{\pi}{2}$.

SOLUTION   The graph is shown in Figure 17, along with the graph of $y = \tan x$ over the same interval. Note that the coefficient 3 affects the graph by making it rise and fall faster than the graph of $y = \tan x$.

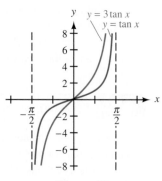

Figure 17

As you can see from the graph, the range for $y = 3 \tan x$ is all real numbers. This means that for any real number y there is a value of x for which $y = 3 \tan x$. In other words, any real number can be written as $3 \tan \theta$, which takes care of the question we had about the substitution we made in Section 1.5.

GETTING READY FOR CLASS

After reading through the preceding section, respond in your own words and in complete sentences.

a. How do you graph the equation $y = \sin x$?

b. What is the period of a function?

c. What is amplitude?

d. How does the coefficient 2 affect the graph of $y = 2 \sin x$?

PROBLEM SET 4.1

Make a table of values for Problems 1 through 6 using multiples of $\pi/4$ for x. Then use the entries in the table to sketch the graph of each function for x between 0 and 2π.

 1. $y = \cos x$ **2.** $y = \cot x$ **3.** $y = \csc x$ **4.** $y = \sin x$
5. $y = \tan x$ **6.** $y = \sec x$

Sketch the graphs of each of the following between $x = -4\pi$ and $x = 4\pi$ by extending the graphs you made in Problems 1 through 6:

7. $y = \sin x$ **8.** $y = \cos x$ **9.** $y = \sec x$ **10.** $y = \csc x$
11. $y = \cot x$ **12.** $y = \tan x$

Find all values of x for which the following are true:

13. $\cos x = 0$ **14.** $\sin x = 0$ **15.** $\sin x = 1$ **16.** $\cos x = 1$
17. $\tan x = 0$ **18.** $\cot x = 0$ **19.** $\csc x = 1$ **20.** $\sec x = 1$

Find all values of x for which the following are true:

21. $\tan x$ is undefined   **22.** $\cot x$ is undefined
23. $\csc x$ is undefined   **24.** $\sec x$ is undefined

Give the amplitude and period of each of the following graphs:

25.

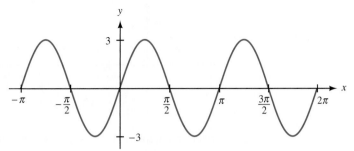

26.

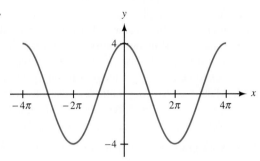

27.

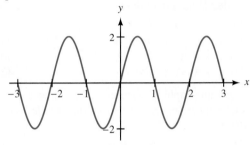

28.

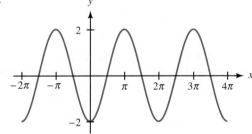

🌐 **29.**

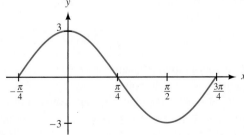

30.

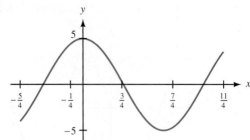

31. Sketch the graph of $y = 2 \sin x$ from $x = 0$ to $x = 2\pi$ by making a table using multiples of $\pi/2$ for x. What is the amplitude of the graph you obtain?

32. Sketch the graph of $y = \frac{1}{2} \cos x$ from $x = 0$ to $x = 2\pi$ by making a table using multiples of $\pi/2$ for x. What is the amplitude of the graph you obtain?

33. Make a table using multiples of $\pi/4$ for x to sketch the graph of $y = \sin 2x$ from $x = 0$ to $x = 2\pi$. After you have obtained the graph, state the number of complete cycles your graph goes through between 0 and 2π.

34. Make a table using multiples of $\pi/6$ for x to sketch the graph of $y = \sin 3x$ from $x = 0$ to $x = 2\pi$. After you have obtained the graph, state the number of complete cycles your graph goes through between 0 and 2π.

Use your graphing calculator to graph each family of functions for $-2\pi \le x \le 2\pi$ together on a single coordinate system. (Make sure your calculator is set to radian mode.) What effect does the value of A have on the graph?

35. $y = A \sin x$ for $A = 1, 2, 3$
36. $y = A \sin x$ for $A = 1, \frac{1}{2}, \frac{1}{3}$
37. $y = A \cos x$ for $A = 1, 0.6, 0.2$
38. $y = A \cos x$ for $A = 1, 3, 5$
39. $y = A \tan x$ for $A = 1, \frac{1}{2}, \frac{1}{3}$
40. $y = A \tan x$ for $A = 1, 2, 3$
41. $y = A \csc x$ for $A = 1, 3, 5$
42. $y = A \csc x$ for $A = 1, 0.6, 0.2$

Use your graphing calculator to graph each pair of functions for $0 \le x \le 4\pi$. (Make sure your calculator is set to radian mode.) What effect does the value of B have on the graph?

43. $y = \sin Bx$ for $B = 1, 2$
44. $y = \sin Bx$ for $B = 1, 4$
45. $y = \cos Bx$ for $B = 1, \frac{1}{2}$
46. $y = \cos Bx$ for $B = 1, \frac{1}{3}$
47. $y = \tan Bx$ for $B = 1, 2$
48. $y = \tan Bx$ for $B = 1, \frac{1}{2}$
49. $y = \csc Bx$ for $B = 1, \frac{1}{2}$
50. $y = \csc Bx$ for $B = 1, 2$

Graph one complete cycle of each of the following. In each case, label the axes accurately and identify the amplitude for each graph (when defined).

51. $y = 6 \sin x$
52. $y = 6 \cos x$
53. $y = \frac{1}{2} \cos x$

54. $y = \frac{1}{3} \sin x$
55. $y = 4 \tan x$
56. $y = 3 \cot x$

57. $y = \frac{1}{3} \cot x$
58. $y = \frac{1}{4} \tan x$
59. $y = 5 \sec x$

60. $y = 4 \csc x$
61. $y = \frac{1}{2} \csc x$
62. $y = \frac{1}{2} \sec x$

Use your graphing calculator to graph each pair of functions for $-2\pi \le x \le 2\pi$ together on a single coordinate system. (Make sure your calculator is set to radian mode.) What effect does the negative sign have on the graph?

63. $y = 3 \sin x, y = -3 \sin x$
64. $y = 4 \cos x, y = -4 \cos x$
65. $y = 2 \tan x, y = -2 \tan x$
66. $y = 6 \cot x, y = -6 \cot x$
67. $y = \frac{1}{2} \sec x, y = -\frac{1}{2} \sec x$
68. $y = \frac{1}{3} \csc x, y = -\frac{1}{3} \csc x$

REVIEW PROBLEMS

The problems that follow review material we covered in Sections 1.5, 3.2, and 3.3.

Prove the following identities.

69. $\cos \theta \tan \theta = \sin \theta$

70. $\sin \theta \tan \theta + \cos \theta = \sec \theta$

71. $(1 + \sin \theta)(1 - \sin \theta) = \cos^2 \theta$

72. $(\sin \theta + \cos \theta)^2 = 1 + 2 \sin \theta \cos \theta$

73. $\csc \theta + \sin (-\theta) = \dfrac{\cos^2 \theta}{\sin \theta}$

74. $\sec \theta - \cos (-\theta) = \dfrac{\sin^2 \theta}{\cos \theta}$

Write each of the following in degrees.

75. $\dfrac{\pi}{3}$

76. $\dfrac{\pi}{6}$

77. $\dfrac{\pi}{4}$

78. $\dfrac{\pi}{2}$

79. $\dfrac{2\pi}{3}$

80. $\dfrac{5\pi}{3}$

81. $\dfrac{11\pi}{6}$

82. $\dfrac{7\pi}{6}$

SECTION 4.2 | PERIOD, REFLECTION, AND VERTICAL TRANSLATION

In Section 4.1, the graphs of $y = \sin x$ and $y = \cos x$ were shown to have a period of 2π. In this section, we will extend our work with graphing to include a more detailed look at period, as well as some additional changes that affect the position of the graph.

 EXAMPLE 1   Graph $y = \sin 2x$ if $0 \le x \le 2\pi$.

SOLUTION   To see how the coefficient 2 in $y = \sin 2x$ affects the graph, we can make a table in which the values of x are multiples of $\pi/4$. (Multiples of $\pi/4$ are convenient because the coefficient 2 divides the 4 in $\pi/4$ exactly.) Table 1 shows the values of x and y, while Figure 1 contains the graphs of $y = \sin x$ and $y = \sin 2x$.

TABLE 1

x	$y = \sin 2x$	(x, y)
0	$y = \sin 2 \cdot 0 = 0$	$(0, 0)$
$\dfrac{\pi}{4}$	$y = \sin 2 \cdot \dfrac{\pi}{4} = \sin \dfrac{\pi}{2} = 1$	$\left(\dfrac{\pi}{4}, 1\right)$
$\dfrac{\pi}{2}$	$y = \sin 2 \cdot \dfrac{\pi}{2} = \sin \pi = 0$	$\left(\dfrac{\pi}{2}, 0\right)$
$\dfrac{3\pi}{4}$	$y = \sin 2 \cdot \dfrac{3\pi}{4} = \sin \dfrac{3\pi}{2} = -1$	$\left(\dfrac{3\pi}{4}, -1\right)$
π	$y = \sin 2 \cdot \pi = 0$	$(\pi, 0)$
$\dfrac{5\pi}{4}$	$y = \sin 2 \cdot \dfrac{5\pi}{4} = \sin \dfrac{5\pi}{2} = 1$	$\left(\dfrac{5\pi}{4}, 1\right)$
$\dfrac{3\pi}{2}$	$y = \sin 2 \cdot \dfrac{3\pi}{2} = \sin 3\pi = 0$	$\left(\dfrac{3\pi}{2}, 0\right)$
$\dfrac{7\pi}{4}$	$y = \sin 2 \cdot \dfrac{7\pi}{4} = \sin \dfrac{7\pi}{2} = -1$	$\left(\dfrac{7\pi}{4}, -1\right)$
2π	$y = \sin 2 \cdot 2\pi = \sin 4\pi = 0$	$(2\pi, 0)$

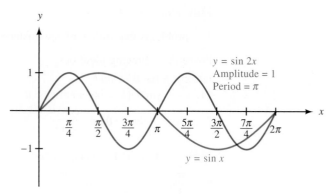

Figure 1

The graph of $y = \sin 2x$ has a period of π. It goes through two complete cycles in 2π units on the x-axis.

EXAMPLE 2   Graph $y = \sin 3x$ from $x = 0$ to $x = 2\pi$.

SOLUTION   To see the effect of the coefficient 3 on the graph, it is convenient to use a table in which the values of x are multiples of $\pi/6$, because 3 divides 6 exactly.

TABLE 2		
x	$y = \sin 3x$	(x, y)
0	$y = \sin 3 \cdot 0 = \sin 0 = 0$	$(0, 0)$
$\dfrac{\pi}{6}$	$y = \sin 3 \cdot \dfrac{\pi}{6} = \sin \dfrac{\pi}{2} = 1$	$\left(\dfrac{\pi}{6}, 1\right)$
$\dfrac{\pi}{3}$	$y = \sin 3 \cdot \dfrac{\pi}{3} = \sin \pi = 0$	$\left(\dfrac{\pi}{3}, 0\right)$
$\dfrac{\pi}{2}$	$y = \sin 3 \cdot \dfrac{\pi}{2} = \sin \dfrac{3\pi}{2} = -1$	$\left(\dfrac{\pi}{2}, -1\right)$
$\dfrac{2\pi}{3}$	$y = \sin 3 \cdot \dfrac{2\pi}{3} = \sin 2\pi = 0$	$\left(\dfrac{2\pi}{3}, 0\right)$

The information in Table 2 indicates the period of $y = \sin 3x$ is $2\pi/3$. Notice that the product $3x$ will vary from 0 to 2π when x takes on values from 0 to $2\pi/3$. That is,

$$0 \leq 3x \leq 2\pi$$

$$\text{if} \quad 0 \leq x \leq \frac{2\pi}{3} \qquad \text{Divide by 3}$$

The graph will go through three complete cycles in 2π units on the x-axis. Figure 2 shows the graph of $y = \sin 3x$ and the graph of $y = \sin x$, on the interval $0 \leq x \leq 2\pi$.

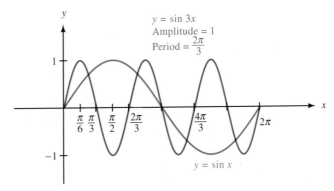

Figure 2

Table 3 summarizes the information obtained from Examples 1 and 2.

TABLE 3		
Equation	**Number of Cycles Every 2π Units**	**Period**
$y = \sin x$	1	2π
$y = \sin 2x$	2	π
$y = \sin 3x$	3	$\dfrac{2\pi}{3}$
$y = \sin Bx$	B	$\dfrac{2\pi}{B}$ B is positive

In order for $y = \sin Bx$ to complete one cycle, the product Bx must vary in value from 0 to 2π. So,

$$0 \leq Bx \leq 2\pi \quad \text{when} \quad 0 \leq x \leq \frac{2\pi}{B}$$

The period is the length of this interval, which is $2\pi/B$.

EXAMPLE 3 Graph one complete cycle of $y = \cos \frac{1}{2} x$.

SOLUTION   The coefficient of x is $\frac{1}{2}$. The graph will go through $\frac{1}{2}$ of a complete cycle every 2π units. The period will be

$$\text{Period} = \frac{2\pi}{1/2} = 4\pi$$

Figure 3 shows the graph.

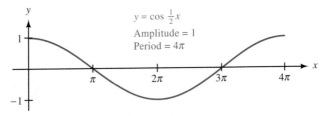

Figure 3

Continuing to generalize from the examples, we have the following summary.

AMPLITUDE AND PERIOD FOR SINE AND COSINE CURVES

If B is a positive number, the graphs of $y = A \sin Bx$ and $y = A \cos Bx$ will have

$$\text{Amplitude} = |A|$$

$$\text{Period} = \frac{2\pi}{B}$$

In the next two examples, we use this information about amplitude and period to graph one complete cycle of a sine and cosine curve, then we extend these graphs to cover more than one complete cycle.

 EXAMPLES   Graph one complete cycle of each of the following equations and then extend the graph to cover the given interval.

4. $y = 4 \cos \dfrac{1}{2}x, \; -4\pi \le x \le 4\pi$

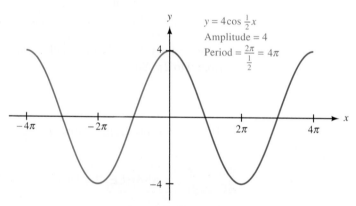

$y = 4\cos \frac{1}{2}x$
Amplitude $= 4$
Period $= \frac{2\pi}{\frac{1}{2}} = 4\pi$

Figure 4

5. $y = 2 \sin \pi x, \; -3 \le x \le 3$

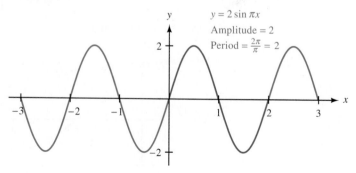

$y = 2 \sin \pi x$
Amplitude $= 2$
Period $= \frac{2\pi}{\pi} = 2$

Figure 5

Note that, on the graphs in Figures 4 and 5, the axes have not been labeled proportionally. Instead, they are labeled so that the amplitude and period are easy to read. As you can see, once we have the graph of one complete cycle of a curve, it is easy to extend the curve to any interval of interest.

EXAMPLE 6 Graph $y = 3 \csc 2x$ from $x = 0$ to $x = 2\pi$.

SOLUTION Since $\csc 2x$ is the reciprocal of $\sin 2x$, the graph of $y = \csc 2x$ will have the same period as $y = \sin 2x$. That period is $2\pi/2 = \pi$. Also, because of this reciprocal relationship, the graph of $y = \csc 2x$ will be undefined, and a vertical asymptote will occur, whenever the graph of $y = \sin 2x$ is 0. Since all the y values on the graph of $y = 3 \csc 2x$ will be three times as large or small as the y values on $y = \csc 2x$, it is reasonable to assume the range of y values on the graph of $y = 3 \csc 2x$ will be $y \geq 3$ and $y \leq -3$. Following this kind of reasoning, we have the graph shown in Figure 6.

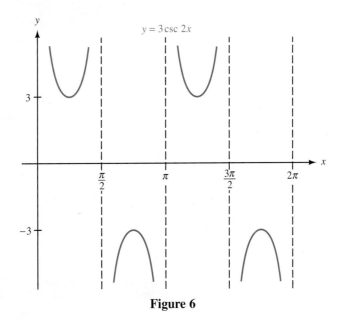

Figure 6

Since the graphs of $y = A \csc Bx$ and $y = A \sec Bx$ are similar, we use the information from Example 6 to generalize as follows:

RANGE AND PERIOD FOR SECANT AND COSECANT GRAPHS

If B is a positive number, the graphs of $y = A \sec Bx$ and $y = A \csc Bx$ will have

$$\text{Period} = \frac{2\pi}{B}$$

$$\text{Range: } y \geq |A| \quad \text{and} \quad y \leq -|A|$$

For our next example, we look at the graph of one of the equations we found in Example 4 of Section 3.5. Example 7 gives the main facts from that example.

EXAMPLE 7 Figure 7 shows a fire truck parked on the shoulder of a freeway next to a long block wall. The red light on the top of the truck is 10 feet from the wall and rotates through one complete revolution every 2 seconds. Graph the equation that gives the length d in terms of time t from $t = 0$ to $t = 2$.

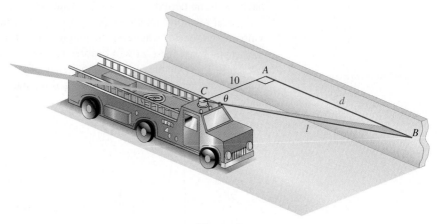

Figure 7

SOLUTION From Example 4 in Section 3.5 we know that

$$d = 10 \tan \pi t$$

To graph this equation between $t = 0$ and $t = 2$, we construct a table of values in which t assumes all multiples of $\frac{1}{4}$ from $t = 0$ to $t = 2$. Plotting the points given in Table 4 and then connecting them with a smooth tangent curve, we have the graph in Figure 8.

t	$d = 10 \tan \pi t$	d
0	$d = 10 \tan \pi \cdot 0 = 10 \tan 0 = 0$	0
$\frac{1}{4}$	$d = 10 \tan \pi \cdot \frac{1}{4} = 10 \tan \frac{\pi}{4} = 10$	10
$\frac{1}{2}$	$d = 10 \tan \pi \cdot \frac{1}{2} = 10 \tan \frac{\pi}{2}$	undefined
$\frac{3}{4}$	$d = 10 \tan \pi \cdot \frac{3}{4} = 10 \tan \frac{3\pi}{4} = -10$	-10
1	$d = 10 \tan \pi \cdot 1 = 10 \tan \pi = 0$	0
$\frac{5}{4}$	$d = 10 \tan \pi \cdot \frac{5}{4} = 10 \tan \frac{5\pi}{4} = 10$	10
$\frac{3}{2}$	$d = 10 \tan \pi \cdot \frac{3}{2} = 10 \tan \frac{3\pi}{2}$	undefined
$\frac{7}{4}$	$d = 10 \tan \pi \cdot \frac{7}{4} = 10 \tan \frac{7\pi}{4} = -10$	-10
2	$d = 10 \tan \pi \cdot 2 = 10 \tan 2\pi = 0$	0

TABLE 4

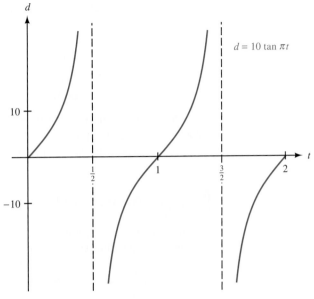

Figure 8

Since the period of $y = \tan x$ is π, the graph will complete one cycle when πt varies between 0 and π. That is,

$$0 \le \pi t \le \pi$$

$$\text{if} \quad 0 \le t \le 1 \qquad \text{Divide by } \pi$$

So, the new period can be found by dividing the normal period π by π to get 1. Also, the number 10 in the equation $d = 10 \tan \pi t$ causes the tangent graph to rise faster above the horizontal axis and fall faster below it. ■

We generalize the results of Example 7 as follows.

PERIOD AND RATE OF INCREASE FOR TANGENT AND COTANGENT GRAPHS

If B is a positive number, the graphs of $y = A \tan Bx$ and $y = A \cot Bx$ will have

$$\text{Period} = \frac{\pi}{B}$$

Each graph will rise and fall at a faster rate than the corresponding graphs of $y = \tan x$ and $y = \cot x$ if $|A|$ is greater than 1 and at a slower rate if $|A|$ is between 0 and 1.

REFLECTING ABOUT THE X-AXIS

So far in this section, all of the coefficients A and B we have encountered have been positive. If we are given an equation to graph in which B is negative, we can use the properties of even and odd functions to rewrite the equation with B positive. For example,

$$y = 3 \sin (-2x) \text{ is equivalent to } y = -3 \sin 2x$$
$$\text{because sine is an odd function.}$$

$$y = 3 \cos (-2x) \text{ is equivalent to } y = 3 \cos 2x$$
$$\text{because cosine is an even function.}$$

So we do not need to worry about negative values of B; we simply make them positive by using the properties of even and odd functions and then graph as usual. To see how a negative value of A affects graphing, we will graph $y = -2 \cos x$.

EXAMPLE 8 Graph $y = -2 \cos x$, from $x = -2\pi$ to $x = 4\pi$.

SOLUTION Each value of y on the graph of $y = -2 \cos x$ will be the opposite of the corresponding value of y on the graph of $y = 2 \cos x$. The result is that the graph of $y = -2 \cos x$ is the reflection of the graph of $y = 2 \cos x$ about the x-axis. Figure 9 shows the extension of one complete cycle of $y = -2 \cos x$ to the interval $-2\pi \leq x \leq 4\pi$.

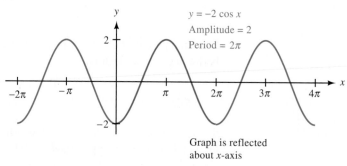

$y = -2 \cos x$
Amplitude = 2
Period = 2π

Graph is reflected
about x-axis

Figure 9

EXAMPLE 9 Graph $y = 10 \tan (-\pi x)$ for $0 \leq x \leq 2$.

SOLUTION Because tangent is an odd function, $y = 10 \tan (-\pi x)$ is equivalent to $y = -10 \tan (\pi x)$. To graph $y = -10 \tan (\pi x)$, we reflect the graph of $y = 10 \tan \pi x$ about the x-axis. Using the graph of $y = 10 \tan \pi x$ in Figure 8 of Example 7 (replacing d with y and t with x), the reflection is shown in Figure 10.

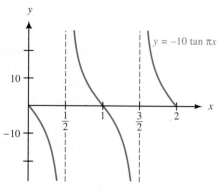

$y = -10 \tan \pi x$

Figure 10

SUMMARY

The graphs of $y = A \sin Bx$ and $y = A \cos Bx$, where B is a positive number, will be reflected about the x-axis if A is negative. The same is true for the tangent, cotangent, secant, and cosecant functions.

VERTICAL TRANSLATIONS

Recall from algebra the relationship between the graphs of $y = x^2$ and $y = x^2 - 3$. Figures 11 and 12 show the graphs. The graph of $y = x^2 - 3$ has the same shape as the graph of $y = x^2$ but with its vertex (and all other points) moved down three units. If we were to graph $y = x^2 + 2$, the graph would have the same shape as the graph of $y = x^2$, but it would be moved up two units. In general, the graph of $y = f(x) + k$ is the graph of $y = f(x)$ translated k units vertically. If k is a positive number, the translation is up. If k is a negative number, the translation is down.

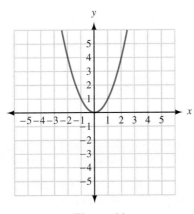

Figure 11

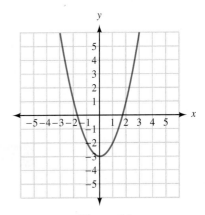

Figure 12

EXAMPLE 10 Use the results of Examples 5 and 6 to graph each of the following equations.

a. $y = 3 + 2 \sin \pi x$
b. $y = -1 + 3 \csc 2x$

SOLUTION Each equation has the form $y = k + f(x)$, indicating that the vertical translation is k. (Since addition is a commutative operation, it makes no difference whether our equations have the form $y = k + f(x)$ or $y = f(x) + k$; both have the same meaning.) The graph of $y = 3 + 2 \sin \pi x$ is the graph of $y = 2 \sin \pi x$ with all points moved up three units. Likewise, the graph of $y = -1 + 3 \csc 2x$ is the graph of $y = 3 \csc 2x$ with all points moved down one unit. The graphs are shown in Figures 13 and 14.

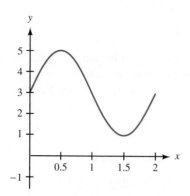

Figure 13

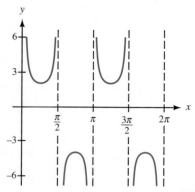

Figure 14

GETTING READY FOR CLASS

After reading through the preceding section, respond in your own words and in complete sentences.

a. How does the coefficient 2 affect the graph of $y = \sin 2x$?

b. How does the coefficient $\dfrac{1}{2}$ affect the graph of $y = \sin \dfrac{1}{2}x$?

c. What is the period of $y = \cos Bx$?

d. How does the coefficient -2 affect the graph of $y = -2 \cos x$?

PROBLEM SET 4.2

Graph one complete cycle of each of the following. In each case, label the axes accurately and identify the period for each graph.

 1. $y = \sin 2x$ **2.** $y = \sin \dfrac{1}{2}x$

3. $y = \cos \dfrac{1}{3}x$ **4.** $y = \cos 3x$

5. $y = \sin \pi x$ **6.** $y = \cos \pi x$

7. $y = \sin \dfrac{\pi}{2}x$ **8.** $y = \cos \dfrac{\pi}{2}x$

 9. $y = \csc 3x$ **10.** $y = \csc 4x$

11. $y = \sec \dfrac{\pi}{4}x$ **12.** $y = \sec \dfrac{1}{4}x$

13. $y = \tan \dfrac{1}{2}x$ **14.** $y = \tan \dfrac{1}{3}x$

15. $y = \cot 4x$ **16.** $y = \cot \pi x$

Graph one complete cycle for each of the following. In each case, label the axes so that the amplitude (when defined) and period are easy to read.

17. $y = 4 \sin 2x$ **18.** $y = 2 \sin 4x$

19. $y = 3 \sin \dfrac{1}{2}x$ **20.** $y = 2 \sin \dfrac{1}{3}x$

21. $y = \dfrac{1}{2} \cos 3x$

22. $y = \dfrac{1}{2} \sin 3x$

23. $y = \dfrac{1}{2} \sin \dfrac{\pi}{2} x$

24. $y = 2 \sin \dfrac{\pi}{2} x$

25. $y = \dfrac{1}{2} \csc 3x$

26. $y = \dfrac{1}{2} \sec 3x$

27. $y = 3 \sec \dfrac{1}{2} x$

28. $y = 3 \csc \dfrac{1}{2} x$

29. $y = 2 \tan 3x$

30. $y = 3 \tan 2x$

31. $y = \dfrac{1}{2} \cot \dfrac{\pi}{2} x$

32. $y = \dfrac{1}{3} \cot \dfrac{1}{2} x$

Use your answers for Problems 17 through 32 for reference, and graph one complete cycle of each of the following equations.

33. $y = 4 + 4 \sin 2x$

34. $y = -2 + 2 \sin 4x$

35. $y = -1 + \dfrac{1}{2} \cos 3x$

36. $y = 1 + \dfrac{1}{2} \sin 3x$

37. $y = -2 + 3 \sec \dfrac{1}{2} x$

38. $y = 3 + 3 \csc \dfrac{1}{2} x$

39. $y = 3 + \dfrac{1}{2} \cot \dfrac{\pi}{2} x$

40. $y = -4 + 3 \tan 2x$

Graph each of the following over the given interval. Label the axes so that the amplitude (when defined) and period are easy to read.

41. $y = 2 \sin \pi x, \ -4 \leq x \leq 4$

42. $y = 3 \cos \pi x, \ -2 \leq x \leq 4$

43. $y = 3 \sin 2x, \ -\pi \leq x \leq 2\pi$

44. $y = -3 \sin 2x, \ -2\pi \leq x \leq 2\pi$

45. $y = -3 \cos \dfrac{1}{2} x, \ -2\pi \leq x \leq 6\pi$

46. $y = 3 \cos \dfrac{1}{2} x, \ -4\pi \leq x \leq 4\pi$

47. $y = -2 \sin (-3x), \ 0 \leq x \leq 2\pi$

48. $y = -2 \cos (-3x), \ 0 \leq x \leq 2\pi$

49. $y = -2 \csc 3x, \ 0 \leq x \leq 2\pi$

50. $y = -2 \sec 3x, \ 0 \leq x \leq 2\pi$

51. $y = -\cot 2x, \ 0 \leq x \leq \pi$

52. $y = -\tan 4x, \ 0 \leq x \leq \pi$

53. Electric Current The current in an alternating circuit varies in intensity with time. If I represents the intensity of the current and t represents time, then the relationship between I and t is given by

$$I = 20 \sin 120 \pi t$$

where I is measured in amperes and t is measured in seconds. Find the maximum value of I and the time it takes for I to go through one complete cycle.

54. Maximum Velocity and Distance A weight is hung from a spring and set in motion so that it moves up and down continuously. The velocity v of the weight at any time t is given by the equation

$$v = 3.5 \cos 2\pi t$$

where v is measured in meters per second and t is measured in seconds. Find the maximum velocity of the weight and the amount of time it takes for the weight to move from its lowest position to its highest position.

55. Rotating Light Figure 15 shows a lighthouse that is 100 feet from a long straight wall on the beach. The light in the lighthouse rotates through one complete rotation once every 4 seconds. In Problem 21 of Problem Set 3.5, you found the equation that gives d in terms of t to be $d = 100 \tan \frac{\pi}{2}t$. Graph this equation by making a table in which t assumes all multiples of $\frac{1}{2}$ from $t = 0$ to $t = 4$.

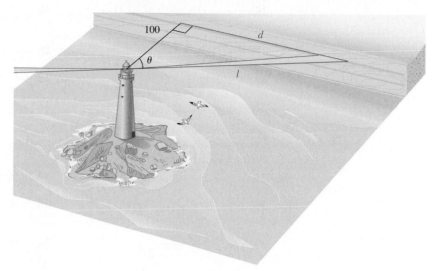

Figure 15

56. Rotating Light In Figure 15, the equation that gives l in terms of time t is $l = 100 \sec \frac{\pi}{2}t$. Graph this equation from $t = 0$ to $t = 4$.

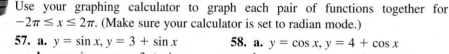 Use your graphing calculator to graph each pair of functions together for $-2\pi \leq x \leq 2\pi$. (Make sure your calculator is set to radian mode.)

57. a. $y = \sin x$, $y = 3 + \sin x$
  b. $y = \sin x$, $y = -3 + \sin x$
  c. $y = \sin x$, $y = -\sin x$
59. a. $y = \tan x$, $y = 2 + \tan x$
  b. $y = \tan x$, $y = -2 + \tan x$
  c. $y = \tan x$, $y = -\tan x$
61. a. $y = \sec x$, $y = 1 + \sec x$
  b. $y = \sec x$, $y = -1 + \sec x$
  c. $y = \sec x$, $y = -\sec x$

58. a. $y = \cos x$, $y = 4 + \cos x$
  b. $y = \cos x$, $y = -4 + \cos x$
  c. $y = \cos x$, $y = -\cos x$
60. a. $y = \cot x$, $y = 5 + \cot x$
  b. $y = \cot x$, $y = -5 + \cot x$
  c. $y = \cot x$, $y = -\cot x$
62. a. $y = \csc x$, $y = 3 + \csc x$
  b. $y = \csc x$, $y = -3 + \csc x$
  c. $y = \csc x$, $y = -\csc x$

Use your graphing calculator to graph each pair of functions together for $-2\pi \leq x \leq 2\pi$. (Make sure your calculator is set to radian mode.) How does the value of C affect the graph in each case?

63. a. $y = \sin x$, $y = \sin (x + C)$ for $C = \dfrac{\pi}{4}$

  b. $y = \sin x$, $y = \sin (x + C)$ for $C = -\dfrac{\pi}{4}$

64. a. $y = \cos x$, $y = \cos (x + C)$ for $C = \dfrac{\pi}{3}$

  b. $y = \cos x$, $y = \cos (x + C)$ for $C = -\dfrac{\pi}{3}$

65. a. $y = \tan x$, $y = \tan (x + C)$ for $C = \dfrac{\pi}{6}$

    b. $y = \tan x$, $y = \tan (x + C)$ for $C = -\dfrac{\pi}{6}$

66. a. $y = \csc x$, $y = \csc (x + C)$ for $C = \dfrac{\pi}{4}$

    b. $y = \csc x$, $y = \csc (x + C)$ for $C = -\dfrac{\pi}{4}$

REVIEW PROBLEMS

The problems that follow review material we covered in Section 3.2. Reviewing these problems will help you with the next section.

Evaluate each of the following if x is $\pi/2$ and y is $\pi/6$.

67. $\sin \left(x + \dfrac{\pi}{2} \right)$ **68.** $\sin \left(x - \dfrac{\pi}{2} \right)$

69. $\cos \left(y - \dfrac{\pi}{6} \right)$ **70.** $\cos \left(y + \dfrac{\pi}{6} \right)$

71. $\sin (x + y)$ **72.** $\cos (x + y)$

73. $\sin x + \sin y$ **74.** $\cos x + \cos y$

Convert each of the following to radians without using a calculator.

75. $45°$ **76.** $30°$ **77.** $60°$ **78.** $90°$

79. $150°$ **80.** $300°$ **81.** $225°$ **82.** $120°$

SECTION 4.3 | PHASE SHIFT

In this section, we will consider equations of the form

$$y = A \sin (Bx + C) \qquad B > 0$$

and

$$y = A \cos (Bx + C) \qquad B > 0$$

We already know how the coefficients A and B affect the graphs of these equations. The only thing we have left to do is discover what effect C has on the graphs. We will start our investigation with a couple of equations in which A and B are equal to 1.

EXAMPLE 1 Graph $y = \sin \left(x + \dfrac{\pi}{2} \right)$, if $-\dfrac{\pi}{2} \le x \le \dfrac{3\pi}{2}$.

SOLUTION Since we have not graphed an equation of this form before, it is a good idea to begin by making a table (Table 1). In this case, multiples of $\pi/2$ will be the most convenient replacements for x in the table. Also, if we start with $x = -\pi/2$, our first value of y will be 0.

TABLE 1

x	$y = \sin\left(x + \dfrac{\pi}{2}\right)$	(x, y)
$-\dfrac{\pi}{2}$	$y = \sin\left(-\dfrac{\pi}{2} + \dfrac{\pi}{2}\right) = \sin 0 = 0$	$\left(-\dfrac{\pi}{2}, 0\right)$
0	$y = \sin\left(0 + \dfrac{\pi}{2}\right) = \sin \dfrac{\pi}{2} = 1$	$(0, 1)$
$\dfrac{\pi}{2}$	$y = \sin\left(\dfrac{\pi}{2} + \dfrac{\pi}{2}\right) = \sin \pi = 0$	$\left(\dfrac{\pi}{2}, 0\right)$
π	$y = \sin\left(\pi + \dfrac{\pi}{2}\right) = \sin \dfrac{3\pi}{2} = -1$	$(\pi, -1)$
$\dfrac{3\pi}{2}$	$y = \sin\left(\dfrac{3\pi}{2} + \dfrac{\pi}{2}\right) = \sin 2\pi = 0$	$\left(\dfrac{3\pi}{2}, 0\right)$

Graphing these points and then drawing the sine curve that connects them gives us the graph of $y = \sin(x + \pi/2)$, as shown in Figure 1. Figure 1 also includes the graph of $y = \sin x$ for reference; we are trying to discover how the graphs of $y = \sin(x + \pi/2)$ and $y = \sin x$ differ.

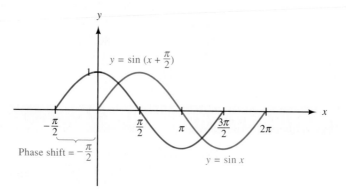

Figure 1

It seems that the graph of $y = \sin(x + \pi/2)$ is shifted $\pi/2$ units to the left of the graph of $y = \sin x$. We say the graph of $y = \sin(x + \pi/2)$ has a *phase shift* of $-\pi/2$, where the negative sign indicates the shift is to the left (in the negative direction). As an x-coordinate, the phase shift indicates the beginning of a cycle.

From the results in Example 1, we would expect the graph of $y = \sin(x - \pi/2)$ to have a phase shift of $+\pi/2$. That is, we expect the graph of $y = \sin(x - \pi/2)$ to be shifted $\pi/2$ units to the *right* of the graph of $y = \sin x$. We will consider this situation in our next example.

EXAMPLE 2 Graph one complete cycle of $y = \sin\left(x - \dfrac{\pi}{2}\right)$.

SOLUTION Proceeding as we did in Example 1, we make a table (Table 2) using multiples of $\pi/2$ for x, and then use the information in the table to sketch the graph. In this example, we start with $x = \pi/2$, since this value of x will give us $y = 0$.

TABLE 2

x	$y = \sin\left(x - \dfrac{\pi}{2}\right)$	(x, y)
$\dfrac{\pi}{2}$	$y = \sin\left(\dfrac{\pi}{2} - \dfrac{\pi}{2}\right) = \sin 0 = 0$	$\left(\dfrac{\pi}{2}, 0\right)$
π	$y = \sin\left(\pi - \dfrac{\pi}{2}\right) = \sin\dfrac{\pi}{2} = 1$	$(\pi, 1)$
$\dfrac{3\pi}{2}$	$y = \sin\left(\dfrac{3\pi}{2} - \dfrac{\pi}{2}\right) = \sin\pi = 0$	$\left(\dfrac{3\pi}{2}, 0\right)$
2π	$y = \sin\left(2\pi - \dfrac{\pi}{2}\right) = \sin\dfrac{3\pi}{2} = -1$	$(2\pi, -1)$
$\dfrac{5\pi}{2}$	$y = \sin\left(\dfrac{5\pi}{2} - \dfrac{\pi}{2}\right) = \sin 2\pi = 0$	$\left(\dfrac{5\pi}{2}, 0\right)$

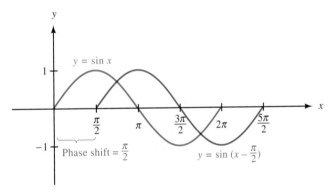

Figure 2

The graph of $y = \sin(x - \pi/2)$, as we expected, is a sine curve shifted $\pi/2$ units to the right of the graph of $y = \sin x$ (Figure 2). The phase shift, in this case, is $+\pi/2$.

Before we write any conclusions about phase shift, we should look at another example in which A and B are not 1.

EXAMPLE 3 Graph $y = 3\sin\left(2x + \dfrac{\pi}{2}\right)$, if $-\dfrac{\pi}{4} \le x \le \dfrac{3\pi}{4}$.

SOLUTION We know the coefficient $B = 2$ will change the period from 2π to π. Because the period is smaller (π instead of 2π), we should use values of x in our table (Table 3) that are closer together, like multiples of $\pi/4$ instead of $\pi/2$.

TABLE 3

x	$y = 3 \sin\left(2x + \dfrac{\pi}{2}\right)$	(x, y)
$-\dfrac{\pi}{4}$	$y = 3 \sin\left[2\left(-\dfrac{\pi}{4}\right) + \dfrac{\pi}{2}\right] = 3 \sin 0 = 0$	$\left(-\dfrac{\pi}{4}, 0\right)$
0	$y = 3 \sin\left(2 \cdot 0 + \dfrac{\pi}{2}\right) = 3 \sin \dfrac{\pi}{2} = 3$	$(0, 3)$
$\dfrac{\pi}{4}$	$y = 3 \sin\left(2 \cdot \dfrac{\pi}{4} + \dfrac{\pi}{2}\right) = 3 \sin \pi = 0$	$\left(\dfrac{\pi}{4}, 0\right)$
$\dfrac{\pi}{2}$	$y = 3 \sin\left(2 \cdot \dfrac{\pi}{2} + \dfrac{\pi}{2}\right) = 3 \sin \dfrac{3\pi}{2} = -3$	$\left(\dfrac{\pi}{2}, -3\right)$
$\dfrac{3\pi}{4}$	$y = 3 \sin\left(2 \cdot \dfrac{3\pi}{4} + \dfrac{\pi}{2}\right) = 3 \sin 2\pi = 0$	$\left(\dfrac{3\pi}{4}, 0\right)$

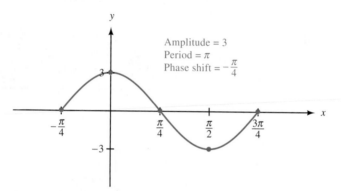

Amplitude = 3
Period = π
Phase shift = $-\dfrac{\pi}{4}$

Figure 3

The amplitude and period are as we would expect. The phase shift, however, is half of $-\pi/2$. The phase shift, $-\pi/4$, comes from the ratio $-C/B$, or in this case

$$\frac{-\pi/2}{2} = -\frac{\pi}{4}$$

The phase shift in the equation $y = A \sin(Bx + C)$ depends on both B and C. In this example, the period is half of the period of $y = \sin x$, and the phase shift is half of the phase shift of $y = \sin(x + \pi/2)$, which was found in Example 1 (see Figure 3). ▮

Notice in Example 3 that a cycle begins when the expression $2x + \pi/2$ assumes a value of zero. Solving $2x + \pi/2 = 0$ for x, we obtain the phase shift of $x = -\pi/4$. Since a cycle of $y = \sin x$ begins when $x = 0$, a cycle of $y = \sin(Bx + C)$ begins when

$$Bx + C = 0$$

$$Bx = -C$$

$$x = -\frac{C}{B}$$

Although all of the examples we have completed so far in this section have been sine curves, the results also apply to cosine curves. Here is a summary.

AMPLITUDE, PERIOD, AND PHASE SHIFT

The graphs of $y = A \sin (Bx + C)$ and $y = A \cos (Bx + C)$, where $B > 0$, will have the following characteristics:

1. Amplitude $= |A|$

2. Period $= \dfrac{2\pi}{B}$

3. Phase shift $= -\dfrac{C}{B}$

If $A < 0$, the graphs will be reflected about the x-axis.

The information on amplitude, period, and phase shift allows us to sketch sine and cosine curves without having to make tables.

EXAMPLE 4 Graph one complete cycle of $y = 2 \sin (3x + \pi)$.

SOLUTION Here is a detailed list of steps to use in graphing sine and cosine curves for which B is positive.

Step 1: Use A, B, and C to find the amplitude, period, and phase shift.

$$\text{Amplitude} = |A| = 2 \qquad \text{Period} = \frac{2\pi}{B} = \frac{2\pi}{3} \qquad \text{Phase shift} = -\frac{C}{B} = -\frac{\pi}{3}$$

Step 2: On the x-axis, label the starting point, ending point, and the point halfway between them for each cycle of the curve in question. The starting point is the phase shift. The ending point is the phase shift plus the period. It is also a good idea to label the points one-fourth and three-fourths of the way between the starting point and ending point.

Step 3: Label the y-axis with the amplitude and the opposite of the amplitude. We have drawn a faint rectangle to help visualize the dimensions of one cycle. It is okay if the units on the x-axis and the y-axis are not proportional. That is, one unit on the y-axis can be a different length from one unit on the x-axis. The idea is to make the graph easy to read (Figure 4).

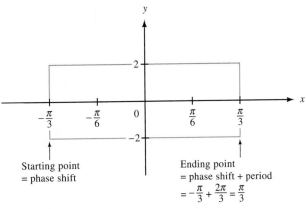

Figure 4

Step 4: Sketch in the curve in question, keeping in mind that the graph will be reflected about the *x*-axis if *A* is negative. In this case, we want a sine curve that will be 0 at the starting point and 0 at the ending point (Figure 5).

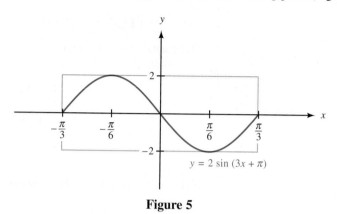

$$y = 2 \sin (3x + \pi)$$

Figure 5

The steps listed in Example 4 may seem complicated at first. With a little practice they do not take much time at all, especially when compared with the time it would take to make a table. Also, once we have graphed one complete cycle of the curve, it would be fairly easy to extend the graph in either direction.

USING TECHNOLOGY

VERIFYING A GRAPH

A graphing calculator can be used to verify the graph of a trigonometric function that has been drawn by hand. For example, to verify the graph in Figure 5, set your calculator to radian mode, define

$$Y_1 = 2 \sin (3x + \pi)$$

and set the window variables to match the characteristics for the cycle that was drawn (set the viewing rectangle to agree with the rectangle shown in the figure):

$$-\pi/3 \le x \le \pi/3, \text{ scale} = \pi/6; \ -2 \le y \le 2$$

If our drawing is correct, we should see one cycle of the sine function that exactly fits the screen. Figure 6 shows the result.

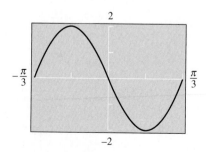

Figure 6

EXAMPLE 5 Graph $y = -2 \cos (3x + \pi)$ from $x = -2\pi/3$ to $x = 2\pi/3$.

SOLUTION A, B, and C are the same here as they were in Example 4, except that A is now negative. We use the same labeling on the axes as we used in Example 4, but we draw in a cosine curve that is reflected about the x-axis instead of a sine curve, and then we extend it to cover the interval $-2\pi/3 \leq x \leq 2\pi/3$ (Figure 7).

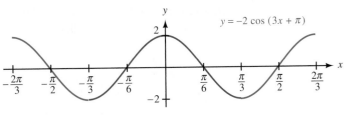

Figure 7

MORE ON LABELING THE X-AXIS

Now that we have been through a few examples in detail, we can be more specific about how we label the x-axis. Suppose the line in Figure 8 represents the x-axis, and the five points we have labeled with the letters a through e are the five points we will use to graph one complete cycle of a sine or cosine curve. Because these points are equally spaced along the x-axis, we can find their coordinates in the following manner (and the following order).

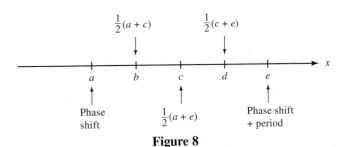

Figure 8

1. The coordinate of the first point, a, is always the phase shift.
2. The coordinate of the last point, e, is the sum of the phase shift and the period.
3. The center point, c, is the average of the first point and the last point. That is,

$$c = \frac{a + e}{2} = \frac{1}{2}(a + e)$$

4. The point b is the average of the first point and the center point. That is,

$$b = \frac{a + c}{2} = \frac{1}{2}(a + c)$$

5. The point d is the average of the center point and the last point. In symbols,

$$d = \frac{c + e}{2} = \frac{1}{2}(c + e)$$

EXAMPLE 6   Graph one complete cycle of $y = 4 \sin\left(2x - \dfrac{\pi}{3}\right)$.

SOLUTION   In this case, $A = 4$, $B = 2$, and $C = -\dfrac{\pi}{3}$. These values give us a sine curve with

$$\text{Amplitude} = 4 \qquad \text{Period} = \frac{2\pi}{2} = \pi \qquad \text{Phase Shift} = \frac{-(-\pi/3)}{2} = \frac{\pi}{6}$$

Because the amplitude is 4, we label the y-axis with 4 and -4. The x-axis is labeled according to the discussion that preceded this example, as shown in Figure 9.

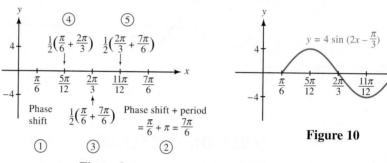

Figure 9

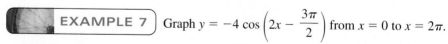

Figure 10

We finish the graph by drawing in a sine curve that starts at $x = \pi/6$ and ends at $x = 7\pi/6$, as shown in Figure 10.

As an alternative to steps 3 through 5 on page 195, we can find points b, c, and d by dividing the period by 4. This gives us the spacing between all five points used to draw the cycle.

$$\text{spacing} = \frac{1}{4}\,\text{period} = \frac{1}{4} \cdot \frac{2\pi}{B} = \frac{\pi}{2B}$$

Then

$$b = a + \frac{\pi}{2B},\, c = b + \frac{\pi}{2B}, \quad \text{and} \quad d = c + \frac{\pi}{2B}$$

EXAMPLE 7   Graph $y = -4 \cos\left(2x - \dfrac{3\pi}{2}\right)$ from $x = 0$ to $x = 2\pi$.

SOLUTION   We use A, B, and C to help us sketch one complete cycle of the curve, then we extend the resulting graph to cover the interval $0 \le x \le 2\pi$.

$$\text{Amplitude} = 4 \qquad \text{Period} = \frac{2\pi}{2} = \pi \qquad \text{Phase Shift} = -\frac{-3\pi/2}{2} = \frac{3\pi}{4}$$

One complete cycle will start at $3\pi/4$ on the x-axis. It will end π units later at $3\pi/4 + \pi = 7\pi/4$. The spacing of the five points will be $\pi/4$, so we have the following:

$$a = 3\pi/4$$

$$b = 3\pi/4 + \pi/4 = 4\pi/4 = \pi$$

$$c = \pi + \pi/4 = 5\pi/4$$

$$d = 5\pi/4 + \pi/4 = 6\pi/4 = 3\pi/2$$

$$e = 7\pi/4$$

Since A is negative, the graph is reflected about the x-axis. Figure 11 shows the five points that will define the cycle. The completed graph is shown in Figure 12.

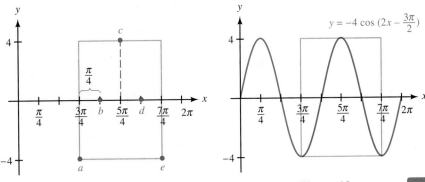

Figure 11

Figure 12

EXAMPLE 8   Graph one complete cycle of $y = 3 + 5 \sin\left(\pi x + \dfrac{\pi}{4}\right)$.

SOLUTION   In this example, $A = 5$, $B = \pi$, $C = \pi/4$, and $k = 3$. The graph will have the following characteristics:

$$\text{Amplitude} = 5 \qquad\qquad \text{Period} = \frac{2\pi}{\pi} = 2$$

$$\text{Phase shift} = -\frac{\pi/4}{\pi} = -\frac{1}{4} \qquad\qquad \text{Spacing} = \frac{2}{4} = \frac{1}{2}$$

$$\text{Vertical translation} = 3$$

Figure 13 shows the five points used to define the cycle. The dashed line reminds us that the graph must be translated upward 3 units. The completed graph is shown in Figure 14.

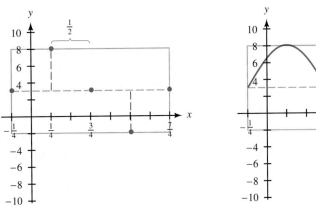

Figure 13

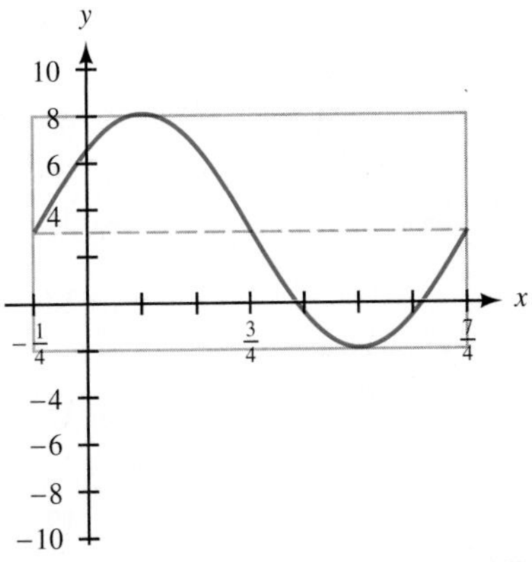

Figure 14

GETTING READY FOR CLASS

After reading through the preceding section, respond in your own words and in complete sentences.

a. How does the term $\frac{\pi}{2}$ affect the graph of $y = \sin\left(x + \frac{\pi}{2}\right)$?

b. How does the term $\frac{\pi}{2}$ affect the graph of $y = \sin\left(x - \frac{\pi}{2}\right)$?

c. How do you find the period for the graph of $y = A \sin (Bx + C)$?

d. How do you find the phase shift for the graph of $y = A \sin (Bx + C)$?

PROBLEM SET 4.3

For each equation, first identify the phase shift and then sketch one complete cycle of the graph. In each case, graph $y = \sin x$ on the same coordinate system.

1. $y = \sin\left(x + \frac{\pi}{4}\right)$

2. $y = \sin\left(x + \frac{\pi}{6}\right)$

3. $y = \sin\left(x - \frac{\pi}{4}\right)$

4. $y = \sin\left(x - \frac{\pi}{6}\right)$

5. $y = \sin\left(x + \frac{\pi}{3}\right)$

6. $y = \sin\left(x - \frac{\pi}{3}\right)$

For each equation, identify the phase shift and then sketch one complete cycle of the graph. In each case, graph $y = \cos x$ on the same coordinate system.

7. $y = \cos\left(x - \frac{\pi}{2}\right)$

8. $y = \cos\left(x + \frac{\pi}{2}\right)$

9. $y = \cos\left(x + \frac{\pi}{3}\right)$

10. $y = \cos\left(x - \frac{\pi}{4}\right)$

For each equation, identify the amplitude, period, and phase shift. Then label the axes accordingly and sketch one complete cycle of the curve.

11. $y = \sin (2x - \pi)$

12. $y = \sin (2x + \pi)$

13. $y = \sin\left(\pi x + \frac{\pi}{2}\right)$

14. $y = \sin\left(\pi x - \frac{\pi}{2}\right)$

15. $y = -\cos\left(2x + \frac{\pi}{2}\right)$

16. $y = -\cos\left(2x - \frac{\pi}{2}\right)$

17. $y = 2 \sin\left(\frac{1}{2}x + \frac{\pi}{2}\right)$

18. $y = 3 \cos\left(\frac{1}{2}x + \frac{\pi}{3}\right)$

19. $y = \frac{1}{2} \cos\left(3x - \frac{\pi}{2}\right)$

20. $y = \frac{4}{3} \cos\left(3x + \frac{\pi}{2}\right)$

21. $y = 3 \sin\left(\frac{\pi}{3}x - \frac{\pi}{3}\right)$

22. $y = 3 \cos\left(\frac{\pi}{3}x - \frac{\pi}{3}\right)$

Use your answers for Problems 11 through 20 for reference, and graph one complete cycle of each of the following equations.

23. $y = 1 + \sin (2x - \pi)$

24. $y = -1 + \sin (2x + \pi)$

25. $y = -3 + \sin \left(\pi x + \dfrac{\pi}{2} \right)$

26. $y = 3 + \sin \left(\pi x - \dfrac{\pi}{2} \right)$

27. $y = 2 - \cos \left(2x + \dfrac{\pi}{2} \right)$

28. $y = -2 - \cos \left(2x - \dfrac{\pi}{2} \right)$

29. $y = -2 + 2 \sin \left(\dfrac{1}{2} x + \dfrac{\pi}{2} \right)$

30. $y = 3 + 3 \cos \left(\dfrac{1}{2} x + \dfrac{\pi}{3} \right)$

31. $y = \dfrac{3}{2} + \dfrac{1}{2} \cos \left(3x - \dfrac{\pi}{2} \right)$

32. $y = \dfrac{2}{3} + \dfrac{4}{3} \cos \left(3x + \dfrac{\pi}{2} \right)$

Graph each of the following equations over the given interval. In each case, be sure to label the axes so that the amplitude, period, and phase shift are easy to read.

33. $y = 4 \cos \left(2x - \dfrac{\pi}{2} \right), \ -\dfrac{\pi}{4} \le x \le \dfrac{3\pi}{2}$

34. $y = 3 \sin \left(2x - \dfrac{\pi}{3} \right), \ -\dfrac{5\pi}{6} \le x \le \dfrac{7\pi}{6}$

35. $y = -4 \cos \left(2x - \dfrac{\pi}{2} \right), \ -\dfrac{\pi}{4} \le x \le \dfrac{3\pi}{2}$

36. $y = -3 \sin \left(2x - \dfrac{\pi}{3} \right), \ -\dfrac{5\pi}{6} \le x \le \dfrac{7\pi}{6}$

37. $y = \dfrac{2}{3} \sin \left(3x + \dfrac{\pi}{2} \right), \ -\pi \le x \le \pi$

38. $y = \dfrac{3}{4} \sin \left(3x - \dfrac{\pi}{2} \right), \ -\dfrac{\pi}{2} \le x \le \dfrac{3\pi}{2}$

39. $y = -\dfrac{2}{3} \sin \left(3x + \dfrac{\pi}{2} \right), \ -\pi \le x \le \pi$

40. $y = -\dfrac{3}{4} \sin \left(3x - \dfrac{\pi}{2} \right), \ -\dfrac{\pi}{2} \le x \le \dfrac{3\pi}{2}$

Sketch one complete cycle of each of the following by first graphing the appropriate sine or cosine curve and then using the reciprocal relationships. In each case, be sure to include the asymptotes on your graph.

41. $y = \csc \left(x + \dfrac{\pi}{4} \right)$

42. $y = \sec \left(x + \dfrac{\pi}{4} \right)$

43. $y = 2 \sec \left(2x - \dfrac{\pi}{2} \right)$

44. $y = 2 \csc \left(2x - \dfrac{\pi}{2} \right)$

45. $y = 3 \csc \left(2x + \dfrac{\pi}{3} \right)$

46. $y = 3 \sec \left(2x - \dfrac{\pi}{3} \right)$

The periods for $y = \tan x$ and $y = \cot x$ are π. The graphs of $y = \tan (Bx + C)$ and $y = \cot (Bx + C)$ will have periods $= \pi/B$ and phase shifts $= -C/B$, for $B > 0$. Sketch the graph of each equation below. Be sure to show the asymptotes with each graph.

47. $y = \tan \left(x + \dfrac{\pi}{4} \right)$ **48.** $y = \tan \left(x - \dfrac{\pi}{4} \right)$

49. $y = \cot \left(x - \dfrac{\pi}{4} \right)$ **50.** $y = \cot \left(x + \dfrac{\pi}{4} \right)$

51. $y = \tan \left(2x - \dfrac{\pi}{2} \right)$ **52.** $y = \tan \left(2x + \dfrac{\pi}{2} \right)$

REVIEW PROBLEMS

The following problems review material we covered in Section 3.4.

53. Arc Length Find the length of arc cut off by a central angle of $\pi/6$ radians in a circle of radius 10 centimeters.

54. Arc Length How long is the arc cut off by a central angle of $90°$ in a circle with radius 22 centimeters?

55. Arc Length The minute hand of a clock is 2.6 centimeters long. How far does the tip of the minute hand travel in 30 minutes?

56. Arc Length The hour hand of a clock is 3 inches long. How far does the tip of the hand travel in 5 hours?

57. Radius of a Circle Find the radius of a circle if a central angle of 6 radians cuts off an arc of length 4 feet.

58. Radius of a Circle In a circle, a central angle of $135°$ cuts off an arc of length 75 meters. Find the radius of the circle.

SECTION 4.4 | FINDING AN EQUATION FROM ITS GRAPH

In this section, we will reverse what we have done in the previous sections of this chapter and produce an equation that describes a graph, rather than a graph that describes an equation. Let's start with an example from algebra.

EXAMPLE 1 Find the equation of the line shown in Figure 1.

SOLUTION From algebra we know that the equation of any straight line (except a vertical one) can be written in slope-intercept form as

$$y = mx + b$$

where m is the slope of the line, and b is its y-intercept.

Since the line in Figure 1 crosses the y-axis at 3, we know the y-intercept b is 3. To find the slope of the line, we find the ratio of the vertical change to the horizontal change between any two points on the line (sometimes called rise/run). From Figure 1 we see that this ratio is $-1/2$. Therefore, $m = -1/2$. The equation of our line must be

$$y = -\dfrac{1}{2}x + 3$$

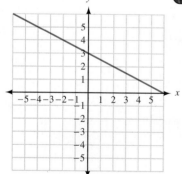

Figure 1

 **EXAMPLE 2** One cycle of the graph of a trigonometric function is shown in Figure 2. Find an equation to match the graph.

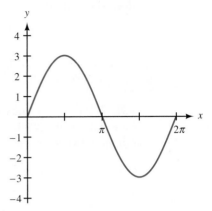

Figure 2

SOLUTION   The graph is a sine curve with an amplitude of 3, period 2π, and no phase shift. The equation is

$$y = 3 \sin x \qquad 0 \le x \le 2\pi$$

 **EXAMPLE 3** Find an equation of the graph shown in Figure 3.

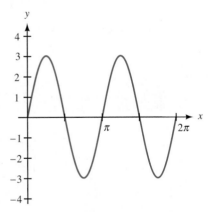

Figure 3

SOLUTION   Again, we have a sine curve, so we know the equation will have the form

$$y = k + A \sin (Bx + C)$$

From Figure 3 we see that the amplitude is 3, which means that $A = 3$. There is no phase shift, nor is there any vertical translation of the graph. Therefore, both C and k are 0.

To find B, we notice that the period is π. Since the formula for the period is $2\pi/B$, we have

$$\pi = \frac{2\pi}{B}$$

which means that B is 2. Our equation must be

$$y = 0 + 3 \sin (2x + 0)$$

which simplifies to

$$y = 3 \sin 2x \quad \text{for} \quad 0 \le x \le 2\pi \quad \blacksquare$$

 EXAMPLE 4 Find an equation of the graph shown in Figure 4.

SOLUTION The graph in Figure 4 has the same shape (amplitude, period, and phase shift) as the graph shown in Figure 3. In addition, it has undergone a vertical shift up of two units; therefore, the equation is

$$y = 2 + 3 \sin 2x \quad \text{for} \quad 0 \le x \le 2\pi$$

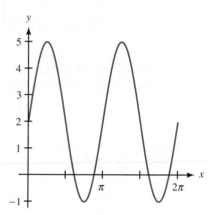

Figure 4

 EXAMPLE 5 Find an equation of the graph shown in Figure 5.

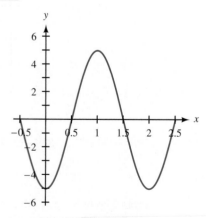

Figure 5

SOLUTION If we look at the graph from $x = 0$ to $x = 2$, it looks like a cosine curve that has been reflected about the x-axis. The general form for a cosine curve is

$$y = k + A \cos (Bx + C)$$

From Figure 5 we see the amplitude is 5. Since the graph has been reflected about the *x*-axis, $A = -5$. The period is 2, giving us an equation to solve for *B*:

$$\text{Period} = \frac{2\pi}{B} = 2 \Rightarrow B = \pi$$

There is no horizontal or vertical translation of the curve (if we assume it is a cosine curve), so *C* and *k* are both 0. An equation that describes this graph is

$$y = -5 \cos \pi x \qquad -0.5 \leq x \leq 2.5 \quad \blacksquare$$

EXAMPLE 6   Find an equation of the curve shown in Figure 6.

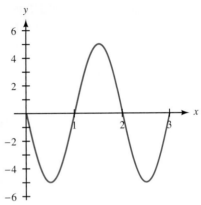

Figure 6

SOLUTION   The graph has the same shape as the graph shown in Figure 5 except it has been moved to the right 0.5 units. Using the results from Example 5, we know that $B = \pi$. To find *C*, we use the formula for phase shift:

$$\text{Phase shift} = -\frac{C}{B}$$

$$0.5 = -\frac{C}{\pi}$$

Solving for *C*, we have $C = -\dfrac{\pi}{2}$.

The equation is

$$y = -5 \cos\left(\pi x - \frac{\pi}{2}\right) \qquad 0 \leq x \leq 3$$

Another possible approach to this problem is to use a reflected sine curve. Then no phase shift is necessary since a cycle begins at $x = 0$. Using $A = -5$ and $B = \pi$, we have

$$y = -5 \sin \pi x \qquad 0 \leq x \leq 3$$

which is actually a bit simpler than the equation we obtained when using a cosine function. ■

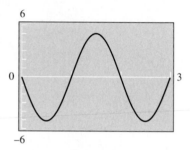

USING TECHNOLOGY

VERIFYING TRIGONOMETRIC MODELS

With a graphing calculator, we can easily verify that our equation, or model, for the graph in Figure 6 is correct. First, set your calculator to radian mode, and then define

$$Y_1 = -5 \cos (\pi x - \pi/2)$$

Set the window variables to match the given graph:

$$0 \le x \le 3; \ -6 \le y \le 6$$

If our equation is correct, the graph drawn by our calculator should be identical to the graph provided. Since the graph shown in Figure 7 is the same as the graph in Figure 6, our equation must be correct.

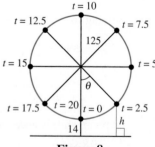

Figure 7

In our next example, we use a table of values for two variables to obtain a graph. From the graph, we find the equation.

EXAMPLE 7 The Ferris wheel built by George Ferris that we encountered in Chapters 2 and 3 is shown in Figure 8. Recall that the diameter is 250 feet and it rotates through one complete revolution every 20 minutes.

Figure 8

First, make a table that shows the rider's height h above the ground for each of the nine values of t shown in Figure 8. Then graph the ordered pairs indicated by the table. Finally, use the graph to find an equation that gives h as a function of t.

TABLE 1

t	h
0 min	14 ft
2.5 min	51 ft
5 min	139 ft
7.5 min	227 ft
10 min	264 ft
12.5 min	227 ft
15 min	139 ft
17.5 min	51 ft
20 min	14 ft

SOLUTION Here is the reasoning we use to find the value of h for each of the given values of t: When $t = 0$ and $t = 20$, the rider is at the bottom of the wheel, 14 feet above the ground. When $t = 10$, the rider is at the top of the wheel, which is 264 feet above the ground. At times $t = 5$ and $t = 15$, the rider is even with the center of the wheel, which is 139 feet above the ground. The other four values of h are found using right triangle trigonometry, as we did in Chapter 2. Table 1 shows our corresponding values of t and h.

If we graph the points (t, h) on a rectangular coordinate system and then connect them with a smooth curve, we produce the diagram shown in Figure 9. The curve is a cosine curve that has been reflected about the t-axis and then shifted up vertically.

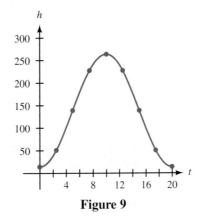

Figure 9

Since the curve is a cosine curve, the equation will have the form

$$h = k + A \cos (Bt + C)$$

To find the equation for the curve in Figure 9, we must find values for k, A, B, and C. We begin by finding C.

Since the curve starts at $t = 0$, the phase shift is 0, which gives us

$$C = 0$$

The amplitude, half the difference between the highest point and the lowest point on the graph, must be 125 (also equal to the radius of the wheel). However, since the graph is a *reflected* cosine curve, we have

$$A = -125$$

The period is 20 minutes, so we find B with the equation

$$20 = \frac{2\pi}{B} \quad \Rightarrow \quad B = \frac{\pi}{10}$$

The amount of the vertical translation, k, is the distance the center of the wheel is off the ground. Therefore,

$$k = 139$$

The equation that gives the height of the rider at any time t during the ride is

$$h = 139 - 125 \cos \frac{\pi}{10} t \quad \blacksquare$$

Many real-life phenomena are periodic in nature and can therefore be modeled very effectively using trigonometric functions. The group project at the end of this chapter provides an opportunity to further explore the use of trigonometric functions as models.

GETTING READY FOR CLASS

After reading through the preceding section, respond in your own words and in complete sentences.

a. How do you find the equation of a line from its graph?

b. How do you find the coefficient A by looking at the graph of
$y = A \sin (Bx + C)$?

c. How do you find the period for a sine function by looking at its graph?

d. How do you find the number k by looking at the graph of
$y = k + A \sin (Bx + C)$?

PROBLEM SET 4.4

Find the equation of each of the following lines. Write your answers in slope-intercept form, $y = mx + b$.

1.

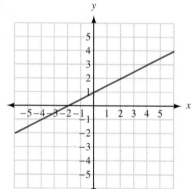

2.

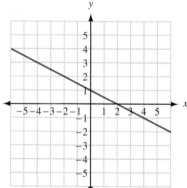

3.

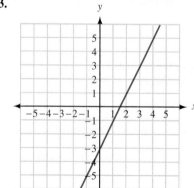

4.

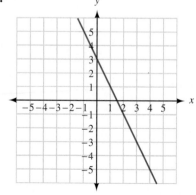

Each graph below is one complete cycle of the graph of an equation containing a trigonometric function. In each case, find an equation to match the graph. If you are using a graphing calculator, graph your equation to verify that it is correct.

5.

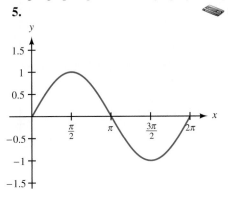

6.

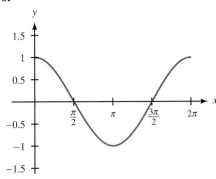

7.

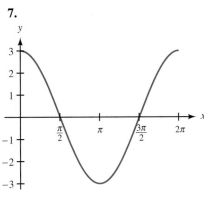

8.

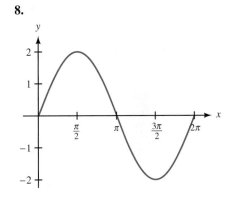

9.

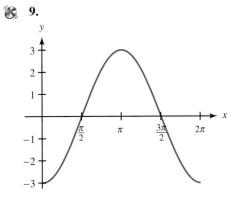

10.

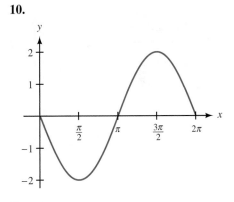

11.

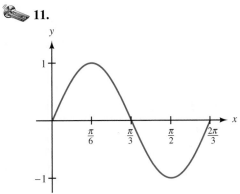

12.

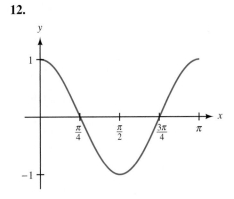

13.

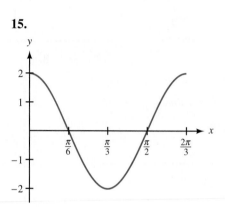

14.

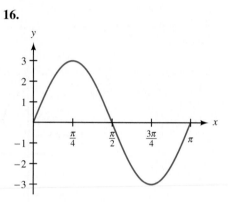

15.

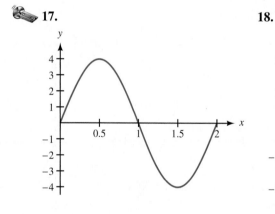

16.

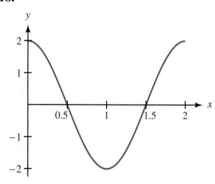

17.

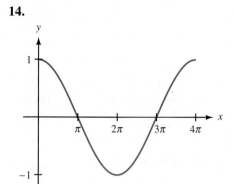

18.

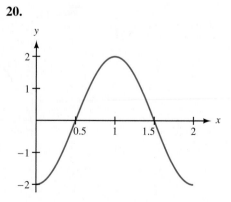

19.

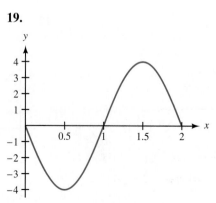

20.

21.

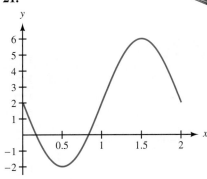

22.

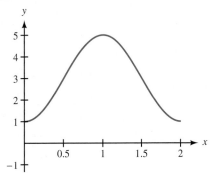

23.

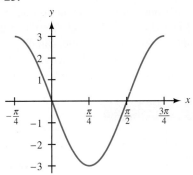

24.

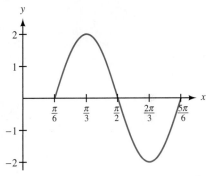

25.

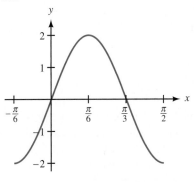

26.

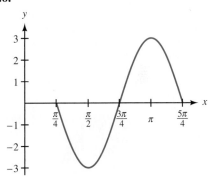

27.

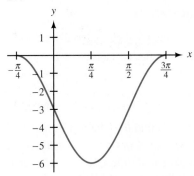

28.

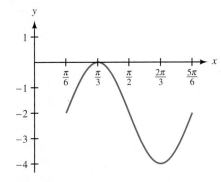

29. **30.**

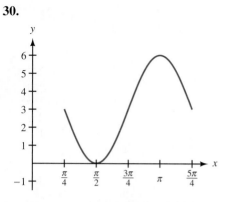

31. Ferris Wheel Figure 10 is a model of the Ferris wheel known as the Riesenrad that we encountered in Chapters 2 and 3. Recall that the diameter of the wheel is 197 feet, and one complete revolution takes 15 minutes. The bottom of the wheel is 12 feet above the ground. Complete the table next to Figure 10, then plot the points (t, h) from the table. Finally, connect the points with a smooth curve, and use the curve to find an equation that will give a passenger's height above the ground at any time t during the ride.

TABLE 2

t	h
0 min	
1.875 min	
3.75 min	
5.625 min	
7.5 min	
9.375 min	
11.25 min	
13.125 min	
15 min	

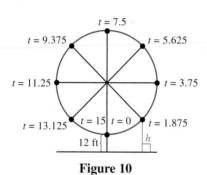

Figure 10

32. Ferris Wheel In Chapters 2 and 3, we worked some problems involving the Ferris wheel called Colossus that was built in St. Louis in 1986. The diameter of the wheel is 165 feet, it rotates at 1.5 revolutions per minute, and the bottom of the wheel is 9 feet above the ground. Find an equation that gives a passenger's height above the ground at any time t during the ride. Assume the passenger starts the ride at the bottom of the wheel.

REVIEW PROBLEMS

The following problems review material we covered in Section 3.1.

Name the reference angle for each angle below.

33. $321°$ **34.** $148°$

35. $236°$ **36.** $-125°$

37. $-276°$ **38.** $450°$

Use your calculator to find θ to the nearest tenth of a degree if $0° < \theta < 360°$ and

39. $\sin \theta = 0.7455$ with θ in QII **40.** $\cos \theta = 0.7455$ with θ in QIV

41. $\csc \theta = -2.3228$ with θ in QIII **42.** $\sec \theta = -3.1416$ with θ in QIII

43. $\cot \theta = -0.2089$ with θ in QIV **44.** $\tan \theta = 0.8156$ with θ in QIII

SECTION 4.5 | GRAPHING COMBINATIONS OF FUNCTIONS

In this section, we will graph equations of the form $y = y_1 + y_2$, where y_1 and y_2 are algebraic or trigonometric functions of x. For instance, the equation $y = 1 + \sin x$ can be thought of as the sum of the two functions $y_1 = 1$ and $y_2 = \sin x$. That is,

$$\text{if} \qquad y_1 = 1 \qquad \text{and} \qquad y_2 = \sin x,$$

$$\text{then} \qquad y = y_1 + y_2$$

Using this kind of reasoning, the graph of $y = 1 + \sin x$ is obtained by adding each value of y_2 in $y_2 = \sin x$ to the corresponding value of y_1 in $y_1 = 1$. Graphically, we can show this by adding the values of y from the graph of y_2 to the corresponding values of y from the graph of y_1 (Figure 1). If $y_2 > 0$, then $y_1 + y_2$ will be *above* y_1 by a distance equal to y_2. If $y_2 < 0$, then $y_1 + y_2$ will be *below* y_1 by a distance equal to the absolute value of y_2.

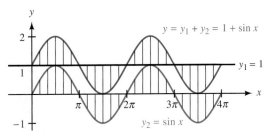

Figure 1

Although in actual practice you may not draw in the little vertical lines we have shown here, they do serve the purpose of allowing us to visualize the idea of adding the y-coordinates on one graph to the corresponding y-coordinates on another graph.

EXAMPLE 1) Graph $y = \frac{1}{3}x - \sin x$ between $x = 0$ and $x = 4\pi$.

SOLUTION   We can think of the equation $y = \frac{1}{3}x - \sin x$ as the sum of the equations $y_1 = \frac{1}{3}x$ and $y_2 = -\sin x$. Graphing each of these two equations on the same set of axes and then adding the values of y_2 to the corresponding values of y_1, we have the graph shown in Figure 2.

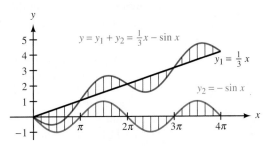

Figure 2

For the rest of the examples in this section, we will not show the vertical lines used to visualize the process of adding y-coordinates. Sometimes the graphs become too confusing to read when the vertical lines are included. It is the idea behind the vertical lines that is important, not the lines themselves. (And remember, the alternative to graphing these types of equations by adding y-coordinates is to make a table. If you try using a table on some of the examples that follow, you will see that the method being presented here is much faster.)

EXAMPLE 2 Graph $y = 2 \sin x + \cos 2x$ for x between 0 and 4π.

SOLUTION We can think of y as the sum of y_1 and y_2, where

$$y_1 = 2 \sin x \qquad \text{(amplitude 2, period } 2\pi)$$

and

$$y_2 = \cos 2x \qquad \text{(amplitude 1, period } \pi)$$

The graphs of y_1, y_2, and $y = y_1 + y_2$ are shown in Figure 3.

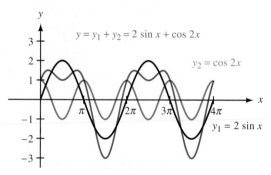

Figure 3

USING TECHNOLOGY

COMBINATIONS OF FUNCTIONS

Although a graphing calculator can easily graph the function $y = 2 \sin x + \cos 2x$ in Example 2 without using the methods described in this section, it can also be used to reinforce them. Set your calculator to radian mode and enter the following three functions:

$$Y_1 = 2 \sin (x), Y_2 = \cos (2x), \qquad \text{and} \qquad Y_3 = Y_1 + Y_2$$

If your calculator has the ability to set different graphing styles, set the first function Y_1 to graph in the normal style, the second function Y_2 in the dot style, and the third function Y_3 in the bold style. Figure 4 shows how the functions would be defined on a TI-83. Now set your window variables so that

$$0 \le x \le 4\pi, \text{ scale} = \pi; \ -4 \le y \le 4$$

When you graph the functions, your calculator screen should look similar to Figure 5. For any x, the dotted graph (Y_2) indicates the distance we should travel above or below the normal graph (Y_1) in order to locate the bold graph (Y_3).

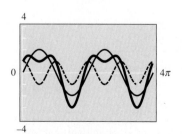

```
Plot1  Plot2  Plot3
\Y1=2sin(X)
:Y2=cos(2X)
\Y3=Y1+Y2
\Y4=
\Y5=
\Y6=
\Y7=
```

Figure 4

Figure 5

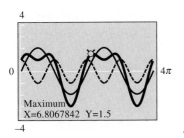

Figure 6

Notice that the period of the function $y = 2 \sin x + \cos 2x$ is 2π. We can also see that the function is at a minimum when the individual component functions $y_1 = 2 \sin x$ and $y_2 = \cos 2x$ are both at their minimum values of -2 and -1. Since $y = y_1 + y_2$, this minimum value will be $(-2) + (-1) = -3$. However, the maximum values of y_1 and y_2 do not occur simultaneously, so the maximum of $y = 2 \sin x + \cos 2x$ is not so easily determined. Using the appropriate command on the calculator we find that the maximum value is 1.5, as shown in Figure 6.

EXAMPLE 3   Graph $y = \cos x + \cos 2x$ for $0 \leq x \leq 4\pi$.

SOLUTION   We let $y = y_1 + y_2$, where

$$y_1 = \cos x \qquad \text{(amplitude 1, period } 2\pi)$$

and

$$y_2 = \cos 2x \qquad \text{(amplitude 1, period } \pi)$$

Figure 7 illustrates the graphs of y_1, y_2, and $y = y_1 + y_2$.

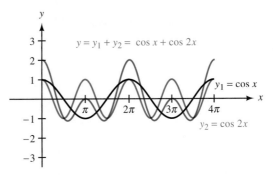

Figure 7

EXAMPLE 4   Graph $y = \sin x + \cos x$ for x between 0 and 4π.

SOLUTION   We let $y_1 = \sin x$ and $y_2 = \cos x$ and graph y_1, y_2, and $y = y_1 + y_2$. Figure 8 illustrates these graphs.

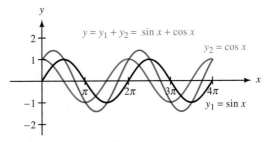

Figure 8

The graph of $y = \sin x + \cos x$ has amplitude $\sqrt{2}$. If we were to extend the graph to the left, we would find it crossed the x-axis at $-\pi/4$. It would then be apparent that the graph of $y = \sin x + \cos x$ is the same as the graph of $y = \sqrt{2} \sin (x + \pi/4)$; both are sine curves with amplitude $\sqrt{2}$ and phase shift $-\pi/4$.

One application of combining trigonometric functions can be seen with Fourier series. Fourier series are used in physics and engineering to represent certain waveforms as an infinite sum of sine and/or cosine functions.

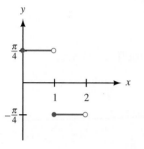

Figure 9

EXAMPLE 5 The square wave function $f(x) = \begin{cases} \dfrac{\pi}{4} & 0 \le x < 1 \\ -\dfrac{\pi}{4} & 1 \le x < 2 \end{cases}$

shown in Figure 9 can be represented by the Fourier series

$$f(x) = \sin(\pi x) + \frac{\sin(3\pi x)}{3} + \frac{\sin(5\pi x)}{5} + \frac{\sin(7\pi x)}{7} + \cdots, 0 \le x < 2$$

Graph the second partial sum for the Fourier series, consisting of the first two terms of this sum:

$$y = \sin(\pi x) + \frac{\sin(3\pi x)}{3}$$

SOLUTION We let $y_1 = \sin(\pi x)$ and $y_2 = \dfrac{\sin(3\pi x)}{3}$ and graph y_1, y_2, and $y = y_1 + y_2$. Figure 10 illustrates these graphs.

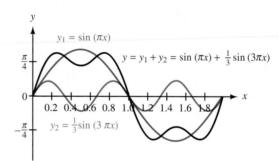

Figure 10

In Figure 11 we have graphed the first, second, third, and fourth partial sums for the Fourier series. You can see that as we include more terms from the series, the resulting curve gives a better approximation of the square wave graph.

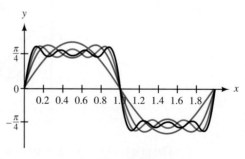

Figure 11

GETTING READY FOR CLASS

After reading through the preceding section, respond in your own words and in complete sentences.

a. We can think of the function $y = 1 + \sin x$ as the sum of what two functions?

b. How do you graph the function $y = 1 + \sin x$?

c. How do you graph the function $y = 2 \sin x + \cos 2x$?

d. What is the period of the function $y = 2 \sin x + \cos 2x$?

PROBLEM SET 4.5

Use addition of y-coordinates to sketch the graph of each of the following between $x = 0$ and $x = 4\pi$.

1. $y = 1 + \sin x$

2. $y = 1 + \cos x$

3. $y = 2 - \cos x$

4. $y = 2 - \sin x$

5. $y = 4 + 2 \sin x$

6. $y = 4 + 2 \cos x$

7. $y = \dfrac{1}{3}x - \cos x$

8. $y = \dfrac{1}{2}x - \sin x$

9. $y = \dfrac{1}{2}x - \cos x$

10. $y = \dfrac{1}{3}x + \cos x$

Sketch the graph of each equation from $x = 0$ to $x = 8$.

11. $y = x + \sin \pi x$

12. $y = x + \cos \pi x$

Sketch the graph from $x = 0$ to $x = 4\pi$.

13. $y = 3 \sin x + \cos 2x$

14. $y = 3 \cos x + \sin 2x$

15. $y = 2 \sin x - \cos 2x$

16. $y = 2 \cos x - \sin 2x$

17. $y = \sin x + \sin \dfrac{x}{2}$

18. $y = \cos x + \cos \dfrac{x}{2}$

19. $y = \sin x + \sin 2x$

20. $y = \cos x + \cos 2x$

21. $y = \cos x + \dfrac{1}{2} \sin 2x$

22. $y = \sin x + \dfrac{1}{2} \cos 2x$

23. $y = \sin x - \cos x$

24. $y = \cos x - \sin x$

25. Make a table using multiples of $\pi/2$ for x between 0 and 4π to help sketch the graph of $y = x \sin x$.

26. Sketch the graph of $y = x \cos x$.

Use your graphing calculator to graph each of the following between $x = 0$ and $x = 4\pi$. In each case, show the graph of y_1, y_2, and $y = y_1 + y_2$. (Make sure your calculator is set to radian mode.)

27. $y = 2 + \sin x$

28. $y = 2 + \cos x$

29. $y = x + \cos x$

30. $y = x - \sin x$

31. $y = \sin x - 2 \cos x$

32. $y = \cos x + 2 \sin x$

33. $y = \cos 2x + 2 \cos 3x$

34. $y = \sin 2x - 2 \sin 3x$

35. In Example 5 we stated that a square wave can be represented by the Fourier series

$$f(x) = \sin(\pi x) + \frac{\sin(3\pi x)}{3} + \frac{\sin(5\pi x)}{5} + \frac{\sin(7\pi x)}{7} + \cdots, \ 0 \le x < 2$$

Use your graphing calculator to graph the following partial sums of this series.

a. $f(x) = \sin(\pi x) + \dfrac{\sin(3\pi x)}{3}$

b. $f(x) = \sin(\pi x) + \dfrac{\sin(3\pi x)}{3} + \dfrac{\sin(5\pi x)}{5}$

c. $f(x) = \sin(\pi x) + \dfrac{\sin(3\pi x)}{3} + \dfrac{\sin(5\pi x)}{5} + \dfrac{\sin(7\pi x)}{7}$

d. $f(x) = \sin(\pi x) + \dfrac{\sin(3\pi x)}{3} + \dfrac{\sin(5\pi x)}{5} + \dfrac{\sin(7\pi x)}{7} + \dfrac{\sin(9\pi x)}{9}$

36. The waveform shown in Figure 12, called a *sawtooth wave,* can be represented by the Fourier series

$$f(x) = \sin(\pi x) + \frac{\sin(2\pi x)}{2} + \frac{\sin(3\pi x)}{3} + \frac{\sin(4\pi x)}{4} + \cdots, \ 0 \le x < 4$$

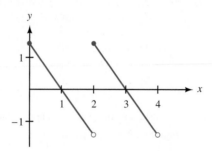

Figure 12

Use your graphing calculator to graph the following partial sums of this series.

a. $f(x) = \sin(\pi x) + \dfrac{\sin(2\pi x)}{2}$

b. $f(x) = \sin(\pi x) + \dfrac{\sin(2\pi x)}{2} + \dfrac{\sin(3\pi x)}{3}$

c. $f(x) = \sin(\pi x) + \dfrac{\sin(2\pi x)}{2} + \dfrac{\sin(3\pi x)}{3} + \dfrac{\sin(4\pi x)}{4}$

d. $f(x) = \sin(\pi x) + \dfrac{\sin(2\pi x)}{2} + \dfrac{\sin(3\pi x)}{3} + \dfrac{\sin(4\pi x)}{4} + \dfrac{\sin(5\pi x)}{5}$

REVIEW PROBLEMS

The following problems review material we covered in Section 3.5.

37. Linear Velocity A point moving on the circumference of a circle covers 5 feet every 20 seconds. Find the linear velocity of the point.

38. Linear Velocity A point is moving on the circumference of a circle. Every 30 seconds the point covers 15 centimeters. Find the linear velocity of the point.

39. Arc Length A point is moving with a linear velocity of 20 feet per second on the circumference of a circle. How far does the point move in 1 minute?

40. **Arc Length** A point moves at 65 meters per second on the circumference of a circle. How far does the point travel in 1 minute?

41. **Arc Length** A point is moving with an angular velocity of 3 radians per second on a circle of radius 6 meters. How far does the point travel in 10 seconds?

42. **Angular Velocity** Convert 30 revolutions per minute (rpm) to angular velocity in radians per second.

43. **Angular Velocity** Convert 120 revolutions per minute to angular velocity in radians per second.

44. **Linear Velocity** A point is rotating at 5 revolutions per minute on a circle of radius 6 inches. What is the linear velocity of the point?

45. **Arc Length** How far does the tip of a 10-centimeter minute hand on a clock travel in 2 hours?

46. **Arc Length** How far does the tip of an 8-centimeter hour hand on a clock travel in 1 day?

SECTION 4.6 | INVERSE TRIGONOMETRIC FUNCTIONS

In Chapter 2 we encountered situations in which we needed to find an angle given a value of one of the trigonometric functions. This is the reverse of what a trigonometric function is designed to do. When we try to use a function in the reverse direction, we are really using what is called the *inverse* of the function. We begin this section with a brief review of the inverse of a function. If this is a new topic for you, a more thorough presentation of functions and inverse functions is provided in Appendix A.

First, let us review the definition of a function and its inverse.

DEFINITION

A *function* is a rule or correspondence that pairs each element of the domain with exactly one element from the range. That is, a function is a set of ordered pairs in which no two different ordered pairs have the same first coordinate.

The *inverse* of a function is found by interchanging the coordinates in each ordered pair that is an element of the function.

With the inverse, the domain and range of the function have switched roles. To find the equation of the inverse of a function, we simply exchange x and y in the equation and then solve for y. For example, to find the inverse of the function $y = x^2 - 4$, we would proceed like this:

The inverse of $\qquad y = x^2 - 4$

is $\qquad\qquad x = y^2 - 4 \qquad$ Exchange x and y

or $\qquad\quad y^2 - 4 = x$

$\qquad\qquad\qquad y^2 = x + 4 \qquad$ Add 4 to both sides

$\qquad\qquad\qquad y = \pm\sqrt{x + 4} \qquad$ Take the square root of both sides

The inverse of the function $y = x^2 - 4$ is given by the equation $y = \pm\sqrt{x + 4}$.

The graph of $y = x^2 - 4$ is a parabola that crosses the x-axis at -2 and 2 and has its vertex at $(0, -4)$. To graph the inverse, we take each point on the graph of $y = x^2 - 4$, interchange the x- and y-coordinates, and then plot the resulting point. Figure 1 shows both graphs. Notice that the graph of the inverse is a reflection of the graph of the original function about the line $y = x$.

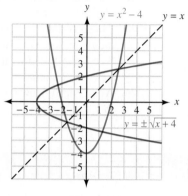

Figure 1

From Figure 1 we see that the inverse of $y = x^2 - 4$ is not a function because the graph of $y = \pm\sqrt{x + 4}$ does not pass the vertical line test. In order for the inverse to also be a function, the graph of the original function must pass the *horizontal* line test. Functions with this property are called *one-to-one* functions. If a function is one-to-one, then we know its inverse will be a function as well.

INVERSE FUNCTION NOTATION

If $y = f(x)$ is a one-to-one function, then the inverse of f is also a function and can be denoted by $y = f^{-1}(x)$.

Because the graphs of all six trigonometric functions do not pass the horizontal line test, the inverse relations for these functions will not be functions themselves. However, we will see that it is possible to define an inverse that is a function if we restrict the original trigonometric function to certain angles. In this section, we will limit our discussion of inverse trigonometric functions to the inverses of the three major functions: sine, cosine, and tangent. The other three inverse trigonometric functions can be handled with the use of reciprocal identities.

THE INVERSE SINE RELATION

To find the inverse of $y = \sin x$, we interchange x and y to obtain

$$x = \sin y$$

This is the equation of the inverse sine relation.

To graph $x = \sin y$, we simply reflect the graph of $y = \sin x$ about the line $y = x$, as shown in Figure 2.

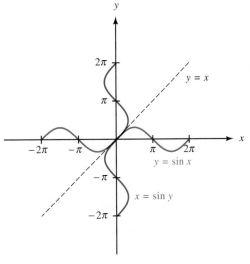

Figure 2

As you can see from the graph, $x = \sin y$ is a relation but not a function. For every value of x in the domain, there are many values of y. The graph of $x = \sin y$ fails the vertical line test.

THE INVERSE SINE FUNCTION

In order that the function $y = \sin x$ have an inverse that is also a function, it is necessary to restrict the values that x can assume so that we may satisfy the horizontal line test. The interval we restrict it to is $-\pi/2 \le x \le \pi/2$. Figure 3 displays the graph of $y = \sin x$ with the restricted interval showing. Notice that this segment of the sine graph passes the horizontal line test, and it maintains the full range of the function $-1 \le y \le 1$. Figure 4 shows the graph of the inverse relation $x = \sin y$ with the restricted interval after the sine curve has been reflected about the line $y = x$.

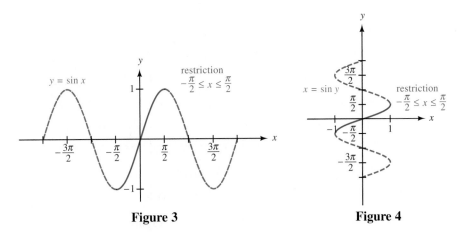

Figure 3   **Figure 4**

It is apparent from Figure 4 that if $x = \sin y$ is restricted to the interval $-\pi/2 \le y \le \pi/2$, then each value of x between -1 and 1 is associated with exactly one value of y, and we have a function rather than just a relation. The equation $x = \sin y$, together with the restriction $-\pi/2 \leq y \leq \pi/2$, forms the inverse sine function. To designate this function, we use the following notation.

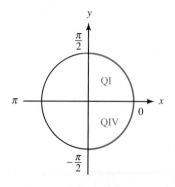

NOTATION

The notation used to indicate the inverse sine function is as follows:

Notation	Meaning
$y = \sin^{-1} x$ or $y = \arcsin x$	$x = \sin y$ and $-\dfrac{\pi}{2} \leq y \leq \dfrac{\pi}{2}$

In words: y is the angle between $-\pi/2$ and $\pi/2$, inclusive, whose sine is x.

The inverse sine function will return an angle between $-\pi/2$ and $\pi/2$, inclusive, corresponding to QIV or QI.

NOTE The notation $\sin^{-1} x$ is not to be interpreted as meaning the reciprocal of $\sin x$. That is,

$$\sin^{-1} x \neq \frac{1}{\sin x}$$

If we want the reciprocal of $\sin x$, we use $\csc x$ or $(\sin x)^{-1}$, but never $\sin^{-1} x$.

THE INVERSE COSINE FUNCTION

Just as we did for the sine function, we must restrict the values that x can assume with the cosine function in order to satisfy the horizontal line test. The interval we restrict it to is $0 \leq x \leq \pi$. Figure 5 shows the graph of $y = \cos x$ with the restricted interval. Figure 6 shows the graph of the inverse relation $x = \cos y$ with the restricted interval after the cosine curve has been reflected about the line $y = x$.

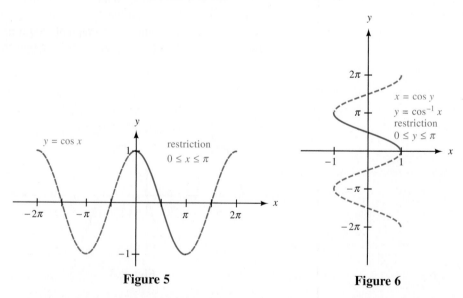

Figure 5

Figure 6

The equation $x = \cos y$, together with the restriction $0 \leq y \leq \pi$, forms the inverse cosine function. To designate this function we use the following notation.

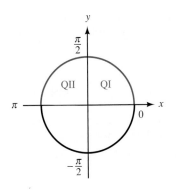

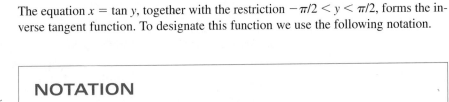

NOTATION

The notation used to indicate the inverse cosine function is as follows:

Notation	Meaning
$y = \cos^{-1} x$ or $y = \arccos x$	$x = \cos y$ and $0 \le y \le \pi$

In words: y is the angle between 0 and π, inclusive, whose cosine is x.

The inverse cosine function will return an angle between 0 and π, inclusive, corresponding to QI or QII.

THE INVERSE TANGENT FUNCTION

For the tangent function, we restrict the values that x can assume to the interval $-\pi/2 < x < \pi/2$. Figure 7 shows the graph of $y = \tan x$ with the restricted interval. Figure 8 shows the graph of the inverse relation $x = \tan y$ with the restricted interval after it has been reflected about the line $y = x$.

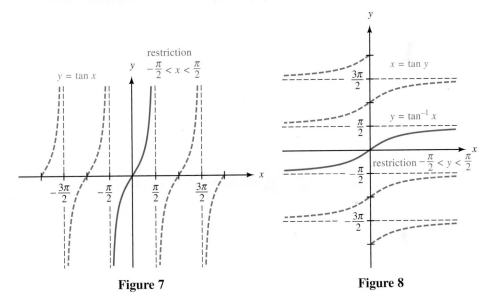

Figure 7   **Figure 8**

The equation $x = \tan y$, together with the restriction $-\pi/2 < y < \pi/2$, forms the inverse tangent function. To designate this function we use the following notation.

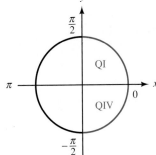

NOTATION

The notation used to indicate the inverse tangent function is as follows:

Notation	Meaning
$y = \tan^{-1} x$ or $y = \arctan x$	$x = \tan y$ and $-\dfrac{\pi}{2} < y < \dfrac{\pi}{2}$

In words: y is the angle between $-\pi/2$ and $\pi/2$ whose tangent is x.

The inverse tangent function will return an angle between $-\pi/2$ and $\pi/2$, corresponding to QIV or QI.

To summarize, here are the definitions of the three inverse trigonometric functions we have presented, along with the domain, range, and graph for each.

INVERSE TRIGONOMETRIC FUNCTIONS

Inverse sine	Inverse cosine	Inverse tangent
$y = \sin^{-1} x = \arcsin x$	$y = \cos^{-1} x = \arccos x$	$y = \tan^{-1} x = \arctan x$

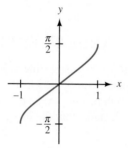
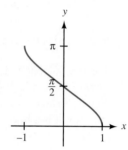
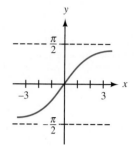

Domain: $-1 \leq x \leq 1$ Domain: $-1 \leq x \leq 1$ Domain: all real numbers

Range: $-\dfrac{\pi}{2} \leq y \leq \dfrac{\pi}{2}$ Range: $0 \leq y \leq \pi$ Range: $-\dfrac{\pi}{2} < y < \dfrac{\pi}{2}$

EXAMPLE 1 Evaluate in radians without using a calculator or tables.

a. $\sin^{-1} \dfrac{1}{2}$ **b.** $\arccos\left(-\dfrac{\sqrt{3}}{2}\right)$ **c.** $\tan^{-1}(-1)$

SOLUTION
a. The angle between $-\pi/2$ and $\pi/2$ whose sine is $\dfrac{1}{2}$ is $\pi/6$.

$$\sin^{-1} \frac{1}{2} = \frac{\pi}{6}$$

b. The angle between 0 and π with a cosine of $-\sqrt{3}/2$ is $5\pi/6$.

$$\arccos\left(-\frac{\sqrt{3}}{2}\right) = \frac{5\pi}{6}$$

c. The angle between $-\pi/2$ and $\pi/2$ the tangent of which is -1 is $-\pi/4$.

$$\tan^{-1}(-1) = -\frac{\pi}{4}$$

NOTE In part c of Example 1, it would be incorrect to give the answer as $7\pi/4$. It is true that $\tan 7\pi/4 = -1$, but $7\pi/4$ is not between $-\pi/2$ and $\pi/2$. There is a difference.

USING TECHNOLOGY

GRAPHING THE INVERSE SINE FUNCTION

A graphing calculator can be used to graph and evaluate an inverse trigonometric function. To graph the inverse sine function, set your calculator to degree mode and enter the function

$$Y_1 = \sin^{-1} x$$

Set your window variables so that

$$-1.5 \le x \le 1.5; \ -120 \le y \le 120, \text{ scale} = 45$$

The graph of the inverse sine function is shown in Figure 9.

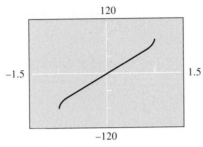

Figure 9

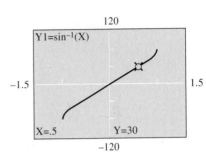

Figure 10

To find $\sin^{-1} \dfrac{1}{2}$ in Example 1a, use the appropriate command to evaluate the function for $x = 0.5$. As Figure 10 illustrates, the result is $y = 30°$. The angle between $-90°$ and $90°$ having a sine of 1/2 is 30°, or $\pi/6$ in radians.

EXAMPLE 2   Use a calculator to evaluate each expression to the nearest tenth of a degree.

a. arcsin (0.5075)   **b.** arcsin (−0.5075)
c. $\cos^{-1}$ (0.6428)   **d.** $\cos^{-1}$ (−0.6428)
e. arctan (4.474)   **f.** arctan (−4.474)

SOLUTION   The easiest method of evaluating these expressions is to use a calculator. Make sure the calculator is set to degree mode, and then enter the number and press the appropriate key. Scientific and graphing calculators are programmed so that the restrictions on the inverse trigonometric functions are automatic.

a. arcsin (0.5075) = 30.5°
b. arcsin (−0.5075) = −30.5°   } Reference angle 30.5°

c. $\cos^{-1}$ (0.6428) = 50.0°
d. $\cos^{-1}$ (−0.6428) = 130.0°   } Reference angle 50°

e. arctan (4.474) = 77.4°
f. arctan (−4.474) = −77.4°   } Reference angle 77.4°

In Example 5 of Section 1.5, we simplified the expression $\sqrt{x^2 + 9}$ using the trigonometric substitution $x = 3\tan\theta$ to eliminate the square root. We will now see how to simplify the result of that example further by removing the absolute value symbol.

EXAMPLE 3 Simplify $3\left|\sec\theta\right|$ if $\theta = \tan^{-1}\dfrac{x}{3}$ for some real number x.

SOLUTION Since $\theta = \tan^{-1}\dfrac{x}{3}$, we know from the definition of the inverse tangent function that $-\dfrac{\pi}{2} < \theta < \dfrac{\pi}{2}$. For any angle θ within this interval, $\sec\theta$ will be a positive value. Therefore, $\left|\sec\theta\right| = \sec\theta$ and we can simplify the expression further as

$$3\left|\sec\theta\right| = 3\sec\theta \quad \blacksquare$$

EXAMPLE 4 Evaluate each expression.

a. $\sin\left(\sin^{-1}\dfrac{1}{2}\right)$

b. $\sin^{-1}(\sin 135°)$

SOLUTION

a. From Example 1a we know that $\sin^{-1}\dfrac{1}{2} = \dfrac{\pi}{6}$. Therefore,

$$\sin\left(\sin^{-1}\dfrac{1}{2}\right) = \sin\left(\dfrac{\pi}{6}\right) = \dfrac{1}{2}$$

b. Since $\sin 135° = \dfrac{1}{\sqrt{2}}$, $\sin^{-1}(\sin 135°) = \sin^{-1}\left(\dfrac{1}{\sqrt{2}}\right)$ will be the angle y,

$-90° \leq y \leq 90°$, for which $\sin y = \dfrac{1}{\sqrt{2}}$. The angle satisfying this requirement is $y = 45°$. So,

$$\sin^{-1}(\sin 135°) = \sin^{-1}\left(\dfrac{1}{\sqrt{2}}\right) = 45° \quad \blacksquare$$

NOTE Notice in Example 4a that the result of 1/2 is the same as the value that appeared in the original expression. Because $y = \sin x$ and $y = \sin^{-1} x$ are inverse functions, the one function will "undo" the action performed by the other. (For more details on this property of inverse functions, see Appendix A.) The reason this same process did not occur in Example 4b is that the original angle 135° is not within the restricted domain of the sine function and is therefore outside the range of the inverse sine function. In a sense, the functions in Example 4b are not really inverses because we did not choose an input within the agreed-upon interval for the sine function.

EXAMPLE 5   Simplify $\tan^{-1}(\tan x)$ if $-\dfrac{\pi}{2} < x < \dfrac{\pi}{2}$.

SOLUTION   Because x is within the restricted domain for the tangent function, the two functions are inverses. Whatever value the tangent function assigns to x, the inverse tangent function will reverse this association and return the original value of x. So, by the property of inverse functions,

$$\tan^{-1}(\tan x) = x$$

EXAMPLE 6   Evaluate $\sin\left(\tan^{-1}\dfrac{3}{4}\right)$ without using a calculator.

SOLUTION   We begin by letting $\theta = \tan^{-1}\left(\dfrac{3}{4}\right)$. (Remember, $\tan^{-1}x$ is the angle whose tangent is x.) Then we have

$$\text{If } \theta = \tan^{-1}\frac{3}{4}, \text{ then } \tan\theta = \frac{3}{4} \text{ and } 0° < \theta < 90°$$

We can draw a triangle in which one of the acute angles is θ (Figure 11). Since $\tan\theta = \dfrac{3}{4}$, we label the side opposite θ with 3 and the side adjacent to θ with 4. The hypotenuse is found by applying the Pythagorean Theorem.

From Figure 11 we find $\sin\theta$ using the ratio of the side opposite θ to the hypotenuse.

$$\sin\left(\tan^{-1}\frac{3}{4}\right) = \sin\theta = \frac{3}{5}$$

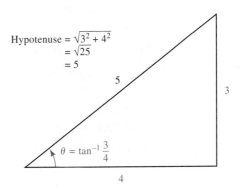

$$\text{Hypotenuse} = \sqrt{3^2 + 4^2}$$
$$= \sqrt{25}$$
$$= 5$$

$$\theta = \tan^{-1}\frac{3}{4}$$

Figure 11

CALCULATOR NOTE   If we were to do the same problem with the aid of a calculator, the sequence would look like this:

Scientific Calculator

3 ÷ 4 = $\tan^{-1}$ sin

Graphing Calculator

sin ($\tan^{-1}$ (3 ÷ 4)) ENTER

The display would read 0.6, which is $\dfrac{3}{5}$.

Although it is a lot easier to use a calculator on problems like the one in Example 6, solving it without a calculator will be of more use to you in the future.

EXAMPLE 7 Write the expression $\sin(\cos^{-1} x)$ as an equivalent expression in x only.

SOLUTION We let $\theta = \cos^{-1} x$. Then $\cos \theta = x = \dfrac{x}{1}$ and $0 \leq \theta \leq \pi$. We can visualize the problem by drawing θ in standard position with terminal side in either quadrant I or quadrant II (Figure 12). Let $P = (x, y)$ be a point on the terminal side of θ. By Definition I, $\cos \theta = \dfrac{x}{r}$, so r must be equal to 1. We can find y by applying the Pythagorean Theorem. Notice that y will be a positive value in either quadrant.

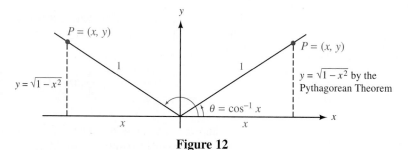

Figure 12

Since $\sin \theta = \dfrac{y}{r}$,

$$\sin(\cos^{-1} x) = \sin \theta = \frac{\sqrt{1-x^2}}{1} = \sqrt{1-x^2}$$

This result is valid whether x is positive (θ terminates in QI) or negative (θ terminates in QII).

GETTING READY FOR CLASS

After reading through the preceding section, respond in your own words and in complete sentences.

a. Why must the graph of $y = \sin x$ be restricted to $-\dfrac{\pi}{2} \leq x \leq \dfrac{\pi}{2}$ in order to have an inverse function?

b. What restriction is made on the values of x for $y = \cos x$ so that it will have an inverse that is a function?

c. What is the meaning of the notation $y = \tan^{-1} x$?

d. If $y = \arctan x$, then what restriction is placed on the value of y?

PROBLEM SET 4.6

1. Graph $y = \cos x$ between -2π and 2π, and then reflect the graph about the line $y = x$ to obtain the graph of $x = \cos y$.

2. Graph $y = \sin x$ between $-\pi/2$ and $\pi/2$, and then reflect the graph about the line $y = x$ to obtain the graph of $y = \sin^{-1} x$ between $-\pi/2$ and $\pi/2$.

3. Graph $y = \tan x$ for x between $-3\pi/2$ and $3\pi/2$, and then reflect the graph about the line $y = x$ to obtain the graph of $x = \tan y$.

4. Graph $y = \cot x$ for x between 0 and 2π, and then reflect the graph about the line $y = x$ to obtain the graph of $x = \cot y$.

Evaluate each expression without using a calculator, and write your answers in radians.

5. $\sin^{-1}\left(\dfrac{\sqrt{3}}{2}\right)$ **6.** $\cos^{-1}\left(\dfrac{1}{2}\right)$

7. $\cos^{-1}(-1)$ **8.** $\cos^{-1}(0)$

9. $\tan^{-1}(1)$ **10.** $\tan^{-1}(0)$

11. $\arccos\left(-\dfrac{1}{\sqrt{2}}\right)$ **12.** $\arccos(1)$

13. $\sin^{-1}\left(-\dfrac{1}{2}\right)$ **14.** $\sin^{-1}\left(\dfrac{1}{\sqrt{2}}\right)$

15. $\arctan(\sqrt{3})$ **16.** $\arctan\left(\dfrac{1}{\sqrt{3}}\right)$

17. $\arcsin(0)$ **18.** $\arcsin\left(-\dfrac{\sqrt{3}}{2}\right)$

19. $\tan^{-1}\left(-\dfrac{1}{\sqrt{3}}\right)$ **20.** $\tan^{-1}(-\sqrt{3})$

21. $\cos^{-1}\left(-\dfrac{1}{2}\right)$ **22.** $\sin^{-1}(1)$

23. $\arccos\left(\dfrac{\sqrt{3}}{2}\right)$ **24.** $\arcsin(-1)$

Use a calculator to evaluate each expression to the nearest tenth of a degree.

25. $\sin^{-1}(0.1702)$ **26.** $\sin^{-1}(-0.1702)$
27. $\cos^{-1}(-0.8425)$ **28.** $\cos^{-1}(0.8425)$
29. $\tan^{-1}(0.3799)$ **30.** $\tan^{-1}(-0.3799)$
31. $\arcsin(0.9627)$ **32.** $\arccos(0.9627)$
33. $\cos^{-1}(-0.4664)$ **34.** $\sin^{-1}(-0.4664)$
35. $\arctan(-2.748)$ **36.** $\arctan(-0.3640)$
37. $\sin^{-1}(-0.7660)$ **38.** $\cos^{-1}(-0.7660)$

39. Use your graphing calculator to graph $y = \sin^{-1} x$ in degree mode. Use the graph with the appropriate command to evaluate each expression.

   a. $\sin^{-1}\left(-\dfrac{1}{2}\right)$ **b.** $\sin^{-1}\left(\dfrac{\sqrt{3}}{2}\right)$ **c.** $\arcsin\left(-\dfrac{1}{\sqrt{2}}\right)$

40. Use your graphing calculator to graph $y = \cos^{-1} x$ in degree mode. Use the graph with the appropriate command to evaluate each expression.

   a. $\cos^{-1}\left(\dfrac{1}{2}\right)$ **b.** $\cos^{-1}\left(-\dfrac{\sqrt{3}}{2}\right)$ **c.** $\arccos\left(\dfrac{1}{\sqrt{2}}\right)$

41. Use your graphing calculator to graph $y = \tan^{-1} x$ in degree mode. Use the graph with the appropriate command to evaluate each expression.

   a. $\tan^{-1}(-1)$ **b.** $\tan^{-1}(\sqrt{3})$ **c.** $\arctan\left(-\dfrac{1}{\sqrt{3}}\right)$

42. Simplify $4\left|\cos\theta\right|$ if $\theta = \sin^{-1}\dfrac{x}{4}$ for some real number x.

43. Simplify $2\left|\sin\theta\right|$ if $\theta = \cos^{-1}\dfrac{x}{2}$ for some real number x.

44. Simplify $5 \left| \sec \theta \right|$ if $\theta = \tan^{-1} \dfrac{x}{5}$ for some real number x.

Evaluate without using a calculator.

45. $\sin \left(\sin^{-1} \dfrac{3}{5} \right)$ **46.** $\cos \left(\cos^{-1} \dfrac{3}{5} \right)$

47. $\cos \left(\cos^{-1} \dfrac{1}{2} \right)$ **48.** $\sin \left(\sin^{-1} \dfrac{1}{\sqrt{2}} \right)$

49. $\tan \left(\tan^{-1} \dfrac{1}{2} \right)$ **50.** $\tan \left(\tan^{-1} \dfrac{3}{4} \right)$

51. $\sin^{-1} (\sin 225°)$ **52.** $\sin^{-1} (\sin 330°)$

53. $\sin^{-1} \left(\sin \dfrac{\pi}{3} \right)$ **54.** $\sin^{-1} \left(\sin \dfrac{\pi}{4} \right)$

55. $\cos^{-1} (\cos 120°)$ **56.** $\cos^{-1} (\cos\ 45°)$

57. $\cos^{-1} \left(\cos \dfrac{7\pi}{4} \right)$ **58.** $\cos^{-1} \left(\cos \dfrac{7\pi}{6} \right)$

59. $\tan^{-1} (\tan 45°)$ **60.** $\tan^{-1} (\tan 60°)$

61. $\tan^{-1} \left(\tan \dfrac{5\pi}{6} \right)$ **62.** $\tan^{-1} \left(\tan \dfrac{2\pi}{3} \right)$

Evaluate without using a calculator.

63. $\cos \left(\tan^{-1} \dfrac{3}{4} \right)$ **64.** $\csc \left(\tan^{-1} \dfrac{3}{4} \right)$

65. $\tan \left(\sin^{-1} \dfrac{3}{5} \right)$ **66.** $\tan \left(\cos^{-1} \dfrac{3}{5} \right)$

67. $\sec \left(\cos^{-1} \dfrac{1}{\sqrt{5}} \right)$ **68.** $\sin \left(\cos^{-1} \dfrac{1}{\sqrt{5}} \right)$

69. $\sin \left(\cos^{-1} \dfrac{1}{2} \right)$ **70.** $\cos \left(\sin^{-1} \dfrac{1}{2} \right)$

71. $\cot \left(\tan^{-1} \dfrac{1}{2} \right)$ **72.** $\cot \left(\tan^{-1} \dfrac{1}{3} \right)$

73. Simplify $\sin^{-1} (\sin x)$ if $-\pi/2 \le x \le \pi/2$.
74. Simplify $\cos^{-1} (\cos x)$ if $0 \le x \le \pi$.

For each expression below, write an equivalent expression that involves x only. (For Problems 81 through 84, assume x is positive.)

75. $\cos (\cos^{-1} x)$ **76.** $\sin (\sin^{-1} x)$
77. $\cos (\sin^{-1} x)$ **78.** $\tan (\cos^{-1} x)$
79. $\sin (\tan^{-1} x)$ **80.** $\cos (\tan^{-1} x)$

81. $\sin \left(\cos^{-1} \dfrac{1}{x} \right)$ **82.** $\cos \left(\sin^{-1} \dfrac{1}{x} \right)$

83. $\sec \left(\cos^{-1} \dfrac{1}{x} \right)$ **84.** $\csc \left(\sin^{-1} \dfrac{1}{x} \right)$

REVIEW PROBLEMS

The problems that follow review material we covered in Sections 4.2 and 4.3.

Graph each of the following equations over the indicated interval. Be sure to label the x- and y-axes so that the amplitude and period are easy to see.

85. $y = 4 \sin 2x$, one complete cycle
86. $y = 2 \sin 4x$, one complete cycle
87. $y = 2 \sin \pi x$, $-4 \le x \le 4$
88. $y = 3 \cos \pi x$, $-2 \le x \le 4$
89. $y = -3 \cos \dfrac{1}{2}x$, $-2\pi \le x \le 6\pi$
90. $y = -3 \sin 2x$, $-2\pi \le x \le 2\pi$

Graph one complete cycle of each of the following equations. Be sure to label the x- and y-axes so that the amplitude, period, and phase shift for each graph are easy to see.

91. $y = \sin\left(x - \dfrac{\pi}{4}\right)$ **92.** $y = \sin\left(x + \dfrac{\pi}{6}\right)$

93. $y = \cos\left(2x + \dfrac{\pi}{2}\right)$ **94.** $y = \sin(2x + \pi)$

95. $y = 3 \sin\left(2x - \dfrac{\pi}{3}\right)$ **96.** $y = 3 \cos\left(2x - \dfrac{\pi}{3}\right)$

CHAPTER 4 SUMMARY

EXAMPLES

1. Since

$$\sin(x + 2\pi) = \sin x,$$

the function $y = \sin x$ is periodic with period 2π. Likewise, since

$$\tan(x + \pi) = \tan x,$$

the function $y = \tan x$ is periodic with period π.

Periodic Functions [4.1]

A function $y = f(x)$ is said to be periodic with period p if p is the smallest positive number such that $f(x + p) = f(x)$ for all x in the domain of f.

Basic Graphs [4.1]

The graphs of $y = \sin x$ and $y = \cos x$ are both periodic with period 2π. The amplitude of each graph is 1. The sine curve passes through 0 on the y-axis, while the cosine curve passes through 1 on the y-axis.

The graphs of $y = \csc x$ and $y = \sec x$ are also periodic with period 2π. We graph them by using the fact that they are reciprocals of sine and cosine. Since there is no largest or smallest value of y, we say the secant and cosecant curves have no amplitude.

The graphs of $y = \tan x$ and $y = \cot x$ are periodic with period π. The tangent curve passes through the origin, while the cotangent is undefined when x is 0. There is no amplitude for either graph.

2.

a.

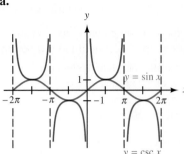

$y = \csc x$

b.

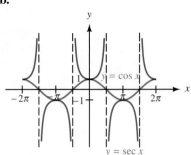

$y = \sec x$

c.

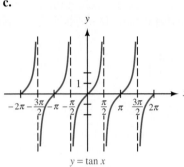

$y = \tan x$

d.

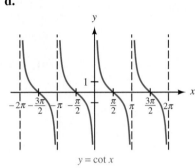

$y = \cot x$

3.

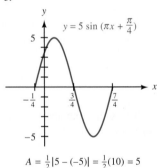

$A = \frac{1}{2}|5 - (-5)| = \frac{1}{2}(10) = 5$

Amplitude [4.1]

The *amplitude* A of a curve is half the absolute value of the difference between the largest value of y, denoted by M, and the smallest value of y, denoted by m.

$$A = \frac{1}{2}|M - m|$$

4. The phase shift for the graph in Example 3 is $-\frac{1}{4}$.

Phase Shift [4.3]

The *phase shift* for a sine or cosine curve is the distance the curve has moved right or left from the curve $y = \sin x$ or $y = \cos x$. For example, we usually think of the graph of $y = \sin x$ as starting at the origin. If we graph another sine curve that starts at $\pi/4$, then we say this curve has a phase shift of $\pi/4$.

5.

a.

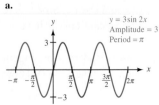

Graphing Sine and Cosine Curves [4.2, 4.3]

The graphs of $y = A \sin(Bx + C)$ and $y = A \cos(Bx + C)$, where $B > 0$, will have the following characteristics:

$$\text{Amplitude} = |A| \qquad \text{Period} = \frac{2\pi}{B} \qquad \text{Phase shift} = -\frac{C}{B}$$

b.

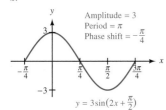

To graph one of these curves, we first find the phase shift and label that point on the x-axis (this will be our starting point). We then add the period to the phase shift and mark the result on the x-axis (this is our ending point). We mark the y-axis with the amplitude. Finally, we sketch in one complete cycle of the curve in question, keeping in mind that, if A is negative, the graph must be reflected about the x-axis.

6.

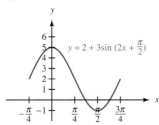

Vertical Translations [4.2]

Adding a constant k to a trigonometric function translates the graph vertically up or down. For example, the graph of $y = k + A \sin (Bx + C)$ will have the same shape (amplitude, period, phase shift, and reflection, if indicated) as $y = A \sin (Bx + C)$ but will be translated k units vertically from the graph of $y = A \sin (Bx + C)$.

7.

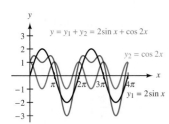

Graphing by Addition of y-Coordinates [4.5]

To graph equations of the form $y = y_1 + y_2$, where y_1 and y_2 are algebraic or trigonometric functions of x, we graph y_1 and y_2 separately on the same coordinate system and then add the two graphs to obtain the graph of y.

8. Evaluate in radians without using a calculator.

a. $\sin^{-1} \dfrac{1}{2}$

The angle between $-\pi/2$ and $\pi/2$ whose sine is $\frac{1}{2}$ is $\pi/6$.

$$\sin^{-1} \frac{1}{2} = \frac{\pi}{6}$$

b. $\arccos \left(-\dfrac{\sqrt{3}}{2} \right)$

The angle between 0 and π with a cosine of $-\sqrt{3}/2$ is $5\pi/6$.

$$\arccos \left(-\frac{\sqrt{3}}{2} \right) = \frac{5\pi}{6}$$

Inverse Trigonometric Functions [4.6]

Inverse Function	Meaning
$y = \sin^{-1} x$ or $y = \arcsin x$	$x = \sin y$ and $-\dfrac{\pi}{2} \leq y \leq \dfrac{\pi}{2}$
In words: y is the angle between $-\pi/2$ and $\pi/2$, inclusive, whose sine is x.	
$y = \cos^{-1} x$ or $y = \arccos x$	$x = \cos y$ and $0 \leq y \leq \pi$
In words: y is the angle between 0 and π, inclusive, whose cosine is x.	
$y = \tan^{-1} x$ or $y = \arctan x$	$x = \tan y$ and $-\dfrac{\pi}{2} < y < \dfrac{\pi}{2}$
In words: y is the angle between $-\pi/2$ and $\pi/2$ whose tangent is x.	

CHAPTER 4 TEST

Graph each of the following between $x = -4\pi$ and $x = 4\pi$.

1. $y = \sin x$ **2.** $y = \cos x$

3. $y = \tan x$ **4.** $y = \sec x$

5. How many complete cycles of the graph of the equation $y = \sin x$ are shown in your answer to Problem 1?

6. How many complete cycles of the graph of the equation $y = \tan x$ are shown in your answer to Problem 3?

7. Use your answer to Problem 2 to find all values of x between -4π and 4π for which $\sec x = -1$.

8. Use your answer to Problem 2 to find all values of x between -4π and 4π for which $\sec x$ is undefined.

For each equation below, first identify the amplitude and period and then use this information to sketch one complete cycle of the graph.

9. $y = \cos \pi x$ **10.** $y = -3 \cos x$

Graph each of the following on the given interval.

11. $y = 2 + 3 \sin 2x, \ -\pi \le x \le 2\pi$ **12.** $y = 2 \sin \pi x, \ -4 \le x \le 4$

For each equation below, identify the amplitude, period, and phase shift and then use this information to sketch one complete cycle of the graph.

13. $y = \sin\left(x + \dfrac{\pi}{4}\right)$ **14.** $y = \cos\left(x - \dfrac{\pi}{2}\right)$

15. $y = 3 \sin\left(2x - \dfrac{\pi}{3}\right)$ **16.** $y = -3 + 3 \sin\left(\dfrac{\pi}{3}x - \dfrac{\pi}{3}\right)$

17. $y = \csc\left(x + \dfrac{\pi}{4}\right)$ **18.** $y = \tan\left(2x - \dfrac{\pi}{2}\right)$

Graph each of the following on the given interval.

19. $y = 2 \sin (3x - \pi), \ -\dfrac{\pi}{3} \le x \le \dfrac{5\pi}{3}$

20. $y = 2 \sin\left(\dfrac{\pi}{2}x - \dfrac{\pi}{4}\right), \ -\dfrac{1}{2} \le x \le \dfrac{13}{2}$

Find an equation for each of the following graphs.

21.   **22.**

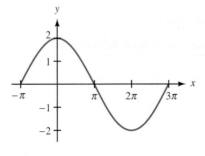

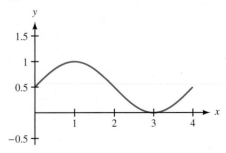

Sketch the following between $x = 0$ and $x = 4\pi$.

23. $y = \dfrac{1}{2}x - \sin x$

24. $y = \sin x + \cos 2x$

Graph each of the following.

25. $y = \cos^{-1} x$

26. $y = \arcsin x$

Evaluate each expression without using a calculator and write your answer in radians.

27. $\sin^{-1}\left(\dfrac{1}{2}\right)$

28. $\cos^{-1}\left(-\dfrac{\sqrt{3}}{2}\right)$

29. $\arctan(-1)$

30. $\arcsin(1)$

Use a calculator to evaluate each expression to the nearest tenth of a degree.

31. $\arcsin(0.5934)$

32. $\arctan(-0.8302)$

33. $\arccos(-0.6981)$

34. $\arcsin(-0.2164)$

Evaluate without using a calculator.

35. $\tan\left(\cos^{-1}\dfrac{2}{3}\right)$

36. $\cos\left(\tan^{-1}\dfrac{2}{3}\right)$

37. $\cos^{-1}(\cos 30°)$

38. $\tan^{-1}\left(\tan\dfrac{7\pi}{6}\right)$

For each expression below, write an equivalent expression that involves x only.

39. $\sin(\cos^{-1} x)$

40. $\tan(\sin^{-1} x)$

CHAPTER 4 GROUP PROJECT

MODELING THE SUNSPOT CYCLE

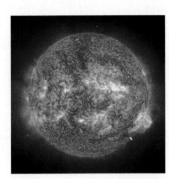

"From my earlier observations, which I have reported every year in this journal, it appears that there is a certain periodicity in the appearance of sunspots and this theory seems more and more probable from the results of this year . . . "

Heinrich Schwabe, 1843

Objective: To find a sinusoidal model for the average annual sunspot number using recent data.

Sunspots have been observed and recorded for thousands of years. Some of the earliest observations were made by Chinese astronomers. Further observations were made following the invention of the telescope in the 17th century by a number of astronomers, including Galileo and William Herschel. However, it was the German astronomer Heinrich Schwabe who, in searching for evidence of other planets between Mercury and the Sun, first observed an apparent periodic cycle in sunspot activity.

The following table provides the average annual sunspot number for each year between 1978 and 1998.

Year	Sunspot Number	Year	Sunspot Number
1978	92.5	1989	157.6
1979	155.4	1990	142.6
1980	154.6	1991	145.7
1981	140.4	1992	94.3
1982	115.9	1993	54.6
1983	66.6	1994	29.9
1984	45.9	1995	17.5
1985	17.9	1996	8.6
1986	13.4	1997	21.5
1987	29.4	1998	64.3
1988	100.2		

1. Sketch a graph of the data, listing the year along the horizontal axis and the sunspot number along the vertical axis.
2. Using the data and your graph from Question 1, find an equation of the form $y = k + A \sin (Bx + C)$ or $y = k + A \cos (Bx + C)$ to match the graph, where x is the year and y is the sunspot number for that year.
3. According to your model, what is the length of the sunspot cycle (in years)? In 1848, Rudolph Wolf determined the length of the sunspot cycle to be 11.1 years. How does your value compare with his?
4. Use your model to predict the sunspot number for 2001. How does your prediction compare with the actual value of 111?
5. Using your graphing calculator, make a scatter plot of the data from the table. Then graph your model from Question 2 along with the data. How well does your model fit the data? What could you do to try to improve your model?
6. If your graphing calculator is capable of computing a least-squares sinusoidal regression model, use it to find a second model for the data. Graph this new equation along with your first model. How do they compare?

CHAPTER 4 RESEARCH PROJECT
THE SUNSPOT CYCLE

Johann Rudolph Wolf

Although many astronomers made regular observations of sunspots, it was the Swiss astronomer Rudolph Wolf that devised the first universal method for counting sunspots. The daily sunspot number, sometimes called the Wolf number, follows a periodic cycle. Wolf calculated the length of this cycle to be about 11.1 years.

Research the astronomer Rudolph Wolf and the sunspot cycle. Why do modern astronomers consider the length of the sunspot cycle to be 22 years? What are some of the effects that the sunspot cycle causes here on Earth? Write a paragraph or two about your findings.

IDENTITIES AND FORMULAS

Mathematics, rightly viewed, possesses not only truth,
but supreme beauty.

Bertrand Russell

INTRODUCTION

Although it doesn't look like it, Figure 1 shows the graphs of two functions, namely

$$y = \cos^2 x \quad \text{and} \quad y = \frac{1 - \sin^4 x}{1 + \sin^2 x}$$

Although these two functions look quite different from one another, they are in fact the same function. This means that, for all values of x,

$$\cos^2 x = \frac{1 - \sin^4 x}{1 + \sin^2 x}$$

This last expression is an *identity,* and identities are one of the topics we will study in this chapter.

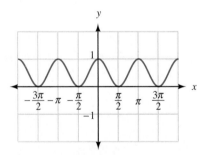

Figure 1

STUDY SKILLS FOR CHAPTER 5

The study skills for this chapter focus on the way you approach new situations in mathematics. The first study skill is a point of view you hold about your natural instincts for what does and doesn't work in mathematics. The second study skill gives you a way of testing your instincts.

1. **Don't Let Your Intuition Fool You**   As you become more experienced and more successful in mathematics you will be able to trust your mathematical intuition. For now though, it can get in the way of your success. For example, if you ask a beginning algebra student to expand $(a + b)^2$, many will write $a^2 + b^2$, which is incorrect. In trigonometry, at first glance it may seem that a statement such as $\sin (A + B) = \sin A + \sin B$ is true. However, it too is false, as you will see in this chapter.

2. **Test Properties You are Unsure Of**   From time to time you will be in a situation where you would like to apply a property or rule, but you are not sure it is true. You can always test a property or statement by substituting numbers for variables. For instance, we always have students that rewrite $(x + 3)^2$ as $x^2 + 9$, thinking that the two expressions are equivalent. The fact that the two expressions are not equivalent becomes obvious when we substitute 10 for x in each one.

When $x = 10$, the expression $(x + 3)^2$ is $(10 + 3)^2 = 13^2 = 169$

When $x = 10$, the expression $x^2 + 9 = 10^2 + 9 = 100 + 9 = 109$

Similarly, there may come a time when you are wondering if $\sin 2A$ is the same as $2 \sin A$. If you try $A = 30°$ in each expression, you will find out quickly that the two expressions are not the same.

When $A = 30°$, the expression $\sin 2A = \sin 2(30°) = \sin 60° = \dfrac{\sqrt{3}}{2}$

When $A = 30°$, the expression $2 \sin A = 2 \sin 30° = 2 \cdot \dfrac{1}{2} = 1$

When you test the equivalence of expressions by substituting numbers for the variable, make it easy on yourself by choosing numbers that are easy to work with, as we did above.

It is not good practice to trust your intuition or instincts in every new situation in mathematics. If you have any doubt about generalizations you are making, test them by replacing variables with numbers and simplifying.

SECTION 5.1 | PROVING IDENTITIES

We began proving identities in Chapter 1. In this section, we will extend the work we did in Chapter 1 to include proving more complicated identities. For review, Table 1 lists the basic identities and some of their more important equivalent forms.

TABLE 1

	Basic Identities	Common Equivalent Forms
Reciprocal	$\csc \theta = \dfrac{1}{\sin \theta}$	$\sin \theta = \dfrac{1}{\csc \theta}$
	$\sec \theta = \dfrac{1}{\cos \theta}$	$\cos \theta = \dfrac{1}{\sec \theta}$
	$\cot \theta = \dfrac{1}{\tan \theta}$	$\tan \theta = \dfrac{1}{\cot \theta}$
Ratio	$\tan \theta = \dfrac{\sin \theta}{\cos \theta}$	
	$\cot \theta = \dfrac{\cos \theta}{\sin \theta}$	
Pythagorean	$\cos^2 \theta + \sin^2 \theta = 1$	$\sin^2 \theta = 1 - \cos^2 \theta$
		$\sin \theta = \pm\sqrt{1 - \cos^2 \theta}$
		$\cos^2 \theta = 1 - \sin^2 \theta$
		$\cos \theta = \pm\sqrt{1 - \sin^2 \theta}$
	$1 + \tan^2 \theta = \sec^2 \theta$	
	$1 + \cot^2 \theta = \csc^2 \theta$	

NOTE The last two Pythagorean identities can be derived from $\cos^2 \theta + \sin^2 \theta = 1$ by dividing each side by $\cos^2 \theta$ and $\sin^2 \theta$, respectively. For example, if we divide each side of $\cos^2 \theta + \sin^2 \theta = 1$ by $\cos^2 \theta$, we have

$$\cos^2 \theta + \sin^2 \theta = 1$$

$$\frac{\cos^2 \theta + \sin^2 \theta}{\cos^2 \theta} = \frac{1}{\cos^2 \theta}$$

$$\frac{\cos^2 \theta}{\cos^2 \theta} + \frac{\sin^2 \theta}{\cos^2 \theta} = \frac{1}{\cos^2 \theta}$$

$$1 + \tan^2 \theta = \sec^2 \theta$$

To derive the last Pythagorean identity, we would need to divide both sides of $\cos^2 \theta + \sin^2 \theta = 1$ by $\sin^2 \theta$ to obtain $1 + \cot^2 \theta = \csc^2 \theta$.

The rest of this section is concerned with using the basic identities (or their equivalent forms) listed above, along with our knowledge of algebra, to prove other identities.

Recall that an identity in trigonometry is a statement that two expressions are equal for all replacements of the variable for which each expression is defined. To prove (or verify) a trigonometric identity, we use trigonometric substitutions and algebraic manipulations to either

1. Transform the right side of the identity into the left side, or
2. Transform the left side of the identity into the right side.

The main thing to remember in proving identities is to work on each side of the identity separately. We do not want to use properties from algebra that involve both sides of the identity—like the addition property of equality. We prove identities in order to develop the ability to transform one trigonometric expression into another.

When we encounter problems in other courses that require the use of the techniques used to verify identities, we usually find that the solution to these problems hinges on transforming an expression containing trigonometric functions into less complicated expressions. In these cases, we do not usually have an equal sign to work with.

EXAMPLE 1 Prove $\sin \theta \cot \theta = \cos \theta$.

PROOF To prove this identity, we transform the left side into the right side.

$$\sin \theta \cot \theta = \sin \theta \cdot \frac{\cos \theta}{\sin \theta} \qquad \text{Ratio identity}$$

$$= \frac{\sin \theta \cos \theta}{\sin \theta} \qquad \text{Multiply}$$

$$= \cos \theta \qquad \text{Divide out common factor } \sin \theta$$

In this example, we have transformed the left side into the right side. Remember, we verify identities by transforming one expression into another. ■

EXAMPLE 2 Prove $\tan x + \cos x = \sin x (\sec x + \cot x)$.

PROOF We can begin by applying the distributive property to the right side to multiply through by $\sin x$. Then we can change each expression on the right side to an equivalent expression involving only $\sin x$ and $\cos x$.

$$\sin x (\sec x + \cot x) = \sin x \sec x + \sin x \cot x \qquad \text{Multiply}$$

$$= \sin x \cdot \frac{1}{\cos x} + \sin x \cdot \frac{\cos x}{\sin x} \qquad \begin{array}{l}\text{Reciprocal and ratio} \\ \text{identities}\end{array}$$

$$= \frac{\sin x}{\cos x} + \cos x \qquad \text{Multiply}$$

$$= \tan x + \cos x \qquad \text{Ratio identity}$$

In this case, we transformed the right side into the left side. ■

Before we go on to the next example, let's list some guidelines that may be useful in learning how to prove identities.

GUIDELINES FOR PROVING IDENTITIES

1. It is usually best to work on the more complicated side first.

2. Look for trigonometric substitutions involving the basic identities that may help simplify things.

3. Look for algebraic operations, such as adding fractions, the distributive property, or factoring, that may simplify the side you are working with or that will at least lead to an expression that will be easier to simplify.

4. If you cannot think of anything else to do, change everything to sines and cosines and see if that helps.

5. Always keep an eye on the side you are not working with to be sure you are working toward it. There is a certain sense of direction that accompanies a successful proof.

Probably the best advice is to remember that these are simply guidelines. The best way to become proficient at proving trigonometric identities is to practice. The more identities you prove, the more you will be able to prove and the more confident you will become. *Don't be afraid to stop and start over if you don't seem to be getting anywhere.* With most identities, there are a number of different proofs that will lead to the same result. Some of the proofs will be longer than others.

EXAMPLE 3 Prove $\dfrac{\cos^4 t - \sin^4 t}{\cos^2 t} = 1 - \tan^2 t.$

PROOF In this example, factoring the numerator on the left side will reduce the exponents there from 4 to 2.

$$\frac{\cos^4 t - \sin^4 t}{\cos^2 t} = \frac{(\cos^2 t + \sin^2 t)(\cos^2 t - \sin^2 t)}{\cos^2 t} \qquad \text{Factor}$$

$$= \frac{1\,(\cos^2 t - \sin^2 t)}{\cos^2 t} \qquad \text{Pythagorean identity}$$

$$= \frac{\cos^2 t}{\cos^2 t} - \frac{\sin^2 t}{\cos^2 t} \qquad \text{Separate into two fractions}$$

$$= 1 - \tan^2 t \qquad \text{Ratio identity} \quad ■$$

EXAMPLE 4 Prove $1 + \cos \theta = \dfrac{\sin^2 \theta}{1 - \cos \theta}.$

PROOF We begin this proof by applying an alternate form of the Pythagorean identity to the right side to write $\sin^2 \theta$ as $1 - \cos^2 \theta$. Then we factor $1 - \cos^2 \theta$ as the difference of two squares and reduce to lowest terms.

$$\frac{\sin^2 \theta}{1 - \cos \theta} = \frac{1 - \cos^2 \theta}{1 - \cos \theta} \qquad \text{Pythagorean identity}$$

$$= \frac{(1 - \cos \theta)(1 + \cos \theta)}{1 - \cos \theta} \qquad \text{Factor}$$

$$= 1 + \cos \theta \qquad \text{Reduce} \quad ■$$

USING TECHNOLOGY

VERIFYING IDENTITIES

You can use your graphing calculator to decide if an equation is an identity or not. If the two expressions are indeed equal for all defined values of the variable, then they should produce identical graphs. Although this does not constitute a proof, it does give strong evidence that the identity is true.

We can verify the identity in Example 4 by defining the expression on the left as a function Y1 and the expression on the right as a second function Y2. If your calculator is equipped with different graphing styles, set the style of Y2 so that you will be able to distinguish the second graph from the first. (In Figure 1, we have used the *path* style on a TI-83 for the second function.) Also, be sure your calculator is set to radian mode. Set your window variables so that

$$-2\pi \leq x \leq 2\pi, \text{ scale } = \pi/2; \; -4 \leq y \leq 4, \text{ scale } = 1$$

```
Plot1  Plot2  Plot3
\Y1■1+cos(X)
-₀Y2▤(sin(X))²/(1−cos(X))
\Y3=
\Y4=
\Y5=
\Y6=
\Y7=
```

Figure 1

When you graph the functions, your calculator screen should look similar to Figure 2 (the small circle is a result of the path style in action). Observe that the two graphs are identical.

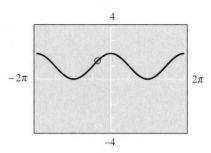

Figure 2

If your calculator is not equipped with different graphing styles, it may be difficult to tell if the second graph really coincides with the first. In this case you can trace the graph, and switch between the two functions at several points to convince yourself that the two graphs are indeed the same.

 EXAMPLE 5 Prove $\tan x + \cot x = \sec x \csc x$.

PROOF We begin this proof by writing the left side in terms of $\sin x$ and $\cos x$. Then we simplify the left side by finding a common denominator in order to add the resulting fractions.

$$\tan x + \cot x = \frac{\sin x}{\cos x} + \frac{\cos x}{\sin x} \qquad \text{Changes to sines and cosines}$$

$$= \frac{\sin x}{\cos x} \cdot \frac{\sin x}{\sin x} + \frac{\cos x}{\sin x} \cdot \frac{\cos x}{\cos x} \qquad \text{LCD}$$

$$= \frac{\sin^2 x + \cos^2 x}{\cos x \sin x} \qquad \text{Add fractions}$$

$$= \frac{1}{\cos x \sin x} \qquad \text{Pythagorean identity}$$

$$= \frac{1}{\cos x} \cdot \frac{1}{\sin x} \qquad \text{Write as separate fractions}$$

$$= \sec x \csc x \qquad \text{Reciprocal identities} \quad \blacksquare$$

 **EXAMPLE 6** Prove $\dfrac{\sin \alpha}{1 + \cos \alpha} + \dfrac{1 + \cos \alpha}{\sin \alpha} = 2 \csc \alpha$.

PROOF The common denominator for the left side of the equation is $\sin \alpha \, (1 + \cos \alpha)$. We multiply the first fraction by $(\sin \alpha)/(\sin \alpha)$ and the second fraction by $(1 + \cos \alpha)/(1 + \cos \alpha)$ to produce two equivalent fractions with the same denominator.

$$\frac{\sin \alpha}{1 + \cos \alpha} + \frac{1 + \cos \alpha}{\sin \alpha}$$

$$= \frac{\mathbf{\sin \alpha}}{\mathbf{\sin \alpha}} \cdot \frac{\sin \alpha}{1 + \cos \alpha} + \frac{1 + \cos \alpha}{\sin \alpha} \cdot \frac{\mathbf{1 + \cos \alpha}}{\mathbf{1 + \cos \alpha}} \qquad \text{LCD}$$

$$= \frac{\sin^2 \alpha + (1 + \cos \alpha)^2}{\sin \alpha \,(1 + \cos \alpha)} \qquad \text{Add numerators}$$

$$= \frac{\sin^2 \alpha + 1 + 2 \cos \alpha + \cos^2 \alpha}{\sin \alpha \,(1 + \cos \alpha)} \qquad \text{Expand } (1 + \cos \alpha)^2$$

$$= \frac{2 + 2 \cos \alpha}{\sin \alpha \,(1 + \cos \alpha)} \qquad \text{Pythagorean identity}$$

$$= \frac{2(1 + \cos \alpha)}{\sin \alpha \,(1 + \cos \alpha)} \qquad \text{Factor out a 2}$$

$$= \frac{2}{\sin \alpha} \qquad \text{Reduce}$$

$$= 2 \csc \alpha \qquad \text{Reciprocal identity} \qquad ■$$

EXAMPLE 7 Prove $\dfrac{1 + \sin t}{\cos t} = \dfrac{\cos t}{1 - \sin t}$.

PROOF   The trick to proving this identity requires that we multiply the numerator and denominator on the right side by $1 + \sin t$. (This is similar to rationalizing the denominator.)

$$\frac{\cos t}{1 - \sin t} = \frac{\cos t}{1 - \sin t} \cdot \frac{\mathbf{1 + \sin t}}{\mathbf{1 + \sin t}} \qquad \begin{array}{l}\text{Multiply numerator and} \\ \text{denominator by } 1 + \sin t\end{array}$$

$$= \frac{\cos t \,(1 + \sin t)}{1 - \sin^2 t} \qquad \text{Multiply out the denominator}$$

$$= \frac{\cos t \,(1 + \sin t)}{\cos^2 t} \qquad \text{Pythagorean identity}$$

$$= \frac{1 + \sin t}{\cos t} \qquad \text{Reduce}$$

Note that it would have been just as easy for us to verify this identity by multiplying the numerator and denominator on the left side by $1 - \sin t$.   ■

GETTING READY FOR CLASS

After reading through the preceding section, respond in your own words and in complete sentences.

a. What is an identity?

b. In trigonometry, how do we prove an identity?

c. What is a first step in simplifying the expression $\dfrac{\cos^4 t - \sin^4 t}{\cos^2 t}$?

d. What is a first step in simplifying the expression $\dfrac{\sin \alpha}{1 + \cos \alpha} + \dfrac{1 + \cos \alpha}{\sin \alpha}$?

PROBLEM SET 5.1

Prove that each of the following identities is true:

1. $\cos \theta \tan \theta = \sin \theta$

2. $\sec \theta \cot \theta = \csc \theta$

3. $\csc \theta \tan \theta = \sec \theta$

4. $\tan \theta \cot \theta = 1$

5. $\dfrac{\tan A}{\sec A} = \sin A$

6. $\dfrac{\cot A}{\csc A} = \cos A$

7. $\sec \theta \cot \theta \sin \theta = 1$

8. $\tan \theta \csc \theta \cos \theta = 1$

9. $\cos x \, (\csc x + \tan x) = \cot x + \sin x$

10. $\sin x \, (\sec x + \csc x) = \tan x + 1$

11. $\cot x - 1 = \cos x \, (\csc x - \sec x)$

12. $\tan x \, (\cos x + \cot x) = \sin x + 1$

13. $\cos^2 x \, (1 + \tan^2 x) = 1$

14. $\sin^2 x \, (\cot^2 x + 1) = 1$

15. $(1 - \sin x)(1 + \sin x) = \cos^2 x$

16. $(1 - \cos x)(1 + \cos x) = \sin^2 x$

17. $\dfrac{\cos^4 t - \sin^4 t}{\sin^2 t} = \cot^2 t - 1$

18. $\dfrac{\sin^4 t - \cos^4 t}{\sin^2 t \cos^2 t} = \sec^2 t - \csc^2 t$

19. $1 + \sin \theta = \dfrac{\cos^2 \theta}{1 - \sin \theta}$

20. $1 - \sin \theta = \dfrac{\cos^2 \theta}{1 + \sin \theta}$

21. $\dfrac{1 - \sin^4 \theta}{1 + \sin^2 \theta} = \cos^2 \theta$

22. $\dfrac{1 - \cos^4 \theta}{1 + \cos^2 \theta} = \sin^2 \theta$

23. $\sec^2 \theta - \tan^2 \theta = 1$

24. $\csc^2 \theta - \cot^2 \theta = 1$

25. $\sec^4 \theta - \tan^4 \theta = \dfrac{1 + \sin^2 \theta}{\cos^2 \theta}$

26. $\csc^4 \theta - \cot^4 \theta = \dfrac{1 + \cos^2 \theta}{\sin^2 \theta}$

27. $\tan \theta - \cot \theta = \dfrac{\sin^2 \theta - \cos^2 \theta}{\sin \theta \cos \theta}$

28. $\sec \theta - \csc \theta = \dfrac{\sin \theta - \cos \theta}{\sin \theta \cos \theta}$

29. $\csc B - \sin B = \cot B \cos B$

30. $\sec B - \cos B = \tan B \sin B$

31. $\cot \theta \cos \theta + \sin \theta = \csc \theta$

32. $\tan \theta \sin \theta + \cos \theta = \sec \theta$

33. $\dfrac{\cos x}{1 + \sin x} + \dfrac{1 + \sin x}{\cos x} = 2 \sec x$

34. $\dfrac{\cos x}{1 + \sin x} - \dfrac{1 - \sin x}{\cos x} = 0$

35. $\dfrac{1}{1 + \cos x} + \dfrac{1}{1 - \cos x} = 2 \csc^2 x$

36. $\dfrac{1}{1 - \sin x} + \dfrac{1}{1 + \sin x} = 2 \sec^2 x$

37. $\dfrac{1 - \sec x}{1 + \sec x} = \dfrac{\cos x - 1}{\cos x + 1}$

38. $\dfrac{\csc x - 1}{\csc x + 1} = \dfrac{1 - \sin x}{1 + \sin x}$

39. $\dfrac{\cos t}{1 + \sin t} = \dfrac{1 - \sin t}{\cos t}$

40. $\dfrac{\sin t}{1 + \cos t} = \dfrac{1 - \cos t}{\sin t}$

41. $\dfrac{(1 - \sin t)^2}{\cos^2 t} = \dfrac{1 - \sin t}{1 + \sin t}$

42. $\dfrac{\sin^2 t}{(1 - \cos t)^2} = \dfrac{1 + \cos t}{1 - \cos t}$

43. $\dfrac{\sec \theta + 1}{\tan \theta} = \dfrac{\tan \theta}{\sec \theta - 1}$

44. $\dfrac{\csc \theta - 1}{\cot \theta} = \dfrac{\cot \theta}{\csc \theta + 1}$

45. $\dfrac{1 - \sin x}{1 + \sin x} = (\sec x - \tan x)^2$

46. $\dfrac{1 + \cos x}{1 - \cos x} = (\csc x + \cot x)^2$

47. $\sec x + \tan x = \dfrac{1}{\sec x - \tan x}$

48. $\dfrac{1}{\csc x - \cot x} = \csc x + \cot x$

49. $\dfrac{\sin x + 1}{\cos x + \cot x} = \tan x$

50. $\dfrac{\cos x + 1}{\cot x} = \sin x + \tan x$

51. $\sin^4 A - \cos^4 A = 1 - 2 \cos^2 A$

52. $\cos^4 A - \sin^4 A = 1 - 2 \sin^2 A$

53. $\dfrac{\sin^2 B - \tan^2 B}{1 - \sec^2 B} = \sin^2 B$

54. $\dfrac{\cot^2 B - \cos^2 B}{\csc^2 B - 1} = \cos^2 B$

55. $\dfrac{\sec^4 y - \tan^4 y}{\sec^2 y + \tan^2 y} = 1$

56. $\dfrac{\csc^2 y + \cot^2 y}{\csc^4 y - \cot^4 y} = 1$

57. $\dfrac{\sin^3 A - 8}{\sin A - 2} = \sin^2 A + 2 \sin A + 4$

58. $\dfrac{1 - \cos^3 A}{1 - \cos A} = \cos^2 A + \cos A + 1$

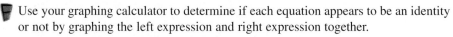 **59.** $\dfrac{1 - \tan^3 t}{1 - \tan t} = \sec^2 t + \tan t$

60. $\dfrac{1 + \cot^3 t}{1 + \cot t} = \csc^2 t - \cot t$

61. $\dfrac{\tan x}{\sin x - \cos x} = \dfrac{\sin^2 x + \sin x \cos x}{\cos x - 2 \cos^3 x}$

62. $\dfrac{\cot^2 x}{\sin x + \cos x} = \dfrac{\cos^2 x \sin x - \cos^3 x}{2 \sin^4 x - \sin^2 x}$

The following identities are from the book *Plane and Spherical Trigonometry with Tables* by Rosenbach, Whitman, and Moskovitz, and published by Ginn and Company in 1937. Verify each identity.

63. $(\tan \theta + \cot \theta)^2 = \sec^2 \theta + \csc^2 \theta$

64. $\dfrac{\tan^2 \psi + 2}{1 + \tan^2 \psi} = 1 + \cos^2 \psi$

65. $\dfrac{1 + \sin \phi}{1 - \sin \phi} - \dfrac{1 - \sin \phi}{1 + \sin \phi} = 4 \tan \phi \sec \phi$

66. $\dfrac{\cos \beta}{1 - \tan \beta} + \dfrac{\sin \beta}{1 - \cot \beta} = \sin \beta + \cos \beta$

Use your graphing calculator to determine if each equation appears to be an identity or not by graphing the left expression and right expression together.

67. $(\sec B - 1)(\sec B + 1) = \tan^2 B$

68. $\dfrac{1 - \sec \theta}{\cos \theta} = \dfrac{\cos \theta}{1 + \sec \theta}$

69. $\sec x + \cos x = \tan x \sin x$

70. $\dfrac{\tan t}{\sec t + 1} = \dfrac{\sec t - 1}{\tan t}$

71. $\sec A - \csc A = \dfrac{\cos A - \sin A}{\cos A \sin A}$

72. $\cos^4 \theta - \sin^4 \theta = 2 \cos^2 \theta - 1$

73. $\dfrac{1}{1 - \sin x} + \dfrac{1}{1 + \sin x} = 2 \sec^2 x$

74. $\cot^4 t - \tan^4 t = \dfrac{\sin^2 t + 1}{\cos^2 t}$

Show that each of the following statements is not an identity by finding a value of θ that makes the statement false.

75. $\sin \theta = \sqrt{1 - \cos^2 \theta}$

76. $\sin \theta + \cos \theta = 1$

77. $\sin \theta = \dfrac{1}{\cos \theta}$

78. $\tan^2 \theta + \cot^2 \theta = 1$

79. $\sqrt{\sin^2 \theta + \cos^2 \theta} = \sin \theta + \cos \theta$

80. $\sin \theta \cos \theta = 1$

81. Show that sin $(A + B)$ is not, in general, equal to sin A + sin B by substituting $30°$ for A and $60°$ for B in both expressions and simplifying. (An example like this, which shows that a statement is not true for certain values of the variables, is called a *counterexample* for the statement.)

82. Show that sin $2x \neq 2$ sin x by substituting $30°$ for x and then simplifying both sides.

REVIEW PROBLEMS

The problems that follow review material we covered in Sections 1.4 and 3.2. Reviewing these problems will help you with some of the material in the next section.

83. If sin $A = 3/5$ and A terminates in quadrant I, find cos A and tan A.

84. If cos $B = -5/13$ with B in quadrant III, find sin B and tan B.

Give the exact value of each of the following:

85. $\sin \dfrac{\pi}{3}$ **86.** $\cos \dfrac{\pi}{3}$ **87.** $\cos \dfrac{\pi}{6}$ **88.** $\sin \dfrac{\pi}{6}$

Convert to degrees.

89. $\pi/12$ **90.** $5\pi/12$ **91.** $7\pi/12$ **92.** $11\pi/12$

SECTION 5.2 | SUM AND DIFFERENCE FORMULAS

The expressions sin $(A + B)$ and cos $(A + B)$ occur frequently enough in mathematics that it is necessary to find expressions equivalent to them that involve sines and cosines of single angles. The most obvious question to begin with is

Is sin $(A + B)$ = sin A + sin B?

The answer is no. Substituting almost any pair of numbers for A and B in the formula will yield a false statement. As a counterexample, we can let $A = 30°$ and $B = 60°$ in the above formula and then simplify each side. (A counterexample is an example that shows that a statement is not, in general, true.)

$$\sin (30° + 60°) \overset{?}{=} \sin 30° + \sin 60°$$

$$\sin 90° \overset{?}{=} \frac{1}{2} + \frac{\sqrt{3}}{2}$$

$$1 \neq \frac{1 + \sqrt{3}}{2}$$

The formula just doesn't work. The next question is, what are the formulas for sin $(A + B)$ and cos $(A + B)$? The answer to that question is what this section is all about. Let's start by deriving the formula for cos $(A + B)$.

We begin by drawing angle A in standard position and then adding B and $-B$ to it. Figure 1 shows these angles in relation to the unit circle. Note that the points on the unit circle through which the terminal sides of the angles A, $A + B$, and $-B$ pass have been labeled with the sines and cosines of those angles.

To derive the formula for cos $(A + B)$, we simply have to see that line segment P_1P_3 is equal to line segment P_2P_4. (From geometry, they are chords cut off by equal central angles.)

$$P_1P_3 = P_2P_4$$

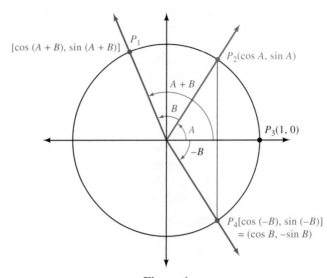

Figure 1

Squaring both sides gives us

$$(P_1 P_3)^2 = (P_2 P_4)^2$$

Now, applying the distance formula, we have

$$[\cos (A + B) - 1]^2 + [\sin (A + B) - 0]^2 = (\cos A - \cos B)^2 + (\sin A + \sin B)^2$$

Let's call this Equation 1. Taking the left side of Equation 1, expanding it, and then simplifying by using the first Pythagorean identity gives us

Left Side of Equation 1

$$\cos^2 (A + B) - 2 \cos (A + B) + 1 + \sin^2 (A + B) \qquad \text{Expand squares}$$

$$= -2 \cos (A + B) + 2 \qquad \text{Pythagorean identity}$$

Applying the same two steps to the right side of Equation 1 gives us

Right Side of Equation 1

$$\cos^2 A - 2 \cos A \cos B + \cos^2 B + \sin^2 A + 2 \sin A \sin B + \sin^2 B$$

$$= -2 \cos A \cos B + 2 \sin A \sin B + 2$$

Equating the simplified versions of the left and right sides of Equation 1 we have

$$-2 \cos (A + B) + 2 = -2 \cos A \cos B + 2 \sin A \sin B + 2$$

Adding -2 to both sides and then dividing both sides by -2 gives us the formula we are after:

$$\boxed{\cos (A + B) = \cos A \cos B - \sin A \sin B}$$

This is the first formula in a series of formulas for trigonometric functions of the sum or difference of two angles. It must be memorized. Before we derive the others, let's look at some of the ways we can use our first formula.

EXAMPLE 1   Find the exact value for cos 75°.

SOLUTION   We write 75° as 45° + 30° and then apply the formula for cos $(A + B)$.

$$\cos 75° = \cos (45° + 30°)$$

$$= \cos 45° \cos 30° - \sin 45° \sin 30°$$

$$= \frac{\sqrt{2}}{2} \cdot \frac{\sqrt{3}}{2} - \frac{\sqrt{2}}{2} \cdot \frac{1}{2}$$

$$= \frac{\sqrt{6} - \sqrt{2}}{4}$$

NOTE   If you completed the Chapter 2 Group Project, compare your value of cos 75° from the project with our result in Example 1. Convince yourself that the two values are the same.

 EXAMPLE 2   Show that cos $(x + 2\pi) = \cos x$.

SOLUTION   Applying the formula for cos $(A + B)$, we have

$$\cos (x + 2\pi) = \cos x \cos 2\pi - \sin x \sin 2\pi$$

$$= \cos x \cdot 1 - \sin x \cdot 0$$

$$= \cos x$$

Notice that this is not a new relationship. We already know that if two angles are coterminal, then their cosines are equal— and $x + 2\pi$ and x are coterminal. What we have done here is shown this to be true with a formula instead of the definition of cosine.

 EXAMPLE 3   Write cos 3x cos 2x − sin 3x sin 2x as a single cosine.

SOLUTION   We apply the formula for cos $(A + B)$ in the reverse direction from the way we applied it in the first two examples.

$$\cos 3x \cos 2x - \sin 3x \sin 2x = \cos (3x + 2x)$$

$$= \cos 5x$$

Here is the derivation of the formula for cos $(A - B)$. It involves the formula for cos $(A + B)$ and the formulas for even and odd functions.

$\cos (A - B) = \cos [A + (-B)]$	Write $A - B$ as a sum
$= \cos A \cos (-B) - \sin A \sin (-B)$	Sum formula
$= \cos A \cos B - \sin A(-\sin B)$	Cosine is an even function, sine is odd
$= \cos A \cos B + \sin A \sin B$	

The only difference in the formulas for the expansion of $\cos(A + B)$ and $\cos(A - B)$ is the sign between the two terms. Here are both formulas again.

$$\cos(A + B) = \cos A \cos B - \sin A \sin B$$
$$\cos(A - B) = \cos A \cos B + \sin A \sin B$$

Again, both formulas are important and should be memorized.

 **EXAMPLE 4** Show that $\cos(90° - A) = \sin A$.

SOLUTION We will need this formula when we derive the formula for $\sin(A + B)$.

$$\cos(90° - A) = \cos 90° \cos A + \sin 90° \sin A$$
$$= 0 \cdot \cos A + 1 \cdot \sin A$$
$$= \sin A \quad \blacksquare$$

Note that the formula we just derived is not a new formula. The angles $90° - A$ and A are complementary angles, and we already know the sine of an angle is always equal to the cosine of its complement. We could also state it this way:

$$\sin(90° - A) = \cos A$$

We can use this information to derive the formula for $\sin(A + B)$. To understand this derivation, you must recognize that $A + B$ and $90° - (A + B)$ are complementary angles.

$$\sin(A + B) = \cos[90° - (A + B)] \qquad \text{The sine of an angle is the}$$
$$\text{cosine of its complement}$$

$$= \cos[90° - A - B] \qquad \text{Remove parentheses}$$

$$= \cos[(90° - A) - B] \qquad \text{Regroup within brackets}$$

Now we expand using the formula for the cosine of a difference.

$$= \cos(90° - A) \cos B + \sin(90° - A) \sin B$$
$$= \sin A \cos B + \cos A \sin B$$

This gives us an expansion formula for $\sin(A + B)$.

$$\sin(A + B) = \sin A \cos B + \cos A \sin B$$

This is the formula for the sine of a sum. To find the formula for $\sin(A - B)$, we write $A - B$ as $A + (-B)$ and proceed as follows:

$$\sin(A - B) = \sin[A + (-B)]$$
$$= \sin A \cos(-B) + \cos A \sin(-B)$$
$$= \sin A \cos B - \cos A \sin B$$

This gives us the formula for the sine of a difference.

$$\sin(A - B) = \sin A \cos B - \cos A \sin B$$

EXAMPLE 5 Graph $y = 4 \sin 5x \cos 3x - 4 \cos 5x \sin 3x$ from $x = 0$ to $x = 2\pi$.

SOLUTION   To write the equation in the form $y = A \sin Bx$, we factor 4 from each term on the right and then apply the formula for $\sin (A - B)$ to the remaining expression to write it as a single trigonometric function.

$$y = 4 \sin 5x \cos 3x - 4 \cos 5x \sin 3x$$

$$= 4 (\sin 5x \cos 3x - \cos 5x \sin 3x)$$

$$= 4 \sin (5x - 3x)$$

$$= 4 \sin 2x$$

The graph of $y = 4 \sin 2x$ will have an amplitude of 4 and a period of $2\pi/2 = \pi$. The graph is shown in Figure 2.

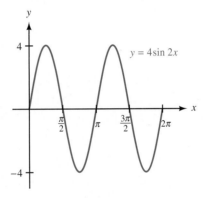

Figure 2

EXAMPLE 6 Find the exact value of $\sin \dfrac{\pi}{12}$.

SOLUTION   We have to write $\pi/12$ in terms of two numbers the exact values of which are known. The numbers $\pi/3$ and $\pi/4$ will work since their difference is $\pi/12$.

$$\sin \frac{\pi}{12} = \sin \left(\frac{\pi}{3} - \frac{\pi}{4} \right)$$

$$= \sin \frac{\pi}{3} \cos \frac{\pi}{4} - \cos \frac{\pi}{3} \sin \frac{\pi}{4}$$

$$= \frac{\sqrt{3}}{2} \cdot \frac{\sqrt{2}}{2} - \frac{1}{2} \cdot \frac{\sqrt{2}}{2}$$

$$= \frac{\sqrt{6} - \sqrt{2}}{4}$$

This is the same answer we obtained in Example 1 when we found the exact value of $\cos 75°$. It should be, though, because $\pi/12 = 15°$, which is the complement of $75°$, and the cosine of an angle is equal to the sine of its complement.

EXAMPLE 7 If $\sin A = \frac{3}{5}$ with A in QI and $\cos B = -\frac{5}{13}$ with B in QIII, find $\sin (A + B)$, $\cos (A + B)$, and $\tan (A + B)$.

SOLUTION We have $\sin A$ and $\cos B$. We need to find $\cos A$ and $\sin B$ before we can apply any of our formulas. Some equivalent forms of our first Pythagorean identity will help here.

If $\sin A = \frac{3}{5}$ with A in QI, then

$$\cos A = \sqrt{1 - \sin^2 A} = \sqrt{1 - \left(\frac{3}{5}\right)^2} = \frac{4}{5}$$

If $\cos B = -\frac{5}{13}$ with B in QIII, then

$$\sin B = -\sqrt{1 - \left(-\frac{5}{13}\right)^2} = -\frac{12}{13}$$

We have

$$\sin A = \frac{3}{5} \qquad \sin B = -\frac{12}{13}$$

$$\cos A = \frac{4}{5} \qquad \cos B = -\frac{5}{13}$$

Therefore,

$$\sin (A + B) = \sin A \cos B + \cos A \sin B$$

$$= \frac{3}{5}\left(-\frac{5}{13}\right) + \frac{4}{5}\left(-\frac{12}{13}\right)$$

$$= -\frac{63}{65}$$

$$\cos (A + B) = \cos A \cos B - \sin A \sin B$$

$$= \frac{4}{5}\left(-\frac{5}{13}\right) - \frac{3}{5}\left(-\frac{12}{13}\right)$$

$$= \frac{16}{65}$$

$$\tan (A + B) = \frac{\sin (A + B)}{\cos (A + B)}$$

$$= \frac{-63/65}{16/65}$$

$$= -\frac{63}{16}$$

Notice also that $A + B$ must terminate in quadrant IV because

$$\sin (A + B) < 0 \qquad \text{and} \qquad \cos (A + B) > 0$$

While working through the last part of Example 7, you may have wondered if there is a separate formula for tan $(A + B)$. (More likely, you are hoping there isn't.) There is, and it is derived from the formulas we already have.

$$\tan (A + B) = \frac{\sin (A + B)}{\cos (A + B)}$$

$$= \frac{\sin A \cos B + \cos A \sin B}{\cos A \cos B - \sin A \sin B}$$

To be able to write this last line in terms of tangents only, we must divide numerator and denominator by $\cos A \cos B$.

$$= \frac{\dfrac{\sin A \cos B}{\cos A \cos B} + \dfrac{\cos A \sin B}{\cos A \cos B}}{\dfrac{\cos A \cos B}{\cos A \cos B} - \dfrac{\sin A \sin B}{\cos A \cos B}}$$

$$= \frac{\tan A + \tan B}{1 - \tan A \tan B}$$

The formula for tan $(A + B)$ is

$$\boxed{\tan (A + B) = \frac{\tan A + \tan B}{1 - \tan A \tan B}}$$

Since tangent is an odd function, the formula for tan $(A - B)$ will look like this:

$$\boxed{\tan (A - B) = \frac{\tan A - \tan B}{1 + \tan A \tan B}}$$

EXAMPLE 8  If $\sin A = \frac{3}{5}$ with A in QI and $\cos B = -\frac{5}{13}$ with B in QIII, find tan $(A + B)$ by using the formula

$$\tan (A + B) = \frac{\tan A + \tan B}{1 - \tan A \tan B}$$

SOLUTION   The angles A and B as given here are the same ones used previously in Example 7. Looking over Example 7 again, we find that

$$\tan A = \frac{3}{4} \qquad \text{and} \qquad \tan B = \frac{12}{5}$$

Therefore,

$$\tan (A + B) = \frac{\tan A + \tan B}{1 - \tan A \tan B}$$

$$= \frac{\dfrac{3}{4} + \dfrac{12}{5}}{1 - \dfrac{3}{4} \cdot \dfrac{12}{5}}$$

$$= \frac{\dfrac{15}{20} + \dfrac{48}{20}}{1 - \dfrac{9}{5}}$$

$$= \frac{\dfrac{63}{20}}{-\dfrac{4}{5}}$$

$$= -\frac{63}{16}$$

which is the same result we obtained previously.

 GETTING READY FOR CLASS

After reading through the preceding section, respond in your own words and in complete sentences.

a. Why is it necessary to have sum and difference formulas for sine, cosine, and tangent?

b. Write both the sum and the difference formulas for cosine.

c. Write both the sum and the difference formulas for sine.

d. Write both the sum and the difference formulas for tangent.

PROBLEM SET 5.2

Find exact values for each of the following:

1. $\sin 15°$

2. $\sin 75°$

3. $\tan 15°$

4. $\tan 75°$

5. $\sin \dfrac{7\pi}{12}$

6. $\cos \dfrac{7\pi}{12}$

7. $\cos 105°$

8. $\sin 105°$

Show that each of the following is true:

9. $\sin (x + 2\pi) = \sin x$

10. $\cos (x - 2\pi) = \cos x$

11. $\cos \left(x - \dfrac{\pi}{2}\right) = \sin x$

12. $\sin \left(x - \dfrac{\pi}{2}\right) = -\cos x$

13. $\cos(180° - \theta) = -\cos\theta$

14. $\sin(180° - \theta) = \sin\theta$

15. $\sin(90° + \theta) = \cos\theta$

16. $\cos(90° + \theta) = -\sin\theta$

17. $\tan\left(x + \dfrac{\pi}{4}\right) = \dfrac{1 + \tan x}{1 - \tan x}$

18. $\tan\left(x - \dfrac{\pi}{4}\right) = \dfrac{\tan x - 1}{\tan x + 1}$

19. $\sin\left(\dfrac{3\pi}{2} - x\right) = -\cos x$

20. $\cos\left(x - \dfrac{3\pi}{2}\right) = -\sin x$

Write each expression as a single trigonometric function.

21. $\sin 3x \cos 2x + \cos 3x \sin 2x$

22. $\cos 3x \cos 2x + \sin 3x \sin 2x$

23. $\cos 5x \cos x - \sin 5x \sin x$

24. $\sin 8x \cos x - \cos 8x \sin x$

25. $\cos 15° \cos 75° - \sin 15° \sin 75°$

26. $\cos 15° \cos 75° + \sin 15° \sin 75°$

Graph each of the following from $x = 0$ to $x = 2\pi$.

27. $y = \sin 5x \cos 3x - \cos 5x \sin 3x$

28. $y = \sin x \cos 2x + \cos x \sin 2x$

29. $y = 3\cos 7x \cos 5x + 3\sin 7x \sin 5x$

30. $y = 2\cos 4x \cos x + 2\sin 4x \sin x$

31. Graph one complete cycle of $y = \sin x \cos\dfrac{\pi}{4} + \cos x \sin\dfrac{\pi}{4}$ by first rewriting the right side in the form $\sin(A + B)$.

32. Graph one complete cycle of $y = \sin x \cos\dfrac{\pi}{6} - \cos x \sin\dfrac{\pi}{6}$ by first rewriting the right side in the form $\sin(A - B)$.

33. Graph one complete cycle of $y = 2(\sin x \cos\dfrac{\pi}{3} + \cos x \sin\dfrac{\pi}{3})$ by first rewriting the right side in the form $2\sin(A + B)$.

34. Graph one complete cycle of $y = 2(\sin x \cos\dfrac{\pi}{3} - \cos x \sin\dfrac{\pi}{3})$ by first rewriting the right side in the form $2\sin(A - B)$.

35. Let $\sin A = \dfrac{3}{5}$ with A in QII and $\sin B = -\dfrac{5}{13}$ with B in QIII. Find $\sin(A + B)$, $\cos(A + B)$, and $\tan(A + B)$. In what quadrant does $A + B$ terminate?

36. Let $\cos A = -\dfrac{5}{13}$ with A in QII and $\sin B = \dfrac{3}{5}$ with B in QI. Find $\sin(A - B)$, $\cos(A - B)$, and $\tan(A - B)$. In what quadrant does $A - B$ terminate?

37. If $\sin A = 1/\sqrt{5}$ with A in QI and $\tan B = \dfrac{3}{4}$ with B in QI, find $\tan(A + B)$ and $\cot(A + B)$. In what quadrant does $A + B$ terminate?

38. If $\sec A = \sqrt{5}$ with A in QI and $\sec B = \sqrt{10}$ with B in QI, find $\sec(A + B)$. [First find $\cos(A + B)$.]

39. If $\tan(A + B) = 3$ and $\tan B = \dfrac{1}{2}$, find $\tan A$.

40. If $\tan(A + B) = 2$ and $\tan B = \dfrac{1}{3}$, find $\tan A$.

41. Write a formula for $\sin 2x$ by writing $\sin 2x$ as $\sin(x + x)$ and using the formula for the sine of a sum.

42. Write a formula for $\cos 2x$ by writing $\cos 2x$ as $\cos(x + x)$ and using the formula for the cosine of a sum.

Prove each identity.

43. $\sin(90° + x) + \sin(90° - x) = 2\cos x$

44. $\sin(90° + x) - \sin(90° - x) = 0$

45. $\cos(x - 90°) - \cos(x + 90°) = 2\sin x$

46. $\cos(x + 90°) + \cos(x - 90°) = 0$

47. $\sin\left(\dfrac{\pi}{6} + x\right) + \sin\left(\dfrac{\pi}{6} - x\right) = \cos x$

48. $\cos\left(\dfrac{\pi}{3} + x\right) + \cos\left(\dfrac{\pi}{3} - x\right) = \cos x$

49. $\cos\left(x + \dfrac{\pi}{4}\right) + \cos\left(x - \dfrac{\pi}{4}\right) = \sqrt{2}\,\cos x$

50. $\sin\left(\dfrac{\pi}{4} + x\right) + \sin\left(\dfrac{\pi}{4} - x\right) = \sqrt{2}\,\cos x$

51. $\sin\left(\dfrac{3\pi}{2} + x\right) + \sin\left(\dfrac{3\pi}{2} - x\right) = -2\cos x$

52. $\cos\left(x + \dfrac{3\pi}{2}\right) + \cos\left(x - \dfrac{3\pi}{2}\right) = 0$

53. $\sin(A + B) + \sin(A - B) = 2\sin A \cos B$
54. $\cos(A + B) + \cos(A - B) = 2\cos A \cos B$

55. $\dfrac{\sin(A - B)}{\cos A \cos B} = \tan A - \tan B$ **56.** $\dfrac{\cos(A + B)}{\sin A \cos B} = \cot A - \tan B$

57. $\sec(A + B) = \dfrac{\cos(A - B)}{\cos^2 A - \sin^2 B}$ **58.** $\sec(A - B) = \dfrac{\cos(A + B)}{\cos^2 A - \sin^2 B}$

Use your graphing calculator to determine if each equation appears to be an identity by graphing the left expression and right expression together. If so, then prove the identity.

59. $\sin x = \cos\left(\dfrac{\pi}{2} - x\right)$ **60.** $\cos x = \sin\left(\dfrac{\pi}{2} - x\right)$

61. $-\cos x = \sin\left(\dfrac{\pi}{2} + x\right)$ **62.** $\sin x = \cos\left(\dfrac{\pi}{2} + x\right)$

63. $-\sin x = \cos\left(\dfrac{\pi}{2} + x\right)$ **64.** $-\cos x = \cos(\pi - x)$

REVIEW PROBLEMS

The problems that follow review material we covered in Section 4.2. Graph one complete cycle of each of the following:

65. $y = 4\sin 2x$ **66.** $y = 2\sin 4x$

67. $y = 3\sin\dfrac{1}{2}x$ **68.** $y = 5\sin\dfrac{1}{3}x$

69. $y = 2\cos \pi x$ **70.** $y = \cos 2\pi x$
71. $y = \csc 3x$ **72.** $y = \sec 3x$

73. $y = \dfrac{1}{2}\cos 3x$ **74.** $y = \dfrac{1}{2}\sin 3x$

75. $y = \dfrac{1}{2}\sin\dfrac{\pi}{2}x$ **76.** $y = 2\sin\dfrac{\pi}{2}x$

SECTION 5.3 | DOUBLE-ANGLE FORMULAS

We will begin this section by deriving the formulas for sin 2*A* and cos 2*A* using the formulas for sin (*A* + *B*) and cos (*A* + *B*). The formulas we derive for sin 2*A* and cos 2*A* are called *double-angle* formulas. Here is the derivation of the formula for sin 2*A*.

$$\sin 2A = \sin (A + A) \qquad \text{Write } 2A \text{ as } A + A$$

$$= \sin A \cos A + \cos A \sin A \qquad \text{Sum formula}$$

$$= \sin A \cos A + \sin A \cos A \qquad \text{Commutative property}$$

$$= 2 \sin A \cos A$$

The last line gives us our first double-angle formula.

$$\boxed{\sin 2A = 2 \sin A \cos A}$$

The first thing to notice about this formula is that it indicates the 2 in sin 2*A* *cannot* be factored out and written as a coefficient. That is,

$$\sin 2A \neq 2 \sin A$$

For example, if *A* = 30°, sin 2 · 30° = sin 60° = $\sqrt{3}/2$, which is not the same as 2 sin 30° = 2($\frac{1}{2}$) = 1.

EXAMPLE 1 If sin *A* = $\frac{3}{5}$ with *A* in QII, find sin 2*A*.

SOLUTION In order to apply the formula for sin 2*A*, we must first find cos *A*. Since *A* terminates in QII, cos *A* is negative.

$$\cos A = -\sqrt{1 - \sin^2 A} = -\sqrt{1 - \left(\frac{3}{5}\right)^2} = -\sqrt{\frac{16}{25}} = -\frac{4}{5}$$

Now we can apply the formula for sin 2*A*.

$$\sin 2A = 2 \sin A \cos A$$

$$= 2 \left(\frac{3}{5}\right) \left(-\frac{4}{5}\right)$$

$$= -\frac{24}{25} \quad \blacksquare$$

We can also use our new formula to expand the work we did previously with identities.

EXAMPLE 2 Prove (sin θ + cos θ)² = 1 + sin 2θ.

PROOF

$$(\sin \theta + \cos \theta)^2 = \sin^2 \theta + 2 \sin \theta \cos \theta + \cos^2 \theta \qquad \text{Expand}$$

$$= 1 + 2 \sin \theta \cos \theta \qquad \text{Pythagorean identity}$$

$$= 1 + \sin 2\theta \qquad \text{Double-angle identity} \quad \blacksquare$$

EXAMPLE 3   Prove $\sin 2x = \dfrac{2 \cot x}{1 + \cot^2 x}$.

PROOF

$$\frac{2 \cot x}{1 + \cot^2 x} = \frac{2 \cdot \dfrac{\cos x}{\sin x}}{1 + \dfrac{\cos^2 x}{\sin^2 x}} \qquad \text{Ratio identity}$$

$$= \frac{2 \sin x \cos x}{\sin^2 x + \cos^2 x} \qquad \begin{array}{l} \text{Multiply numerator and} \\ \text{denominator by } \sin^2 x \end{array}$$

$$= 2 \sin x \cos x \qquad \text{Pythagorean identity}$$

$$= \sin 2x \qquad \text{Double-angle identity} \quad \blacksquare$$

There are three forms of the double-angle formula for $\cos 2A$. The first involves both $\sin A$ and $\cos A$, the second involves only $\cos A$, and the third involves only $\sin A$. Here is how we obtain the three formulas.

$$\cos 2A = \cos (A + A) \qquad \text{Write } 2A \text{ as } A + A$$

$$= \cos A \cos A - \sin A \sin A \qquad \text{Sum formula}$$

$$= \cos^2 A - \sin^2 A$$

To write this last formula in terms of $\cos A$ only, we substitute $1 - \cos^2 A$ for $\sin^2 A$.

$$\cos 2A = \cos^2 A - (1 - \cos^2 A)$$

$$= \cos^2 A - 1 + \cos^2 A$$

$$= 2 \cos^2 A - 1$$

To write the formula in terms of $\sin A$ only, we substitute $1 - \sin^2 A$ for $\cos^2 A$ in the last line above.

$$\cos 2A = 2 \cos^2 A - 1$$

$$= 2(1 - \sin^2 A) - 1$$

$$= 2 - 2 \sin^2 A - 1$$

$$= 1 - 2 \sin^2 A$$

Here are the three forms of the double-angle formula for $\cos 2A$.

$\cos 2A = \cos^2 A - \sin^2 A$	First form
$= 2 \cos^2 A - 1$	Second form
$= 1 - 2 \sin^2 A$	Third form

Which form we choose will depend on the circumstances of the problem, as the next three examples illustrate.

EXAMPLE 4   If $\sin A = 1/\sqrt{5}$, find $\cos 2A$.

SOLUTION   In this case, since we are given $\sin A$, applying the third form of the formula for $\cos 2A$ will give us the answer more quickly than applying either of the other two forms.

$$\cos 2A = 1 - 2 \sin^2 A$$

$$= 1 - 2 \left(\frac{1}{\sqrt{5}} \right)^2$$

$$= 1 - \frac{2}{5}$$

$$= \frac{3}{5} \quad \text{}$$

EXAMPLE 5   Prove $\cos 4x = 8 \cos^4 x - 8 \cos^2 x + 1$.

PROOF   We can write $\cos 4x$ as $\cos (2 \cdot 2x)$ and apply our double-angle formula. Since the right side is written in terms of $\cos x$ only, we will choose the second form of our double-angle formula for $\cos 2A$.

$$\cos 4x = \cos (2 \cdot 2x)$$

$$= 2 \cos^2 2x - 1 \qquad \text{Double-angle formula}$$

$$= 2 (2 \cos^2 x - 1)^2 - 1 \qquad \text{Double-angle formula}$$

$$= 2 (4 \cos^4 x - 4 \cos^2 x + 1) - 1 \qquad \text{Square}$$

$$= 8 \cos^4 x - 8 \cos^2 x + 2 - 1 \qquad \text{Distribute}$$

$$= 8 \cos^4 x - 8 \cos^2 x + 1 \qquad \text{Simplify} \quad \text{}$$

EXAMPLE 6   Graph $y = 3 - 6 \sin^2 x$ from $x = 0$ to $x = 2\pi$.

SOLUTION   To write the equation in the form $y = A \cos Bx$, we factor 3 from each term on the right side and then apply the formula for $\cos 2A$ to the remaining expression to write it as a single trigonometric function.

$$y = 3 - 6 \sin^2 x$$

$$= 3(1 - 2 \sin^2 x) \qquad \text{Factor 3 from each term}$$

$$= 3 \cos 2x \qquad \text{Double-angle formula}$$

The graph of $y = 3 \cos 2x$ will have an amplitude of 3 and a period of $2\pi/2 = \pi$. The graph is shown in Figure 1.

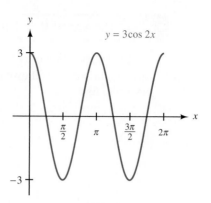

Figure 1

 EXAMPLE 7 Prove $\tan\theta = \dfrac{1-\cos 2\theta}{\sin 2\theta}$.

PROOF

$$\frac{1-\cos 2\theta}{\sin 2\theta} = \frac{1-(1-2\sin^2\theta)}{2\sin\theta\cos\theta} \qquad \text{Double-angle formulas}$$

$$= \frac{2\sin^2\theta}{2\sin\theta\cos\theta} \qquad \text{Simplify numerator}$$

$$= \frac{\sin\theta}{\cos\theta} \qquad \text{Divide out common factor } 2\sin\theta$$

$$= \tan\theta \qquad \text{Ratio identity} \quad \blacksquare$$

We end this section by deriving the formula for $\tan 2A$.

$$\tan 2A = \tan(A+A)$$

$$= \frac{\tan A + \tan A}{1 - \tan A \tan A}$$

$$= \frac{2\tan A}{1 - \tan^2 A}$$

Our double-angle formula for $\tan 2A$ is

$$\boxed{\tan 2A = \frac{2\tan A}{1-\tan^2 A}}$$

 EXAMPLE 8 Simplify $\dfrac{2\tan 15°}{1-\tan^2 15°}$.

SOLUTION The expression has the same form as the right side of our double-angle formula for $\tan 2A$. Therefore,

$$\frac{2\tan 15°}{1-\tan^2 15°} = \tan(2\cdot 15°)$$

$$= \tan 30°$$

$$= \frac{1}{\sqrt{3}} \quad \blacksquare$$

EXAMPLE 9   If $x = 3 \tan \theta$, write the expression below in terms of just x.

$$\frac{\theta}{2} + \frac{\sin 2\theta}{4}$$

SOLUTION   To substitute for the first term above, we need to write θ in terms of x. To do so, we solve the equation $x = 3 \tan \theta$ for θ.

If $3 \tan \theta = x$

then $\tan \theta = \dfrac{x}{3}$

and $\theta = \tan^{-1} \dfrac{x}{3}$

Next, since the inverse tangent function can take on values only between $-\pi/2$ and $\pi/2$, we can visualize θ by drawing a right triangle in which θ is one of the acute angles. Since $\tan \theta = x/3$, we label the side opposite θ with x and the side adjacent to θ with 3.

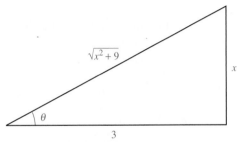

Figure 2

By the Pythagorean Theorem, the hypotenuse of the triangle in Figure 2 must be $\sqrt{x^2 + 9}$, which means $\sin \theta = x/\sqrt{x^2 + 9}$ and $\cos \theta = 3/\sqrt{x^2 + 9}$. Now we are ready to simplify and substitute to solve our problem.

$$\frac{\theta}{2} + \frac{\sin 2\theta}{4} = \frac{\theta}{2} + \frac{2 \sin \theta \cos \theta}{4}$$

$$= \frac{\theta}{2} + \frac{\sin \theta \cos \theta}{2}$$

$$= \frac{1}{2}(\theta + \sin \theta \cos \theta)$$

$$= \frac{1}{2}\left(\tan^{-1}\frac{x}{3} + \frac{x}{\sqrt{x^2 + 9}} \cdot \frac{3}{\sqrt{x^2 + 9}}\right)$$

$$= \frac{1}{2}\left(\tan^{-1}\frac{x}{3} + \frac{3x}{x^2 + 9}\right)$$

Note that we do not have to use absolute value symbols when we multiply and simplify the square roots in the second to the last line above because we know that $x^2 + 9$ is always positive.

GETTING READY FOR CLASS

After reading through the preceding section, respond in your own words and in complete sentences.

a. What is the formula for $\sin 2A$?

b. What are the three formulas for $\cos 2A$?

c. What is the formula for $\tan 2A$?

d. As a general rule, when do we use θ to indicate an angle, and when do we use A, B, or other non-Greek symbols? (*Hint:* Think in terms of degrees and radians.)

PROBLEM SET 5.3

Let $\sin A = -\frac{3}{5}$ with A in QIII and find

 1. $\sin 2A$ **2.** $\cos 2A$

3. $\tan 2A$ **4.** $\cot 2A$

Let $\cos x = 1/\sqrt{10}$ with x in QIV and find

5. $\cos 2x$ **6.** $\sin 2x$

7. $\cot 2x$ **8.** $\tan 2x$

Let $\tan \theta = \frac{5}{12}$ with θ in QI and find

9. $\sin 2\theta$ **10.** $\cos 2\theta$

11. $\csc 2\theta$ **12.** $\sec 2\theta$

Let $\csc t = \sqrt{5}$ with t in QII and find

13. $\cos 2t$ **14.** $\sin 2t$

15. $\sec 2t$ **16.** $\csc 2t$

Graph each of the following from $x = 0$ to $x = 2\pi$.

17. $y = 4 - 8 \sin^2 x$ **18.** $y = 2 - 4 \sin^2 x$

19. $y = 6 \cos^2 x - 3$ **20.** $y = 4 \cos^2 x - 2$

21. $y = 1 - 2 \sin^2 2x$ **22.** $y = 2 \cos^2 2x - 1$

Use exact values to show that each of the following is true.

23. $\sin 60° = 2 \sin 30° \cos 30°$ **24.** $\cos 60° = 1 - 2 \sin^2 30°$

25. $\cos 120° = \cos^2 60° - \sin^2 60°$ **26.** $\sin 90° = 2 \sin 45° \cos 45°$

27. If $\tan A = \frac{3}{4}$, find $\tan 2A$.   **28.** If $\tan A = -\sqrt{3}$, find $\tan 2A$.

Simplify each of the following.

29. $2 \sin 15° \cos 15°$ **30.** $\cos^2 15° - \sin^2 15°$

31. $1 - 2 \sin^2 75°$ **32.** $2 \cos^2 105° - 1$

33. $\sin \dfrac{\pi}{12} \cos \dfrac{\pi}{12}$ **34.** $\sin \dfrac{\pi}{8} \cos \dfrac{\pi}{8}$

35. $\dfrac{\tan 22.5°}{1 - \tan^2 22.5°}$ **36.** $\dfrac{\tan \dfrac{3\pi}{8}}{1 - \tan^2 \dfrac{3\pi}{8}}$

Prove each of the following identities.

37. $(\sin x - \cos x)^2 = 1 - \sin 2x$

38. $(\cos x - \sin x)(\cos x + \sin x) = \cos 2x$

39. $\cos^2 \theta = \dfrac{1 + \cos 2\theta}{2}$

40. $\sin^2 \theta = \dfrac{1 - \cos 2\theta}{2}$

41. $\cot \theta = \dfrac{\sin 2\theta}{1 - \cos 2\theta}$

42. $\cos 2\theta = \dfrac{1 - \tan^2 \theta}{1 + \tan^2 \theta}$

43. $2 \csc 2x = \tan x + \cot x$

44. $2 \cot 2x = \cot x - \tan x$

45. $\sin 3\theta = 3 \sin \theta - 4 \sin^3 \theta$

46. $\cos 3\theta = 4 \cos^3 \theta - 3 \cos \theta$

47. $\cos^4 x - \sin^4 x = \cos 2x$

48. $2 \sin^4 x + 2 \sin^2 x \cos^2 x = 1 - \cos 2x$

49. $\cot \theta - \tan \theta = \dfrac{\cos 2\theta}{\sin \theta \cos \theta}$

50. $\csc \theta - 2 \sin \theta = \dfrac{\cos 2\theta}{\sin \theta}$

51. $\sin 4A = 4 \sin A \cos^3 A - 4 \sin^3 A \cos A$

52. $\cos 4A = \cos^4 A - 6 \cos^2 A \sin^2 A + \sin^4 A$

53. $\dfrac{1 - \tan x}{1 + \tan x} = \dfrac{1 - \sin 2x}{\cos 2x}$

54. $\dfrac{2 - 2 \cos 2x}{\sin 2x} = \sec x \csc x - \cot x + \tan x$

Use your graphing calculator to determine if each equation appears to be an identity by graphing the left expression and right expression together. If so, then prove the identity.

55. $\cot 2x = \dfrac{\cos x - \sin x \tan x}{\sec x \sin 2x}$

56. $\sec 2x = \dfrac{\sec^2 x \csc^2 x}{\csc^2 x - \sec^2 x}$

57. $\sec 2x = \dfrac{\sec^2 x \csc^2 x}{\csc^2 x + \sec^2 x}$

58. $\csc 2x = \dfrac{\sec x + \csc x}{2 \sin x - 2 \cos x}$

59. $\csc 2x = \dfrac{\sec x + \csc x}{2 \sin x + 2 \cos x}$

60. $\tan 2x = \dfrac{2 \cot x}{\csc^2 x - 2}$

61. If $x = 5 \tan \theta$, write the expression $\dfrac{\theta}{2} - \dfrac{\sin 2\theta}{4}$ in terms of just x.

62. If $x = 4 \sin \theta$, write the expression $\dfrac{\theta}{2} - \dfrac{\sin 2\theta}{4}$ in terms of just x.

63. If $x = 3 \sin \theta$, write the expression $\dfrac{\theta}{2} - \dfrac{\sin 2\theta}{4}$ in terms of just x.

64. If $x = 2 \sin \theta$, write the expression $2\theta - \tan 2\theta$ in terms of just x.

REVIEW PROBLEMS

The problems that follow review material we covered in Section 4.5. Graph each of the following from $x = 0$ to $x = 4\pi$.

65. $y = 2 - 2 \cos x$

66. $y = 2 + 2 \cos x$

67. $y = 3 - 3 \cos x$

68. $y = 3 + 3 \cos x$

69. $y = \cos x + \dfrac{1}{2} \sin 2x$

70. $y = \sin x + \dfrac{1}{2} \cos 2x$

Graph each of the following from $x = 0$ to $x = 8$.

71. $y = \dfrac{1}{2}x + \sin \pi x$

72. $y = x + \sin \dfrac{\pi}{2}x$

SECTION 5.4 | HALF-ANGLE FORMULAS

In this section, we will derive formulas for $\sin \dfrac{A}{2}$ and $\cos \dfrac{A}{2}$. These formulas are called *half-angle* formulas and are derived from the double-angle formulas for $\cos 2A$.

In Section 5.3, we developed three ways to write the formula for $\cos 2A$, two of which were

$$\cos 2A = 1 - 2 \sin^2 A \quad \text{and} \quad \cos 2A = 2 \cos^2 A - 1$$

Since the choice of the letter we use to denote the angles in these formulas is arbitrary, we can use an x instead of A.

$$\cos 2x = 1 - 2 \sin^2 x \quad \text{and} \quad \cos 2x = 2 \cos^2 x - 1$$

Let us exchange sides in the first formula and solve for $\sin x$.

$$1 - 2 \sin^2 x = \cos 2x \qquad \qquad \text{Exchange sides}$$

$$-2 \sin^2 x = -1 + \cos 2x \qquad \qquad \text{Add } -1 \text{ to both sides}$$

$$\sin^2 x = \frac{1 - \cos 2x}{2} \qquad \qquad \text{Divide both sides by } -2$$

$$\sin x = \pm \sqrt{\frac{1 - \cos 2x}{2}} \qquad \qquad \text{Take the square root of both sides}$$

Since every value of x can be written as $\frac{1}{2}$ of some other number A, we can replace x with $A/2$. This is equivalent to saying $2x = A$.

$$\boxed{\sin \frac{A}{2} = \pm \sqrt{\frac{1 - \cos A}{2}}}$$

This last expression is the half-angle formula for $\sin \dfrac{A}{2}$. To find the half-angle formula for $\cos \dfrac{A}{2}$, we solve $\cos 2x = 2 \cos^2 x - 1$ for $\cos x$ and then replace x with $A/2$ (and $2x$ with A). Without showing the steps involved in this process, here is the result:

$$\boxed{\cos \frac{A}{2} = \pm \sqrt{\frac{1 + \cos A}{2}}}$$

In both half-angle formulas, the sign in front of the radical, $+$ or $-$, is determined by the quadrant in which $A/2$ terminates.

EXAMPLE 1 If $\cos A = \frac{3}{5}$ with $270° < A < 360°$, find $\sin \dfrac{A}{2}$, $\cos \dfrac{A}{2}$, and $\tan \dfrac{A}{2}$.

SOLUTION First of all, we determine the quadrant in which $A/2$ terminates.

$$270° < A < 360° \Rightarrow \frac{270°}{2} < \frac{A}{2} < \frac{360°}{2}$$

$$\text{or} \qquad 135° < \frac{A}{2} < 180° \Rightarrow \frac{A}{2} \in \text{QII}$$

In quadrant II, sine is positive, and cosine and tangent are negative.

$$\sin\frac{A}{2} = \sqrt{\frac{1-\cos A}{2}} \qquad \cos\frac{A}{2} = -\sqrt{\frac{1+\cos A}{2}}$$

$$= \sqrt{\frac{1-3/5}{2}} \qquad = -\sqrt{\frac{1+3/5}{2}}$$

$$= \sqrt{\frac{1}{5}} \qquad = -\sqrt{\frac{4}{5}}$$

$$= \frac{1}{\sqrt{5}} \qquad = -\frac{2}{\sqrt{5}}$$

$$\tan\frac{A}{2} = \frac{\sin\dfrac{A}{2}}{\cos\dfrac{A}{2}} = \frac{\dfrac{1}{\sqrt{5}}}{-\dfrac{2}{\sqrt{5}}} = -\frac{1}{2} \quad$$

EXAMPLE 2 If $\sin A = -\frac{12}{13}$ with $180° < A < 270°$, find the six trigonometric functions of $A/2$.

SOLUTION   To use the half-angle formulas, we need to find $\cos A$.

Because $A \in$ QIII

$$\cos A = -\sqrt{1-\sin^2 A} = -\sqrt{1-\left(-\frac{12}{13}\right)^2} = -\sqrt{\frac{25}{169}} = -\frac{5}{13}$$

Also, $A/2$ terminates in QII because

$$180° < A < 270°$$

$$\frac{180°}{2} < \frac{A}{2} < \frac{270°}{2}$$

$$90° < \frac{A}{2} < 135° \Rightarrow \frac{A}{2} \in \text{QII}$$

In quadrant II, sine is positive and cosine is negative.

$$\sin\frac{A}{2} = \sqrt{\frac{1-(-5/13)}{2}} \qquad \cos\frac{A}{2} = -\sqrt{\frac{1+(-5/13)}{2}}$$

$$= \sqrt{\frac{9}{13}} \qquad = -\sqrt{\frac{4}{13}}$$

$$= \frac{3}{\sqrt{13}} \qquad = -\frac{2}{\sqrt{13}}$$

Now that we have sine and cosine of $A/2$, we can apply the ratio identity for tangent to find $\tan\dfrac{A}{2}$.

$$\tan\frac{A}{2} = \frac{\sin\dfrac{A}{2}}{\cos\dfrac{A}{2}} = \frac{\dfrac{3}{\sqrt{13}}}{-\dfrac{2}{\sqrt{13}}} = -\frac{3}{2}$$

Next, we apply our reciprocal identities to find cosecant, secant, and cotangent of $A/2$.

$$\csc \frac{A}{2} = \frac{1}{\sin \dfrac{A}{2}} = \frac{\sqrt{13}}{3} \qquad \sec \frac{A}{2} = \frac{1}{\cos \dfrac{A}{2}} = -\frac{\sqrt{13}}{2}$$

$$\cot \frac{A}{2} = \frac{1}{\tan \dfrac{A}{2}} = -\frac{2}{3} \quad \blacksquare$$

In the previous two examples, we found $\tan \dfrac{A}{2}$ by using the ratio of $\sin \dfrac{A}{2}$ to $\cos \dfrac{A}{2}$. There are formulas that allow us to find $\tan \dfrac{A}{2}$ directly from $\sin A$ and $\cos A$. In Example 7 of Section 5.3, we proved the following identity:

$$\tan \theta = \frac{1 - \cos 2\theta}{\sin 2\theta}$$

If we let $\theta = \dfrac{A}{2}$ in this identity, we obtain a formula for $\tan \dfrac{A}{2}$ that involves only $\sin A$ and $\cos A$. Here it is.

$$\boxed{\tan \frac{A}{2} = \frac{1 - \cos A}{\sin A}}$$

If we multiply the numerator and denominator of the right side of this formula by $1 + \cos A$ and simplify the result, we have a second formula for $\tan \dfrac{A}{2}$.

$$\boxed{\tan \frac{A}{2} = \frac{\sin A}{1 + \cos A}}$$

 EXAMPLE 3   Find $\tan 15°$.

SOLUTION   Since $15° = 30°/2$, we can use a half-angle formula to find $\tan 15°$.

$$\tan 15° = \tan \frac{30°}{2}$$

$$= \frac{1 - \cos 30°}{\sin 30°}$$

$$= \frac{1 - \dfrac{\sqrt{3}}{2}}{\dfrac{1}{2}}$$

$$= 2 - \sqrt{3} \quad \blacksquare$$

EXAMPLE 4 Graph $y = 4 \cos^2 \dfrac{x}{2}$ from $x = 0$ to $x = 4\pi$.

SOLUTION   Applying our half-angle formula for $\cos \dfrac{x}{2}$ to the right side, we have

$$y = 4 \cos^2 \frac{x}{2}$$

$$= 4 \left(\pm \sqrt{\frac{1 + \cos x}{2}} \right)^2$$

$$= 4 \left(\frac{1 + \cos x}{2} \right)$$

$$= 2 + 2 \cos x$$

We graph $y = 2 + 2 \cos x$ using the method developed in Section 4.5. The graph is shown in Figure 1.

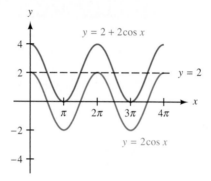

Figure 1

EXAMPLE 5 Prove $\sin^2 \dfrac{x}{2} = \dfrac{\tan x - \sin x}{2 \tan x}$.

PROOF   We can use a half-angle formula on the left side. In this case, since we have $\sin^2 (x/2)$, we write the half-angle formula without the square root sign. After that, we multiply the numerator and denominator on the left side by $\tan x$ because the right side has $\tan x$ in both the numerator and the denominator.

$$\sin^2 \frac{x}{2} = \frac{1 - \cos x}{2} \qquad \text{Square of half-angle formula}$$

$$= \frac{\tan x}{\tan x} \cdot \frac{1 - \cos x}{2} \qquad \begin{array}{l}\text{Multiply numerator and}\\ \text{denominator by } \tan x\end{array}$$

$$= \frac{\tan x - \tan x \cos x}{2 \tan x} \qquad \text{Distributive property}$$

$$= \frac{\tan x - \sin x}{2 \tan x} \qquad \tan x \cos x \text{ is } \sin x$$

GETTING READY FOR CLASS

After reading through the preceding section, respond in your own words and in complete sentences.

a. From what other formulas are half-angle formulas derived?

b. What is the formula for $\sin \dfrac{A}{2}$?

c. What is the formula for $\cos \dfrac{A}{2}$?

d. What are the two formulas for $\tan \dfrac{A}{2}$?

PROBLEM SET 5.4

NOTE For the following problems, assume that all the given angles are in simplest form, so that if A is in QIV you may assume that $270° < A < 360°$.

If $\cos A = \frac{1}{2}$ with A in QIV, find

1. $\sin \dfrac{A}{2}$ **2.** $\cos \dfrac{A}{2}$

3. $\csc \dfrac{A}{2}$ **4.** $\sec \dfrac{A}{2}$

If $\sin A = -\frac{3}{5}$ with A in QIII, find

5. $\cos \dfrac{A}{2}$ **6.** $\sin \dfrac{A}{2}$

7. $\sec \dfrac{A}{2}$ **8.** $\csc \dfrac{A}{2}$

If $\sin B = -\frac{1}{3}$ with B in QIII, find

9. $\sin \dfrac{B}{2}$ **10.** $\csc \dfrac{B}{2}$

11. $\cos \dfrac{B}{2}$ **12.** $\sec \dfrac{B}{2}$

13. $\tan \dfrac{B}{2}$ **14.** $\cot \dfrac{B}{2}$

If $\sin A = \frac{4}{5}$ with A in QII, and $\sin B = \frac{3}{5}$ with B in QI, find

15. $\sin \dfrac{A}{2}$ **16.** $\cos \dfrac{A}{2}$

17. $\cos 2A$ **18.** $\sin 2A$

19. $\sec 2A$ **20.** $\csc 2A$

21. $\cos \dfrac{B}{2}$ **22.** $\sin \dfrac{B}{2}$

23. $\sin (A + B)$ **24.** $\cos (A + B)$

25. $\cos (A - B)$ **26.** $\sin (A - B)$

Graph each of the following from $x = 0$ to $x = 4\pi$.

27. $y = 4 \sin^2 \dfrac{x}{2}$

28. $y = 6 \cos^2 \dfrac{x}{2}$

29. $y = 2 \cos^2 \dfrac{x}{2}$

30. $y = 2 \sin^2 \dfrac{x}{2}$

Use half-angle formulas to find exact values for each of the following:

31. $\cos 15°$

32. $\tan 15°$

33. $\sin 75°$

34. $\cos 75°$

35. $\cos 105°$

36. $\sin 105°$

Prove the following identities.

37. $\sin^2 \dfrac{\theta}{2} = \dfrac{\csc \theta - \cot \theta}{2 \csc \theta}$

38. $2 \cos^2 \dfrac{\theta}{2} = \dfrac{\sin^2 \theta}{1 - \cos \theta}$

39. $\sec^2 \dfrac{A}{2} = \dfrac{2 \sec A}{\sec A + 1}$

40. $\csc^2 \dfrac{A}{2} = \dfrac{2 \sec A}{\sec A - 1}$

41. $\tan \dfrac{B}{2} = \csc B - \cot B$

42. $\tan \dfrac{B}{2} = \dfrac{\sec B}{\sec B \csc B + \csc B}$

43. $\tan \dfrac{x}{2} + \cot \dfrac{x}{2} = 2 \csc x$

44. $\tan \dfrac{x}{2} - \cot \dfrac{x}{2} = -2 \cot x$

45. $\cos^2 \dfrac{\theta}{2} = \dfrac{\tan \theta + \sin \theta}{2 \tan \theta}$

46. $2 \sin^2 \dfrac{\theta}{2} = \dfrac{\sin^2 \theta}{1 + \cos \theta}$

47. $\cos^4 \theta = \dfrac{1}{4} + \dfrac{\cos 2\theta}{2} + \dfrac{\cos^2 2\theta}{4}$

48. $4 \sin^4 \theta = 1 - 2 \cos 2\theta + \cos^2 2\theta$

REVIEW PROBLEMS

The following problems review material we covered in Section 4.6. Reviewing these problems will help you with the next section.

Evaluate without using a calculator or tables.

49. $\sin \left(\arcsin \dfrac{3}{5} \right)$

50. $\cos \left(\arcsin \dfrac{3}{5} \right)$

51. $\cos (\arctan 2)$

52. $\sin (\arctan 2)$

Write an equivalent expression that involves x only.

53. $\sin (\tan^{-1} x)$

54. $\cos (\tan^{-1} x)$

55. $\tan (\sin^{-1} x)$

56. $\tan (\cos^{-1} x)$

57. Graph $y = \sin^{-1} x$

58. Graph $y = \cos^{-1} x$

SECTION 5.5 | ADDITIONAL IDENTITIES

There are two main parts to this section, both of which rely on the work we have done previously with identities and formulas. In the first part of this section, we will extend our work on identities to include problems that involve inverse trigonometric functions. In the second part, we will use the formulas we obtained for the sine and cosine of a sum or difference to write some new formulas involving sums and products.

IDENTITIES AND FORMULAS INVOLVING INVERSE FUNCTIONS

The solution to our first example combines our knowledge of inverse trigonometric functions with our formula for sin $(A + B)$.

 **EXAMPLE 1** Evaluate sin $(\arcsin \frac{3}{5} + \arctan 2)$ without using a calculator.

SOLUTION We can simplify things somewhat if we let $\alpha = \arcsin \frac{3}{5}$ and $\beta = \arctan 2$.

$$\sin\left(\arcsin \frac{3}{5} + \arctan 2\right) = \sin(\alpha + \beta)$$

$$= \sin \alpha \cos \beta + \cos \alpha \sin \beta$$

Drawing and labeling a triangle for α and another for β, we have

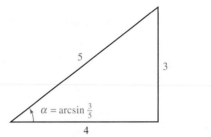
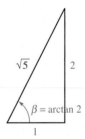

Figure 1

From the triangles in Figure 1, we have

$$\sin \alpha = \frac{3}{5} \qquad \sin \beta = \frac{2}{\sqrt{5}}$$

$$\cos \alpha = \frac{4}{5} \qquad \cos \beta = \frac{1}{\sqrt{5}}$$

Substituting these numbers into

$$\sin \alpha \cos \beta + \cos \alpha \sin \beta$$

gives us

$$\frac{3}{5} \cdot \frac{1}{\sqrt{5}} + \frac{4}{5} \cdot \frac{2}{\sqrt{5}} = \frac{11}{5\sqrt{5}}$$

CALCULATOR NOTE To check our answer for Example 1 using a calculator, we would use the following sequence:

Scientific Calculator

$$3 \boxed{\div} 5 \boxed{=} \boxed{\sin^{-1}} \boxed{+} 2 \boxed{\tan^{-1}} \boxed{=} \boxed{\sin}$$

Graphing Calculator

The display would show 0.9839 to four decimal places, which is the decimal approximation of $\dfrac{11}{5\sqrt{5}}$. It is appropriate to check your work on problems like this by using your calculator. The concepts are best understood, however, by working through the problems without using a calculator.

Here is a similar example involving inverse trigonometric functions and a double-angle identity.

 **EXAMPLE 2** Write $\sin(2\tan^{-1}x)$ as an equivalent expression involving only x. (Assume x is positive.)

SOLUTION We begin by letting $\theta = \tan^{-1}x$ and then draw a right triangle with an acute angle of θ. If we label the opposite side with x and the adjacent side with 1, the ratio of the side opposite θ to the side adjacent to θ is $x/1 = x$.

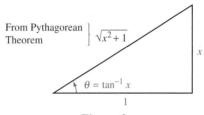

From Pythagorean Theorem $\sqrt{x^2 + 1}$

x

$\theta = \tan^{-1}x$

1

Figure 2

From Figure 2, we have $\sin\theta = x/\sqrt{x^2 + 1}$ and $\cos\theta = 1/\sqrt{x^2 + 1}$. Therefore,

$$\sin(2\tan^{-1}x) = \sin 2\theta \qquad\qquad \text{Substitute } \theta \text{ for } \tan^{-1}x$$

$$= 2\sin\theta\cos\theta \qquad\qquad \text{Double-angle identity}$$

$$= 2 \cdot \frac{x}{\sqrt{x^2 + 1}} \cdot \frac{1}{\sqrt{x^2 + 1}} \qquad \text{From Figure 2}$$

$$= \frac{2x}{x^2 + 1} \qquad\qquad \text{Multiplication}$$

To conclude our work with identities in this chapter, we will derive some additional formulas that contain sums and products of sines and cosines.

PRODUCT TO SUM FORMULAS

If we add the formula for $\sin(A - B)$ to the formula for $\sin(A + B)$, we will eventually arrive at a formula for the product $\sin A \cos B$.

$$\sin A \cos B + \cos A \sin B = \sin(A + B)$$
$$\underline{\sin A \cos B - \cos A \sin B = \sin(A - B)}$$
$$2\sin A \cos B \qquad\qquad = \sin(A + B) + \sin(A - B)$$

Dividing both sides of this result by 2 gives us

$$\boxed{\sin A \cos B = \frac{1}{2}[\sin(A + B) + \sin(A - B)]} \qquad (1)$$

By similar methods, we can derive the formulas that follow.

$$\cos A \sin B = \frac{1}{2}[\sin (A + B) - \sin (A - B)] \qquad (2)$$

$$\cos A \cos B = \frac{1}{2}[\cos (A + B) + \cos (A - B)] \qquad (3)$$

$$\sin A \sin B = \frac{1}{2}[\cos (A - B) - \cos (A + B)] \qquad (4)$$

These four product formulas are of use in calculus. The reason they are useful is that they indicate how we can convert a product into a sum. In calculus, it is sometimes much easier to work with sums of trigonometric functions than it is to work with products.

 EXAMPLE 3 Verify product formula (3) for $A = 30°$ and $B = 120°$.

SOLUTION Substituting $A = 30°$ and $B = 120°$ into

$$\cos A \cos B = \frac{1}{2} [\cos (A + B) + \cos (A - B)]$$

we have

$$\cos 30° \cos 120° = \frac{1}{2} [\cos 150° + \cos (-90°)]$$

$$\frac{\sqrt{3}}{2} \cdot \left(-\frac{1}{2}\right) = \frac{1}{2}\left(-\frac{\sqrt{3}}{2} + 0\right)$$

$$-\frac{\sqrt{3}}{4} = -\frac{\sqrt{3}}{4} \qquad \text{A true statement} \quad \blacksquare$$

 EXAMPLE 4

Write $10 \cos 5x \sin 3x$ as a sum or difference.

SOLUTION Product formula (2) is appropriate for an expression of the form $\cos A \sin B$.

$$10 \cos 5x \sin 3x = 10 \cdot \frac{1}{2} [\sin (5x + 3x) - \sin (5x - 3x)]$$

$$= 5 (\sin 8x - \sin 2x) \quad \blacksquare$$

SUM TO PRODUCT FORMULAS

By some simple manipulations we can change our product formulas into sum formulas. If we take the formula for $\sin A \cos B$, exchange sides, and then multiply through by 2 we have

$$\sin (A + B) + \sin (A - B) = 2 \sin A \cos B$$

If we let $\alpha = A + B$ and $\beta = A - B$, then we can solve for A by adding the left sides and the right sides.

$$A + B = \alpha$$
$$\underline{A - B = \beta}$$
$$2A \quad\quad = \alpha + \beta$$
$$A \quad\quad = \frac{\alpha + \beta}{2}$$

By subtracting the expression for β from the expression for α, we have

$$B = \frac{\alpha - \beta}{2}$$

Writing the equation $\sin(A + B) + \sin(A - B) = 2 \sin A \cos B$ in terms of α and β gives us our first sum formula:

$$\sin \alpha + \sin \beta = 2 \sin \frac{\alpha + \beta}{2} \cos \frac{\alpha - \beta}{2} \tag{5}$$

Similarly, the following sum formulas can be derived from the other product formulas:

$$\sin \alpha - \sin \beta = 2 \cos \frac{\alpha + \beta}{2} \sin \frac{\alpha - \beta}{2} \tag{6}$$

$$\cos \alpha + \cos \beta = 2 \cos \frac{\alpha + \beta}{2} \cos \frac{\alpha - \beta}{2} \tag{7}$$

$$\cos \alpha - \cos \beta = -2 \sin \frac{\alpha + \beta}{2} \sin \frac{\alpha - \beta}{2} \tag{8}$$

EXAMPLE 5   Verify sum formula (7) for $\alpha = 30°$ and $\beta = 90°$.

SOLUTION   We substitute $\alpha = 30°$ and $\beta = 90°$ into sum formula (7) and simplify each side of the resulting equation.

$$\cos 30° + \cos 90° = 2 \cos \frac{30° + 90°}{2} \cos \frac{30° - 90°}{2}$$

$$\cos 30° + \cos 90° = 2 \cos 60° \cos(-30°)$$

$$\frac{\sqrt{3}}{2} + 0 = 2 \left(\frac{1}{2}\right)\left(\frac{\sqrt{3}}{2}\right)$$

$$\frac{\sqrt{3}}{2} = \frac{\sqrt{3}}{2} \quad\quad \text{A true statement}$$

EXAMPLE 6 Verify the identity.

$$-\tan x = \frac{\cos 3x - \cos x}{\sin 3x + \sin x}$$

PROOF Applying the formulas for $\cos \alpha - \cos \beta$ and $\sin \alpha + \sin \beta$ to the right side and then simplifying, we arrive at $-\tan x$.

$$\frac{\cos 3x - \cos x}{\sin 3x + \sin x} = \frac{-2 \sin \dfrac{3x + x}{2} \sin \dfrac{3x - x}{2}}{2 \sin \dfrac{3x + x}{2} \cos \dfrac{3x - x}{2}} \qquad \text{Sum to product formulas}$$

$$= \frac{-2 \sin 2x \sin x}{2 \sin 2x \cos x} \qquad \text{Simplify}$$

$$= -\frac{\sin x}{\cos x} \qquad \text{Divide out common factors}$$

$$= -\tan x \qquad \text{Ratio identity} \quad \blacksquare$$

GETTING READY FOR CLASS

After reading through the preceding section, respond in your own words and in complete sentences.

a. How would you describe $\arcsin \dfrac{3}{5}$ to a classmate?

b. How do we arrive at a formula for the product $\sin A \cos B$?

c. Explain the difference between product to sum formulas and sum to product formulas.

d. Write $\sin \alpha + \sin \beta$ as a product using only sine and cosine.

PROBLEM SET 5.5

Evaluate each expression below without using a calculator. (Assume any variables represent positive numbers.)

 1. $\sin \left(\arcsin \dfrac{3}{5} - \arctan 2 \right)$

2. $\cos \left(\arcsin \dfrac{3}{5} - \arctan 2 \right)$

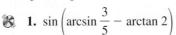 **3.** $\cos \left(\tan^{-1} \dfrac{1}{2} + \sin^{-1} \dfrac{1}{2} \right)$

4. $\sin \left(\tan^{-1} \dfrac{1}{2} - \sin^{-1} \dfrac{1}{2} \right)$

5. $\sin \left(2 \cos^{-1} \dfrac{1}{\sqrt{5}} \right)$

6. $\sin \left(2 \tan^{-1} \dfrac{3}{4} \right)$

Write each expression as an equivalent expression involving only x. (Assume x is positive.)

7. $\tan (\sin^{-1} x)$

8. $\tan (\cos^{-1} x)$

9. $\sin (2 \sin^{-1} x)$

10. $\sin (2 \cos^{-1} x)$

11. $\cos (2 \cos^{-1} x)$

12. $\cos (2 \sin^{-1} x)$

13. Verify product formula (4) for $A = 30°$ and $B = 120°$.

14. Verify product formula (1) for $A = 120°$ and $B = 30°$.

Rewrite each expression as a sum or difference, then simplify if possible.

15. $10 \sin 5x \cos 3x$ **16.** $10 \sin 5x \sin 3x$

17. $\cos 8x \cos 2x$ **18.** $\cos 2x \sin 8x$

19. $\sin 60° \cos 30°$ **20.** $\cos 90° \cos 180°$

21. $\sin 4\pi \sin 2\pi$ **22.** $\cos 3\pi \sin \pi$

23. Verify sum formula (6) for $\alpha = 30°$ and $\beta = 90°$.

24. Verify sum formula (8) for $\alpha = 90°$ and $\beta = 30°$.

Rewrite each expression as a product. Simplify if possible.

25. $\sin 7x + \sin 3x$ **26.** $\cos 5x - \cos 3x$

27. $\cos 45° + \cos 15°$ **28.** $\sin 75° - \sin 15°$

29. $\sin \dfrac{7\pi}{12} - \sin \dfrac{\pi}{12}$ **30.** $\cos \dfrac{\pi}{12} + \cos \dfrac{7\pi}{12}$

Verify each identity.

31. $-\cot x = \dfrac{\sin 3x + \sin x}{\cos 3x - \cos x}$ **32.** $\cot x = \dfrac{\cos 3x + \cos x}{\sin 3x - \sin x}$

33. $\cot x = \dfrac{\sin 4x + \sin 6x}{\cos 4x - \cos 6x}$ **34.** $-\tan 4x = \dfrac{\cos 3x - \cos 5x}{\sin 3x - \sin 5x}$

35. $\tan 4x = \dfrac{\sin 5x + \sin 3x}{\cos 3x + \cos 5x}$ **36.** $\cot 2x = \dfrac{\sin 3x - \sin x}{\cos x - \cos 3x}$

REVIEW PROBLEMS

The problems that follow review material we covered in Section 4.3. Graph one complete cycle.

37. $y = \sin\left(x + \dfrac{\pi}{4}\right)$ **38.** $y = \sin\left(x - \dfrac{\pi}{4}\right)$

39. $y = \cos\left(x - \dfrac{\pi}{3}\right)$ **40.** $y = \cos\left(x + \dfrac{\pi}{3}\right)$

41. $y = \sin\left(2x - \dfrac{\pi}{2}\right)$ **42.** $y = \sin\left(2x - \dfrac{\pi}{3}\right)$

43. $y = \dfrac{1}{2}\cos\left(3x - \dfrac{\pi}{2}\right)$ **44.** $y = \dfrac{4}{3}\cos\left(3x - \dfrac{\pi}{2}\right)$

45. $y = 3\sin\left(\pi x - \dfrac{\pi}{2}\right)$ **46.** $y = 4\sin\left(2\pi x - \dfrac{\pi}{2}\right)$

> ## CHAPTER 5 SUMMARY

Basic Identities [5.1]

	Basic Identities	Common Equivalent Forms
Reciprocal	$\csc \theta = \dfrac{1}{\sin \theta}$	$\sin \theta = \dfrac{1}{\csc \theta}$
	$\sec \theta = \dfrac{1}{\cos \theta}$	$\cos \theta = \dfrac{1}{\sec \theta}$
	$\cot \theta = \dfrac{1}{\tan \theta}$	$\tan \theta = \dfrac{1}{\cot \theta}$
Ratio	$\tan \theta = \dfrac{\sin \theta}{\cos \theta}$	
	$\cot \theta = \dfrac{\cos \theta}{\sin \theta}$	
Pythagorean	$\cos^2 \theta + \sin^2 \theta = 1$	$\sin^2 \theta = 1 - \cos^2 \theta$
		$\sin \theta = \pm\sqrt{1 - \cos^2 \theta}$
		$\cos^2 \theta = 1 - \sin^2 \theta$
		$\cos \theta = \pm\sqrt{1 - \sin^2 \theta}$
	$1 + \tan^2 \theta = \sec^2 \theta$	
	$1 + \cot^2 \theta = \csc^2 \theta$	

1. To prove

$\tan x + \cos x = \sin x\,(\sec x + \cot x)$,

we can multiply through by $\sin x$ on the right side and then change to sines and cosines.

$\sin x\,(\sec x + \cot x)$

$= \sin x \sec x + \sin x \cot x$

$= \sin x \cdot \dfrac{1}{\cos x} + \sin x \cdot \dfrac{\cos x}{\sin x}$

$= \dfrac{\sin x}{\cos x} + \cos x$

$= \tan x + \cos x$

Proving Identities [5.1]

An identity in trigonometry is a statement that two expressions are equal for all replacements of the variable for which each expression is defined. To prove a trigonometric identity, we use trigonometric substitutions and algebraic manipulations to either

1. Transform the right side into the left side, or
2. Transform the left side into the right side.

Remember to work on each side separately. We do not want to use properties from algebra that involve both sides of the identity—like the addition property of equality.

2. To find the exact value for $\cos 75°$, we write $75°$ as $45° + 30°$ and then apply the formula for $\cos(A + B)$.

$\cos 75°$

$= \cos(45° + 30°)$

$= \cos 45° \cos 30°$
  $- \sin 45° \sin 30°$

$= \dfrac{\sqrt{2}}{2} \cdot \dfrac{\sqrt{3}}{2} - \dfrac{\sqrt{2}}{2} \cdot \dfrac{1}{2}$

$= \dfrac{\sqrt{6} - \sqrt{2}}{4}$

Sum and Difference Formulas [5.2]

$\sin(A + B) = \sin A \cos B + \cos A \sin B$

$\sin(A - B) = \sin A \cos B - \cos A \sin B$

$\cos(A + B) = \cos A \cos B - \sin A \sin B$

$\cos(A - B) = \cos A \cos B + \sin A \sin B$

$\tan(A + B) = \dfrac{\tan A + \tan B}{1 - \tan A \tan B}$

$\tan(A - B) = \dfrac{\tan A - \tan B}{1 + \tan A \tan B}$

3. If $\sin A = \frac{3}{5}$ with A in QII, then

$$\cos 2A = 1 - 2 \sin^2 A$$
$$= 1 - 2\left(\frac{3}{5}\right)^2$$
$$= \frac{7}{25}$$

Double-Angle Formulas [5.3]

$$\sin 2A = 2 \sin A \cos A$$

$$\cos 2A = \cos^2 A - \sin^2 A \qquad \text{First form}$$
$$= 2 \cos^2 A - 1 \qquad \text{Second form}$$
$$= 1 - 2 \sin^2 A \qquad \text{Third form}$$

$$\tan 2A = \frac{2 \tan A}{1 - \tan^2 A}$$

4. We can use a half-angle formula to find the exact value of $\sin 15°$ by writing $15°$ as $30°/2$.

$$\sin 15° = \sin \frac{30°}{2}$$
$$= \sqrt{\frac{1 - \cos 30°}{2}}$$
$$= \sqrt{\frac{1 - \sqrt{3}/2}{2}}$$
$$= \sqrt{\frac{2 - \sqrt{3}}{4}}$$
$$= \frac{\sqrt{2 - \sqrt{3}}}{2}$$

Half-Angle Formulas [5.4]

$$\sin \frac{A}{2} = \pm\sqrt{\frac{1 - \cos A}{2}}$$

$$\cos \frac{A}{2} = \pm\sqrt{\frac{1 + \cos A}{2}}$$

$$\tan \frac{A}{2} = \frac{1 - \cos A}{\sin A} = \frac{\sin A}{1 + \cos A}$$

5. We can write the product

$$10 \cos 5x \sin 3x$$

as a difference by applying the second product to sum formula:

$$10 \cos 5x \sin 3x$$
$$= 10 \cdot \frac{1}{2}[\sin (5x + 3x) - \sin (5x - 3x)]$$
$$= 5 (\sin 8x - \sin 2x)$$

Product to Sum Formulas [5.5]

$$\sin A \cos B = \frac{1}{2}[\sin (A + B) + \sin (A - B)]$$

$$\cos A \sin B = \frac{1}{2}[\sin (A + B) - \sin (A - B)]$$

$$\cos A \cos B = \frac{1}{2}[\cos (A + B) + \cos (A - B)]$$

$$\sin A \sin B = \frac{1}{2}[\cos (A - B) - \cos (A + B)]$$

6. Prove

$$-\tan x = \frac{\cos 3x - \cos x}{\sin 3x + \sin x}$$

Proof

$$\frac{\cos 3x - \cos x}{\sin 3x + \sin x}$$

$$= \frac{-2 \sin \dfrac{3x + x}{2} \sin \dfrac{3x - x}{2}}{2 \sin \dfrac{3x + x}{2} \cos \dfrac{3x - x}{2}}$$

$$= \frac{-2 \sin 2x \sin x}{2 \sin 2x \cos x}$$

$$= -\frac{\sin x}{\cos x}$$

$$= -\tan x$$

Sum to Product Formulas [5.5]

$$\sin \alpha + \sin \beta = 2 \sin \frac{\alpha + \beta}{2} \cos \frac{\alpha - \beta}{2}$$

$$\sin \alpha - \sin \beta = 2 \cos \frac{\alpha + \beta}{2} \sin \frac{\alpha - \beta}{2}$$

$$\cos \alpha + \cos \beta = 2 \cos \frac{\alpha + \beta}{2} \cos \frac{\alpha - \beta}{2}$$

$$\cos \alpha - \cos \beta = -2 \sin \frac{\alpha + \beta}{2} \sin \frac{\alpha - \beta}{2}$$

> ## CHAPTER 5 TEST

Prove each identity.

1. $\tan \theta = \sin \theta \sec \theta$

2. $\dfrac{\cot \theta}{\csc \theta} = \cos \theta$

3. $(\sec x - 1)(\sec x + 1) = \tan^2 x$

4. $\sec \theta - \cos \theta = \tan \theta \sin \theta$

5. $\dfrac{\cos t}{1 - \sin t} = \dfrac{1 + \sin t}{\cos t}$

6. $\dfrac{1}{1 - \sin t} + \dfrac{1}{1 + \sin t} = 2 \sec^2 t$

7. $\sin (\theta - 90°) = -\cos \theta$

8. $\cos \left(\dfrac{\pi}{2} + \theta \right) = -\sin \theta$

9. $\cos^4 A - \sin^4 A = \cos 2A$

10. $\cot A = \dfrac{\sin 2A}{1 - \cos 2A}$

11. $\cot x - \tan x = \dfrac{\cos 2x}{\sin x \cos x}$

12. $\tan \dfrac{x}{2} = \dfrac{\tan x}{\sec x + 1}$

Use your graphing calculator to determine if each equation appears to be an identity by graphing the left expression and right expression together. If so, then prove the identity. (Some of these identities are from the book *Plane and Spherical Trigonometry* written by Leonard M. Passano and published by The Macmillan Company in 1918.)

13. $\dfrac{\sec^2 \alpha}{\sin^2 \alpha} = \csc^2 \alpha + \sec^2 \alpha$

14. $(1 + \sec x)(1 - \cos x) = \tan^2 x \cos x$

15. $\dfrac{1}{\tan \theta + \cot \theta} = \csc \theta \sec \theta$

16. $\dfrac{\sin \theta + \cot \theta}{1 - \cos \theta} = \tan \theta$

17. $\cot^2 \theta - \cos^2 \theta = \cot^2 \theta \cos^2 \theta$

18. $\sec^2 x \csc^2 x = \sec^2 x + \csc^2 x$

Let $\sin A = -\dfrac{3}{5}$ with $270° \le A \le 360°$ and $\sin B = \dfrac{12}{13}$ with $90° \le B \le 180°$ and find

19. $\sin (A + B)$

20. $\cos (A - B)$

21. $\cos 2B$

22. $\sin 2B$

23. $\sin \dfrac{A}{2}$

24. $\cos \dfrac{A}{2}$

Find exact values for each of the following:

25. $\sin 75°$

26. $\cos 15°$

27. $\tan \dfrac{\pi}{12}$

28. $\cot \dfrac{\pi}{12}$

Write each expression as a single trigonometric function.

29. $\cos 4x \cos 5x - \sin 4x \sin 5x$

30. $\sin 15° \cos 75° + \cos 15° \sin 75°$

31. If $\sin A = -\dfrac{1}{\sqrt{5}}$ with $180° \le A \le 270°$, find $\cos 2A$ and $\cos \dfrac{A}{2}$.

32. If $\sec A = \sqrt{10}$ with $0° \le A \le 90°$, find $\sin 2A$ and $\sin \dfrac{A}{2}$.

33. Find $\tan A$ if $\tan B = \dfrac{1}{2}$ and $\tan (A + B) = 3$.

34. Find $\cos x$ if $\cos 2x = \dfrac{1}{2}$.

Evaluate each expression below without using a calculator. (Assume any variables represent positive numbers.)

35. $\cos \left(\arcsin \dfrac{4}{5} - \arctan 2 \right)$ **36.** $\sin \left(\arccos \dfrac{4}{5} + \arctan 2 \right)$

37. $\cos (2 \sin^{-1} x)$ **38.** $\sin (2 \cos^{-1} x)$

39. Rewrite the product $\sin 6x \sin 4x$ as a sum or difference.

40. Rewrite the sum $\cos 15° + \cos 75°$ as a product and simplify.

CHAPTER 5 GROUP PROJECT

COMBINATIONS OF FUNCTIONS

Figure 1

Objective: To write $y = a \sin Bx + b \cos Bx$ as a single trigonometric function $y = A \sin (Bx + C)$.

The GPS (global positioning system) allows a person to locate their position anywhere in the world. In order to determine position, a GPS receiver downloads a signal that contains navigation data from a NAVSTAR satellite (Figure 1) and performs a series of calculations using this data. Several of the calculations require writing the sum of a sine and cosine function as a single sine function. In this project you will learn how this is done.

In Example 4 of Section 4.5 we showed how we could graph the function $y = \sin x + \cos x$ by adding y-coordinates. We observed that the resulting graph appeared to have the shape of a sine function.

 1. Graph $y = \sin x + \cos x$ with your graphing calculator. Use your calculator to find the coordinates of a high point on the graph and a low point (you can use your $\boxed{\text{TRACE}}$ key, or your calculator may have special commands for finding maximum and minimum points). Find an equation of the graph in the form $y = A \sin (Bx + C)$ by using these coordinates to estimate values of A, B, and C. Check your equation by graphing it. The two graphs should be nearly identical.

To find an equation for this graph using an algebraic method, we need to find a way to write $y = a \sin Bx + b \cos Bx$ as $y = A \sin (Bx + C)$. That is, we would like to find values of A, B and C so that the equation $a \sin Bx + b \cos Bx = A \sin (Bx + C)$ is an identity. Since the value of B will be the same for both functions, we only need to find values for A and C.

2. First, we will assume $a \sin Bx + b \cos Bx = A \sin (Bx + C)$. Express the right side of this equation as a sum using a sum identity. In order for the resulting equation to be an identity, explain why it must be true that $A \cos C = a$ and $A \sin C = b$.

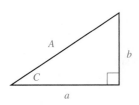

Figure 2

3. Solve for A in terms of a and b using the Pythagorean identity

$$\sin^2 C + \cos^2 C = 1$$

Assume A is positive. To find the value of C, we simply need to find an angle satisfying the criteria:

$$\sin C = \frac{b}{A} \quad \text{and} \quad \cos C = \frac{a}{A}$$

Figure 2 shows a visual way to remember the relationships between a, b, A, and C. You just have to remember to choose angle C so that its terminal side lies in the correct quadrant.

4. Write $y = \sin x + \cos x$ in the form $y = A \sin (Bx + C)$. First identify the values of a, b, and B. Then solve for A and C using your results from Question 3.

5. Graph your equation from Question 4 and compare it with the graph shown in Figure 8 of Section 4.5. The two graphs should be identical!

CHAPTER 5 RESEARCH PROJECT
PTOLEMY

Claudius Ptolemy

Many of the identities presented in this chapter were known to the Greek astronomer and mathematician Claudius Ptolemy (A.D. 85–165). In his work *Almagest,* he was able to find the sine of sums and differences of angles and half-angles.

Research Ptolemy and his use of chords and the chord function. What was the notation that he used, and how is it different from modern notation? What other significant contributions did Ptolemy make to science? Write a paragraph or two about your findings.

CHAPTER **6**

EQUATIONS

Use, use your powers: what now costs you effort will in the end become mechanical.

Georg C. Lichtenberg

INTRODUCTION

The human cannonball is launched from the cannon with an initial velocity of 64 feet per second at an angle of θ from the horizontal.

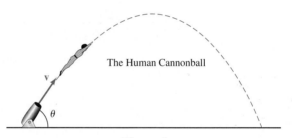

Figure 1

The maximum height attained by the human cannonball is a function of both his initial velocity and the angle that the cannon is inclined from the horizontal. His height t seconds after the cannon is fired is given by the equation

$$h(t) = -16t^2 + 64t \sin \theta$$

which you can see involves a trigonometric function. Equations that contain trigonometric functions are what we will study in this chapter.

STUDY SKILLS FOR CHAPTER 6

This is the last chapter where we will mention study skills. You should know by now what works best for you and what you have to do to achieve your goals for this course. From now on it is simply a matter of sticking with the things that work for you and avoiding the things that do not. It seems simple, but as with anything that takes effort, it is up to you to see that you maintain the skills that get you where you want to be in the course.

SECTION 6.1 | SOLVING TRIGONOMETRIC EQUATIONS

The solution set for an equation is the set of all numbers which, when used in place of the variable, make the equation a true statement. For example, the solution set for the equation $4x^2 - 9 = 0$ is $\left\{-\frac{3}{2}, \frac{3}{2}\right\}$ since these are the only two numbers that, when used in place of x, turn the equation into a true statement.

In algebra, the first kind of equations you learned to solve were linear (or first-degree) equations in one variable. Solving these equations was accomplished by applying two important properties: the *addition property of equality* and the *multiplication property of equality*. These two properties were stated as follows:

ADDITION PROPERTY OF EQUALITY

For any three algebraic expressions A, B, and C

$$\text{If } A = B$$

$$\text{then } A + C = B + C$$

In Words: Adding the same quantity to both sides of an equation will not change the solution set.

MULTIPLICATION PROPERTY OF EQUALITY

For any three algebraic expressions A, B, and C, with $C \neq 0$,

$$\text{If } A = B$$

$$\text{then } AC = BC$$

In Words: Multiplying both sides of an equation by the same nonzero quantity will not change the solution set.

Here is an example that shows how we use these two properties to solve a linear equation in one variable.

 **EXAMPLE 1** Solve for x: $5x + 7 = 2x - 5$.

SOLUTION

$$5x + 7 = 2x - 5$$

$$3x + 7 = -5 \qquad \text{Add } -2x \text{ to each side}$$

$$3x = -12 \qquad \text{Add } -7 \text{ to each side}$$

$$x = -4 \qquad \text{Multiply each side by } \frac{1}{3}$$

Notice in the last step we could just as easily have divided both sides by 3 instead of multiplying both sides by $\frac{1}{3}$. Division by a number and multiplication by its reciprocal are equivalent operations. ■

EXAMPLE 4   Find all degree solutions to $\sin(2A - 50°) = \dfrac{\sqrt{3}}{2}$.

SOLUTION   The expression $2A - 50°$ must be coterminal with $60°$ or $120°$, since $\sin 60° = \sqrt{3}/2$ and $\sin 120° = \sqrt{3}/2$. Therefore,

$$2A - 50° = 60° + 360°k \quad \text{or} \quad 2A - 50° = 120° + 360°k$$

where k is an integer. To solve each of these equations for A, we first add $50°$ to each side of the equation and then divide each side by 2. Here are the steps involved in doing so:

$$2A - 50° = 60° + 360°k \quad \text{or} \quad 2A - 50° = 120° + 360°k$$

$$2A = 110° + 360°k \qquad\qquad 2A = 170° + 360°k \qquad \text{Add } 50° \text{ to each side}$$

$$A = \frac{110° + 360°k}{2} \qquad\qquad A = \frac{170° + 360°k}{2} \qquad \text{Divide each side by 2}$$

$$A = \frac{110°}{2} + \frac{360°k}{2} \qquad\qquad A = \frac{170°}{2} + \frac{360°k}{2}$$

$$A = 55° + 180°k \qquad\qquad\qquad A = 85° + 180°k$$

NOTE   Unless directed otherwise, let's agree to write all degree solutions to our equations in decimal degrees, to the nearest tenth of a degree.

USING TECHNOLOGY

SOLVING EQUATIONS: FINDING INTERSECTION POINTS

We can solve the equation in Example 4 with a graphing calculator by defining the expression on each side of the equation as a function. The solutions to the equation will be the x-values of the points where the two graphs intersect.

Set your calculator to degree mode and define $Y_1 = \sin(2x - 50)$ and $Y_2 = \sqrt{3}/2$. Set your window variables so that

$$0 \le x \le 360, \text{ scale} = 90; \quad -2 \le y \le 2, \text{ scale} = 1$$

Graph both functions and use the appropriate command on your calculator to find the coordinates of the first two intersection points. From Figure 5 we see that the x-coordinates of these points are $x = 55°$ and $x = 85°$. We can also see that the next intersection points occur after $180°$ instead of $360°$. Find the second pair of intersection points and verify that this is true.

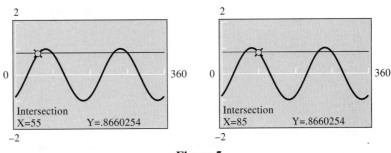

Figure 5

EXAMPLE 5 Solve $3 \sin \theta - 2 = 7 \sin \theta - 1$, if $0° \le \theta \le 360°$.

SOLUTION We can solve for $\sin \theta$ by collecting all the variable terms on the left side and all the constant terms on the right side.

$$3 \sin \theta - 2 = 7 \sin \theta - 1$$

$$-4 \sin \theta - 2 = -1 \qquad \text{Add } -7 \sin \theta \text{ to each side}$$

$$-4 \sin \theta = 1 \qquad \text{Add 2 to each side}$$

$$\sin \theta = -\frac{1}{4} \qquad \text{Divide each side by } -4$$

Since we have not memorized the angle whose sine is $-\frac{1}{4}$, we must convert $-\frac{1}{4}$ to a decimal and use a calculator to find the reference angle.

$$\hat{\theta} = \sin^{-1}(0.2500) = 14.5°$$

We find that the angle whose sine is nearest to 0.2500 is 14.5°. Therefore, the reference angle is 14.5°. Since $\sin \theta$ is negative, θ will terminate in quadrant III or IV (Figure 6).

Figure 6

In quadrant III we have	In quadrant IV we have
$\theta = 180° + 14.5°$	$\theta = 360° - 14.5°$
$= 194.5°$	$= 345.5°$

CALCULATOR NOTE Remember, because of the restricted values on your calculator, if you use the $\boxed{\sin^{-1}}$ key with -0.2500, your calculator will display approximately $-14.5°$, which is not within the desired interval. The best way to proceed is to find the reference angle using $\boxed{\sin^{-1}}$ with the positive value 0.2500. Then do the rest of the calculations as we have here. ■

The next kind of trigonometric equation we will solve is quadratic in form. In algebra, the two most common methods of solving quadratic equations are factoring and applying the quadratic formula. Here is an example that reviews the factoring method.

EXAMPLE 6 Solve $2x^2 - 9x = 5$ for x.

SOLUTION We begin by writing the equation in standard form (0 on one side—decreasing powers of the variable on the other). We then factor the left side and set each factor equal to 0.

$$2x^2 - 9x = 5$$

$$2x^2 - 9x - 5 = 0 \qquad \text{Standard form}$$

$$(2x + 1)(x - 5) = 0 \qquad \text{Factor}$$

$$2x + 1 = 0 \quad \text{or} \quad x - 5 = 0 \qquad \text{Set each factor to 0}$$

$$x = -\frac{1}{2} \quad \text{or} \quad x = 5 \qquad \text{Solving resulting equations}$$

The two solutions, $x = -\frac{1}{2}$ and $x = 5$, are the only two numbers that satisfy the original equation. ■

EXAMPLE 7 Solve $2 \cos^2 t - 9 \cos t = 5$, if $0 \le t < 2\pi$.

SOLUTION This equation is the equation from Example 6 with $\cos t$ in place of x. The fact that $0 \le t < 2\pi$ indicates we are to write our solutions in radians.

$$2 \cos^2 t - 9 \cos t = 5$$

$$2 \cos^2 t - 9 \cos t - 5 = 0 \qquad \text{Standard form}$$

$$(2 \cos t + 1)(\cos t - 5) = 0 \qquad \text{Factor}$$

$$2 \cos t + 1 = 0 \quad \text{or} \quad \cos t - 5 = 0 \qquad \text{Set each factor to 0}$$

$$\cos t = -\frac{1}{2} \quad \text{or} \quad \cos t = 5$$

The first result, $\cos t = -\frac{1}{2}$, gives us a reference angle of $\hat{\theta} = \cos^{-1}(1/2) = \pi/3$. Since $\cos t$ is negative, t must terminate in quadrant II or III (Figure 7). Therefore,

$$t = \pi - \pi/3 = 2\pi/3 \quad \text{or} \quad t = \pi + \pi/3 = 4\pi/3$$

The second result, $\cos t = 5$, has no solution. For any value of t, $\cos t$ must be between -1 and 1. It can never be 5. ▮

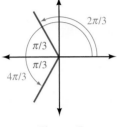

Figure 7

 EXAMPLE 8 Solve $2 \sin^2 \theta + 2 \sin \theta - 1 = 0$, if $0° \le \theta < 360°$.

SOLUTION The equation is already in standard form. If we try to factor the left side, however, we find it does not factor. We must use the quadratic formula. The quadratic formula states that the solutions to the equation

$$ax^2 + bx + c = 0$$

will be

$$x = \frac{-b \pm \sqrt{b^2 - 4ac}}{2a}$$

In our case, the coefficients a, b, and c are

$$a = 2, \qquad b = 2, \qquad c = -1$$

Using these numbers, we can solve for $\sin \theta$ as follows:

$$\sin \theta = \frac{-2 \pm \sqrt{4 - 4(2)(-1)}}{2(2)}$$

$$= \frac{-2 \pm \sqrt{12}}{4}$$

$$= \frac{-2 \pm 2\sqrt{3}}{4}$$

$$= \frac{-1 \pm \sqrt{3}}{2}$$

Using the approximation $\sqrt{3} = 1.7321$, we arrive at the following decimal approximations for $\sin \theta$:

$$\sin \theta = \frac{-1 + 1.7321}{2} \qquad \text{or} \qquad \sin \theta = \frac{-1 - 1.7321}{2}$$

$$\sin \theta = 0.3661 \qquad \text{or} \qquad \sin \theta = -1.3661$$

We will not obtain any solutions from the second expression, $\sin \theta = -1.3661$, since $\sin \theta$ must be between -1 and 1. For $\sin \theta = 0.3661$, we use a calculator to find the angle whose sine is nearest to 0.3661. That angle is $21.5°$, and it is the reference angle for θ. Since $\sin \theta$ is positive, θ must terminate in quadrant I or II (Figure 8). Therefore,

$$\theta = 21.5° \qquad \text{or} \qquad \theta = 180° - 21.5° = 158.5° \quad \text{}$$

Figure 8

GETTING READY FOR CLASS

After reading through the preceding section, respond in your own words and in complete sentences.

a. State the multiplication property of equality.

b. What is the solution set for an equation?

c. How many solutions between $0°$ and $360°$ does the equation $2 \sin x - 1 = 0$ contain?

d. Under what condition is factoring part of the process of solving an equation?

PROBLEM SET 6.1

Solve each equation for θ if $0° \leq \theta < 360°$. Do not use a calculator.

1. $2 \sin \theta = 1$ **2.** $2 \cos \theta = 1$

3. $2 \cos \theta - \sqrt{3} = 0$ **4.** $2 \cos \theta + \sqrt{3} = 0$

5. $2 \tan \theta + 2 = 0$ **6.** $\sqrt{3} \cot \theta - 1 = 0$

Solve each equation for t if $0 \leq t < 2\pi$. Give all answers as exact values in radians. Do not use a calculator.

7. $4 \sin t - \sqrt{3} = 2 \sin t$ **8.** $\sqrt{3} + 5 \sin t = 3 \sin t$

9. $2 \cos t = 6 \cos t - \sqrt{12}$ **10.** $5 \cos t + \sqrt{12} = \cos t$

11. $3 \sin t + 5 = -2 \sin t$ **12.** $3 \sin t + 4 = 4$

Find all solutions in the interval $0° \leq \theta < 360°$. Use a calculator on the last step and write all answers to the nearest tenth of a degree.

13. $4 \sin \theta - 3 = 0$ **14.** $4 \sin \theta + 3 = 0$

15. $2 \cos \theta - 5 = 3 \cos \theta - 2$ **16.** $4 \cos \theta - 1 = 3 \cos \theta + 4$

17. $\sin \theta - 3 = 5 \sin \theta$ **18.** $\sin \theta - 4 = -2 \sin \theta$

Solve for x, if $0 \leq x < 2\pi$. Write your answers in exact values only.

19. $(\sin x - 1)(2 \sin x - 1) = 0$ **20.** $(\cos x - 1)(2 \cos x - 1) = 0$

21. $\tan x (\tan x - 1) = 0$ **22.** $\tan x (\tan x + 1) = 0$

23. $\sin x + 2 \sin x \cos x = 0$ **24.** $\cos x - 2 \sin x \cos x = 0$

25. $2 \sin^2 x - \sin x - 1 = 0$ **26.** $2 \cos^2 x + \cos x - 1 = 0$

Solve for θ, if $0° \leq \theta < 360°$.

27. $(2 \cos \theta + \sqrt{3})(2 \cos \theta + 1) = 0$
28. $(2 \sin \theta - \sqrt{3})(2 \sin \theta - 1) = 0$

29. $\sqrt{3} \tan \theta - 2 \sin \theta \tan \theta = 0$
30. $\tan \theta - 2 \cos \theta \tan \theta = 0$

31. $2 \cos^2 \theta + 11 \cos \theta = -5$
32. $2 \sin^2 \theta - 7 \sin \theta = -3$

Use the quadratic formula to find all solutions in the interval $0° \leq \theta < 360°$ to the nearest tenth of a degree.

33. $2 \sin^2 \theta - 2 \sin \theta - 1 = 0$
34. $2 \cos^2 \theta + 2 \cos \theta - 1 = 0$

35. $\cos^2 \theta + \cos \theta - 1 = 0$
36. $\sin^2 \theta - \sin \theta - 1 = 0$

37. $2 \sin^2 \theta + 1 = 4 \sin \theta$
38. $1 - 4 \cos \theta = -2 \cos^2 \theta$

Write expressions representing all solutions to the equations you solved in the problems below.

39. Problem 1
40. Problem 2

41. Problem 7
42. Problem 8

43. Problem 11
44. Problem 12

45. Problem 13
46. Problem 14

Find all degree solutions to the following equations.

47. $\cos (2A - 50°) = \dfrac{\sqrt{3}}{2}$
48. $\sin (2A + 50°) = \dfrac{\sqrt{3}}{2}$

49. $\sin (3A + 30°) = \dfrac{1}{2}$
50. $\cos (3A + 30°) = \dfrac{1}{2}$

51. $\cos (4A - 20°) = -\dfrac{1}{2}$
52. $\sin (4A - 20°) = -\dfrac{1}{2}$

53. $\sin (5A + 15°) = -\dfrac{1}{\sqrt{2}}$
54. $\cos (5A + 15°) = -\dfrac{1}{\sqrt{2}}$

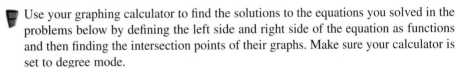 Use your graphing calculator to find the solutions to the equations you solved in the problems below by graphing the function represented by the left side of the equation and then finding its zeros. Make sure your calculator is set to degree mode.

55. Problem 3
56. Problem 4

57. Problem 13
58. Problem 14

59. Problem 29
60. Problem 30

61. Problem 33
62. Problem 34

Use your graphing calculator to find the solutions to the equations you solved in the problems below by defining the left side and right side of the equation as functions and then finding the intersection points of their graphs. Make sure your calculator is set to degree mode.

63. Problem 1
64. Problem 2

65. Problem 17
66. Problem 18

67. Problem 31
68. Problem 32

69. Problem 37
70. Problem 38

71. Problem 47
72. Problem 48

73. Problem 53
74. Problem 54

Motion of a Projectile If a projectile (such as a bullet) is fired into the air with an initial velocity v at an angle of elevation θ (see figure), then the height h of the projectile at time t is given by

$$h = -16t^2 + vt \sin \theta$$

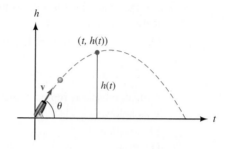

75. Give the equation for the height, if v is 1,500 feet per second and θ is 30°.
76. Give the equation for h, if v is 600 feet per second and θ is 45°. (Leave your answer in exact value form.)
77. Use the equation found in Problem 75 to find the height of the object after 2 seconds.
78. Use the equation found in Problem 76 to find the height of the object after $\sqrt{3}$ seconds to the nearest tenth.
79. Find the angle of elevation θ of a rifle barrel, if a bullet fired at 1,500 feet per second takes 2 seconds to reach a height of 750 feet to the nearest tenth.
80. Find the angle of elevation of a rifle, if a bullet fired at 1,500 feet per second takes 3 seconds to reach a height of 750 feet. Give your answer to the nearest tenth of a degree.

REVIEW PROBLEMS

The problems that follow review material we covered in Sections 5.2 and 5.3. Reviewing these problems will help you with the next section.

81. Write the double-angle formula for $\sin 2A$.
82. Write $\cos 2A$ in terms of $\sin A$ only.
83. Write $\cos 2A$ in terms of $\cos A$ only.
84. Write $\cos 2A$ in terms of $\sin A$ and $\cos A$.
85. Expand $\sin (\theta + 45°)$ and then simplify.
86. Expand $\sin (\theta + 30°)$ and then simplify.
87. Find the exact value of $\sin 75°$.
88. Find the exact value of $\cos 105°$.

89. Prove the identity $\cos 2x = \dfrac{1 - \tan^2 x}{1 + \tan^2 x}$.

90. Prove the identity $\sin 2x = \dfrac{2}{\tan x + \cot x}$.

SECTION 6.2 | MORE ON TRIGONOMETRIC EQUATIONS

In this section, we will use our knowledge of identities to replace some parts of the equations we are solving with equivalent expressions that will make the equations easier to solve. Here are some examples.

 EXAMPLE 1) Solve $2 \cos x - 1 = \sec x$, if $0 \le x < 2\pi$.

SOLUTION   To solve this equation as we have solved the equations in the previous section, we must write each term using the same trigonometric function. To do so, we can use a reciprocal identity to write $\sec x$ in terms of $\cos x$.

$$2 \cos x - 1 = \frac{1}{\cos x}$$

To clear the equation of fractions, we multiply both sides by $\cos x$. (Note that we must assume $\cos x \ne 0$ in order to multiply both sides by it. If we obtain solutions for which $\cos x = 0$, we will have to discard them.)

$$\cos x\,(2 \cos x - 1) = \frac{1}{\cos x} \cdot \cos x$$

$$2 \cos^2 x - \cos x = 1$$

We are left with a quadratic equation that we write in standard form and then solve.

$$2 \cos^2 x - \cos x - 1 = 0 \qquad \text{Standard form}$$

$$(2 \cos x + 1)(\cos x - 1) = 0 \qquad \text{Factor}$$

$$2 \cos x + 1 = 0 \quad \text{or} \quad \cos x - 1 = 0 \qquad \text{Set each factor to 0}$$

$$\cos x = -\frac{1}{2} \quad \text{or} \quad \cos x = 1$$

$$x = \frac{2\pi}{3}, \frac{4\pi}{3} \quad \text{or} \quad x = 0$$

The solutions are 0, $2\pi/3$, and $4\pi/3$.

 EXAMPLE 2) Solve $\sin 2\theta + \sqrt{2} \cos \theta = 0$, $0° \le \theta < 360°$.

SOLUTION   In order to solve this equation, both trigonometric functions must be functions of the same angle. As the equation stands now, one angle is 2θ, while the other is θ. We can write everything as a function of θ by using the double-angle identity $\sin 2\theta = 2 \sin \theta \cos \theta$.

$$\sin 2\theta + \sqrt{2} \cos \theta = 0$$

$$2 \sin \theta \cos \theta + \sqrt{2} \cos \theta = 0 \qquad \sin 2\theta = 2 \sin \theta \cos \theta$$

$$\cos \theta\,(2 \sin \theta + \sqrt{2}) = 0 \qquad \text{Factor out } \cos \theta$$

$$\cos \theta = 0 \quad \text{or} \quad 2 \sin \theta + \sqrt{2} = 0 \qquad \text{Set each factor to 0}$$

$$\sin \theta = -\frac{\sqrt{2}}{2}$$

$$\theta = 90°, 270° \qquad \text{or} \qquad \theta = 225°, 315°$$

EXAMPLE 3 Solve $\cos 2\theta + 3 \sin \theta - 2 = 0$, if $0° \le \theta < 360°$.

SOLUTION We have the same problem with this equation that we did with the equation in Example 2. We must rewrite $\cos 2\theta$ in terms of functions of just θ. Recall that there are three forms of the double-angle identity for $\cos 2\theta$. We choose the double-angle identity that involves $\sin \theta$ only, since the middle term of our equation involves $\sin \theta$, and it is best to have all terms involve the same trigonometric function.

$$\cos 2\theta + 3 \sin \theta - 2 = 0$$

$1 - 2 \sin^2 \theta + 3 \sin \theta - 2 = 0$	$\cos 2\theta = 1 - 2 \sin^2 \theta$
$-2 \sin^2 \theta + 3 \sin \theta - 1 = 0$	Simplify
$2 \sin^2 \theta - 3 \sin \theta + 1 = 0$	Multiply each side by -1
$(2 \sin \theta - 1)(\sin \theta - 1) = 0$	Factor

$$2 \sin \theta - 1 = 0 \quad \text{or} \quad \sin \theta - 1 = 0 \qquad \text{Set factors to 0}$$

$$\sin \theta = \frac{1}{2} \qquad\qquad \sin \theta = 1$$

$$\theta = 30°, 150° \quad \text{or} \quad \theta = 90° \quad \blacksquare$$

EXAMPLE 4 Solve $4 \cos^2 x + 4 \sin x - 5 = 0$, $0 \le x < 2\pi$.

SOLUTION We cannot factor and solve this quadratic equation until each term involves the same trigonometric function. If we change the $\cos^2 x$ in the first term to $1 - \sin^2 x$, we will obtain an equation that involves the sine function only.

$$4 \cos^2 x + 4 \sin x - 5 = 0$$

$4(1 - \sin^2 x) + 4 \sin x - 5 = 0$	$\cos^2 x = 1 - \sin^2 x$
$4 - 4 \sin^2 x + 4 \sin x - 5 = 0$	Distributive property
$-4 \sin^2 x + 4 \sin x - 1 = 0$	Add 4 and -5
$4 \sin^2 x - 4 \sin x + 1 = 0$	Multiply each side by -1
$(2 \sin x - 1)^2 = 0$	Factor
$2 \sin x - 1 = 0$	Set factor to 0

$$\sin x = \frac{1}{2}$$

$$x = \frac{\pi}{6}, \frac{5\pi}{6} \quad \blacksquare$$

EXAMPLE 5 Solve $\sin \theta - \cos \theta = 1$, if $0 \le \theta < 2\pi$.

SOLUTION If we separate $\sin \theta$ and $\cos \theta$ on opposite sides of the equal sign, and then square both sides of the equation, we will be able to use an identity to write the equation in terms of one trigonometric function only.

$$\sin \theta - \cos \theta = 1$$

$\sin \theta = 1 + \cos \theta$	Add $\cos \theta$ to each side
$\sin^2 \theta = (1 + \cos \theta)^2$	Square each side
$\sin^2 \theta = 1 + 2 \cos \theta + \cos^2 \theta$	Expand $(1 + \cos \theta)^2$
$1 - \cos^2 \theta = 1 + 2 \cos \theta + \cos^2 \theta$	$\sin^2 \theta = 1 - \cos^2 \theta$
$0 = 2 \cos \theta + 2 \cos^2 \theta$	Standard form
$0 = 2 \cos \theta (1 + \cos \theta)$	Factor
$2 \cos \theta = 0 \quad$ or $\quad 1 + \cos \theta = 0$	Set factors to 0
$\cos \theta = 0 \qquad\qquad\quad \cos \theta = -1$	
$\theta = \pi/2, 3\pi/2 \quad$ or $\quad \theta = \pi$	

We have three possible solutions, some of which may be extraneous since we squared both sides of the equation in Step 2. Any time we raise both sides of an equation to an even power, we have the possibility of introducing extraneous solutions. We must check each possible solution in our original equation.

Checking $\theta = \pi/2$	*Checking* $\theta = \pi$
$\sin \pi/2 - \cos \pi/2 \overset{?}{=} 1$	$\sin \pi - \cos \pi \overset{?}{=} 1$
$1 - 0 \overset{?}{=} 1$	$0 - (-1) \overset{?}{=} 1$
$1 = 1$	$1 = 1$
$\theta = \pi/2$ is a solution	$\theta = \pi$ is a solution

Checking $\theta = 3\pi/2$

$$\sin 3\pi/2 - \cos 3\pi/2 \overset{?}{=} 1$$

$$-1 - 0 \overset{?}{=} 1$$

$$-1 \neq 1$$

$\theta = 3\pi/2$ is not a solution

All possible solutions, except $\theta = 3\pi/2$, produce true statements when used in place of the variable in the original equation. $\theta = 3\pi/2$ is an extraneous solution produced by squaring both sides of the equation. Our solution set is $\{\pi/2, \pi\}$. ■

USING TECHNOLOGY

VERIFYING SOLUTIONS GRAPHICALLY

We can verify the solutions in Example 5 and confirm that there are only two solutions between 0 and 2π using a graphing calculator. First, write the equation so that all terms are on one side and a zero is on the other.

$$\sin \theta - \cos \theta = 1$$

$\sin \theta - \cos \theta - 1 = 0$	Add -1 to both sides

```
Plot1 Plot2 Plot3
\Y1■sin(X)–cos(X)–1
\Y2=
\Y3=
\Y4=
\Y5=
\Y6=
\Y7=
```

Figure 1

Define the expression on the left side of this equation as function Y₁ (Figure 1). Set your calculator to radian mode and graph the function using the following window settings.

$$0 \le x \le 2\pi, \text{scale} = \pi/2; \; -3 \le y \le 2, \text{scale} = 1$$

The solutions of the equation will be the zeros (x-intercepts) of this function. From the graph, we see that there are only two solutions. Use the feature of your calculator that will allow you to evaluate the function from the graph, and verify that $x = \pi/2$ and $x = \pi$ are x-intercepts (Figure 2). It is clear from the graph that $x = 3\pi/2$ is not a solution.

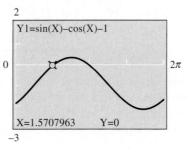 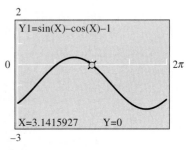

Figure 2

GETTING READY FOR CLASS

After reading through the preceding section, respond in your own words and in complete sentences.

a. What is the first step in solving the equation $2 \cos x - 1 = \sec x$?

b. Why do we need 0 on one side of a quadratic equation in order to solve the equation?

c. How many solutions between 0 and 2π does the equation $\cos x = 0$ contain?

d. How do you factor the left side of the equation $2 \sin \theta \cos \theta + \sqrt{2} \cos \theta = 0$?

PROBLEM SET 6.2

Solve each equation for θ if $0° \le \theta < 360°$. Give your answers in degrees.

1. $\sqrt{3} \sec \theta = 2$ **2.** $\sqrt{2} \csc \theta = 2$

3. $\sqrt{2} \csc \theta + 5 = 3$ **4.** $2\sqrt{3} \sec \theta + 7 = 3$

5. $4 \sin \theta - 2 \csc \theta = 0$ **6.** $4 \cos \theta - 3 \sec \theta = 0$

7. $\sec \theta - 2 \tan \theta = 0$ **8.** $\csc \theta + 2 \cot \theta = 0$

9. $\sin 2\theta - \cos \theta = 0$ **10.** $2 \sin \theta + \sin 2\theta = 0$

11. $2 \cos \theta + 1 = \sec \theta$ **12.** $2 \sin \theta - 1 = \csc \theta$

Solve each equation for x if $0 \le x < 2\pi$. Give your answers in radians using exact values only.

13. $\cos 2x - 3 \sin x - 2 = 0$

14. $\cos 2x - \cos x - 2 = 0$

15. $\cos x - \cos 2x = 0$

16. $\sin x = -\cos 2x$

17. $2 \cos^2 x + \sin x - 1 = 0$

18. $2 \sin^2 x - \cos x - 1 = 0$

19. $4 \sin^2 x + 4 \cos x - 5 = 0$

20. $4 \cos^2 x - 4 \sin x - 5 = 0$

21. $2 \sin x + \cot x - \csc x = 0$

22. $2 \cos x + \tan x = \sec x$

23. $\sin x + \cos x = \sqrt{2}$

24. $\sin x - \cos x = \sqrt{2}$

Solve for θ if $0° \le \theta < 360°$.

25. $\sqrt{3} \sin \theta + \cos \theta = \sqrt{3}$

26. $\sin \theta - \sqrt{3} \cos \theta = \sqrt{3}$

27. $\sqrt{3} \sin \theta - \cos \theta = 1$

28. $\sin \theta - \sqrt{3} \cos \theta = 1$

29. $\sin \dfrac{\theta}{2} - \cos \theta = 0$

30. $\sin \dfrac{\theta}{2} + \cos \theta = 1$

31. $\cos \dfrac{\theta}{2} - \cos \theta = 1$

32. $\cos \dfrac{\theta}{2} - \cos \theta = 0$

For each equation, find all degree solutions in the interval $0° \le \theta < 360°$. If rounding is necessary, round to the nearest tenth of a degree. Use your graphing calculator to verify each solution graphically.

33. $6 \cos \theta + 7 \tan \theta = \sec \theta$

34. $13 \cot \theta + 11 \csc \theta = 6 \sin \theta$

35. $23 \csc^2 \theta - 22 \cot \theta \csc \theta - 15 = 0$

36. $18 \sec^2 \theta - 17 \tan \theta \sec \theta - 12 = 0$

37. $7 \sin^2 \theta - 9 \cos 2\theta = 0$

38. $16 \cos 2\theta - 18 \sin^2 \theta = 0$

Write expressions that give all solutions to the equations you solved in the problems given below.

39. Problem 3

40. Problem 4

41. Problem 23

42. Problem 24

43. Problem 31

44. Problem 32

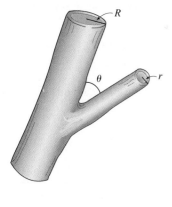

45. Physiology In the human body, the value of θ that makes the following expression 0 is the angle at which an artery of radius r will branch off from a larger artery of radius R in order to minimize the energy loss due to friction. Show that the following expression is 0 when $\cos \theta = r^4/R^4$.

$$r^4 \csc^2 \theta - R^4 \csc \theta \cot \theta$$

46. Physiology Find the value of θ that makes the expression in Problem 45 zero, if $r = 2$ mm and $R = 4$ mm. (Give your answer to the nearest tenth of a degree.)

Solving the following equations will require you to use the quadratic formula. Solve each equation for θ between $0°$ and $360°$, and round your answers to the nearest tenth of a degree.

47. $2 \sin^2 \theta - 2 \cos \theta - 1 = 0$

48. $2 \cos^2 \theta + 2 \sin \theta - 1 = 0$

49. $\cos^2 \theta + \sin \theta = 0$

50. $\sin^2 \theta = \cos \theta$

51. $2 \sin^2 \theta = 3 - 4 \cos \theta$

52. $4 \sin \theta = 3 - 2 \cos^2 \theta$

Use your graphing calculator to find all radian solutions in the interval $0 \le x < 2\pi$ for each of the following equations. Round your answers to four decimal places.

53. $\cos x + 3 \sin x - 2 = 0$

54. $2 \cos x + \sin x + 1 = 0$

55. $\sin^2 x - 3 \sin x - 1 = 0$

56. $\cos^2 x - 3 \cos x + 1 = 0$

57. $\sec x + 2 = \cot x$

58. $\csc x - 3 = \tan x$

REVIEW PROBLEMS

The problems that follow review material we covered in Section 5.4.

If $\sin A = \dfrac{2}{3}$ with A in the interval $0° \le A \le 90°$, find

59. $\sin \dfrac{A}{2}$

60. $\cos \dfrac{A}{2}$

61. $\csc \dfrac{A}{2}$

62. $\sec \dfrac{A}{2}$

63. $\tan \dfrac{A}{2}$

64. $\cot \dfrac{A}{2}$

65. Graph $y = 4 \sin^2 \dfrac{x}{2}$.

66. Graph $y = 6 \cos^2 \dfrac{x}{2}$.

67. Use a half-angle formula to find $\sin 22.5°$.

68. Use a half-angle formula to find $\cos 15°$.

SECTION 6.3 | TRIGONOMETRIC EQUATIONS INVOLVING MULTIPLE ANGLES

In this section, we will consider equations that contain multiple angles. We will use most of the same techniques to solve these equations that we have used in the past. We have to be careful at the last step, however, when our equations contain multiple angles. Here is an example.

 EXAMPLE 1 Solve $\cos 2\theta = \sqrt{3}/2$, if $0° \le \theta < 360°$.

SOLUTION The equation cannot be simplified further. Since we are looking for θ in the interval $0° \le \theta < 360°$, we must first find all values of 2θ in the interval $0° \le 2\theta < 720°$ that satisfy the equation. Here's why:

If $0° \le \theta < 360°$, then $2(0°) \le 2\theta < 2(360°)$

or $0° \le 2\theta < 720°$

To find all values of 2θ between $0°$ and $720°$ that satisfy our original equation, we reason as follows:

$$\text{If } \cos 2\theta = \frac{\sqrt{3}}{2}$$

$$\text{then} \quad 2\theta = 30° \quad \text{or} \quad 2\theta = 330° \quad \text{or} \quad \overbrace{2\theta = 390°}^{30° + 360°} \quad \text{or} \quad \overbrace{2\theta = 690°}^{330° + 360°}$$

Dividing both sides of each equation by 2 gives us all θ between $0°$ and $360°$ that satisfy $\cos 2\theta = \sqrt{3}/2$.

$$\theta = 15° \quad \text{or} \quad \theta = 165° \quad \text{or} \quad \theta = 195° \quad \text{or} \quad \theta = 345°$$

If we had originally been asked to find *all solutions* to the equation in Example 1, instead of only those between 0° and 360°, our work would have been a little less complicated. To find all solutions, we would first find all values of 2θ between 0° and 360° that satisfy the equation, and then we would add on multiples of 360°, since the period of the cosine function is 360°. After that, it is just a matter of dividing everything by 2.

If
$$\cos 2\theta = \frac{\sqrt{3}}{2}$$

then $2\theta = 30° + 360°k$ or $2\theta = 330° + 360°k$ k is an integer

$\theta = 15° + 180°k$ or $\theta = 165° + 180°k$ Divide by 2

The last line gives us *all values* of θ that satisfy $\cos 2\theta = \sqrt{3}/2$. Note that $k = 0$ and $k = 1$ will give us the four solutions between 0° and 360° that we found in Example 1.

USING TECHNOLOGY

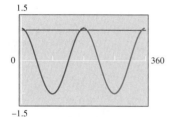

Figure 1

To solve the equation in Example 1 graphically, we can use the intersection of graphs method that was introduced in Section 6.1. Set your graphing calculator to degree mode, and then define each side of the equation as an individual function.

$$Y_1 = \cos(2x), Y_2 = \sqrt{3}/2$$

Graph the two functions using the following window settings:

$$0 \le x \le 360, \text{scale} = 90; -1.5 \le y \le 1.5, \text{scale} = 1$$

Because the period of $y = \cos(2x)$ is $360°/2 = 180°$, we can see that two cycles of the cosine function occur between 0° and 360° (Figure 1). Within the first cycle there are two intersection points. Using the appropriate intersection command, we find that the x-coordinates of these points are $x = 15°$ and $x = 165°$ (Figure 2).

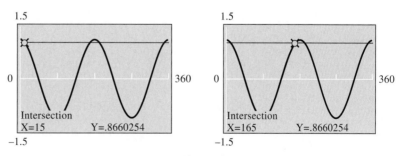

Figure 2

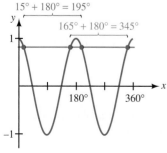

Figure 3

Since the period is 180°, the next pair of intersection points (located on the second cycle) is found by adding 180° to each of the above values (Figure 3). Use the intersection command on your calculator to verify that the next solutions occur at $x = 195°$ and $x = 345°$.

EXAMPLE 2 Find all solutions to tan $3x = 1$, if x is measured in radians with exact values.

SOLUTION First we find all values of $3x$ in the interval $0 \le 3x < \pi$ that satisfy tan $3x = 1$, and then we add on multiples of π because the period of the tangent function is π. After that, we simply divide by 3 to solve for x.

$$\text{If } \tan 3x = 1$$

$$\text{then } 3x = \frac{\pi}{4} + k\pi \qquad k \text{ is an integer}$$

$$x = \frac{\pi}{12} + \frac{k\pi}{3} \qquad \text{Divide by 3}$$

Note that $k = 0, 1,$ and 2 will give us all values of x between 0 and π that satisfy tan $3x = 1$ (Figure 4).

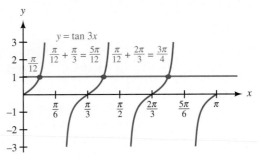

Figure 4

EXAMPLE 3 Solve $\sin 2x \cos x + \cos 2x \sin x = \dfrac{1}{\sqrt{2}}$, if $0 \le x < 2\pi$.

SOLUTION We can simplify the left side by using the formula for $\sin(A + B)$.

$$\sin 2x \cos x + \cos 2x \sin x = \frac{1}{\sqrt{2}}$$

$$\sin(2x + x) = \frac{1}{\sqrt{2}}$$

$$\sin 3x = \frac{1}{\sqrt{2}}$$

First we find *all* possible solutions for x:

$$3x = \frac{\pi}{4} + 2k\pi \qquad \text{or} \qquad 3x = \frac{3\pi}{4} + 2k\pi \qquad k \text{ is an integer}$$

$$x = \frac{\pi}{12} + \frac{2k\pi}{3} \qquad \text{or} \qquad x = \frac{\pi}{4} + \frac{2k\pi}{3} \qquad \text{Divide by 3}$$

To find those solutions that lie in the interval $0 \le x < 2\pi$, we let k take on values of $0, 1,$ and 2. Doing so results in the following solutions (Figure 5):

$$x = \frac{\pi}{12}, \frac{\pi}{4}, \frac{3\pi}{4}, \frac{11\pi}{12}, \frac{17\pi}{12}, \text{ and } \frac{19\pi}{12}$$

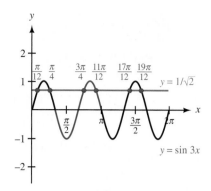

Figure 5

 EXAMPLE 4   Find all solutions to $2 \sin^2 3\theta - \sin 3\theta - 1 = 0$, if θ is measured in degrees.

SOLUTION   We have an equation that is quadratic in $\sin 3\theta$. We factor and solve as usual.

$$2 \sin^2 3\theta - \sin 3\theta - 1 = 0 \qquad \text{Standard form}$$

$$(2 \sin 3\theta + 1)(\sin 3\theta - 1) = 0 \qquad \text{Factor}$$

$$2 \sin 3\theta + 1 = 0 \qquad \text{or} \qquad \sin 3\theta - 1 = 0 \qquad \text{Set factors to 0}$$

$$\sin 3\theta = -\frac{1}{2} \qquad \text{or} \qquad \sin 3\theta = 1$$

$$3\theta = 210° + 360°k \quad \text{or} \quad 3\theta = 330° + 360°k \quad \text{or} \quad 3\theta = 90° + 360°k$$

$$\theta = 70° + 120°k \quad \text{or} \quad \theta = 110° + 120°k \quad \text{or} \quad \theta = 30° + 120°k$$

EXAMPLE 5   Find all solutions, in radians, for $\tan^2 3x = 1$.

SOLUTION   Taking the square root of both sides we have

$$\tan^2 3x = 1$$

$$\tan 3x = \pm 1 \qquad \text{Square root of both sides}$$

Since the period of the tangent function is π, we have

$$3x = \frac{\pi}{4} + k\pi \qquad \text{or} \qquad 3x = \frac{3\pi}{4} + k\pi$$

$$x = \frac{\pi}{12} + \frac{k\pi}{3} \qquad \text{or} \qquad x = \frac{\pi}{4} + \frac{k\pi}{3}$$

EXAMPLE 6 Solve $\sin \theta - \cos \theta = 1$ if $0° \le \theta < 360°$.

SOLUTION We have solved this equation before—in Section 6.2. Here is a second solution.

$$\sin \theta - \cos \theta = 1$$

$$(\sin \theta - \cos \theta)^2 = 1^2 \qquad \text{Square both sides}$$

$$\sin^2 \theta - 2 \sin \theta \cos \theta + \cos^2 \theta = 1 \qquad \text{Expand left side}$$

$$-2 \sin \theta \cos \theta + 1 = 1 \qquad \sin^2 \theta + \cos^2 \theta = 1$$

$$-2 \sin \theta \cos \theta = 0 \qquad \text{Add } -1 \text{ to both sides}$$

$$-\sin 2\theta = 0 \qquad \text{Double-angle identity}$$

$$\sin 2\theta = 0 \qquad \text{Multiply both sides by } -1$$

$$2\theta = 0°, 180°, 360°, 540°$$

$$\theta = 0°, 90°, 180°, 270°$$

Since we squared both sides of the equation in Step 2, we must check all the possible solutions to see if they satisfy the original equation. Doing so gives us solutions $\theta = 90°$ and $180°$. The other two are extraneous, as can be seen in Figure 6.

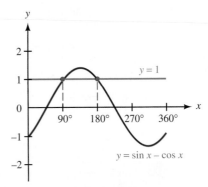

Figure 6

GETTING READY FOR CLASS

After reading through the preceding section, respond in your own words and in complete sentences.

a. If θ is between $0°$ and $360°$, then what can you conclude about 2θ?

b. If $2\theta = 30° + 360°k$, then what can you say about θ?

c. What is an extraneous solution to an equation?

d. Why is it necessary to check solutions to equations that occur after squaring both sides of the equation?

PROBLEM SET 6.3

Find all solutions if $0° \leq \theta < 360°$. Verify your answer graphically.

 **1.** $\sin 2\theta = \sqrt{3}/2$ **2.** $\sin 2\theta = -\sqrt{3}/2$
3. $\tan 2\theta = -1$ **4.** $\cot 2\theta = 1$
5. $\cos 3\theta = -1$ **6.** $\sin 3\theta = -1$

Find all solutions if $0 \leq x < 2\pi$. Use exact values only. Verify your answer graphically.

7. $\sin 2x = 1/\sqrt{2}$ **8.** $\cos 2x = 1/\sqrt{2}$
9. $\sec 3x = -1$ **10.** $\csc 3x = 1$
11. $\tan 2x = \sqrt{3}$ **12.** $\tan 2x = -\sqrt{3}$

Find all degree solutions for each of the following:

13. $\sin 2\theta = 1/2$ **14.** $\sin 2\theta = -\sqrt{3}/2$
15. $\cos 3\theta = 0$ **16.** $\cos 3\theta = -1$
17. $\sin 10\theta = \sqrt{3}/2$ **18.** $\cos 8\theta = 1/2$

Use your graphing calculator to find all degree solutions in the interval $0° \leq x < 360°$ for each of the following equations.

19. $\sin 2x = -\dfrac{1}{\sqrt{2}}$ **20.** $\cos 2x = -\dfrac{1}{2}$

21. $\cos 3x = \dfrac{1}{2}$ **22.** $\sin 3x = \dfrac{\sqrt{3}}{2}$

23. $\tan 2x = \dfrac{1}{\sqrt{3}}$ **24.** $\tan 2x = 1$

Find all solutions if $0 \leq x < 2\pi$. Use exact values only.

25. $\sin 2x \cos x + \cos 2x \sin x = 1/2$
26. $\sin 2x \cos x + \cos 2x \sin x = -1/2$
27. $\cos 2x \cos x - \sin 2x \sin x = -\sqrt{3}/2$
28. $\cos 2x \cos x - \sin 2x \sin x = 1/\sqrt{2}$

Find all solutions in radians using exact values only.

29. $\sin 3x \cos 2x + \cos 3x \sin 2x = 1$
30. $\sin 2x \cos 3x + \cos 2x \sin 3x = -1$
31. $\sin^2 4x = 1$ **32.** $\cos^2 4x = 1$
33. $\cos^3 5x = -1$ **34.** $\sin^3 5x = -1$

Find all degree solutions.

35. $2 \sin^2 3\theta + \sin 3\theta - 1 = 0$ **36.** $2 \sin^2 3\theta + 3 \sin 3\theta + 1 = 0$
37. $2 \cos^2 2\theta + 3 \cos 2\theta + 1 = 0$ **38.** $2 \cos^2 2\theta - \cos 2\theta - 1 = 0$
39. $\tan^2 3\theta = 3$ **40.** $\cot^2 3\theta = 1$

Find all solutions if $0° \leq \theta < 360°$.

41. $\cos \theta - \sin \theta = 1$ **42.** $\sin \theta - \cos \theta = 1$
43. $\sin \theta + \cos \theta = -1$ **44.** $\cos \theta - \sin \theta = -1$
45. $\sin^2 2\theta - 4 \sin 2\theta - 1 = 0$ **46.** $\cos^2 3\theta - 6 \cos 3\theta + 4 = 0$
47. $4 \cos^2 3\theta - 8 \cos 3\theta + 1 = 0$ **48.** $2 \sin^2 2\theta - 6 \sin 2\theta + 3 = 0$
49. $2 \cos^2 4\theta + 2 \sin 4\theta = 1$ **50.** $2 \sin^2 4\theta - 2 \cos 4\theta = 1$

51. Ferris Wheel In Example 7 of Section 4.4, we found the equation that gives the height h of a passenger on a Ferris wheel at any time t during the ride to be

$$h = 139 - 125 \cos \frac{\pi}{10} t$$

where h is given in feet and t is given in minutes. Use this equation to find the times at which a passenger will be 100 feet above the ground. Round your answers to the nearest tenth of a minute. Use your graphing calculator to graph the function and verify your answers.

52. Ferris Wheel In Problem 31 of Problem Set 4.4, you found the equation that gives the height h of a passenger on a Ferris wheel at any time t during the ride to be

$$h = 110.5 - 98.5 \cos \frac{2\pi}{15} t$$

where the units for h are feet and the units for t are minutes. Use this equation to find the times at which a passenger will be 100 feet above the ground. Round your answers to the nearest tenth of a minute. Use your graphing calculator to graph the function and verify your answers.

53. Geometry The formula below gives the relationship between the number of sides n, the radius r, and the length of each side l in a regular polygon. Find n, if $l = r$.

$$l = 2r \sin \frac{180°}{n}$$

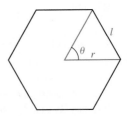

54. Geometry If central angle θ cuts off a chord of length c in a circle of radius r, then the relationship between θ, c, and r is given by

$$2r \sin \frac{\theta}{2} = c$$

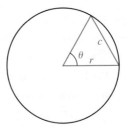

Find θ, if $c = \sqrt{3}r$.

55. Rotating Light In Example 4 of Section 3.5, we found the equation that gives d in terms of t in Figure 7 to be $d = 10 \tan \pi t$. If a person is standing against the wall, 10 feet from point A, how long after the light is at point A will the person see the light? (*Hint:* You must find t when d is 10.)

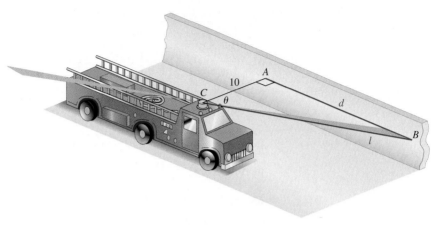

Figure 7

56. Rotating Light In Problem 21 of Problem Set 3.5, you found the equation that gives d in terms of t in Figure 8 to be $d = 100 \tan \frac{1}{2}\pi t$. Two people are sitting on the wall. One of them is directly opposite the lighthouse, while the other person is 100 feet further down the wall. How long after one of them sees the light does the other one see the light? (*Hint:* There are two solutions depending on who sees the light first.)

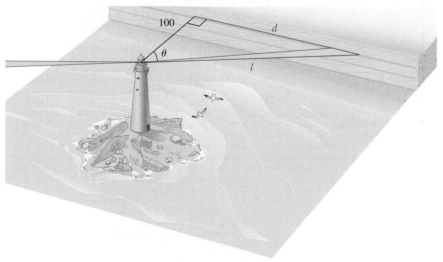

Figure 8

57. Find the smallest positive value of t for which $\sin 2\pi t = \frac{1}{2}$.
58. Find the smallest positive value of t for which $\sin 2\pi t = 1/\sqrt{2}$.

REVIEW PROBLEMS

The problems that follow review material we covered in Sections 5.1 through 5.4.

Prove each identity.

59. $\dfrac{\sin x}{1 + \cos x} = \dfrac{1 - \cos x}{\sin x}$

60. $\dfrac{\sin^2 x}{(1 - \cos x)^2} = \dfrac{1 + \cos x}{1 - \cos x}$

61. $\dfrac{1}{1 + \cos t} + \dfrac{1}{1 - \cos t} = 2\csc^2 t$

62. $\dfrac{1}{1 - \sin t} + \dfrac{1}{1 + \sin t} = 2\sec^2 t$

63. $\tan \dfrac{A}{2} = \dfrac{\sin A}{1 + \cos A}$

64. $\tan \dfrac{A}{2} = \dfrac{1 - \cos A}{\sin A}$

If $\sin A = \dfrac{1}{3}$ with $90° \le A \le 180°$ and $\sin B = \dfrac{3}{5}$ with $0° \le B \le 90°$, find each of the following.

65. $\sin 2A$ **66.** $\cos 2B$ **67.** $\cos \dfrac{A}{2}$ **68.** $\sin \dfrac{B}{2}$

69. $\sin (A + B)$ **70.** $\cos (A - B)$ **71.** $\csc (A + B)$ **72.** $\sec (A - B)$

SECTION 6.4 | PARAMETRIC EQUATIONS AND FURTHER GRAPHING

Up to this point we have had a number of encounters with the Ferris wheel problem. In Section 4.4, we were able to derive an equation that gives the rider's height above the ground at any time t during the ride. We did this by first looking at the graph and then deriving the equation from that graph. Is there a simple way to proceed directly from the diagram of the Ferris wheel problem to an equation that gives height as a function of time? There is, and one of the easiest ways to do so is with *parametric equations,* which we will study in this section.

Let's begin with a model of the giant wheel built by George Ferris. The diameter of this wheel is 250 feet, and the bottom of the wheel sits 14 feet above the ground. We can superimpose a coordinate system on our model, so that the origin of the coordinate system is at the center of the wheel. This is shown in Figure 1.

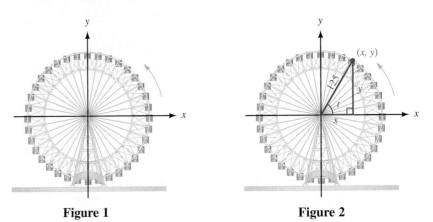

Figure 1   **Figure 2**

Choosing one of the carriages in the first quadrant as the point (x, y), we draw a line from the origin to the carriage. Then we draw the right triangle shown in Figure 2.

From the triangle we have

$$\cos t = \frac{x}{125} \Rightarrow x = 125 \cos t$$

$$\sin t = \frac{y}{125} \Rightarrow y = 125 \sin t$$

The two equations on the right are called *parametric equations*. They show x and y as functions of a third variable, t. That is, each number we substitute for t gives us an ordered pair (x, y). As you would expect, the graph of all such ordered pairs is a circle with a radius of 125 feet, centered at the origin.

USING TECHNOLOGY

GRAPHING PARAMETRIC EQUATIONS

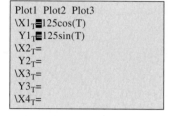

Figure 3

We can use a graphing calculator to confirm that graphing the parametric equations above results in a circle. With our calculator set to degree mode and parametric mode, we input our two equations in the y-variables list, as shown in Figure 3.

Below are the window settings we use depending on whether our calculators are set to degree mode or radian mode. You should try both settings. Generally, degrees give us a more intuitive feel for the problem, while radians are more useful as the problem becomes more complicated:

Degree mode: $0 \leq t \leq 360$,   step $= 5$
$-130 \leq x \leq 130$, scale $= 20$
$-130 \leq y \leq 130$, scale $= 20$

Radian mode: $0 \leq t \leq 2\pi$,   step $= \pi/12$
$-130 \leq x \leq 130$, scale $= 20$
$-130 \leq y \leq 130$, scale $= 20$

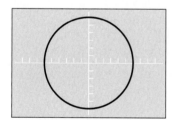

Figure 4

Finally, we graph our parametric equations using the zoom-square command. The result is shown in Figure 4.

Suppose the circle in Figure 4 was found with the calculator in degree mode. If we begin tracing around the circle, we see that when $t = 0°$, $x = 125$ and $y = 0$ (Figure 5). When we trace to $t = 45°$, we see that $x \approx 88$ and $y \approx 88$ (Figure 6).

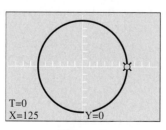

Figure 5

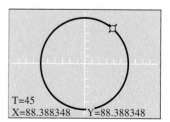

Figure 6

TABLE 1

t	(x, y)
0°	(125, 0)
45°	(88, 88)
90°	(0, 125)
135°	(−88, 88)
180°	(−125, 0)
225°	(−88, −88)
270°	(0, −125)
315°	(88, −88)
360°	(125, 0)

As we continue to trace around the circle, we obtain the values of x and y shown in Table 1.

Before we had graphing calculators and computer software to help us draw these graphs, we had to construct tables, like Table 1, and then use the ordered pairs from the table to plot points. As you can imagine, it could be a long and tedious process.

ELIMINATING THE PARAMETER

Let's go back to our original set of parametric equations and solve for $\cos t$ and $\sin t$:

$$x = 125 \cos t \Rightarrow \cos t = \frac{x}{125}$$

$$y = 125 \sin t \Rightarrow \sin t = \frac{y}{125}$$

Substituting the expressions above for $\cos t$ and $\sin t$ into the Pythagorean identity $\cos^2 t + \sin^2 t = 1$, we have

$$\left(\frac{x}{125}\right)^2 + \left(\frac{y}{125}\right)^2 = 1$$

$$\frac{x^2}{125^2} + \frac{y^2}{125^2} = 1$$

$$x^2 + y^2 = 125^2$$

We recognize this last equation as the equation of a circle with a radius of 125 and a center at the origin. What we have done is eliminate the parameter t to obtain an equation in just x and y whose graph we recognize. This process is called *eliminating the parameter*. Note that it gives us further justification that the graph of our set of parametric equations is a circle.

 **EXAMPLE 1** Eliminate the parameter t from the parametric equations $x = 3 \cos t$ and $y = 2 \sin t$.

SOLUTION Again, we will use the identity $\cos^2 t + \sin^2 t = 1$. Before we do so, however, we must solve the first equation for $\cos t$ and the second equation for $\sin t$.

$$x = 3 \cos t \Rightarrow \cos t = \frac{x}{3}$$

$$y = 2 \sin t \Rightarrow \sin t = \frac{y}{2}$$

Substituting $x/3$ and $y/2$ for $\cos t$ and $\sin t$ into the first Pythagorean identity gives us

$$\left(\frac{x}{3}\right)^2 + \left(\frac{y}{2}\right)^2 = 1$$

$$\frac{x^2}{9} + \frac{y^2}{4} = 1$$

which is the equation of an ellipse. The center is at the origin, the x-intercepts are 3 and -3, and the y-intercepts are 2 and -2. Figure 7 shows the graph.

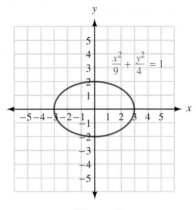

Figure 7

EXAMPLE 2   Eliminate the parameter t from the parametric equations $x = 3 + \sin t$ and $y = \cos t - 2$.

SOLUTION   Solving the first equation for $\sin t$ and the second equation for $\cos t$, we have

$$\sin t = x - 3 \qquad \text{and} \qquad \cos t = y + 2$$

Substituting these expressions into the Pythagorean identity for $\sin t$ and $\cos t$ gives us

$$(x - 3)^2 + (y + 2)^2 = 1$$

which is the equation of a circle with a radius of 1 and center at $(3, -2)$.

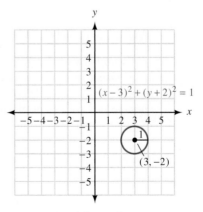

Figure 8

EXAMPLE 3 Eliminate the parameter t from the parametric equations $x = 3 + 2 \sec t$ and $y = 2 + 4 \tan t$.

SOLUTION In this case, we solve for $\sec t$ and $\tan t$ and then use the identity $1 + \tan^2 t = \sec^2 t$.

$$x = 3 + 2 \sec t \Rightarrow \sec t = \frac{x - 3}{2}$$

$$y = 2 + 4 \tan t \Rightarrow \tan t = \frac{y - 2}{4}$$

$$1 + \left(\frac{y - 2}{4} \right)^2 = \left(\frac{x - 3}{2} \right)^2$$

or

$$\frac{(x - 3)^2}{4} - \frac{(y - 2)^2}{16} = 1$$

This is the equation of a hyperbola.

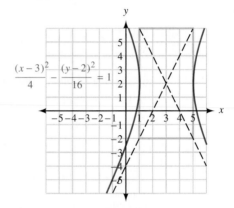

Figure 9

MAKING OUR MODELS MORE REALISTIC

Let's go back to our Ferris wheel model from the beginning of this section. Our wheel has a radius of 125 feet and sits 14 feet above the ground. One trip around the wheel takes 20 minutes. Figure 10 shows this model with a coordinate system superimposed with its origin at the center of the wheel. Also shown are the parametric equations that describe the path of someone riding the wheel.

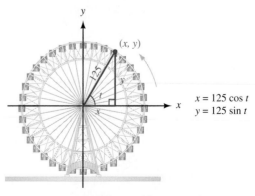

$$x = 125 \cos t$$
$$y = 125 \sin t$$

Figure 10

Our model would be more realistic if the *x*-axis was along the ground, below the wheel. We can accomplish this very easily by moving everything up 139 feet (125 feet for the radius and another 14 feet for its height above the ground). Figure 11 shows our new graph next to a new set of parametric equations.

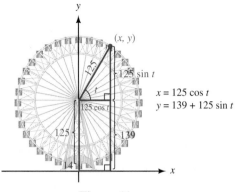

$$x = 125 \cos t$$
$$y = 139 + 125 \sin t$$

Figure 11

Next, let's assume there is a ticket booth 200 feet to the left of the wheel. If we want to use the ticket booth as our starting point, we can move our graph to the right 200 feet. Figure 12 shows this model, along with the corresponding set of parametric equations.

Note that we can solve this last set of equations for cos *t* and sin *t* to get

$$\cos t = \frac{x - 200}{125}$$

$$\sin t = \frac{y - 139}{125}$$

Using these results in our Pythagorean identity, $\cos^2 t + \sin^2 t = 1$, we have $(x - 200)^2 + (y - 139)^2 = 125^2$ which we recognize as the equation of a circle with center at (200, 139) and radius 125.

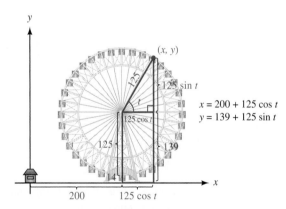

$$x = 200 + 125 \cos t$$
$$y = 139 + 125 \sin t$$

Figure 12

Continuing to improve our model, we would like the rider on the wheel to start their ride at the bottom of the wheel. Assuming the *t* is in radians, we accomplish this by subtracting $\pi/2$ from *t* giving us

$$x = 200 + 125 \cos\left(t - \frac{\pi}{2}\right)$$

$$y = 139 + 125 \sin\left(t - \frac{\pi}{2}\right)$$

Finally, since one trip around the wheel takes 20 minutes, we can write our equations in terms of time *T* by using the proportion

$$\frac{t}{2\pi} = \frac{T}{20}$$

Solving for t we have

$$t = \frac{\pi}{10} T$$

Substituting this expression for t into our parametric equations we have

$$x = 200 + 125 \cos\left(\frac{\pi}{10} T - \frac{\pi}{2}\right)$$

$$y = 139 + 125 \sin\left(\frac{\pi}{10} T - \frac{\pi}{2}\right)$$

These last equations give us a very accurate model of the path taken by someone riding on this Ferris wheel. To graph these equations on our graphing calculator, we can use the following window:

Radian mode: $0 \le T \le 20$,   step $= 1$
$-50 \le x \le 330$, scale $= 20$
$-50 \le y \le 330$, scale $= 20$

Graphing with the zoom-square we have the graph shown in Figure 13. Figure 14 shows a trace of the graph. Tracing around this graph gives us the position of the rider at each minute of the 20-minute ride.

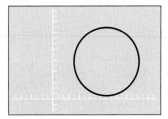

Figure 13

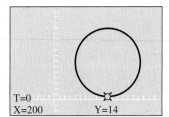

Figure 14

PARAMETRIC EQUATIONS AND THE HUMAN CANNONBALL

In Section 2.5, we found that the human cannonball, shot from a cannon at 50 miles per hour at 60° from the horizontal, will have a horizontal velocity of 25 miles per hour and an initial vertical velocity of 43 miles per hour (Figure 15). We can generalize this so that a human cannonball, shot from a cannon at $|\mathbf{V}_0|$ miles per hour at θ from the horizontal, will have a horizontal speed of $|\mathbf{V}_0| \cos\theta$ miles per hour and an initial vertical speed of $|\mathbf{V}_0| \sin\theta$ miles per hour (Figure 16).

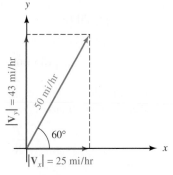

Figure 15

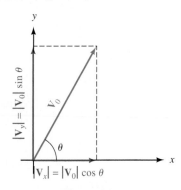

Figure 16

Neglecting the resistance of air, the only force acting on the human cannonball is the force of gravity, which accelerates him at 32 ft/sec² toward the earth. Since the cannonball's horizontal speed is constant, we can find the distance traveled after t seconds by simply multiplying speed and time:

The distance traveled horizontally after t seconds is

$$x = (|\mathbf{V}_0| \cos \theta)t$$

To find the cannonball's vertical distance from the cannon after t seconds, we use a formula from physics:

$$y = (|\mathbf{V}_0| \sin \theta)t - \tfrac{1}{2}gt^2 \quad \text{where } g = 32 \text{ ft/sec}^2 \text{ (the acceleration of gravity on earth)}$$

This gives us the following set of parametric equations:

$$x = (|\mathbf{V}_0| \cos \theta)t$$

$$y = (|\mathbf{V}_0| \sin \theta)t - 16t^2$$

The equations describe the path of a human cannonball shot from a cannon at a speed of $|\mathbf{V}_0|$, at an angle of θ degrees from horizontal. So that the units will agree, $|\mathbf{V}_0|$ must be in feet per second because t is in seconds.

 EXAMPLE 4 Graph the path of the human cannonball if the initial velocity out of the cannon is 50 miles per hour at an angle of 60° from the horizontal.

SOLUTION The position of the cannonball at time t is described parametrically by x and y as given by the equations above. To use these equations, we must first convert 50 miles per hour to feet per second.

$$50 \text{ mi/hr} = \frac{50 \text{ mi}}{\text{hr}} \times \frac{5,280 \text{ ft}}{1 \text{ mi}} \times \frac{1 \text{ hr}}{3,600 \text{ sec}} \approx 73.3 \text{ ft/sec}$$

Substituting 73.3 for $|\mathbf{V}_0|$ and 60° for θ we have parametric equations that together describe the path of the human cannonball.

$$x = (73.3 \cos 60°)t$$

$$y = (73.3 \sin 60°)t - 16t^2$$

We set our calculators to parametric mode, then we set up our function list and window as follows.

```
Plot1  Plot2  Plot3
\X1ₜ■(73.3cos(60))T
 Y1ₜ■(73.3sin(60))T—16T²
\X2ₜ=
 Y2ₜ=
\X3ₜ=
 Y3ₜ=
\X4ₜ=
```

Degree mode: $0 \le t \le 6,$ step $= 0.1$
$0 \le x \le 160,$ scale $= 20$
$0 \le y \le 80,$ scale $= 20$

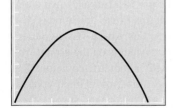

Figure 17   The graph is shown in Figure 17.

GETTING READY FOR CLASS

After reading through the preceding section, respond in your own words and in complete sentences.

a. What are parametric equations?

b. What trigonometric identity do we use to eliminate the parameter t from the equations $x = 3 \cos t$ and $y = 2 \sin t$?

c. What is the first step in eliminating the parameter t from the equations $x = 3 + \sin t$ and $y = \cos t - 2$?

d. How do you convert 50 miles per hour into an equivalent quantity in feet per second?

PROBLEM SET 6.4

Eliminate the parameter t from each of the following and then sketch the graph of each:

1. $x = \sin t \quad y = \cos t$
2. $x = -\sin t \quad y = \cos t$
 3. $x = 3 \cos t \quad y = 3 \sin t$
4. $x = 2 \cos t \quad y = 2 \sin t$
5. $x = 2 \sin t \quad y = 4 \cos t$
6. $x = 3 \sin t \quad y = 4 \cos t$
7. $x = 2 + \sin t \quad y = 3 + \cos t$
8. $x = 3 + \sin t \quad y = 2 + \cos t$
9. $x = \sin t - 2 \quad y = \cos t - 3$
10. $x = \cos t - 3 \quad y = \sin t + 2$
11. $x = 3 + 2 \sin t \quad y = 1 + 2 \cos t$
12. $x = 2 + 3 \sin t \quad y = 1 + 3 \cos t$
13. $x = 3 \cos t - 3 \quad y = 3 \sin t + 1$
14. $x = 4 \sin t - 5 \quad y = 4 \cos t - 3$

Eliminate the parameter t in each of the following:

15. $x = \sec t \quad y = \tan t$
16. $x = \tan t \quad y = \sec t$
17. $x = 3 \sec t \quad y = 3 \tan t$
18. $x = 3 \cot t \quad y = 3 \csc t$
19. $x = 2 + 3 \tan t \quad y = 4 + 3 \sec t$
20. $x = 3 + 5 \tan t \quad y = 2 + 5 \sec t$
21. $x = \cos 2t \quad y = \sin t$
22. $x = \cos 2t \quad y = \cos t$
23. $x = \sin t \quad y = \sin t$
24. $x = \cos t \quad y = \cos t$
25. $x = 3 \sin t \quad y = 2 \sin t$
26. $x = 2 \sin t \quad y = 3 \sin t$

 27. Ferris Wheel The Ferris wheel built in Vienna in 1897 has a diameter of 197 feet and sits 12 feet above the ground. It rotates in a counterclockwise direction, making one complete revolution every 15 minutes. Use parametric equations to model the path of a rider on this wheel. Place your coordinate system so that the origin of the coordinate system is on the ground below the bottom of the wheel. You want to end up with parametric equations that will give you the position of the rider every minute of the ride. Graph your results on a graphing calculator.

 28. Ferris Wheel A Ferris wheel named Colossus was built in St. Louis in 1986. It has a diameter of 165 feet and sits 9 feet above the ground. It rotates in a counterclockwise direction, making one complete revolution every 1.5 minutes. Use parametric equations to model the path of a rider on this wheel. Place your coordinate system so that the origin of the coordinate system is on the ground 100 feet to the left of the wheel. You want to end up with parametric equations that will give you the position of the rider every second of the ride. Graph your results on a graphing calculator.

REVIEW PROBLEMS

The problems that follow review material we covered in Sections 4.6 and 5.5. Evaluate each expression.

29. $\sin\left(\cos^{-1}\dfrac{1}{2}\right)$

30. $\tan\left(\sin^{-1}\dfrac{1}{3}\right)$

31. $\sin\left(\tan^{-1}\dfrac{1}{3}+\sin^{-1}\dfrac{1}{4}\right)$

32. $\cos\left(\tan^{-1}\dfrac{2}{3}+\cos^{-1}\dfrac{1}{3}\right)$

33. $\cos\left(\sin^{-1}x\right)$

34. $\sin\left(\cos^{-1}x\right)$

35. $\cos\left(2\tan^{-1}x\right)$

36. $\sin\left(2\tan^{-1}x\right)$

37. Write $8\sin 3x\cos 2x$ as a sum.

38. Write $\sin 8x + \sin 4x$ as a product.

EXTENDING THE CONCEPTS

39. **Human Cannonball** Graph the parametric equations in Example 4 and then find the maximum height of the cannonball, the maximum distance traveled horizontally, and the time at which the cannonball hits the net. (Assume the barrel of the cannon and the net are the same distance above the ground.)

40. **Human Cannonball** A human cannonball is fired from a cannon with an initial velocity of 50 miles per hour. On the same screen on your calculator, graph the paths taken by the cannonball if the angle between the cannon and the horizontal is 20°, 30°, 40°, 50°, 60°, 70°, and 80°.

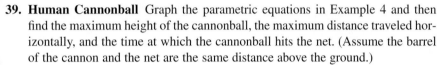

CHAPTER 6 | SUMMARY

EXAMPLES

1. Solve for x:

$$2\cos x - \sqrt{3} = 0$$
$$2\cos x = \sqrt{3}$$
$$\cos x = \dfrac{\sqrt{3}}{2}$$

**SOLUTIONS BETWEEN 0°
AND 360° OR 0 AND 2π**

In Degrees	In Radians
$x = 30°$	$x = \dfrac{\pi}{6}$
or	or
$x = 330°$	$x = \dfrac{11\pi}{6}$

**ALL SOLUTIONS
(k IS AN INTEGER)**

In Degrees	In Radians
$x = 30° + 360°k$	$x = \dfrac{\pi}{6} + 2k\pi$
or	or
$x = 330° + 360°k$	$x = \dfrac{11\pi}{6} + 2k\pi$

Solving Simple Trigonometric Equations [6.1]

We solve linear equations in trigonometry by applying the properties of equality developed in algebra. The two most important properties from algebra are stated as follows:

Addition Property of Equality

For any three algebraic expressions A, B, and C,

If $A = B$

then $A + C = B + C$

In Words: Adding the same quantity to both sides of an equation will not change the solution set.

Multiplication Property of Equality

For any three algebraic expressions A, B, and C, with $C \neq 0$,

If $A = B$

then $AC = BC$

In Words: Multiplying both sides of an equation by the same nonzero quantity will not change the solution set.

To solve a trigonometric equation that is quadratic in $\sin x$ or $\cos x$, we write it in standard form and then factor it or use the quadratic formula.

2. Solve if $0° \leq \theta < 360°$:

$$\cos 2\theta + 3 \sin \theta - 2 = 0$$
$$1 - 2 \sin^2 \theta + 3 \sin \theta - 2 = 0$$
$$2 \sin^2 \theta - 3 \sin \theta + 1 = 0$$
$$(2 \sin \theta - 1)(\sin \theta - 1) = 0$$
$$2 \sin \theta - 1 = 0 \text{ or}$$
$$\sin \theta - 1 = 0$$
$$\sin \theta = \frac{1}{2} \qquad \sin \theta = 1$$
$$\theta = 30°, 150°, 90°$$

Using Identities in Trigonometric Equations [6.2]

Sometimes it is necessary to use identities to make trigonometric substitutions when solving equations. Identities are usually required if the equation contains more than one trigonometric function or if there is more than one angle named in the equation. In the example to the left, we begin by replacing $\cos 2\theta$ with $1 - 2 \sin^2 \theta$. Doing so gives us a quadratic equation in $\sin \theta$, which we put in standard form and solve by factoring.

3. Solve:

$$\sin 2x \cos x + \cos 2x \sin x = \frac{1}{\sqrt{2}}$$
$$\sin (2x + x) = \frac{1}{\sqrt{2}}$$
$$\sin 3x = \frac{1}{\sqrt{2}}$$

$$3x = \frac{\pi}{4} + 2k\pi \quad \text{or} \quad 3x = \frac{3\pi}{4} + 2k\pi$$

$$x = \frac{\pi}{12} + \frac{2k\pi}{3} \quad \text{or} \quad x = \frac{\pi}{4} + \frac{2k\pi}{3}$$

where k is an integer.

Equations Involving Multiple Angles [6.3]

Sometimes the equations we solve in trigonometry reduce to equations that contain multiple angles. When this occurs, we have to be careful in the last step that we do not leave out any solutions. For instance, if we are asked to find all solutions between $x = 0$ and $x = 2\pi$, and our final equation contains $2x$, we must find all values of $2x$ between 0 and 4π in order that x remain between 0 and 2π.

4. Eliminate the parameter t from the equations $x = 3 + \sin t$ and $y = \cos t - 2$.

Solving for $\sin t$ and $\cos t$ we have

$$\sin t = x - 3 \quad \text{and} \quad \cos t = y + 2$$

Substituting these expressions into the Pythagorean identity, we have

$$(x - 3)^2 + (y + 2)^2 = 1$$

which is the equation of a circle with a radius of 1 and center at $(3, -2)$.

Parametric Equations [6.4]

When the coordinates of point (x, y) are described separately by two equations of the form $x = f(t)$ and $y = g(t)$, then the two equations are called *parametric equations* and t is called the *parameter*. One way to graph a set of points (x, y) that are given in terms of the parameter t is to eliminate the parameter and obtain an equation in just x and y that gives the same set of points (x, y).

CHAPTER 6 TEST

Find all solutions in the interval $0° \leq \theta < 360°$. If rounding is necessary, round to the nearest tenth of a degree.

1. $2 \sin \theta - 1 = 0$

2. $\sqrt{3} \tan \theta + 1 = 0$

3. $\cos \theta - 2 \sin \theta \cos \theta = 0$

4. $\tan \theta - 2 \cos \theta \tan \theta = 0$

5. $4 \cos \theta - 2 \sec \theta = 0$

6. $2 \sin \theta - \csc \theta = 1$

7. $\sin \dfrac{\theta}{2} + \cos \theta = 0$

8. $\cos \dfrac{\theta}{2} - \cos \theta = 0$

9. $4 \cos 2\theta + 2 \sin \theta = 1$

10. $\sin (3\theta - 45°) = -\dfrac{\sqrt{3}}{2}$

11. $\sin \theta + \cos \theta = 1$

12. $\sin \theta - \cos \theta = 1$

13. $\cos 3\theta = -\dfrac{1}{2}$

14. $\tan 2\theta = 1$

Find all solutions for the following equations. Write your answers in radians using exact values.

15. $\cos 2x - 3 \cos x = -2$ **16.** $\sqrt{3} \sin x - \cos x = 0$

17. $\sin 2x \cos x + \cos 2x \sin x = -1$ **18.** $\sin^3 4x = 1$

Find all solutions, to the nearest tenth of a degree, in the interval $0° \leq \theta < 360°$.

19. $5 \sin^2 \theta - 3 \sin \theta = 2$ **20.** $4 \cos^2 \theta - 4 \cos \theta = 2$

 Use your graphing calculator to find all radian solutions in the interval $0 \leq x < 2\pi$ for each of the following equations. Round your answers to four decimal places.

21. $3 \sin x - 2 = 0$ **22.** $\cos x + 3 = 4 \sin x$

23. $\sin^2 x + 3 \sin x - 1 = 0$ **24.** $\sin 2x = \dfrac{3}{5}$

 25. Ferris Wheel In Example 7 of Section 4.4, we found the equation that gives the height h of a passenger on a Ferris wheel at any time t during the ride to be

$$h = 139 - 125 \cos \frac{\pi}{10}t$$

where h is given in feet and t is given in minutes. Use this equation to find the times at which a passenger will be 150 feet above the ground. Round your answers to the nearest tenth of a minute.

Eliminate the parameter t from each of the following and then sketch the graph.

26. $x = 3 \cos t$ $y = 3 \sin t$ **27.** $x = \sec t$ $y = \tan t$

28. $x = 3 + 2 \sin t$ $y = 1 + 2 \cos t$

29. $x = 3 \cos t - 3$ $y = 3 \sin t + 1$

30. Ferris Wheel A Ferris wheel has a diameter of 180 feet and sits 8 feet above the ground. It rotates in a counterclockwise direction, making one complete revolution every 3 minutes. Use parametric equations to model the path of a rider on this wheel. Place your coordinate system so that the origin of the coordinate system is on the ground below the bottom of the wheel. You want to end up with parametric equations that will give you the position of the rider every minute of the ride. Graph your results on a graphing calculator.

CHAPTER **6** GROUP PROJECT

DERIVING THE CYCLOID

Objective: To derive parametric equations for the cycloid.

The cycloid is a famous planar curve with a rich history. First studied and named by Galileo, the cycloid is defined as the curve traced by a point on the circumference of a circle that is rolling along a line without slipping (Figure 1).

Figure 1

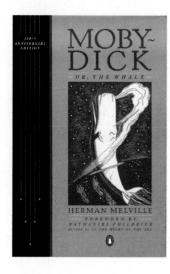

Herman Melville referred to the cycloid in this passage from *Moby Dick:*

> It was in the left-hand try-pot of the Pequod, with the soapstone diligently circling round me, that I was first indirectly struck by the remarkable fact, that in geometry all bodies gliding along a cycloid, my soapstone, for example, will descend from any point in precisely the same time.

To begin, we will assume the circle has radius r and is positioned on a rectangular coordinate system with its center on the positive y-axis and tangent to the x-axis (Figure 2). The x-axis will serve as the line that the circle will roll along. We will choose point P, initially at the origin, to be the fixed point on the circumference of the circle that will trace the cycloid. Figure 3 shows the position of the circle after it has rolled a short distance d along the x-axis. We will use t to represent the angle (in radians) through which the circle has rotated.

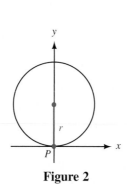

Figure 2   **Figure 3**

1. If d is the distance that the circle has rolled, then what is the length of the arc s? Use this to find a relationship between d and t.
2. Use the lengths a, b, r, and d to find the coordinates of point P. That is, find equations for x and y in terms of these values.
3. Now use right triangle trigonometry to find a and b in terms of r and t.
4. Using your answers to Questions 1 through 3, find equations for x and y in terms of r and t only. Since r is a constant, you now have parametric equations for the cycloid, using the angle t as the parameter!
5. Suppose that $r = 1$. Complete the table by finding x and y for each value of t given. Then use your results to sketch the graph of the cycloid for $0 \le t \le 2\pi$.

t	0	$\pi/4$	$\pi/2$	$3\pi/4$	π	$5\pi/4$	$3\pi/2$	$7\pi/4$	2π
x									
y									

 6. Use your graphing calculator to graph the cycloid for $0 \le t \le 6\pi$. First, put your calculator into parametric mode. Then define x and y using your equations from Question 4 (using $r = 1$). Set your window so that $0 \le t \le 6\pi$, $0 \le x \le 20$, and $0 \le y \le 3$.
7. Research the cycloid. What property of the cycloid is Melville referring to? What other famous property does the cycloid have? What was the connection between the cycloid and navigation at sea? Write a paragraph or two about your findings.

CHAPTER **6** RESEARCH PROJECT

THE WITCH OF AGNESI

Maria Gaetana Agnesi

Maria Gaetana Agnesi (1718–1799) was the author of *Instituzioni Analitiche ad uso Della Gioventu Italiana*, a calculus textbook considered to be the best book of its time and the first surviving mathematical work written by a woman. Within this text Maria Agnesi describes a famous curve that has come to be known as the *Witch of Agnesi*. Figure 1 shows a diagram from the book used to illustrate the derivation of the curve.

Research the Witch of Agnesi. How did this curve get its name? How is the curve defined? What are the parametric equations for the curve? Who was the first to study this curve almost 100 years earlier? What other contributions did Maria Agnesi make in mathematics? Write a paragraph or two about your findings.

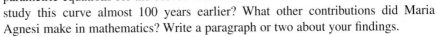

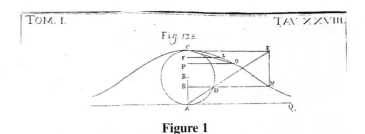

Figure 1

TRIANGLES

Mathematics, the non-empirical science par excellence . . . the science of sciences, delivering the key to those laws of nature and the universe which are concealed by appearances.

Hannah Arendt

INTRODUCTION

The aerial tram in Rio de Janeiro, which carries sightseers up to the top of Sugarloaf Mountain, provides a good example of forces acting in equilibrium. The force representing the weight of the gondola and its passengers must be withstood by the tension in the cable, which can be represented by a pair of forces pointing in the same direction as the cable on either side (Figure 1). Since these three forces are in static equilibrium, they must form a triangle as shown in Figure 2.

Figure 1

Figure 2

In this chapter we will continue our study of triangles and their usefulness as mathematical models in many situations. Specifically, we will consider triangles like the one shown in Figure 2 that are not right triangles.

SECTION 7.1 | THE LAW OF SINES

There are many relationships that exist between the sides and angles in a triangle. One such relationship is called the *law of sines,* which states that the ratio of the sine of an angle to the length of the side opposite that angle is constant in any triangle. Here it is stated in symbols:

Law of Sines

$$\frac{\sin A}{a} = \frac{\sin B}{b} = \frac{\sin C}{c}$$

or, equivalently,

$$\frac{a}{\sin A} = \frac{b}{\sin B} = \frac{c}{\sin C}$$

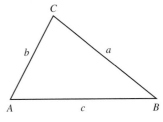

Figure 1

PROOF

The altitude h of the triangle in Figure 2 can be written in terms of $\sin A$ or $\sin B$ depending on which of the two right triangles we are referring to:

$$\sin A = \frac{h}{b} \qquad \sin B = \frac{h}{a}$$

$$h = b \sin A \qquad h = a \sin B$$

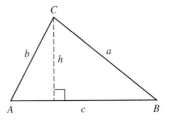

Figure 2

Since h is equal to itself, we have

$$h = h$$

$$b \sin A = a \sin B$$

$$\frac{b \sin A}{ab} = \frac{a \sin B}{ab} \qquad \text{Divide both sides by } ab$$

$$\frac{\sin A}{a} = \frac{\sin B}{b} \qquad \text{Divide out common factors}$$

If we do the same kind of thing with the altitude that extends from A, we will have the third ratio in the law of sines, $\dfrac{\sin C}{c}$, equal to the two ratios above.

Note that the derivation of the law of sines will proceed in the same manner if triangle ABC contains an obtuse angle, as in Figure 3.

In triangle BDC we have

$$\sin (180° - B) = \frac{h}{a}$$

but,

$$\sin (180° - B) = \sin 180° \cos B - \cos 180° \sin B$$

$$= (0) \cos B - (-1) \sin B$$

$$= \sin B$$

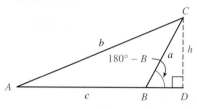

Figure 3

So, $\sin B = h/a$, which is the result we obtained previously. Using triangle ADC, we have $\sin A = h/b$. As you can see, these are the same two expressions we began with when deriving the law of sines for the acute triangle in Figure 2. From this point on, the derivation would match our previous derivation.

We can use the law of sines to find missing parts of triangles for which we are given two angles and a side.

TWO ANGLES AND ONE SIDE

In our first example, we are given two angles and the side opposite one of them. (You may recall that in geometry these were the parts we needed equal in two triangles in order to prove them congruent using the AAS Theorem.)

 **EXAMPLE 1** In triangle ABC, $A = 30°$, $B = 70°$, and $a = 8.0$ cm. Find the length of side c.

SOLUTION We begin by drawing a picture of triangle ABC (it does not have to be accurate) and labeling it with the information we have been given.

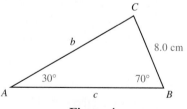

Figure 4

When we use the law of sines, we must have one of the ratios given to us. In this case, since we are given a and A, we have the ratio $\dfrac{a}{\sin A}$. To solve for c, we need first to find angle C. Since the sum of the angles in any triangle is 180°, we have

$$C = 180° - (A + B)$$

$$= 180° - (30° + 70°)$$

$$= 80°$$

To find side c, we use the following two ratios given in the law of sines.

$$\frac{c}{\sin C} = \frac{a}{\sin A}$$

To solve for c, we multiply both sides by $\sin C$ and then substitute.

$c = \dfrac{a \sin C}{\sin A}$ Multiply both sides by $\sin C$

$\quad = \dfrac{8.0 \sin 80°}{\sin 30°}$ Substitute in known values

$\quad = \dfrac{8.0(0.9848)}{0.5000}$ Calculator

$\quad = 16$ cm   Rounded to the nearest integer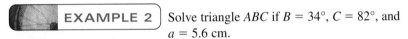

NOTE   The equal sign in the third line above should actually be replaced by the *approximately equal to* symbol, $\approx$, since the decimal 0.9848 is an approximation to $\sin 80°$. (Remember, most of the trigonometric functions are irrational numbers.) In this chapter, we will use an equal sign in the solutions to all of our examples, even when the $\approx$ symbol would be more appropriate, in order to make the examples a little easier to follow.

In our next example, we are given two angles and the side included between them (ASA) and are asked to find all the missing parts.

EXAMPLE 2   Solve triangle ABC if $B = 34°$, $C = 82°$, and $a = 5.6$ cm.

SOLUTION   We begin by finding angle A so that we have one of the ratios in the law of sines completed.

Angle A

$A = 180° - (B + C)$

$\quad = 180° - (34° + 82°)$

$\quad = 64°$

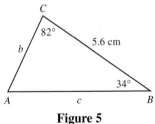

Figure 5

Side b

If $\dfrac{b}{\sin B} = \dfrac{a}{\sin A}$

then $b = \dfrac{a \sin B}{\sin A}$ Multiply both sides by $\sin B$

$\quad = \dfrac{5.6 \sin 34°}{\sin 64°}$ Substitute in known values

$\quad = \dfrac{5.6(0.5592)}{0.8988}$ Calculator

$\quad = 3.5$ cm   To the nearest tenth

Side c

If
$$\frac{c}{\sin C} = \frac{a}{\sin A}$$

then
$$c = \frac{a \sin C}{\sin A} \qquad \text{Multiply both sides by } \sin C$$

$$= \frac{5.6 \sin 82°}{\sin 64°} \qquad \text{Substitute in known values}$$

$$= \frac{5.6(0.9903)}{0.8988} \qquad \text{Calculator}$$

$$= 6.2 \text{ cm} \qquad \text{To the nearest tenth} \quad ■$$

The law of sines, along with some fancy electronic equipment, was used to obtain the results of some of the field events in one of the recent Olympic Games.

Figure 6 is a diagram of a shot put ring. The shot is tossed (put) from the left and lands at *A*. A small electronic device is then placed at *A* (there is usually a dent in the ground where the shot lands, so it is easy to find where to place the device). The device at *A* sends a signal to a booth in the stands that gives the measures of angles *A* and *B*. The distance *a* is found ahead of time. To find the distance *x*, the law of sines is used.

$$\frac{x}{\sin B} = \frac{a}{\sin A}$$

$$\text{or} \qquad x = \frac{a \sin B}{\sin A}$$

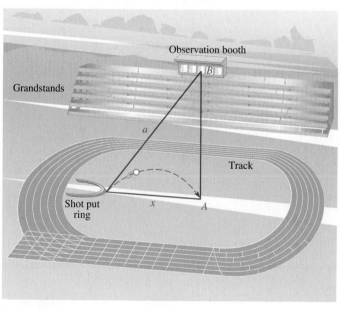

Figure 6

EXAMPLE 3 Find x in Figure 6 if $a = 562$ ft, $B = 5.7°$, and $A = 85.3°$.

SOLUTION

$$x = \frac{a \sin B}{\sin A}$$

$$= \frac{562 \sin 5.7°}{\sin 85.3°}$$

$$= 56.0 \text{ ft} \qquad \text{To three significant digits} \quad \blacksquare$$

EXAMPLE 4 A satellite is circling above the earth as shown in Figure 7. When the satellite is directly above point B, angle A is 75.4°. If the distance between points B and D on the circumference of the earth is 910 miles and the radius of the earth is 3,960 miles, how far above the earth is the satellite?

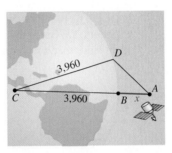

Figure 7

SOLUTION   First we find the radian measure of central angle C by dividing the arc length BD by the radius of the earth. Multiplying this number by $180/\pi$ will give us the degree measure of angle C.

$$C = \underbrace{\frac{910}{3,960}}_{\substack{\text{Angle } C \text{ in} \\ \text{radians}}} \cdot \underbrace{\frac{180}{\pi}}_{\substack{\text{Convert to} \\ \text{degrees}}} = 13.2°$$

Next we find angle CDA.

$$\angle CDA = 180° - (75.4° + 13.2°) = 91.4°$$

To find x, we use the law of sines.

$$\frac{x + 3,960}{\sin 91.4°} = \frac{3,960}{\sin 75.4°}$$

$$x + 3,960 = \frac{3,960 \sin 91.4°}{\sin 75.4°}$$

$$x = \frac{3,960 \sin 91.4°}{\sin 75.4°} - 3,960$$

$$x = 131 \text{ mi} \qquad \text{To three significant digits} \quad \blacksquare$$

EXAMPLE 5 A hot-air balloon is flying over a dry lake when the wind stops blowing. The balloon comes to a stop 450 feet above the ground at point *D* as shown in Figure 8. A jeep following the balloon runs out of gas at point *A*. The nearest service station is due north of the jeep at point *B*. The bearing of the balloon from the jeep at *A* is N 13° E, while the bearing of the balloon from the service station at *B* is S 19° E. If the angle of elevation of the balloon from *A* is 12°, how far will the people in the jeep have to walk to reach the service station at point *B*?

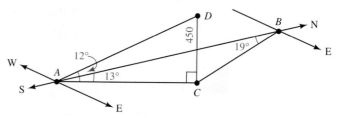

Figure 8

SOLUTION First we find the distance between *C* and *A* using right triangle trigonometry. Since this is an intermediate calculation, which we will use again, we keep more than two significant digits for *AC*.

$$\tan 12° = \frac{450}{AC}$$

$$AC = \frac{450}{\tan 12°}$$

$$= 2{,}117 \text{ ft}$$

Next we find angle *ACB*.

$$\angle ACB = 180° - (13° + 19°)$$

$$= 148°$$

Finally, we find *AB* using the law of sines.

$$\frac{AB}{\sin 148°} = \frac{2{,}117}{\sin 19°}$$

$$AB = \frac{2{,}117 \sin 148°}{\sin 19°}$$

$$= 3{,}400 \text{ ft} \qquad \text{To two significant digits}$$

Since there are 5,280 feet in a mile, the people at *A* will walk approximately 3,400/5,280 = 0.6 miles to get to the service station at *B*.

Our next example involves vectors. It is taken from the text *College Physics* by Miller and Schroeer, published by Saunders College Publishing.

EXAMPLE 6 A traffic light weighing 22 pounds is suspended by two wires as shown in Figure 9. Find the magnitude of the tension in wire *AB*, and the magnitude of the tension in wire *AC*.

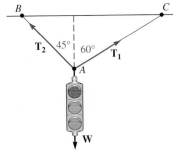

Figure 9

SOLUTION   We assume that the traffic light is not moving and is therefore in the state of *static equilibrium.* When an object is in this state, the sum of the forces acting on the object must be 0. It is because of this fact that we can redraw the vectors from Figure 9 and be sure that they form a closed triangle. Figure 10 shows a convenient redrawing of the two tension vectors $\mathbf{T_1}$ and $\mathbf{T_2}$, and the vector $\mathbf{W}$ that is due to gravity.

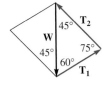

Figure 10

Using the law of sines we have:

$$\frac{|\mathbf{T_1}|}{\sin 45°} = \frac{22}{\sin 75°}$$

$$|\mathbf{T_1}| = \frac{22 \sin 45°}{\sin 75°}$$

$$= 16 \text{ lb} \qquad \text{To two significant figures}$$

$$\frac{|\mathbf{T_2}|}{\sin 60°} = \frac{22}{\sin 75°}$$

$$|\mathbf{T_2}| = \frac{22 \sin 60°}{\sin 75°}$$

$$= 20 \text{ lb} \qquad \text{To two significant figures}$$

GETTING READY FOR CLASS

After reading through the preceding section, respond in your own words and in complete sentences.

a. What is the law of sines?

b. Find $\sin(180° - B)$.

c. Why is it always possible to find the third angle of any triangle once you are given the first two?

d. When an object is in the state of static equilibrium, what do we know about the forces acting upon that object?

PROBLEM SET 7.1

Each problem that follows refers to triangle ABC.

1. If $A = 40°$, $B = 60°$, and $a = 12$ cm, find b.
2. If $A = 80°$, $B = 30°$, and $b = 14$ cm, find a.
3. If $B = 120°$, $C = 20°$, and $c = 28$ inches, find b.
4. If $B = 110°$, $C = 40°$, and $b = 18$ inches, find c.
5. If $A = 10°$, $C = 100°$, and $a = 24$ yd, find c.
6. If $A = 5°$, $C = 125°$, and $c = 510$ yd, find a.
7. If $A = 50°$, $B = 60°$, and $a = 36$ km, find C and then find c.
8. If $B = 40°$, $C = 70°$, and $c = 42$ km, find A and then find a.
9. If $A = 52°$, $B = 48°$, and $c = 14$ cm, find C and then find a.
10. If $A = 33°$, $C = 82°$, and $b = 18$ cm, find B and then find c.

The information below refers to triangle ABC. In each case, find all the missing parts.

11. $A = 42.5°$, $B = 71.4°$, $a = 215$ inches
12. $A = 110.4°$, $C = 21.8°$, $c = 246$ inches
13. $A = 46°$, $B = 95°$, $c = 6.8$ m
14. $B = 57°$, $C = 31°$, $a = 7.3$ m
15. $A = 43° \, 30'$, $C = 120° \, 30'$, $a = 3.48$ ft
16. $B = 14° \, 20'$, $C = 75° \, 40'$, $b = 2.72$ ft
17. $B = 13.4°$, $C = 24.8°$, $a = 315$ cm
18. $A = 105°$, $B = 45°$, $c = 630$ cm

19. In triangle ABC, $A = 30°$, $b = 20$ ft, and $a = 2$ ft. Show that it is impossible to solve this triangle by using the law of sines to find $\sin B$.
20. In triangle ABC, $A = 40°$, $b = 20$ ft, and $a = 18$ ft. Use the law of sines to find $\sin B$ and then give two possible values for B.

Geometry The circle in Figure 11 has a radius of r and center at C. The distance from A to B is x, the distance from A to D is y, and the length of arc BD is s. For Problems 21 through 24, redraw Figure 11, label it as indicated in each problem, and then solve the problem.

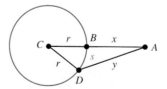

Figure 11

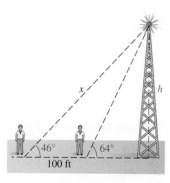

Figure 12

21. If $A = 31°$, $s = 11$, and $r = 12$, find x.
22. If $A = 26°$, $s = 22$, and $r = 20$, find x.
23. If $A = 45°$, $s = 18$, and $r = 15$, find y.
24. If $A = 55°$, $s = 21$, and $r = 22$, find y.

25. **Angle of Elevation** A man standing near a radio station antenna observes that the angle of elevation to the top of the antenna is $64°$. He then walks 100 feet further away and observes that the angle of elevation to the top of the antenna is $46°$. Find the height of the antenna to the nearest foot. (*Hint:* Find x first.)

26. Angle of Elevation A person standing on the street looks up to the top of a building and finds that the angle of elevation is 38°. She then walks one block further away (440 feet) and finds that the angle of elevation to the top of the building is now 28°. How far away from the building is she when she makes her second observation? (See Figure 13.)

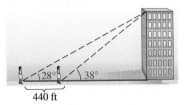

Figure 13

27. Angle of Depression A man is flying in a hot-air balloon in a straight line at a constant rate of 5 feet per second, while keeping it at a constant altitude. As he approaches the parking lot of a market, he notices that the angle of depression from his balloon to a friend's car in the parking lot is 35°. A minute and a half later, after flying directly over this friend's car, he looks back to see his friend getting into the car and observes the angle of depression to be 36°. At that time, what is the distance between him and his friend? (Give your answer to the nearest foot.)

28. Angle of Elevation A woman entering an outside glass elevator on the ground floor of a hotel glances up to the top of the building across the street and notices that the angle of elevation is 48°. She rides the elevator up three floors (60 feet) and finds that the angle of elevation to the top of the building across the street is 32°. How tall is the building across the street? (Give your answer to the nearest foot.)

29. Angle of Elevation From a point on the ground, a person notices that a 110-foot antenna on the top of a hill subtends an angle of 0.5°. If the angle of elevation to the bottom of the antenna is 35°, find the height of the hill. (See Figure 14.)

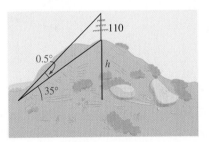

Figure 14

30. Angle of Elevation A 150-foot antenna is on top of a tall building. From a point on the ground, the angle of elevation to the top of the antenna is 28.5°, while the angle of elevation to the bottom of the antenna from the same point is 23.5°. How tall is the building?

31. Height of a Tree Figure 15 is a diagram that shows how Colleen estimates the height of a tree that is on the other side of a stream. She stands at point A facing the tree and finds the angle of elevation from A to the top of the tree to be 51°. Then she turns 105° and walks 25 feet to point B, where she measures the angle between her path and the base of the tree. She finds that angle to be 44°. Use this information to find the height of the tree.

Figure 15

32. Sea Rescue A plane makes a forced landing at sea. The last radio signal received at station C gives the bearing of the plane from C as N 55.4° E at an altitude of 1,050 feet. An observer at C sights the plane and gives $\angle DCB$ as 22.5°. How far will a rescue boat at A have to travel to reach any survivors at B, if the bearing of B from A is S 56.4° E? (See Figure 16.)

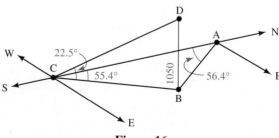

Figure 16

33. Distance to a Ship A ship is anchored off a long straight shoreline that runs north and south. From two observation points 18 miles apart on shore, the bearings of the ship are N 31° E and S 53° E. What is the distance from the ship to each of the observation points?

34. Distance to a Rocket Tom and Fred are 3.5 miles apart watching a rocket being launched from Vandenberg Air Force Base. Tom estimates the bearing of the rocket from his position to be S 75° W, while Fred estimates that the bearing of the rocket from his position is N 65° W. If Fred is due south of Tom, how far is each of them from the rocket?

35. Force A tightrope walker is standing still with one foot on the tightrope as shown in Figure 17. If the tightrope walker weighs 125 pounds, find the magnitudes of the tension in the rope toward each end of the rope.

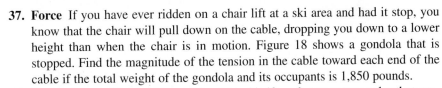

Figure 17

36. Force A tightrope walker weighing 145 pounds is standing still at the center of a tightrope that is 46.5 feet long. The weight of the walker causes the center of the tightrope to move down 14.5 inches. Find the magnitude of the tension in the tightrope.

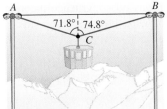

Figure 18

37. Force If you have ever ridden on a chair lift at a ski area and had it stop, you know that the chair will pull down on the cable, dropping you down to a lower height than when the chair is in motion. Figure 18 shows a gondola that is stopped. Find the magnitude of the tension in the cable toward each end of the cable if the total weight of the gondola and its occupants is 1,850 pounds.

38. Force A chair lift at a ski resort is stopped halfway between two poles that support the cable to which the chair is attached. The poles are 215 feet apart and the combined weight of the chair and the three people on the chair is 725 pounds. If the weight of the chair and the people riding it causes the chair to move to a position 15.8 feet below the horizontal line that connects the top of the two poles, find the tension in the cable.

REVIEW PROBLEMS

The problems that follow review material we covered in Sections 3.1 and 6.1.

Solve each equation for θ if $0° \leq \theta < 360°$. If rounding is necessary, round to the nearest tenth of a degree.

39. $2 \sin \theta - \sqrt{2} = 0$ **40.** $5 \cos \theta - 3 = 0$

41. $\sin \theta \cos \theta - 2 \cos \theta = 0$ **42.** $3 \cos \theta - 2 \sin \theta \cos \theta = 0$

43. $2 \sin^2 \theta - 3 \sin \theta = -1$ **44.** $10 \cos^2 \theta + \cos \theta - 3 = 0$

45. $\cos^2 \theta - 4 \cos \theta + 2 = 0$ **46.** $2 \sin^2 \theta - 6 \sin \theta + 3 = 0$

Find all radian solutions to each equation using exact values only.

47. $(\sin x + 1)(2 \sin x - 1) = 0$ **48.** $2 \sin x \cos x - \sqrt{3} \sin x = 0$

Find θ to the nearest tenth of a degree if $0 \leq \theta < 360°$, and

49. $\sin \theta = 0.7380$ **50.** $\sin \theta = 0.7965$

51. $\sin \theta = 0.9668$ **52.** $\sin \theta = 0.2351$

SECTION 7.2 | THE AMBIGUOUS CASE

In this section, we will extend the law of sines to solve triangles for which we are given two sides and the angle opposite one of the given sides.

EXAMPLE 1 Find angle B in triangle ABC if $a = 2$, $b = 6$, and $A = 30°$.

SOLUTION Applying the law of sines we have

$$\sin B = \frac{b \sin A}{a}$$

$$= \frac{6 \sin 30°}{2}$$

$$= \frac{6(0.5000)}{2}$$

$$= 1.5$$

Since $\sin B$ can never be larger than 1, no triangle exists for which $a = 2$, $b = 6$, and $A = 30°$. (You may recall from geometry that there was no congruence theorem SSA.) Figure 1 illustrates what went wrong here. No matter how we orient side a, we cannot complete a triangle.

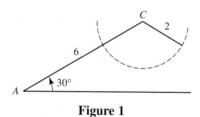

Figure 1

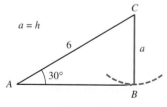

Figure 2

When we are given two sides and an angle opposite one of them (SSA), we have several possibilities for the triangle or triangles that result. As was the case in Example 1, one of the possibilities is that no triangle will fit the given information. If side a in Example 1 had been equal in length to the altitude drawn from vertex C, then we would have had a right triangle that fit the given information, as shown in Figure 2. If side a had been longer than this altitude but shorter than side b, we would have had two triangles that fit the given information, triangle $AB'C$ and triangle ABC as illustrated in Figure 3 and Figure 4. Notice that the two possibilities for side a in Figure 3 create an isosceles triangle $B'CB$.

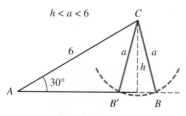

Figure 3

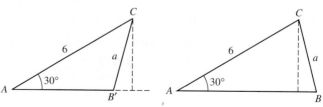

Figure 4

If side *a* in triangle *ABC* of Example 1 had been longer than side *b*, we would have had only one triangle that fit the given information, as shown in Figure 5.

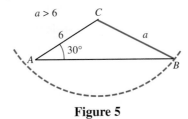

Figure 5

Because of the different possibilities that arise in solving a triangle for which we are given two sides and an angle opposite one of the given sides, we call this situation the *ambiguous case.*

 Find the missing parts in triangle *ABC* if $a = 54$ cm, $b = 62$ cm, and $A = 40°$.

SOLUTION   First we solve for sin *B* with the law of sines.

Angle B

$$\sin B = \frac{b \sin A}{a}$$

$$= \frac{62 \sin 40°}{54}$$

$$= 0.7380$$

Now, since sin *B* is positive for any angle in quadrant I or II, we have two possibilities (Figure 6). We will call one of them *B* and the other *B'*.

$$B = \sin^{-1}(0.7380) = 48° \qquad \text{or} \qquad B' = 180° - 48° = 132°$$

Notice that $B' = 132°$ is the supplement of *B*. We have two different angles that can be found with $a = 54$ cm, $b = 62$ cm, and $A = 40°$. Figure 7 shows both of them. One is labeled *ABC*, while the other is labeled *AB'C*. Also, since triangle *B'BC* is isosceles, we can see that angle *AB'C* must be supplementary to *B*.

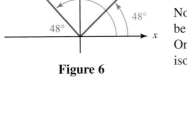

Figure 6

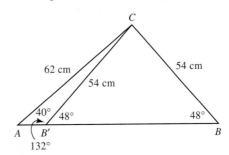

Figure 7

Angles C and C'

Since there are two values for *B*, we have two values for *C*.

$$C = 180 - (A + B) \qquad \text{and} \qquad C' = 180 - (A + B')$$

$$= 180 - (40° + 48°) \qquad\qquad = 180 - (40° + 132°)$$

$$= 92° \qquad\qquad\qquad = 8°$$

Sides c and c′

$$c = \frac{a \sin C}{\sin A} \quad \text{and} \quad c' = \frac{a \sin C'}{\sin A}$$

$$= \frac{54 \sin 92°}{\sin 40°} \qquad\qquad = \frac{54 \sin 8°}{\sin 40°}$$

$$= 84 \text{ cm} \qquad\qquad = 12 \text{ cm} \qquad \text{To two significant digits}$$

Figure 8 shows both triangles.

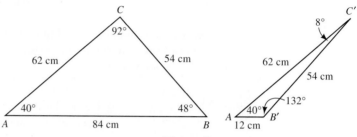

Figure 8

    EXAMPLE 3   Find the missing parts of triangle ABC if $C = 35.4°$, $a = 205$ ft, and $c = 314$ ft.

SOLUTION   Applying the law of sines, we find $\sin A$.

Angle A

$$\sin A = \frac{a \sin C}{c}$$

$$= \frac{205 \sin 35.4°}{314}$$

$$= 0.3782$$

Since $\sin A$ is positive in quadrants I and II, we have two possible values for A.

$$A = \sin^{-1}(0.3782) = 22.2° \quad \text{and} \quad A' = 180° - 22.2° = 157.8°$$

The second possibility, $A' = 157.8°$, will not work, however, since C is $35.4°$ and therefore

$$C + A' = 35.4° + 157.8° = 193.2°$$

which is larger than 180°. This result indicates that there is exactly one triangle that fits the description given in Example 3. In that triangle

$$A = 22.2°$$

Angle B

$$B = 180° - (35.4° + 22.2°) = 122.4°$$

Side b

$$b = \frac{c \sin B}{\sin C}$$

$$b = \frac{314 \sin 122.4°}{\sin 35.4°}$$

$$= 458 \text{ ft} \qquad \text{To three significant digits}$$

Figure 9 is a diagram of this triangle.

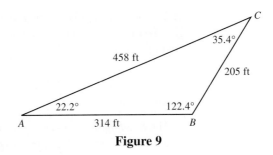

Figure 9

The different cases that can occur when we solve the kinds of triangles given in Examples 1, 2, and 3 become apparent in the process of solving for the missing parts. The following table completes the set of conditions under which we will have 1, 2, or no triangles in the ambiguous case.

In Table 1, we are assuming that we are given angle A and sides a and b in triangle ABC, and that h is the altitude from vertex C.

TABLE 1

Conditions	Number of Triangles	Diagram
$A < 90°$ and $a < h$	0	
$A > 90°$ and $a < b$	0	
$A < 90°$ and $a = h$	1	
$A < 90°$ and $a \geq b$	1	
$A > 90°$ and $a > b$	1	
$A < 90°$ and $h < a < b$	2	

NAVIGATION

In Section 2.4, we used bearing to give the position of one point from another and to give the direction of a moving object. Another way to specify the direction of a moving object is to use *heading*.

> **DEFINITION**
>
> The *heading* of an object is the angle, measured clockwise from due north, to the vector representing the path of the object.

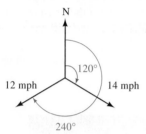

Figure 10

Figure 10 shows two vectors that represent the velocities of two ships—one traveling 14 miles per hour in the direction 120° from due north, and the other traveling 12 miles per hour at 240° from due north. Note that this method of giving the path of a moving object is a little simpler than the bearing method used in Section 2.4. To give the path of the ship traveling 14 miles per hour using the bearing method, we would say it is on a course with bearing S 60° E.

In Example 1 of Section 2.5 we found that although a boat is headed in one direction, its actual course may be in a different direction because of the current of the river. The same type of thing can happen with a plane if the air currents are in a direction different from that of the plane.

To generalize the vocabulary associated with these situations, we say the direction in which a plane or boat is headed is called its *heading* (the direction the plane or boat would travel if there were no wind or current), while the direction in which the plane or boat is actually moving with respect to the ground is called its *true course*. Both heading and true course can be given, as above, by angles measured clockwise from due north. We further notice that the airspeed of a plane is the speed with which it is moving through the air. The ground speed is its speed with respect to the ground below. If there were no wind, then the airspeed and ground speed would be the same. However, if the wind is blowing, then these values will be different.

As illustrated in Figure 11, the vector representing ground speed and true course is the resultant vector found by adding the vector representing airspeed and heading and the vector representing wind speed and direction.

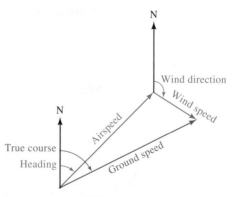

Figure 11

EXAMPLE 4 A plane is flying with an airspeed of 170 miles per hour and a heading of 52.5°. Its true course, however, is at 64.1° from due north. If the wind currents are a constant 40.0 miles per hour, what are the possibilities for the ground speed of the plane?

SOLUTION We represent the velocity of the plane and the velocity of the wind with vectors, the resultant sum of which will be the ground speed and true course of the plane. First, measure an angle of 52.5° from north (clockwise) and draw a vector in that direction with magnitude 170 to represent the heading and airspeed. Then draw a second vector at an angle of 64.1° from north with unknown magnitude to represent the true course and ground speed. Because this second vector must be the resultant, the wind velocity will be some vector originating at the tip of the first vector and ending at the tip of the second vector, and having a length of 40.0. Figure 12 illustrates the situation.

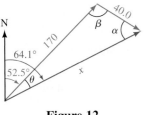

Figure 12

From Figure 12, $\theta = 64.1° - 52.5° = 11.6°$. First we use the law of sines to find angle α.

$$\frac{\sin \alpha}{170} = \frac{\sin 11.6°}{40.0}$$

$$\sin \alpha = \frac{170 \sin 11.6°}{40.0}$$

$$= 0.8546$$

Since $\sin \alpha$ is positive in quadrants I and II, we have two possible values for α.

$$\alpha = \sin^{-1}(0.8546) = 58.7° \qquad \text{and} \qquad \alpha' = 180° - 58.7°$$
$$= 121.3°$$

We have two different triangles that can be formed with the three vectors. Figure 13 shows both of them.

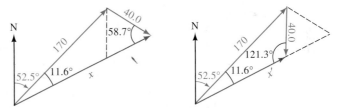

Figure 13

Since there are two values for α, there are two values for β.

$$\beta = 180° - (11.6° + 58.7°) \quad \text{and} \quad \beta' = 180° - (11.6° + 121.3°)$$
$$= 109.7° \qquad\qquad\qquad\qquad = 47.1°$$

Now we can use the law of sines again to find the ground speed.

$$\frac{x}{\sin 109.7°} = \frac{40}{\sin 11.6°} \qquad \text{and} \qquad \frac{x'}{\sin 47.1°} = \frac{40}{\sin 11.6°}$$

$$x = \frac{40 \sin 109.7°}{\sin 11.6°} \qquad\qquad x' = \frac{40 \sin 47.1°}{\sin 11.6°}$$

$$= 187 \text{ mi/hr} \qquad\qquad\qquad = 146 \text{ mi/hr}$$

The two possibilities for the ground speed of the plane are 187 miles per hour and 146 miles per hour.

GETTING READY FOR CLASS

After reading through the preceding section, respond in your own words and in complete sentences.

a. What are the three possibilities for the resulting triangle or triangles when we are given two sides and an angle opposite them?

b. In triangle ABC, $A = 40°$, $B = 48°$, and $B' = 132°$. What are the measures of angles C and C'?

c. How many triangles are possible if angle A is greater than 90° and side a is less than side b?

d. How many triangles are possible if angle A is greater than 90° and side a is greater than side b?

PROBLEM SET 7.2

For each triangle below, solve for B and use the results to explain why the triangle has the given number of solutions.

1. $A = 30°$, $b = 40$ ft, $a = 10$ ft; no solution
2. $A = 150°$, $b = 30$ ft, $a = 10$ ft; no solution
3. $A = 120°$, $b = 20$ cm, $a = 30$ cm; one solution
4. $A = 30°$, $b = 12$ cm, $a = 6$ cm; one solution
5. $A = 60°$, $b = 18$ m, $a = 16$ m; two solutions
6. $A = 20°$, $b = 40$ m, $a = 30$ m; two solutions

Find all solutions to each of the following triangles:

7. $A = 38°$, $a = 41$ ft, $b = 54$ ft
8. $A = 43°$, $a = 31$ ft, $b = 37$ ft
9. $A = 112.2°$, $a = 43.8$ cm, $b = 22.3$ cm
10. $A = 124.3°$, $a = 27.3$ cm, $b = 50.2$ cm
11. $C = 27° 50'$, $c = 347$ m, $b = 425$ m
12. $C = 51° 30'$, $c = 707$ m, $b = 821$ m
13. $B = 45° 10'$, $b = 1.79$ inches, $c = 1.12$ inches
14. $B = 62° 40'$, $b = 6.78$ inches, $c = 3.48$ inches
15. $B = 118°$, $b = 0.68$ cm, $a = 0.92$ cm
16. $B = 30°$, $b = 4.2$ cm, $a = 8.4$ cm
17. $A = 142°$, $b = 2.9$ yd, $a = 1.4$ yd
18. $A = 65°$, $b = 7.6$ yd, $a = 7.1$ yd
19. $C = 26.8°$, $c = 36.8$ km, $b = 36.8$ km
20. $C = 73.4°$, $c = 51.1$ km, $b = 92.4$ km

21. Distance A 50-foot wire running from the top of a tent pole to the ground makes an angle of 58° with the ground. If the length of the tent pole is 44 feet, how far is it from the bottom of the tent pole to the point where the wire is fastened to the ground? (The tent pole is not necessarily perpendicular to the ground.)

22. **Distance** A hot-air balloon is held at a constant altitude by two ropes that are anchored to the ground. One rope is 120 feet long and makes an angle of 65° with the ground. The other rope is 115 feet long. What is the distance between the points on the ground at which the two ropes are anchored?

Draw vectors representing the course of a ship that travels

23. 75 miles on a course with direction 30°
24. 75 miles on a course with direction 330°
25. 25 miles on a course with direction 135°
26. 25 miles on a course with direction 225°

27. **Ground Speed** A plane is headed due east with an airspeed of 340 miles per hour. Its true course, however, is at 98° from due north. If the wind currents are a constant 55 miles per hour, what are the possibilities for the ground speed of the plane?

28. **Current** A ship is headed due north at a constant 16 miles per hour. Because of the ocean current, the true course of the ship is 15°. If the currents are a constant 14 miles per hour, in what direction are the currents running?

29. **Ground Speed** A ship headed due east is moving through the water at a constant speed of 12 miles per hour. However, the true course of the ship is 60°. If the currents are a constant 6 miles per hour, what is the ground speed of the ship?

30. **True Course** A plane headed due east is traveling with an airspeed of 180 miles per hour. The wind currents are moving with constant speed in the direction 240°. If the ground speed of the plane is 90 miles per hour, what is its true course?

31. **Leaning Windmill** After a wind storm, a farmer notices that his 32-foot windmill may be leaning, but he is not sure. From a point on the ground 30 feet from the base of the windmill, he finds that the angle of elevation to the top of the windmill is 48°. Is the windmill leaning? If so, what is the acute angle the windmill makes with the ground?

32. **Distance** A boy is riding his motorcycle on a road that runs east and west. He leaves the road at a service station and rides 5.25 miles in the direction N 15.5° E. Then he turns to his right and rides 6.50 miles back to the road, where his motorcycle breaks down. How far will he have to walk to get back to the service station?

REVIEW PROBLEMS

The problems that follow review material we covered in Section 6.2.

Find all solutions in the interval $0° \leq \theta < 360°$. If rounding is necessary, round to the nearest tenth of a degree.

33. $4 \sin \theta - \csc \theta = 0$
34. $2 \sin \theta - 1 = \csc \theta$
35. $2 \cos \theta - \sin 2\theta = 0$
36. $\cos 2\theta + 3 \cos \theta - 2 = 0$
37. $18 \sec^2 \theta - 17 \tan \theta \sec \theta - 12 = 0$
38. $7 \sin^2 \theta - 9 \cos 2\theta = 0$

Find all radian solutions using exact values only.

39. $2 \cos x - \sec x + \tan x = 0$
40. $2 \cos^2 x - \sin x = 1$
41. $\sin x + \cos x = 0$
42. $\sin x - \cos x = 1$

SECTION 7.3 | THE LAW OF COSINES

In this section, we will derive another relationship that exists between the sides and angles in any triangle. It is called the *law of cosines* and is stated like this:

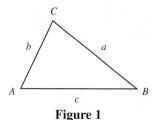

Figure 1

Law of Cosines

$$a^2 = b^2 + c^2 - 2bc \cos A$$

$$b^2 = a^2 + c^2 - 2ac \cos B$$

$$c^2 = a^2 + b^2 - 2ab \cos C$$

DERIVATION

To derive the formulas stated in the law of cosines, we apply the Pythagorean Theorem and some of our basic trigonometric identities. Applying the Pythagorean Theorem to right triangle *BCD* in Figure 2, we have

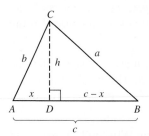

Figure 2

$$a^2 = (c - x)^2 + h^2$$

$$= c^2 - 2cx + x^2 + h^2$$

But from right triangle *ACD*, we have $x^2 + h^2 = b^2$, so

$$a^2 = c^2 - 2cx + b^2$$

$$= b^2 + c^2 - 2cx$$

Now, since $\cos A = x/b$, we have $x = b \cos A$, or

$$a^2 = b^2 + c^2 - 2bc \cos A$$

Applying the same sequence of substitutions and reasoning to the right triangles formed by the altitudes from vertices *A* and *B* will give us the other two formulas listed in the law of cosines.

We can use the law of cosines to solve triangles for which we are given two sides and the angle included between them (SAS) or triangles for which we are given all three sides (SSS).

TWO SIDES AND THE INCLUDED ANGLE

 **EXAMPLE 1** Find the missing parts of triangle *ABC* if $A = 60°$, $b = 20$ inches, and $c = 30$ inches.

SOLUTION The solution process will include the use of both the law of cosines and the law of sines. We begin by using the law of cosines to find *a*.

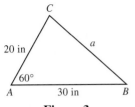

Figure 3

Side a

$$a^2 = b^2 + c^2 - 2bc \cos A \qquad \text{Law of cosines}$$

$$= 20^2 + 30^2 - 2(20)(30) \cos 60° \qquad \text{Substitute in given values}$$

$$= 400 + 900 - 1{,}200(0.5000) \qquad \text{Calculator}$$

$$a^2 = 700$$

$$a = 26 \text{ inches} \qquad \text{To the nearest integer}$$

Now that we have a, we can use the law of sines to solve for either B or C. When we have a choice of angles to solve for, and we are using the law of sines to do so, it is usually best to solve for the smaller angle. Since side b is smaller than side c, angle B will be smaller than angle C.

Angle B

$$\sin B = \frac{b \sin A}{a}$$

$$= \frac{20 \sin 60°}{26}$$

$$\sin B = 0.6662$$

so $\quad B = \sin^{-1}(0.6662) = 42° \qquad$ To the nearest degree

Note that we don't have to check $B' = 180° - 42° = 138°$ because we know B is an acute angle since it is smaller than angles A and C.

Angle C

$$C = 180° - (A + B)$$

$$= 180° - (60° + 42°)$$

$$= 78° \quad \blacksquare$$

EXAMPLE 2 The diagonals of a parallelogram are 24.2 centimeters and 35.4 centimeters and intersect at an angle of 65.5°. Find the length of the shorter side of the parallelogram.

SOLUTION A diagram of the parallelogram is shown in Figure 4. We used the variable x to represent the length of the shorter side. Note also that we labeled half of each diagonal with its length to give us the sides of a triangle. (Recall that the diagonals of a parallelogram bisect each other.)

$$x^2 = (12.1)^2 + (17.7)^2 - 2(12.1)(17.7) \cos 65.5°$$

$$x^2 = 282.07$$

$$x = 16.8 \text{ cm} \qquad \text{To the nearest tenth}$$

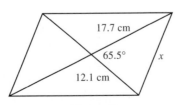

Figure 4

THREE SIDES

To use the law of cosines to solve a triangle for which we are given all three sides, it is convenient to rewrite the equations with the cosines isolated on one side. Here is an equivalent form of the law of cosines.

$$\cos A = \frac{b^2 + c^2 - a^2}{2bc} \qquad \cos B = \frac{a^2 + c^2 - b^2}{2ac} \qquad \cos C = \frac{a^2 + b^2 - c^2}{2ab}$$

Here is how we arrived at the first of these formulas.

$$a^2 = b^2 + c^2 - 2bc \cos A$$

$b^2 + c^2 - 2bc \cos A = a^2$	Exchange sides
$-2bc \cos A = -b^2 - c^2 + a^2$	Add $-b^2$ and $-c^2$ to both sides
$\cos A = \dfrac{b^2 + c^2 - a^2}{2bc}$	Divide both sides by $-2bc$

EXAMPLE 3 Solve triangle ABC if $a = 34$ km, $b = 20$ km, and $c = 18$ km.

SOLUTION We will use the law of cosines to solve for one of the angles and then use the law of sines to find one of the remaining angles. Since there is never any confusion as to whether an angle is acute or obtuse if we have its cosine (the cosine of an obtuse angle is negative), it is best to solve for the largest angle first. Since the longest side is a, we solve for A first.

Angle A

$$\cos A = \frac{b^2 + c^2 - a^2}{2bc} = \frac{20^2 + 18^2 - 34^2}{(2)(20)(18)} = -0.6000$$

so $\quad A = \cos^{-1}(-0.6000) = 127°$ To the nearest degree

There are two ways to find angle C. We can use either the law of sines or the law of cosines.

Angle C

Using the law of sines,

$$\sin C = \frac{c \sin A}{a}$$

$$= \frac{18 \sin 127°}{34}$$

$$\sin C = 0.4228$$

so $\quad C = \sin^{-1}(0.4228)$
$\qquad\ = 25°$ To the nearest degree

Using the law of cosines,

$$\cos C = \frac{a^2 + b^2 - c^2}{2ab}$$

$$= \frac{34^2 + 20^2 - 18^2}{2(34)(20)}$$

$$\cos C = 0.9059$$

so $\quad C = \cos^{-1}(0.9059)$
$\qquad\ = 25°$ To the nearest degree

Angle B

$$B = 180° - (A + C)$$

$$= 180° - (127° + 25°)$$

$$= 28°$$

EXAMPLE 4 A plane is flying with an airspeed of 185 mi/hr with heading 120°. The wind currents are running at a constant 32 mi/hr at 165° clockwise from due north. Find the true course and ground speed of the plane.

SOLUTION Figure 5 is a diagram of the situation with the vector **V** representing the airspeed and direction of the plane and **W** representing the speed and direction of the wind currents. From Figure 5, $\alpha = 180° - 120° = 60°$ and $\theta = 360° - (\alpha + 165°) = 135°$.

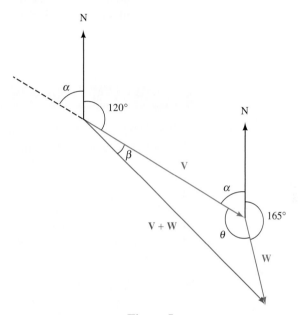

Figure 5

The magnitude of **V** + **W** can be found from the law of cosines.

$$|\mathbf{V} + \mathbf{W}|^2 = |\mathbf{V}|^2 + |\mathbf{W}|^2 - 2|\mathbf{V}||\mathbf{W}|\cos\theta$$

$$= 185^2 + 32^2 - 2(185)(32)\cos 135°$$

$$= 43{,}621$$

so $|\mathbf{V} + \mathbf{W}| = 210$ mph   To two significant digits

To find the direction of **V** + **W**, we first find β using the law of sines.

$$\frac{\sin\beta}{32} = \frac{\sin\theta}{210}$$

$$\sin\beta = \frac{32\sin 135°}{210}$$

$$= 0.1077$$

so $\beta = \sin^{-1}(0.1077) = 6°$ To the nearest degree

The true course is $120° + \beta = 120° + 6° = 126°$. The speed of the plane with respect to the ground is 210 mph. ■

GETTING READY FOR CLASS

After reading through the preceding section, respond in your own words and in complete sentences.

a. State the law of cosines.

b. From what other formula or theorem is the law of cosines derived?

c. What information must we be given in order to use the law of cosines to solve triangles?

d. State the form of the law of cosines used to solve a triangle to which all three sides are given (SSS).

PROBLEM SET 7.3

Each problem below refers to triangle *ABC*.

1. If $a = 120$ inches, $b = 66$ inches, and $C = 60°$, find c.
2. If $a = 120$ inches, $b = 66$ inches, and $C = 120°$, find c.
3. If $a = 22$ yd, $b = 24$ yd, and $c = 26$ yd, find the largest angle.
4. If $a = 13$ yd, $b = 14$ yd, and $c = 15$ yd, find the largest angle.
5. If $b = 4.2$ m, $c = 6.8$ m, and $A = 116°$, find a.
6. If $a = 3.7$ m, $c = 6.4$ m, and $B = 23°$, find b.

7. If $a = 38$ cm, $b = 10$ cm, and $c = 31$ cm, find the largest angle.
8. If $a = 51$ cm, $b = 24$ cm, and $c = 31$ cm, find the largest angle.

Solve each triangle below.

9. $a = 410$ m, $c = 340$ m, $B = 151.5°$
10. $a = 76.3$ m, $c = 42.8$ m, $B = 16.3°$
11. $a = 0.48$ yd, $b = 0.63$ yd, $c = 0.75$ yd
12. $a = 48$ yd, $b = 75$ yd, $c = 63$ yd
13. $b = 0.923$ km, $c = 0.387$ km, $A = 43° \, 20'$
14. $b = 63.4$ km, $c = 75.2$ km, $A = 124° \, 40'$
15. $a = 4.38$ ft, $b = 3.79$ ft, $c = 5.22$ ft
16. $a = 832$ ft, $b = 623$ ft, $c = 345$ ft

17. Use the law of cosines to show that, if $A = 90°$, then $a^2 = b^2 + c^2$.
18. Use the law of cosines to show that, if $a^2 = b^2 + c^2$, then $A = 90°$.

19. **Geometry** The diagonals of a parallelogram are 56 inches and 34 inches and intersect at an angle of 120°. Find the length of the shorter side.

20. **Geometry** The diagonals of a parallelogram are 14 meters and 16 meters and intersect at an angle of 60°. Find the length of the longer side.

21. **Distance Between Two Planes** Two planes leave an airport at the same time. Their speeds are 130 miles per hour and 150 miles per hour, and the angle between their courses is 36°. How far apart are they after 1.5 hours?

22. **Distance Between Two Ships** Two ships leave a harbor entrance at the same time. The first ship is traveling at a constant 18 miles per hour, while the second is traveling at a constant 22 miles per hour. If the angle between their courses is 123°, how far apart are they after 2 hours?

23. Heading and Distance Two planes take off at the same time from an airport. The first plane is flying at 246 miles per hour on a course of 135.0°. The second plane is flying in the direction 175.0° at 357 miles per hour. Assuming there are no wind currents blowing, how far apart are they after 2 hours?

24. Bearing and Distance Two ships leave the harbor at the same time. One ship is traveling at 14 miles per hour on a course with a bearing of S 13° W, while the other is traveling at 12 miles per hour on a course with a bearing of N 75° E. How far apart are they after 3 hours?

25. True Course and Speed A plane is flying with an airspeed of 160 miles per hour and heading of 150°. The wind currents are running at 35 miles per hour at 165° clockwise from due north. Use vectors to find the true course and ground speed of the plane.

26. True Course and Speed A plane is flying with an airspeed of 244 miles per hour with heading 272.7°. The wind currents are running at a constant 45.7 miles per hour in the direction 262.6°. Find the ground speed and true course of the plane.

 27. Speed and Direction A plane has an airspeed of 195 miles per hour and a heading of 30.0°. The ground speed of the plane is 207 miles per hour, and its true course is in the direction of 34.0°. Find the speed and direction of the air currents, assuming they are constants.

28. Speed and Direction The airspeed and heading of a plane are 140 miles per hour and 130°, respectively. If the ground speed of the plane is 135 miles per hour and its true course is 137°, find the speed and direction of the wind currents, assuming they are constants.

Problems 29 and 30 refer to Figure 6, which is a diagram of the YAQUI Mariola bike frame. The frame can be approximated as two triangles that have the seat tube as a common side.

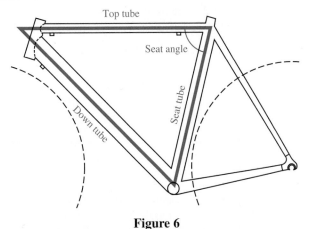

Figure 6

29. Bike Frame Geometry The 47-centimeter Mariola frame has a top tube length of 47.0 centimeters, a seat tube length of 47.5 centimeters, and a seat angle of 78.0°. Find the length of the down tube.

30. Bike Frame Geometry The 57-centimeter Mariola frame has a top tube length of 55.0 centimeters, a seat tube length of 53.5 centimeters, and a seat angle of 78.0°. Find the length of the down tube.

Problems 31 and 32 refer to Figure 7, which is a diagram of the Colnago Dream Plus bike frame. The frame can be approximated as two triangles that have the seat tube as a common side.

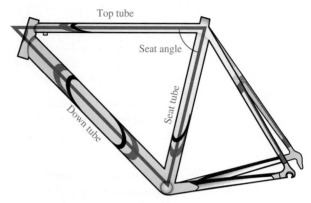

Figure 7

31. **Bike Frame Geometry** The 50-centimeter Colnago frame has a top tube length of 52.3 centimeters, a seat tube length of 48.0 centimeters, and a seat angle of 75.0°. Find the length of the down tube and the angle between the seat tube and the down tube.
32. **Bike Frame Geometry** The 59-centimeter Colnago frame has a top tube length of 56.9 centimeters, a seat tube length of 57.0 centimeters, and a seat angle of 73.0°. Find the length of the down tube and the angle between the seat tube and the down tube.

REVIEW PROBLEMS

The problems that follow review material we covered in Sections 2.5, 6.3, and 7.2. Find all solutions in radians using exact values only.

33. $\sin 3x = 1/2$ 34. $\cos 4x = 0$
35. $\tan^2 3x = 1$ 36. $\tan^2 4x = 1$

Find all degree solutions.

37. $2 \cos^2 3\theta - 9 \cos 3\theta + 4 = 0$
38. $3 \sin^2 2\theta - 2 \sin 2\theta - 5 = 0$
39. $\sin 4\theta \cos 2\theta + \cos 4\theta \sin 2\theta = -1$
40. $\cos 3\theta \cos 2\theta - \sin 3\theta \sin 2\theta = -1$

Solve each equation for θ if $0° \le \theta < 360°$.

41. $\sin \theta + \cos \theta = 1$ 42. $\sin \theta - \cos \theta = 0$

43. **True Course and Speed** A plane has an airspeed of 176 miles per hour and a heading of 40.0°. The air currents are moving at a constant 45.5 miles per hour at 130.0°. Find the ground speed and true course of the plane (Figure 8).
44. **True Course and Speed** A plane has an airspeed of 255 miles per hour and a heading of 130°. The air currents are moving at a constant 25.5 miles per hour at 40.0°. Find the ground speed and true course of the plane.
45. **True Course** A plane has an airspeed of 195 miles per hour and a heading of 30.0°. The air currents are moving at a constant 32.5 miles per hour in the direction N 60.0° W. Find the ground speed and true course of the plane.
46. **True Course** A plane has an airspeed of 140 miles per hour and a heading of 130°. The wind is blowing at a constant 14 miles per hour in the direction N 40° E. Find the ground speed and true course of the plane.

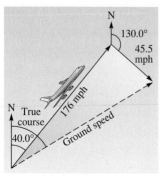

Figure 8

SECTION 7.4 | THE AREA OF A TRIANGLE

In this section, we will derive three formulas for the area S of a triangle. We will start by deriving the formula used to find the area of a triangle for which two sides and the included angle are given.

TWO SIDES AND THE INCLUDED ANGLE

To derive our first formula, we begin with the general formula for the area of a triangle:

$$S = \frac{1}{2}(\text{base})(\text{height})$$

The base of triangle ABC in Figure 1 is c and the height is h. So the formula for S becomes, in this case,

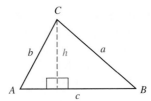

$$S = \frac{1}{2}ch$$

Figure 1

Suppose that, for triangle ABC, we are given the lengths of sides b and c and the measure of angle A. Then we can write $\sin A$ as

$$\sin A = \frac{h}{b}$$

or, by solving for h, $h = b \sin A$

Substituting this expression for h into the formula

$$S = \frac{1}{2}ch$$

we have

$$S = \frac{1}{2}bc \sin A$$

Applying the same kind of reasoning to the heights drawn from A and B, we also have

$$S = \frac{1}{2}ab \sin C$$

$$S = \frac{1}{2}ac \sin B$$

Each of these three formulas indicates that to find the area of a triangle for which we are given two sides and the angle included between them, we multiply half the product of the two sides by the sine of the angle included between them.

EXAMPLE 1 Find the area of triangle ABC if $A = 35.1°$, $b = 2.43$ cm, and $c = 3.57$ cm.

SOLUTION Applying the first formula we derived, we have

$$S = \frac{1}{2} bc \sin A$$

$$= \frac{1}{2}(2.43)(3.57) \sin 35.1°$$

$$= \frac{1}{2}(2.43)(3.57)(0.5750)$$

$$= 2.49 \text{ cm}^2 \qquad \text{To three significant digits}$$

TWO ANGLES AND ONE SIDE

The next area formula we will derive is used to find the area of triangles for which we are given two angles and one side.

Suppose we were given angles A and B and side a in triangle ABC in Figure 1. We could easily solve for C by subtracting the sum of A and B from $180°$.

To find side b, we use the law of sines

$$\frac{b}{\sin B} = \frac{a}{\sin A}$$

Solving this equation for b would give us

$$b = \frac{a \sin B}{\sin A}$$

Substituting this expression for b into the formula

$$S = \frac{1}{2} ab \sin C$$

we have

$$S = \frac{1}{2} a \left(\frac{a \sin B}{\sin A} \right) \sin C$$

$$= \frac{a^2 \sin B \sin C}{2 \sin A}$$

A similar sequence of steps can be used to derive

$$S = \frac{b^2 \sin A \sin C}{2 \sin B}$$

and

$$S = \frac{c^2 \sin A \sin B}{2 \sin C}$$

The formula we use depends on the side we are given.

EXAMPLE 2   Find the area of triangle ABC if $A = 24°\ 10'$, $B = 120°\ 40'$, and $a = 4.25$ ft.

SOLUTION   We begin by finding C.

$$C = 180° - (24°\ 10' + 120°\ 40')$$

$$C = 35°\ 10'$$

Now, applying the formula

$$S = \frac{a^2 \sin B \sin C}{2 \sin A}$$

with $a = 4.25$, $A = 24°\ 10'$, $B = 120°\ 40'$, and $C = 35°\ 10'$, we have

$$S = \frac{(4.25)^2(\sin 120°\ 40')(\sin 35°\ 10')}{2 \sin 24°\ 10'}$$

$$= \frac{(4.25)^2(0.8601)(0.5760)}{2(0.4094)}$$

$$= 10.9 \text{ ft}^2 \qquad \text{To three significant digits}$$

THREE SIDES

The last area formula we will discuss is called Heron's formula. It is used to find the area of a triangle in which all three sides are known.

Heron's formula is attributed to Heron of Alexandria, a geometer of the first century A.D. Heron proved his formula in Book I of his work, *Metrica*.

HERON'S FORMULA

The area of a triangle with sides of length a, b, and c is given by

$$S = \sqrt{s(s-a)(s-b)(s-c)}$$

where s is half the perimeter of the triangle; that is,

$$s = \frac{1}{2}(a + b + c) \quad \text{or} \quad 2s = a + b + c$$

PROOF

We begin our proof by squaring both sides of the formula

$$S = \frac{1}{2}ab \sin C$$

to obtain

$$S^2 = \frac{1}{4}a^2b^2 \sin^2 C$$

Next, we multiply both sides of the equation by $4/a^2b^2$ to isolate $\sin^2 C$ on the right side.

$$\frac{4S^2}{a^2b^2} = \sin^2 C$$

Replacing $\sin^2 C$ with $1 - \cos^2 C$ and then factoring as the difference of two squares, we have

$$\frac{4S^2}{a^2b^2} = 1 - \cos^2 C$$

$$= (1 + \cos C)(1 - \cos C)$$

From the law of cosines we know that $\cos C = (a^2 + b^2 - c^2)/2ab$.

$$= \left[1 + \frac{a^2 + b^2 - c^2}{2ab}\right]\left[1 - \frac{a^2 + b^2 - c^2}{2ab}\right]$$

$$= \left[\frac{2ab + a^2 + b^2 - c^2}{2ab}\right]\left[\frac{2ab - a^2 - b^2 + c^2}{2ab}\right]$$

$$= \left[\frac{(a^2 + 2ab + b^2) - c^2}{2ab}\right]\left[\frac{c^2 - (a^2 - 2ab + b^2)}{2ab}\right]$$

$$= \left[\frac{(a + b)^2 - c^2}{2ab}\right]\left[\frac{c^2 - (a - b)^2}{2ab}\right]$$

Now we factor each numerator as the difference of two squares and multiply the denominators.

$$= \frac{[(a + b + c)(a + b - c)][(c + a - b)(c - a + b)]}{4a^2b^2}$$

Now, since $a + b + c = 2s$, it is also true that

$$a + b - c = a + b + c - 2c = 2s - 2c$$

$$c + a - b = a + b + c - 2b = 2s - 2b$$

$$c - a + b = a + b + c - 2a = 2s - 2a$$

Substituting these expressions into our last equation, we have

$$= \frac{2s(2s - 2c)(2s - 2b)(2s - 2a)}{4a^2b^2}$$

Factoring out a 2 from each term in the numerator and showing the left side of our equation along with the right side, we have

$$\frac{4S^2}{a^2b^2} = \frac{16s(s - a)(s - b)(s - c)}{4a^2b^2}$$

Multiplying both sides by $a^2b^2/4$ we have

$$S^2 = s(s - a)(s - b)(s - c)$$

Taking the square root of both sides of the equation, we have Heron's formula.

$$S = \sqrt{s(s - a)(s - b)(s - c)}$$

EXAMPLE 3 Find the area of triangle ABC if $a = 12$ m, $b = 14$, and $c = 8.0$ m.

SOLUTION We begin by calculating the formula for s, half the perimeter of ABC.

$$s = \frac{1}{2}(12 + 14 + 8) = 17$$

Substituting this value of *s* into Heron's formula along with the given values of *a*, *b*, and *c*, we have

$$S = \sqrt{17(17 - 12)(17 - 14)(17 - 8)}$$

$$= \sqrt{17(5)(3)(9)}$$

$$= \sqrt{2{,}295}$$

$$= 48 \text{ m}^2 \quad \text{To two significant digits} \quad \blacksquare$$

GETTING READY FOR CLASS

After reading through the preceding section, respond in your own words and in complete sentences.

a. State the formula for the area of a triangle given two sides and the included angle.

b. State the formula for the area of a triangle given two angles and one side.

c. State the formula for the area of a triangle given all three sides.

d. What is the name given to this last formula?

PROBLEM SET 7.4

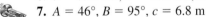

Each problem below refers to triangle *ABC*. In each case, find the area of the triangle. Round to three significant digits.

 1. $a = 50$ cm, $b = 70$ cm, $C = 60°$
 2. $a = 10$ cm, $b = 12$ cm, $C = 120°$
 3. $a = 41.5$ m, $c = 34.5$ m, $B = 151.5°$
 4. $a = 76.3$ m, $c = 42.8$ m, $B = 16.3°$
 5. $b = 0.923$ km, $c = 0.387$ km, $A = 43° \ 20'$
 6. $b = 63.4$ km, $c = 75.2$ km, $A = 124° \ 40'$
 7. $A = 46°$, $B = 95°$, $c = 6.8$ m
 8. $B = 57°$, $C = 31°$, $a = 7.3$ m
 9. $A = 42.5°$, $B = 71.4°$, $a = 210$ inches
10. $A = 110.4°$, $C = 21.8°$, $c = 240$ inches
11. $A = 43° \ 30'$, $C = 120° \ 30'$, $a = 3.48$ ft
12. $B = 14° \ 20'$, $C = 75° \ 40'$, $b = 2.72$ ft
13. $a = 44$ inches, $b = 66$ inches, $c = 88$ inches
14. $a = 23$ inches, $b = 34$ inches, $c = 45$ inches
15. $a = 4.8$ yd, $b = 6.3$ yd, $c = 7.5$ yd
16. $a = 48$ yd, $b = 75$ yd, $c = 63$ yd
17. $a = 4.38$ ft, $b = 3.79$ ft, $c = 5.22$ ft
18. $a = 8.32$ ft, $b = 6.23$ ft, $c = 3.45$ ft

19. Geometry and Area Find the area of a parallelogram if the angle between two of the sides is $120°$ and the two sides are 15 inches and 12 inches.

20. Geometry and Area Find the area of a parallelogram if the two sides measure 24.1 inches and 31.4 inches and the shorter diagonal is 32.4 inches.

21. Geometry and Area The area of a triangle is 40 cm^2. Find the length of the side included between the angles $A = 30°$ and $B = 50°$.

22. Geometry and Area The area of a triangle is 80 in^2. Find the length of the side included between $A = 25°$ and $C = 110°$.

REVIEW PROBLEMS

The problems that follow review material we covered in Section 6.4.

Eliminate the parameter t and graph the resulting equation.

23. $x = \cos t, y = \sin t$
24. $x = \cos t, y = -\sin t$
25. $x = 3 + 2 \sin t, y = 1 + 2 \cos t$
26. $x = \cos t - 3, y = \sin t + 2$

Eliminate the parameter t but do not graph.

27. $x = 2 \tan t, y = 3 \sec t$
28. $x = 4 \cot t, y = 2 \csc t$
29. $x = \sin t, y = \cos 2t$
30. $x = \cos t, y = \cos 2t$

SECTION 7.5 | VECTORS: AN ALGEBRAIC APPROACH

Hermann Grassmann

In this section, we will take a second look at vectors, this time from an algebraic point of view. Much of the credit for this treatment of vectors is attributed to both Irish mathematician William Rowan Hamilton (1805–1865) and German mathematician Hermann Grassmann (1809–1877). Grassmann is famous for his book *Ausdehnungslehre (The Calculus of Extension),* which was first published in 1844. However, it wasn't until 60 years later that this work was fully accepted or the significance realized, when Albert Einstein used it in his theory of relativity.

STANDARD POSITION

As we mentioned in Section 2.5, a vector is in *standard position* when it is placed on a coordinate system so that its tail is located at the origin. If the tip of the vector corresponds to the point (a, b), then the coordinates of this point provide a unique representation for the vector. That is, the point (a, b) determines both the length of the vector and its direction, as shown in Figure 1.

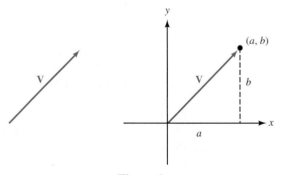

Figure 1

In order to avoid confusion between the point (a, b) and the vector, which is the ray extending from the origin to the point, we will denote the vector as $\mathbf{V} = \langle a, b \rangle$. We refer to this notation as *component form.* The x-coordinate, a, is called the *horizontal component* of $\mathbf{V}$, and the y-coordinate, b, is called the *vertical component* of $\mathbf{V}$.

MAGNITUDE

As you know from Section 2.5, the magnitude of a vector is its length. Referring to Figure 1, we can find the magnitude of the vector $\mathbf{V} = \langle a, b \rangle$ using the Pythagorean Theorem:

$$|\mathbf{V}| = \sqrt{a^2 + b^2}$$

 EXAMPLE 1   Draw the vector $\mathbf{V} = \langle 3, -4 \rangle$ in standard position and find its magnitude.

SOLUTION   We draw the vector by sketching an arrow from the origin to the point $(3, -4)$, as shown in Figure 2. To find the magnitude of $\mathbf{V}$, we find the positive square root of the sum of the squares of the horizontal and vertical components.

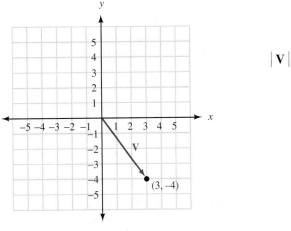

$$|\mathbf{V}| = \sqrt{3^2 + (-4)^2}$$
$$= \sqrt{9 + 16}$$
$$= \sqrt{25}$$
$$= 5$$

Figure 2

ADDITION AND SUBTRACTION WITH ALGEBRAIC VECTORS

Adding and subtracting vectors written in component form is simply a matter of adding (or subtracting) the horizontal components and adding (or subtracting) the vertical components. Figure 3 shows the vector sum of vectors $\mathbf{U} = \langle 6, 2 \rangle$ and $\mathbf{V} = \langle -3, 5 \rangle$.

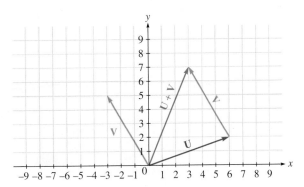

Figure 3

By simply counting squares on the grid you can convince yourself that the sum can be obtained by adding horizontal components and adding vertical components. That is,

$$\begin{aligned} \mathbf{U} + \mathbf{V} &= \langle 6, 2 \rangle + \langle -3, 5 \rangle \\ &= \langle 6 - 3, 2 + 5 \rangle \\ &= \langle 3, 7 \rangle \end{aligned}$$

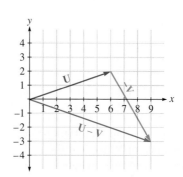

Figure 4

Figure 4 shows the difference of the vectors $\mathbf{U}$ and $\mathbf{V}$. As the diagram in Figure 4 indicates, subtraction of algebraic vectors can be accomplished by subtracting corresponding components. That is,

$$\begin{aligned} \mathbf{U} - \mathbf{V} &= \langle 6, 2 \rangle - \langle -3, 5 \rangle \\ &= \langle 6 - (-3), 2 - 5 \rangle \\ &= \langle 9, -3 \rangle \end{aligned}$$

SCALAR MULTIPLICATION

To multiply a vector in component form by a scalar (real number) we multiply each component of the vector by the scalar. Figure 5 shows the vector $\mathbf{V} = \langle 2, 3 \rangle$ and the vector $3\mathbf{V}$. As you can see,

$$\begin{aligned} 3\mathbf{V} &= 3\langle 2, 3 \rangle \\ &= \langle 3 \cdot 2, 3 \cdot 3 \rangle \\ &= \langle 6, 9 \rangle \end{aligned}$$

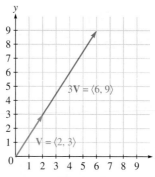

Figure 5

Notice that the vector $3\mathbf{V}$ has the same direction as $\mathbf{V}$ but is three times as long. Multiplying a vector by a positive scalar will preserve the direction of the vector but change its length (assuming the scalar is not equal to 1). Multiplying a vector by a negative scalar will result in a vector with the opposite direction and different length (assuming the scalar is not equal to -1).

EXAMPLE 2 If $\mathbf{U} = \langle 5, -3 \rangle$ and $\mathbf{V} = \langle -6, 4 \rangle$, find

a. $\mathbf{U} + \mathbf{V}$
b. $4\mathbf{U} - 5\mathbf{V}$

SOLUTION
a. $\begin{aligned}[t] \mathbf{U} + \mathbf{V} &= \langle 5, -3 \rangle + \langle -6, 4 \rangle \\ &= \langle 5 - 6, -3 + 4 \rangle \\ &= \langle -1, 1 \rangle \end{aligned}$

b. $\begin{aligned}[t] 4\mathbf{U} - 5\mathbf{V} &= 4\langle 5, -3 \rangle - 5\langle -6, 4 \rangle \\ &= \langle 20, -12 \rangle + \langle 30, -20 \rangle \\ &= \langle 20 + 30, -12 - 20 \rangle \\ &= \langle 50, -32 \rangle \end{aligned}$

COMPONENT VECTOR FORM

Another way to represent a vector algebraically is to express the vector as the sum of a horizontal vector and a vertical vector. As we mentioned in Section 2.5, any vector $\mathbf{V}$ can be written in terms of its horizontal and vertical component vectors, $\mathbf{V}_x$, and $\mathbf{V}_y$, respectively. To do this, we need to define two special vectors.

DEFINITION

The vector that extends from the origin to the point $(1, 0)$ is called the *unit horizontal vector* and is denoted by $\mathbf{i}$. The vector that extends from the origin to the point $(0, 1)$ is called the *unit vertical vector* and is denoted by the vector $\mathbf{j}$. Figure 6 shows the vectors $\mathbf{i}$ and $\mathbf{j}$.

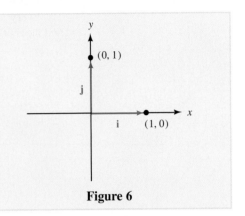

Figure 6

NOTE   A *unit vector* is any vector whose magnitude is 1. Using our previous component notation, we would write $\mathbf{i} = \langle 1, 0 \rangle$ and $\mathbf{j} = \langle 0, 1 \rangle$.

EXAMPLE 3   Write the vector $\mathbf{V} = \langle 3, 4 \rangle$ in terms of the unit vectors $\mathbf{i}$ and $\mathbf{j}$.

SOLUTION   From the origin, we must go three units in the positive x-direction, and then four units in the positive y-direction, to locate the terminal point of $\mathbf{V}$ at $(3, 4)$. Since $\mathbf{i}$ is a vector of length 1 in the positive x-direction, $3\mathbf{i}$ will be a vector of length 3 in that same direction. Likewise, $4\mathbf{j}$ will be a vector of length 4 in the positive y-direction. As shown in Figure 7, $\mathbf{V}$ is the resultant of vectors $3\mathbf{i}$ and $4\mathbf{j}$.

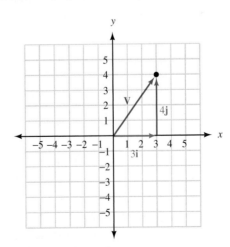

Figure 7

Therefore we can write $\mathbf{V}$ in terms of the unit vectors $\mathbf{i}$ and $\mathbf{j}$ as

$$\mathbf{V} = 3\mathbf{i} + 4\mathbf{j}$$

In Example 3, the vector **3i** is the *horizontal vector component* of **V**, which we have previously referred to as $\mathbf{V}_x$. Likewise, **4j** is the *vertical vector component* of **V**, previously referred to as $\mathbf{V}_y$. Notice that the coefficients of these vectors are simply the coordinates of the terminal point of **V**. That is,

$$\mathbf{V} = \langle 3, 4 \rangle = 3\mathbf{i} + 4\mathbf{j}$$

We refer to the notation **V** = 3**i** + 4**j** as *vector component form.*

Every vector **V** can be written in terms of horizontal and vertical components and the unit vectors **i** and **j**, as Figure 8 illustrates.

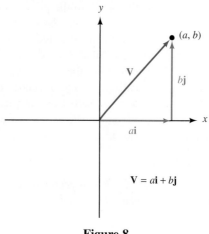

Figure 8

Here is a summary of the information we have developed to this point.

ALGEBRAIC VECTORS

If **i** is the unit vector from (0, 0) to (1, 0), and **j** is the unit vector from (0, 0) to (0, 1), then any vector **V** can be written as

$$\mathbf{V} = a\mathbf{i} + b\mathbf{j} = \langle a, b \rangle$$

where a and b are real numbers. The magnitude of **V** is

$$|\mathbf{V}| = \sqrt{a^2 + b^2}$$

 **EXAMPLE 4** Vector **V** has its tail at the origin, and makes an angle of 35° with the positive *x*-axis. Its magnitude is 12. Write **V** in terms of the unit vectors **i** and **j**.

SOLUTION Figure 9 is a diagram of **V**. The horizontal and vertical components of **V** are a and b, respectively.

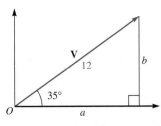

Figure 9

We find a and b using right triangle trigonometry.

$$a = 12 \cos 35° = 9.8$$
$$b = 12 \sin 35° = 6.9$$

Writing **V** in terms of the unit vectors **i** and **j** we have

$$\mathbf{V} = 9.8\mathbf{i} + 6.9\mathbf{j}$$

Working with vectors in vector component form $a\mathbf{i} + b\mathbf{j}$ is no different than in component form $\langle a, b \rangle$, as illustrated in the next example.

EXAMPLE 5 If $\mathbf{U} = 5\mathbf{i} - 3\mathbf{j}$ and $\mathbf{V} = -6\mathbf{i} + 4\mathbf{j}$, find

a. $\mathbf{U} + \mathbf{V}$
b. $4\mathbf{U} - 5\mathbf{V}$

SOLUTION
a. $\mathbf{U} + \mathbf{V} = (5\mathbf{i} - 3\mathbf{j}) + (-6\mathbf{i} + 4\mathbf{j})$

$$= (5 - 6)\mathbf{i} + (-3 + 4)\mathbf{j}$$
$$= -\mathbf{i} + \mathbf{j}$$

b. $4\mathbf{U} - 5\mathbf{V} = 4(5\mathbf{i} - 3\mathbf{j}) - 5(-6\mathbf{i} + 4\mathbf{j})$

$$= (20\mathbf{i} - 12\mathbf{j}) + (30\mathbf{i} - 20\mathbf{j})$$
$$= (20 + 30)\mathbf{i} + (-12 - 20)\mathbf{j}$$
$$= 50\mathbf{i} - 32\mathbf{j}$$

Notice that these results are equivalent to those of Example 2.

As an application of algebraic vectors, let's return to one of the static equilibrium problems we solved in Section 2.5.

EXAMPLE 6 Danny is 5 years old and weighs 42 pounds. He is sitting on a swing when his sister Stacey pulls him and the swing back horizontally through an angle of 30° and then stops. Find the tension in the ropes of the swing. (Figure 10 is a diagram of the situation.)

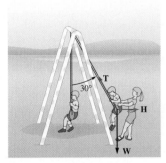

Figure 10

SOLUTION When we solved this problem in Section 2.5, we noted that there are three forces acting on Danny (and the swing), which we have labeled **W**, **H**, and **T**. Recall that the vector **W** is due to the force of gravity: its magnitude is $|\mathbf{W}| = 42$ lb, and its direction is straight down. The vector **H** represents the force with which Stacey is pulling Danny horizontally, and **T** is the force acting on Danny in the direction of the ropes.

If we place the origin of a coordinate system on the point at which the tails of the three vectors intersect, we can write each vector in terms of its magnitude and the unit vectors **i** and **j**. The coordinate system is shown in Figure 11. Since the direction along which **H** acts is in the positive x direction, we can write

$$\mathbf{H} = |\mathbf{H}|\mathbf{i}$$

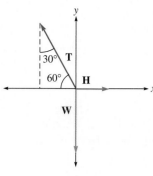

Figure 11

Likewise, since the weight vector **W** is straight down, we can write

$$\mathbf{W} = -|\mathbf{W}|\mathbf{j} = -42\mathbf{j}$$

Using right triangle trigonometry, we write **T** in terms of its horizontal and vertical components:

$$\mathbf{T} = -|\mathbf{T}|\cos 60°\mathbf{i} + |\mathbf{T}|\sin 60°\mathbf{j}$$

Since Danny and the swing are at rest, we have static equilibrium. Therefore, the sum of the vectors is 0.

$$\mathbf{T} + \mathbf{H} + \mathbf{W} = 0$$

$$-|\mathbf{T}|\cos 60°\mathbf{i} + |\mathbf{T}|\sin 60°\mathbf{j} + |\mathbf{H}|\mathbf{i} + (-42\mathbf{j}) = 0$$

Collecting all the **i** components and all the **j** components together, we have

$$(-|\mathbf{T}|\cos 60° + |\mathbf{H}|)\mathbf{i} + (|\mathbf{T}|\sin 60° - 42)\mathbf{j} = 0$$

The only way this can happen is if both components are 0. Setting the coefficient of **j** to 0, we have

$$|\mathbf{T}|\sin 60° - 42 = 0$$

$$|\mathbf{T}| = \frac{42}{\sin 60°} = 48 \text{ lb} \qquad \text{To two significant digits}$$

Setting the coefficient of **i** to 0, and then substituting 48 for $|\mathbf{T}|$, we have

$$-|\mathbf{T}|\cos 60° + |\mathbf{H}| = 0$$

$$-48\cos 60° + |\mathbf{H}| = 0$$

$$|\mathbf{H}| = 48\cos 60° = 24 \text{ lb} \quad \blacksquare$$

GETTING READY FOR CLASS

After reading through the preceding section, respond in your own words and in complete sentences.

a. How can we algebraically represent a vector that is in standard position?

b. Explain how to add or subtract two vectors in component form.

c. What is a scalar and how is it used in our study of vectors?

d. State the definitions for the unit horizontal vector and the unit vertical vector.

PROBLEM SET 7.5

Draw the vector **V** that goes from the origin to the given point. Then write **V** in component form $\langle a, b \rangle$.

1. (4, 1)   **2.** (1, 4)
3. (−5, 2)   **4.** (−2, 5)
5. (3, −3)   **6.** (5, −5)
7. (−6, −4)   **8.** (−4, −6)

Draw the vector **V** that goes from the origin to the given point. Then write **V** in terms of the unit vectors **i** and **j**.

9. (2, 5)   **10.** (5, 2)
11. (−3, 6)   **12.** (−6, 3)
13. (4, −5)   **14.** (5, −4)
15. (−1, −5)   **16.** (−5, −1)

Find the magnitude of each of the following vectors.

17. ⟨−5, 6⟩   **18.** ⟨−3, 7⟩
19. ⟨2, 0⟩   **20.** ⟨0, 5⟩
21. ⟨−2, −5⟩   **22.** ⟨−8, −3⟩
23. **V** = 3**i** + 4**j**   **24.** **V** = 6**i** + 8**j**
25. **U** = 5**i** + 12**j**   **26.** **U** = 20**i** − 21**j**
27. **W** = **i** + 2**j**   **28.** **W** = 3**i** + **j**

For each pair of vectors, find **U** + **V**, **U** − **V**, and 2**U** − 3**V**.

29. **U** = ⟨4, 4⟩, **V** = ⟨4, −4⟩   **30.** **U** = ⟨−4, 4⟩, **V** = ⟨4, 4⟩
31. **U** = ⟨2, 0⟩, **V** = ⟨0, −7⟩   **32.** **U** = ⟨−5, 0⟩, **V** = ⟨0, 1⟩
33. **U** = ⟨4, 1⟩, **V** = ⟨−5, 2⟩   **34.** **U** = ⟨1, 4⟩, **V** = ⟨−2, 5⟩

For each pair of vectors, find **U** + **V**, **U** − **V**, and 3**U** + 2**V**.

35. **U** = **i** + **j**   **V** = **i** − **j**   **36.** **U** = −**i** + **j**   **V** = **i** + **j**
37. **U** = 6**i**   **V** = −8**j**   **38.** **U** = −3**i**   **V** = 5**j**
39. **U** = 2**i** + 5**j**   **V** = 5**i** + 2**j**   **40.** **U** = 5**i** + 3**j**   **V** = 3**i** + 5**j**

41. Vector **V** is in standard position, and makes an angle of 40° with the positive *x*-axis. Its magnitude is 18. Write **V** in component form ⟨*a, b*⟩ and in vector component form *a***i** + *b***j**.

42. Vector **U** is in standard position, and makes an angle of 110° with the positive *x*-axis. Its magnitude is 25. Write **U** in component form ⟨*a, b*⟩ and in vector component form *a***i** + *b***j**.

43. Vector **W** is in standard position, and makes an angle of 230° with the positive *x*-axis. Its magnitude is 8. Write **W** in component form ⟨*a, b*⟩ and in vector component form *a***i** + *b***j**.

44. Vector **F** is in standard position, and makes an angle of 285° with the positive *x*-axis. Its magnitude is 30. Write **F** in component form ⟨*a, b*⟩ and in vector component form *a***i** + *b***j**.

Find the magnitude of each vector and the angle θ, $0° \leq \theta < 360°$, that the vector makes with the positive *x*-axis.

45. **U** = ⟨3, 3⟩   **46.** **V** = ⟨5, −5⟩
47. **W** = −**i** − $\sqrt{3}$**j**   **48.** **F** = −2$\sqrt{3}$**i** + 2**j**

REVIEW PROBLEMS

The problems that follow are problems you may have worked previously in Section 2.5 or Section 7.1. Solve each problem using the methods shown in Example 6 of this section.

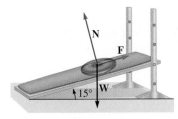

Figure 12

49. **Force** A 10-pound weight is lying on a situp bench at the gym. If the bench is inclined at an angle of 15°, there are three forces acting on the weight, as shown in Figure 12. Find the magnitude of **N** and the magnitude of **F**.

50. **Force** Repeat Problem 49 for a 25-pound weight and a bench inclined at 10°.

51. Force Tyler and his cousin Kelly have attached a rope to the branch of a tree and tied a board to the other end to form a swing. Tyler stands on the board while his cousin pushes him through an angle of 25.5° and holds him there. If Tyler weighs 95.5 pounds, find the magnitude of the force Kelly must push with horizontally to keep Tyler in static equilibrium. See Figure 13.

Figure 13

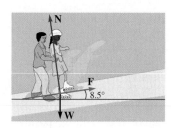

Figure 14

52. Force After they are finished swinging, Tyler and Kelly decide to rollerskate. They come to a hill that is inclined at 8.5°. Tyler pushes Kelly halfway up the hill and then holds her there (Figure 14). If Kelly weighs 58.0 pounds, find the magnitude of the force Tyler must push with to keep Kelly from rolling down the hill. (We are assuming that the rollerskates make the hill into a frictionless surface so that the only force keeping Kelly from rolling backwards down the hill is the force Tyler is pushing with.)

53. Force A traffic light weighing 22 pounds is suspended by two wires as shown in Figure 15. Find the magnitude of the tension in wire AB, and the magnitude of the tension in wire AC.

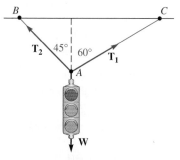

Figure 15

Figure 16

54. Force A tightrope walker is standing still with one foot on the tightrope as shown in Figure 16. If the tightrope walker weighs 125 pounds, find the magnitudes of the tension in the rope toward each end of the rope.

SECTION 7.6 | VECTORS: THE DOT PRODUCT

Now that we have a way to represent vectors algebraically, we can define a type of multiplication between two vectors. The *dot product* (also called the *scalar product*) is a form of multiplication that results in a scalar quantity. For our purposes, it will be useful when finding the angle between two vectors or for finding the work done by a force in moving an object. Here is the definition of the dot product of two vectors.

DEFINITION

The *dot product* of two vectors $\mathbf{U} = a\mathbf{i} + b\mathbf{j}$ and $\mathbf{V} = c\mathbf{i} + d\mathbf{j}$ is written $\mathbf{U} \cdot \mathbf{V}$ and is defined as follows:

$$\mathbf{U} \cdot \mathbf{V} = (a\mathbf{i} + b\mathbf{j}) \cdot (c\mathbf{i} + d\mathbf{j})$$

$$= (ac) + (bd)$$

As you can see, the dot product is a real number (scalar), not a vector.

EXAMPLE 1   Find each of the following dot products.

a. $\mathbf{U} \cdot \mathbf{V}$ when $\mathbf{U} = \langle 3, 4 \rangle$ and $\mathbf{V} = \langle 2, 5 \rangle$
b. $\langle -1, 2 \rangle \cdot \langle 3, -5 \rangle$
c. $\mathbf{S} \cdot \mathbf{W}$ when $\mathbf{S} = 6\mathbf{i} + 3\mathbf{j}$ and $\mathbf{W} = 2\mathbf{i} - 7\mathbf{j}$

SOLUTION   For each problem, we simply multiply the coefficients a and c and add that result to the product of the coefficients b and d.

a. $\mathbf{U} \cdot \mathbf{V} = 3(2) + 4(5)$

$= 6 + 20$

$= 26$

b. $\langle -1, 2 \rangle \cdot \langle 3, -5 \rangle = -1(3) + 2(-5)$

$= -3 + (-10)$

$= -13$

c. $\mathbf{S} \cdot \mathbf{W} = 6(2) + 3(-7)$

$= 12 + (-21)$

$= -9$

FINDING THE ANGLE BETWEEN TWO VECTORS

One application of the dot product is finding the angle between two vectors. To do this, we will use an alternate form of the dot product, shown in the following theorem.

THEOREM 7.1

The dot product of two vectors is equal to the product of their magnitudes multiplied by the cosine of the angle between them. That is, when θ is the angle between two nonzero vectors **U** and **V**, then

$$\mathbf{U} \cdot \mathbf{V} = |\mathbf{U}||\mathbf{V}| \cos \theta$$

The proof of this theorem is derived from the law of cosines and is left as an exercise.

When we are given two vectors and asked to find the angle between them, we rewrite the formula in Theorem 7.1 by dividing each side by $|\mathbf{U}||\mathbf{V}|$. The result is

$$\cos \theta = \frac{\mathbf{U} \cdot \mathbf{V}}{|\mathbf{U}||\mathbf{V}|}$$

This formula is equivalent to our original formula, but is easier to work with when finding the angle between two vectors.

 EXAMPLE 2 Find the angle between the vectors **U** and **V**.

a. $\mathbf{U} = \langle 2, 3 \rangle$ and $\mathbf{V} = \langle -3, 2 \rangle$
b. $\mathbf{U} = 6\mathbf{i} - \mathbf{j}$ and $\mathbf{V} = \mathbf{i} + 4\mathbf{j}$

SOLUTION

a. $\cos \theta = \dfrac{\mathbf{U} \cdot \mathbf{V}}{|\mathbf{U}||\mathbf{V}|}$

$ = \dfrac{2(-3) + 3(2)}{\sqrt{2^2 + 3^2} \cdot \sqrt{(-3)^2 + 2^2}}$

$ = \dfrac{-6 + 6}{\sqrt{13} \cdot \sqrt{13}}$

$ = \dfrac{0}{13}$

$\cos \theta = 0$

$\theta = 90°$

b. $\cos \theta = \dfrac{\mathbf{U} \cdot \mathbf{V}}{|\mathbf{U}||\mathbf{V}|}$

$ = \dfrac{6(1) + (-1)4}{\sqrt{6^2 + (-1)^2} \cdot \sqrt{1^2 + 4^2}}$

$ = \dfrac{6 + (-4)}{\sqrt{37} \cdot \sqrt{17}}$

$ = \dfrac{2}{25.08}$

$\cos \theta = 0.0797$

$\theta = \cos^{-1}(0.0797) = 85.43°$ To the nearest hundredth of a degree

PERPENDICULAR VECTORS

If two nonzero vectors are perpendicular, then the angle between them is 90°. Since the cosine of 90° is always 0, the dot product of two perpendicular vectors must also be 0. This fact gives rise to the following theorem.

THEOREM 7.2

If **U** and **V** are two nonzero vectors, then

$$\mathbf{U} \cdot \mathbf{V} = 0 \Longleftrightarrow \mathbf{U} \perp \mathbf{V}$$

In Words: Two nonzero vectors are perpendicular if and only if their dot product is 0.

 **EXAMPLE 3**   Which of the following vectors are perpendicular to each other?

$$\mathbf{U} = 8\mathbf{i} + 6\mathbf{j}$$

$$\mathbf{V} = 3\mathbf{i} - 4\mathbf{j}$$

$$\mathbf{W} = 4\mathbf{i} + 3\mathbf{j}$$

SOLUTION   Find **U** • **V**, **V** • **W** and **U** • **W**. If the dot product is zero, then the two vectors are perpendicular.

$$\mathbf{U} \cdot \mathbf{V} = 8(3) + 6(-4)$$

$$= 24 - 24$$

$$= 0 \qquad\qquad \text{Therefore, } \mathbf{U} \text{ and } \mathbf{V} \text{ are perpendicular}$$

$$\mathbf{V} \cdot \mathbf{W} = 3(4) + (-4)3$$

$$= 12 - 12$$

$$= 0 \qquad\qquad \text{Therefore, } \mathbf{V} \text{ and } \mathbf{W} \text{ are perpendicular}$$

$$\mathbf{U} \cdot \mathbf{W} = 8(4) + (6)3$$

$$= 32 + 18$$

$$= 50 \qquad\qquad \text{Therefore, } \mathbf{U} \text{ and } \mathbf{W} \text{ are not perpendicular}$$

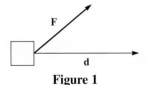

Figure 1

WORK

In Section 2.5 we introduced the concept of work. Recall that work is performed when a constant force **F** is used to move an object a certain distance. We can represent the movement of the object using a displacement vector, **d**, as shown in Figure 1.

In Figure 2 we let **V** represent the component of **F** that is oriented in the same direction as **d**, since only the amount of the force in the direction of movement can be used in calculating work. **V** is sometimes called the *projection of* **F** *onto* **d**. We can find the magnitude of **V** using right triangle trigonometry:

$$|\mathbf{V}| = |\mathbf{F}| \cos \theta$$

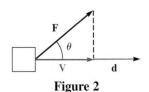

Figure 2

Since $|\mathbf{d}|$ represents the distance the object is moved, the work performed by the force is

$$\text{Work} = |\mathbf{V}||\mathbf{d}|$$

$$= |\mathbf{F}||\mathbf{d}|\cos\theta$$

$$= \mathbf{F}\cdot\mathbf{d} \qquad \text{by Theorem 7.1}$$

We have just established the following theorem.

THEOREM 7.3

If a constant force $\mathbf{F}$ is applied to an object, and the resulting movement of the object is represented by the displacement vector $\mathbf{d}$, then the work performed by the force is

$$\text{Work} = \mathbf{F}\cdot\mathbf{d}$$

 EXAMPLE 4 A force $\mathbf{F} = 35\mathbf{i} - 12\mathbf{j}$ (in pounds) is used to push an object up a ramp. The resulting movement of the object is represented by the displacement vector $\mathbf{d} = 15\mathbf{i} + 4\mathbf{j}$ (in feet), as illustrated in Figure 3. Find the work done by the force.

SOLUTION By Theorem 7.3,

$$\text{Work} = \mathbf{F}\cdot\mathbf{d}$$

$$= 35(15) + (-12)(4)$$

$$= 477 \text{ ft-lb} \quad \blacksquare$$

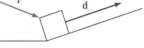

Figure 3

GETTING READY FOR CLASS

After reading through the preceding section, respond in your own words and in complete sentences.

a. Explain how to find the dot product of two vectors.

b. Explain how the dot product is used to find the angle between two vectors.

c. How do we know if two nonzero vectors are perpendicular?

d. Explain how vectors can be used to calculate work.

PROBLEM SET 7.6

Find each of the following dot products.

1. $\langle 6, 6\rangle \cdot \langle 3, 5\rangle$ **2.** $\langle 3, 4\rangle \cdot \langle 5, 5\rangle$

3. $\langle -23, 4\rangle \cdot \langle 15, -6\rangle$ **4.** $\langle 11, -8\rangle \cdot \langle 4, -7\rangle$

For each pair of vectors, find $\mathbf{U}\cdot\mathbf{V}$.

5. $\mathbf{U} = \mathbf{i} + \mathbf{j}$ $\mathbf{V} = \mathbf{i} - \mathbf{j}$ **6.** $\mathbf{U} = -\mathbf{i} + \mathbf{j}$ $\mathbf{V} = \mathbf{i} + \mathbf{j}$

7. $\mathbf{U} = 6\mathbf{i}$ $\mathbf{V} = -8\mathbf{j}$ **8.** $\mathbf{U} = -3\mathbf{i}$ $\mathbf{V} = 5\mathbf{j}$

9. $\mathbf{U} = 2\mathbf{i} + 5\mathbf{j}$ $\mathbf{V} = 5\mathbf{i} + 2\mathbf{j}$ **10.** $\mathbf{U} = 5\mathbf{i} + 3\mathbf{j}$ $\mathbf{V} = 3\mathbf{i} + 5\mathbf{j}$

Find the angle θ between the given vectors to the nearest tenth of a degree.

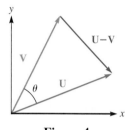

Figure 4

11. $\mathbf{U} = 13\mathbf{i}$ $\mathbf{V} = -6\mathbf{j}$ **12.** $\mathbf{U} = -4\mathbf{i}$ $\mathbf{V} = 17\mathbf{j}$
13. $\mathbf{U} = 4\mathbf{i} + 5\mathbf{j}$ $\mathbf{V} = 7\mathbf{i} - 4\mathbf{j}$ **14.** $\mathbf{U} = -3\mathbf{i} + 5\mathbf{j}$ $\mathbf{V} = 6\mathbf{i} + 3\mathbf{j}$
15. $\mathbf{U} = 13\mathbf{i} - 8\mathbf{j}$ $\mathbf{V} = 2\mathbf{i} + 11\mathbf{j}$ **16.** $\mathbf{U} = 11\mathbf{i} + 7\mathbf{j}$ $\mathbf{V} = -4\mathbf{i} + 6\mathbf{j}$

Show that each pair of vectors is perpendicular.

17. $\mathbf{i}$ and $\mathbf{j}$ **18.** $\mathbf{i} + \mathbf{j}$ and $\mathbf{i} - \mathbf{j}$
19. $-\mathbf{i}$ and $\mathbf{j}$ **20.** $2\mathbf{i} + \mathbf{j}$ and $\mathbf{i} - 2\mathbf{j}$

21. In general, show that the vectors $\mathbf{V} = a\mathbf{i} + b\mathbf{j}$ and $\mathbf{W} = -b\mathbf{i} + a\mathbf{j}$ are always perpendicular.
22. Find the value of a so that vectors $\mathbf{U} = a\mathbf{i} + 6\mathbf{j}$ and $\mathbf{V} = 9\mathbf{i} + 12\mathbf{j}$ are perpendicular.

Find the work performed when the given force $\mathbf{F}$ is applied to an object, whose resulting motion is represented by the displacement vector $\mathbf{d}$. Assume the force is in pounds and the displacement is measured in feet.

23. $\mathbf{F} = 22\mathbf{i} + 9\mathbf{j}$ $\mathbf{d} = 30\mathbf{i} + 4\mathbf{j}$ **24.** $\mathbf{F} = 45\mathbf{i} - 12\mathbf{j}$ $\mathbf{d} = 70\mathbf{i} + 15\mathbf{j}$
25. $\mathbf{F} = -67\mathbf{i} + 39\mathbf{j}$ $\mathbf{d} = -96\mathbf{i} - 28\mathbf{j}$ **26.** $\mathbf{F} = -6\mathbf{i} + 19\mathbf{j}$ $\mathbf{d} = 8\mathbf{i} + 55\mathbf{j}$
27. $\mathbf{F} = 85\mathbf{i}$ $\mathbf{d} = 6\mathbf{i}$ **28.** $\mathbf{F} = 54\mathbf{i}$ $\mathbf{d} = 20\mathbf{i}$
29. $\mathbf{F} = 39\mathbf{j}$ $\mathbf{d} = 72\mathbf{i}$ **30.** $\mathbf{F} = 13\mathbf{j}$ $\mathbf{d} = 44\mathbf{i}$

31. Use the diagram shown in Figure 4 along with the law of cosines to prove Theorem 7.1. (Begin by writing $\mathbf{U} = a\mathbf{i} + b\mathbf{j}$ and $\mathbf{V} = c\mathbf{i} + d\mathbf{j}$.)
32. Use Theorem 7.1 to prove Theorem 7.2.

REVIEW PROBLEMS

The problems that follow are problems you may have worked previously in Section 2.5. Solve each problem by first expressing the force and the displacement of the object as vectors in terms of the unit vectors $\mathbf{i}$ and $\mathbf{j}$. Then use Theorem 7.3 to find the work done.

33. **Work** A package is pushed across a floor a distance of 75 feet by exerting a force of 40 pounds downward at an angle of 20° with the horizontal. How much work is done?
34. **Work** A package is pushed across a floor a distance of 50 feet by exerting a force of 15 pounds downward at an angle of 25° with the horizontal. How much work is done?
35. **Work** An automobile is pushed down a level street by exerting a force of 85 pounds at an angle of 15° with the horizontal (Figure 5). How much work is done in pushing the car 100 feet?

Figure 5

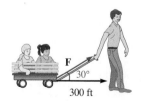

Figure 6

36. **Work** Mark pulls Allison and Mattie in a wagon by exerting a force of 25 pounds on the handle at an angle of 30° with the horizontal (Figure 6). How much work is done by Mark in pulling the wagon 300 feet?

CHAPTER 7 SUMMARY

EXAMPLES

1. If $A = 30°$, $B = 70°$, and $a = 8.0$ cm in triangle ABC, then, by the law of sines,

$$b = \frac{a \sin B}{\sin A} = \frac{8 \sin 70°}{\sin 30°}$$

$$= 15 \text{ cm}$$

The Law of Sines [7.1]

For any triangle ABC, the following relationships are always true:

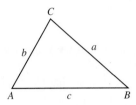

$$\frac{\sin A}{a} = \frac{\sin B}{b} = \frac{\sin C}{c}$$

or, equivalently,

$$\frac{a}{\sin A} = \frac{b}{\sin B} = \frac{c}{\sin C}$$

2. In triangle ABC, if $a = 54$ cm, $b = 62$ cm, and $A = 40°$, then

$$\sin B = \frac{b \sin A}{a} = \frac{62 \sin 40°}{54}$$

$$= 0.7380$$

Since $\sin B$ is positive for any angle in quadrant I or II, we have two possibilities for B:

$$B = 48° \quad \text{or} \quad B' = 180° - 48°$$
$$= 132°$$

This indicates that two triangles exist, both of which fit the given information.

The Ambiguous Case [7.2]

When we are given two sides and an angle opposite one of them (SSA), we have several possibilities for the triangle or triangles that result. One of the possibilities is that no triangle will fit the given information. Another possibility is that two different triangles can be obtained from the given information, and a third possibility is that exactly one triangle will fit the given information. Because of these different possibilities, we call the situation where we are solving a triangle in which we are given two sides and the angle opposite one of them the *ambiguous case*.

3.

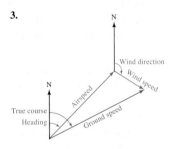

Navigation [7.2]

The *heading* of an object is the angle measured clockwise from due north to the vector representing the path of the object. If the object is subject to wind or currents, then the actual direction of the object is called its *true course*.

4. In triangle ABC, if $a = 34$ km, $b = 20$ km, and $c = 18$ km, then we can find A using the law of cosines.

$$\cos A = \frac{b^2 + c^2 - a^2}{2bc}$$

$$= \frac{20^2 + 18^2 - 34^2}{(2)(20)(18)}$$

$$\cos A = -0.6000$$
$$\text{so} \quad A = 127°$$

The Law of Cosines [7.3]

In any triangle ABC, the following relationships are always true:

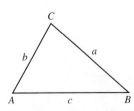

$$a^2 = b^2 + c^2 - 2bc \cos A$$

$$b^2 = a^2 + c^2 - 2ac \cos B$$

$$c^2 = a^2 + b^2 - 2ab \cos C$$

or, equivalently,

$$\cos A = \frac{b^2 + c^2 - a^2}{2bc}$$

$$\cos B = \frac{a^2 + c^2 - b^2}{2ac}$$

$$\cos C = \frac{a^2 + b^2 - c^2}{2ab}$$

5. For triangle ABC,

a. If $a = 12$ cm, $b = 15$ cm, and $C = 20°$, then the area of ABC is

$$S = \frac{1}{2}(12)(15) \sin 20°$$

$$= 30.8 \text{ cm}^2 \text{ to the nearest tenth}$$

The Area of a Triangle [7.4]

The area of a triangle for which we are given two sides and the included angle is given by

$$S = \frac{1}{2}ab \sin C$$

$$S = \frac{1}{2}ac \sin B$$

$$S = \frac{1}{2}bc \sin A$$

b. If $a = 24$ inches, $b = 14$ inches, and $c = 18$ inches, then the area of ABC is

$$S = \sqrt{28(28 - 24)(28 - 14)(28 - 18)}$$

$$= \sqrt{28(4)(14)(10)}$$

$$= \sqrt{15,680}$$

$$= 125.2 \text{ inches}^2 \text{ to the}$$
$$\text{nearest tenth}$$

The area of a triangle for which we are given all three sides is given by the formula

$$S = \sqrt{s(s - a)(s - b)(s - c)}$$

where $s = \frac{1}{2}(a + b + c)$

c. If $A = 40°$, $B = 72°$, and $c = 45$ m, then the area of ABC is

$$S = \frac{45^2 \sin 40° \sin 72°}{2 \sin 68°}$$

$$S = \frac{2,025(0.6428)(0.9511)}{2(0.9272)}$$

$$= 667.6 \text{ m}^2 \text{ to the nearest}$$
$$\text{tenth}$$

The area of a triangle for which we are given two angles and a side is given by

$$S = \frac{a^2 \sin B \sin C}{2 \sin A}$$

$$S = \frac{b^2 \sin C \sin A}{2 \sin B}$$

$$S = \frac{c^2 \sin A \sin B}{2 \sin C}$$

6. The vector **V** that extends from the origin to the point $(-3, 4)$ is

$$\mathbf{V} = -3\mathbf{i} + 4\mathbf{j} = \langle -3, 4 \rangle$$

Algebraic Vectors [7.5]

The vector that extends from the origin to the point $(1, 0)$ is called the *unit horizontal vector* and is denoted by **i**. The vector that extends from the origin to the point $(0, 1)$ is called the *unit vertical vector* and is denoted by **j**. Any nonzero vector **V** can be written

1. In terms of unit vectors as

$$\mathbf{V} = a\mathbf{i} + b\mathbf{j}$$

2. In component form as

$$\mathbf{V} = \langle a, b \rangle$$

where a and b are real numbers.

7. The magnitude of $\mathbf{V} = -3\mathbf{i} + 4\mathbf{j}$ is

$$|\mathbf{V}| = \sqrt{(-3)^2 + 4^2}$$

$$= \sqrt{25}$$

$$= 5$$

Magnitude [7.5]

The magnitude of $\mathbf{V} = a\mathbf{i} + b\mathbf{j} = \langle a, b \rangle$ is

$$|\mathbf{V}| = \sqrt{a^2 + b^2}$$

8. If $\mathbf{U} = 6\mathbf{i} + 2\mathbf{j}$ and
$\mathbf{V} = -3\mathbf{i} + 5\mathbf{j}$, then

$$\begin{aligned}
\mathbf{U} + \mathbf{V} &= (6\mathbf{i} + 2\mathbf{j}) + (-3\mathbf{i} + 5\mathbf{j}) \\
&= (6 - 3)\mathbf{i} + (2 + 5)\mathbf{j} \\
&= 3\mathbf{i} + 7\mathbf{j} \\
\mathbf{U} - \mathbf{V} &= (6\mathbf{i} + 2\mathbf{j}) - (-3\mathbf{i} + 5\mathbf{j}) \\
&= [6 - (-3)]\mathbf{i} + (2 - 5)\mathbf{j} \\
&= 9\mathbf{i} - 3\mathbf{j}
\end{aligned}$$

9. If $\mathbf{U} = -3\mathbf{i} + 4\mathbf{j}$ and
$\mathbf{V} = 4\mathbf{i} + 3\mathbf{j}$, then

$$\mathbf{U} \cdot \mathbf{V} = -3(4) + 4(3) = 0$$

10. The vectors $\mathbf{U}$ and $\mathbf{V}$ in Example 9 above are perpendicular because their dot product is 0.

11. If $\mathbf{F} = 35\mathbf{i} - 12\mathbf{j}$ and
$\mathbf{d} = 15\mathbf{i} + 4\mathbf{j}$, then

$$\begin{aligned}
\text{Work} &= \mathbf{F} \cdot \mathbf{d} \\
&= 35(15) + (-12)(4) \\
&= 477
\end{aligned}$$

Addition and Subtraction with Algebraic Vectors [7.5]

If $\mathbf{U} = a\mathbf{i} + b\mathbf{j} = \langle a, b \rangle$ and $\mathbf{V} = c\mathbf{i} + d\mathbf{j} = \langle c, d \rangle$, then vector addition and subtraction are defined as follows:

Addition: $\quad \mathbf{U} + \mathbf{V} = (a + c)\mathbf{i} + (b + d)\mathbf{j} = \langle a + c, b + d \rangle$

Subtraction: $\quad \mathbf{U} - \mathbf{V} = (a - c)\mathbf{i} + (b - d)\mathbf{j} = \langle a - c, b - d \rangle$

Dot Product [7.6]

The *dot product* of two vectors $\mathbf{U} = a\mathbf{i} + b\mathbf{j}$ and $\mathbf{V} = c\mathbf{i} + d\mathbf{j}$ is written $\mathbf{U} \cdot \mathbf{V}$ and is defined as follows:

$$\mathbf{U} \cdot \mathbf{V} = ac + bd$$

If θ is the angle between the two vectors, then it is also true that

$$\mathbf{U} \cdot \mathbf{V} = |\mathbf{U}||\mathbf{V}| \cos \theta$$

which we can solve for $\cos \theta$ to obtain the formula that allows us to find angle θ:

$$\cos \theta = \frac{\mathbf{U} \cdot \mathbf{V}}{|\mathbf{U}||\mathbf{V}|}$$

Perpendicular Vectors [7.6]

If $\mathbf{U}$ and $\mathbf{V}$ are two nonzero vectors, then

$$\mathbf{U} \cdot \mathbf{V} = 0 \Longleftrightarrow \mathbf{U} \perp \mathbf{V}$$

In Words: Two nonzero vectors are perpendicular if and only if their dot product is 0.

Work [7.6]

If a constant force $\mathbf{F}$ is applied to an object, and the resulting movement of the object is represented by the displacement vector $\mathbf{d}$, then the work performed by the force is

$$\text{Work} = \mathbf{F} \cdot \mathbf{d}$$

CHAPTER 7 TEST

Problems 1 through 14 refer to triangle ABC, which is not necessarily a right triangle.

1. If $A = 32°$, $B = 70°$, and $a = 3.8$ inches, use the law of sines to find b.
2. If $B = 118°$, $C = 37°$, and $c = 2.9$ inches, use the law of sines to find b.
3. If $A = 38.2°$, $B = 63.4°$, and $c = 42.0$ cm, find all the missing parts.
4. If $A = 24.7°$, $C = 106.1°$, and $b = 34.0$ cm, find all the missing parts.
5. Use the law of sines to show that no triangle exists for which $A = 60°$, $a = 12$ inches, and $b = 42$ inches.
6. Use the law of sines to show that exactly one triangle exists for which $A = 42°$, $a = 29$ inches, and $b = 21$ inches.
7. Find two triangles for which $A = 51°$, $a = 6.5$ ft, and $b = 7.9$ ft.

8. Find two triangles for which $A = 26°$, $a = 4.8$ ft, and $b = 9.4$ ft.
9. If $C = 60°$, $a = 10$ cm, and $b = 12$ cm, use the law of cosines to find c.
10. If $C = 120°$, $a = 10$ cm, and $b = 12$ cm, use the law of cosines to find c.
11. If $a = 5$ km, $b = 7$ km, and $c = 9$ km, use the law of cosines to find C to the nearest tenth of a degree.
12. If $a = 10$ km, $b = 12$ km, and $c = 11$ km, use the law of cosines to find B to the nearest tenth of a degree.
13. Find all the missing parts if $a = 6.4$ m, $b = 2.8$ m, and $C = 119°$.
14. Find all the missing parts if $b = 3.7$ m, $c = 6.2$ m, and $A = 35°$.

15. Find the area of the triangle in Problem 3.
16. Find the area of the triangle in Problem 4.
17. Find the area of the triangle in Problem 9.
18. Find the area of the triangle in Problem 10.
19. Find the area of the triangle in Problem 11.
20. Find the area of the triangle in Problem 12.

21. **Geometry** The two equal sides of an isosceles triangle are each 38 centimeters. If the base measures 48 centimeters, find the measure of the two equal angles.
22. **Angle of Elevation** A lamp pole casts a shadow 53 feet long when the angle of elevation of the sun is 48°. Find the height of the lamp pole to the nearest foot.
23. **Angle of Elevation** A man standing near a building notices that the angle of elevation to the top of the building is 64°. He then walks 240 feet farther away from the building and finds the angle of elevation to the top to be 43°. How tall is the building?
24. **Geometry** The diagonals of a parallelogram are 26.8 meters and 39.4 meters. If they meet at an angle of 134.5°, find the length of the shorter side of the parallelogram.
25. **Arc Length** Suppose Figure 1 is an exaggerated diagram of a plane flying above the earth. When the plane is 4.55 miles above point B, the pilot finds angle A to be 90.8°. Assuming that the radius of the earth is 3,960 miles, what is the distance from point A to point B along the circumference of the earth? (*Hint:* Find angle C first, then use the formula for arc length.)

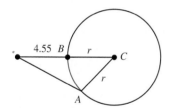

Figure 1

26. **Distance and Bearing** A man wandering in the desert walks 3.3 miles in the direction S 44° W. He then turns and walks 2.2 miles in the direction N 55° W. At that time, how far is he from his starting point, and what is his bearing from his starting point?
27. **Distance** Two guy wires from the top of a tent pole are anchored to the ground on each side of the pole by two stakes so that the two stakes and the tent pole lie along the same line. One of the wires is 56 feet long and makes an angle of 47° with the ground. The other wire is 65 feet long and makes an angle of 37° with the ground. How far apart are the stakes that hold the wires to the ground?
28. **Ground Speed** A plane is headed due east with an airspeed of 345 miles per hour. Its true course, however, is at 95.5° from due north. If the wind currents are a constant 55.0 miles per hour, what are the possibilities for the ground speed of the plane?
29. **Height of a Tree** To estimate the height of a tree, two people position themselves 25 feet apart. From the first person, the bearing of the tree is N 48° E and the angle of elevation to the top of the tree is 73°. If the bearing of the tree from the second person is N 38° W, estimate the height of the tree to the nearest foot.
30. **True Course and Speed** A plane flying with an airspeed of 325 miles per hour is headed in the direction 87.6°. The wind currents are running at a constant 65.4 miles per hour at 262.6°. Find the ground speed and true course of the plane.

Let $\mathbf{U} = 5\mathbf{i} + 12\mathbf{j}$, $\mathbf{V} = -4\mathbf{i} + \mathbf{j}$, and $\mathbf{W} = \mathbf{i} - 4\mathbf{j}$, and find

31. $|\mathbf{U}|$

32. $3\mathbf{U} + 5\mathbf{V}$

33. $3\mathbf{U} - 5\mathbf{V}$

34. $|2\mathbf{V} - \mathbf{W}|$

35. $\mathbf{V} \cdot \mathbf{W}$

36. The angle between $\mathbf{U}$ and $\mathbf{V}$ to the nearest tenth of a degree.

37. Show that the slope of the line that contains $\mathbf{V} = a\mathbf{i} + b\mathbf{j}$ is $m = \dfrac{b}{a}$.

38. Show that $\mathbf{V} = 3\mathbf{i} + 6\mathbf{j}$ and $\mathbf{W} = -8\mathbf{i} + 4\mathbf{j}$ are perpendicular.

39. Find the value of b so that vectors $\mathbf{U} = 5\mathbf{i} + 12\mathbf{j}$ and $\mathbf{V} = 4\mathbf{i} + b\mathbf{j}$ are perpendicular.

40. Find the work performed by the force $\mathbf{F} = 33\mathbf{i} - 4\mathbf{j}$ in moving an object, whose resulting motion is represented by the displacement vector $\mathbf{d} = 56\mathbf{i} + 10\mathbf{j}$.

CHAPTER 7 GROUP PROJECT

MEASURING THE DISTANCE TO MARS

Objective: To find the distance of Mars from the sun using two Earth-based observations.

The planet Mars orbits the sun once every 687 days. This is referred to as its sidereal period. When observing Mars from the earth, the angle that Mars makes with the sun is called the solar elongation (Figure 1). On November 13, 2000, Mars had a solar elongation of 45.5° W. On October 1, 2002, 687 days later, the solar elongation of Mars was 17.1° W. Because the number of days between these observations is equal to the sidereal period of Mars, the planet was in the same position on both occasions. Figure 2 is an illustration of the situation. From this perspective the earth would be traveling counterclockwise.

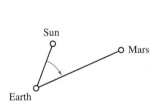

Figure 1

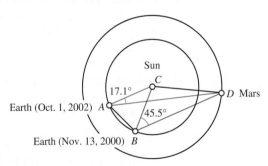

Figure 2

The distance from the sun to the earth is defined as 1 astronomical unit (1 AU), which is about 93 million miles. Because both AC and BC have a length of 1 AU, triangle ABC is an isosceles triangle.

1. Use the fact that the earth completes one revolution about the sun every 365 days to determine how many degrees the earth travels in one day. On November 13, 2002, the earth would again be at point B. Find the number of days it would take the earth to move from point A to point B. Now, using both of these results, compute the number of degrees for central angle ACB.
2. Find the distance of AB using triangle ABC.
3. Since triangle ABC is isosceles, angles CAB and CBA are equal. Find the measure of these two angles. Use this result to find angles DAB and DBA.
4. Use triangle ABD to find the length of AD.
5. Find the length of CD. Give your answer both in terms of AU and in miles.

CHAPTER 7 RESEARCH PROJECT

QUATERNIONS

We mentioned in Section 7.5 that our algebraic treatment of vectors could be attributed, in part, to the Irish mathematician William Rowan Hamilton. Hamilton considered his greatest achievement to be the discovery of *quaternions,* which he (incorrectly) predicted would revolutionize physics.

Research the subject of quaternions. What are they? Why did quaternions fail to be as useful for physics as Hamilton predicted? How are quaternions useful, instead, for 3-D computer graphics? Write a paragraph or two about your findings.

COMPLEX NUMBERS AND POLAR COORDINATES

Mathematical discoveries, small or great, are never born of spontaneous generation. They always presuppose a soil seeded with preliminary knowledge and well prepared by labour, both conscious and subconscious.

Henri Poincaré

INTRODUCTION

In 1545 Jerome Cardan published the mathematical work *Ars Magna*, the title page of which is shown below. The title translates to *The Great Art of Solving Algebraic Equations.* In this book Cardan posed a number of problems, such as solving certain cubic and quartic equations.

HIERONYMI CAR
DANI, PRÆSTANTISSIMI MATHE
MATIC̄, PHILOSOPHI, AC MEDICI,
ARTIS MAGNÆ,
SIVE DE REGVLIS ALGEBRAICIS,
Lib. unus. Qui & totius operis de Arithmetica, quod
OPVS PERFECTVM
inscripsit, est in ordine Decimus.

In our present-day notation, many of the problems posed by Cardan can be solved with some simple algebra, but the solutions require taking square roots of negative numbers. Square roots of negative numbers are handled with the complex number system, which had not been developed when Cardan wrote his book. The foundation of the complex number system is contained in this chapter, along with the very interesting and useful connection between complex numbers and trigonometric functions.

SECTION 8.1 | COMPLEX NUMBERS

One of the problems posed by Cardan in *Ars Magna* is the following:

> *If someone says to you, divide 10 into two parts, one of which multiplied into the other shall produce 40, it is evident that this case or question is impossible.*

Using our present-day notation, this problem can be solved with a system of equations:

$$x + y = 10$$
$$xy = 40$$

Solving the first equation for y, we have $y = 10 - x$. Substituting this value of y into the second equation, we have

$$x(10 - x) = 40$$

The equation is quadratic. We write it in standard form and apply the quadratic formula.

$$0 = x^2 - 10x + 40$$

$$x = \frac{10 \pm \sqrt{100 - 4(1)(40)}}{2}$$

$$= \frac{10 \pm \sqrt{100 - 160}}{2}$$

$$= \frac{10 \pm \sqrt{-60}}{2}$$

$$= \frac{10 \pm 2\sqrt{-15}}{2}$$

$$= 5 \pm \sqrt{-15}$$

This is as far as Cardan could take the problem because he did not know what to do with the square root of a negative number. We handle this situation by using *complex numbers*. Our work with complex numbers is based on the following definition.

DEFINITION

The number i is such that $i^2 = -1$. (That is, i is the number whose square is -1.)

The number i is not a real number. We can use it to write square roots of negative numbers without a negative sign. To do so, we reason that if $a > 0$, then $\sqrt{-a} = \sqrt{ai^2} = i\sqrt{a}$.

EXAMPLE 1 Write each expression in terms of i.

a. $\sqrt{-9}$ **b.** $\sqrt{-12}$ **c.** $\sqrt{-17}$

SOLUTION

a. $\sqrt{-9} = i\sqrt{9} = 3i$

b. $\sqrt{-12} = i\sqrt{12} = 2i\sqrt{3}$

c. $\sqrt{-17} = i\sqrt{17}$

NOTE In order to simplify expressions that contain square roots of negative numbers by using the properties of radicals developed in algebra, it is necessary to write each square root in terms of i before applying the properties of radicals. For example,

this is correct: $\sqrt{-4}\sqrt{-9} = (i\sqrt{4})(i\sqrt{9}) = (2i)(3i) = 6i^2 = -6$

this is incorrect: $\sqrt{-4}\sqrt{-9} = \sqrt{-4(-9)} = \sqrt{36} = 6$

Remember, the properties of radicals you developed in algebra hold only for expressions in which the numbers under the radical sign are nonnegative. When the radicals contain negative numbers, you must first write each radical in terms of i and then simplify.

Next, we use i to write a definition for complex numbers.

DEFINITION

A *complex number* is any number that can be written in the form

$$a + bi$$

where a and b are real numbers and $i^2 = -1$. The form $a + bi$ is called *standard form* for complex numbers. The number a is called the *real part* of the complex number. The number b is called the *imaginary part* of the complex number. If $b = 0$, then $a + bi = a$, which is a real number. If $a = 0$ and $b \neq 0$, then $a + bi = bi$, which is called an *imaginary number.*

EXAMPLE 2

a. The number $3 + 2i$ is a complex number in standard form. The number 3 is the real part, and the number 2 (not $2i$) is the imaginary part.

b. The number $-7i$ is a complex number because it can be written as $0 + (-7)i$. The real part is 0. The imaginary part is -7. The number $-7i$ is also an imaginary number since $a = 0$ and $b \neq 0$.

c. The number 4 is a complex number because it can be written as $4 + 0i$. The real part is 4 and the imaginary part is 0.

From part c in Example 2, it is apparent that real numbers are also complex numbers. The real numbers are a subset of the complex numbers.

EQUALITY FOR COMPLEX NUMBERS

> **DEFINITION**
>
> Two complex numbers are equal if and only if their real parts are equal and their imaginary parts are equal. That is, for real numbers a, b, c, and d,
>
> $$a + bi = c + di \quad \text{if and only if} \quad a = c \text{ and } b = d$$

 EXAMPLE 3   Find x and y if $(-3x - 9) + 4i = 6 + (3y - 2)i$.

SOLUTION   The real parts are $-3x - 9$ and 6. The imaginary parts are 4 and $3y - 2$.

$$
\begin{array}{ll}
-3x - 9 = 6 \quad \text{and} & 4 = 3y - 2 \\
-3x = 15 & 6 = 3y \\
x = -5 & y = 2
\end{array}
$$

ADDITION AND SUBTRACTION OF COMPLEX NUMBERS

> **DEFINITION**
>
> If $z_1 = a_1 + b_1 i$ and $z_2 = a_2 + b_2 i$ are complex numbers, then the sum and difference of z_1 and z_2 are defined as follows:
>
> $$z_1 + z_2 = (a_1 + b_1 i) + (a_2 + b_2 i)$$
> $$= (a_1 + a_2) + (b_1 + b_2)i$$
> $$z_1 - z_2 = (a_1 + b_1 i) - (a_2 + b_2 i)$$
> $$= (a_1 - a_2) + (b_1 - b_2)i$$

As you can see, we add and subtract complex numbers in the same way we would add and subtract polynomials: by combining similar terms.

  EXAMPLE 4   If $z_1 = 3 - 5i$ and $z_2 = -6 - 2i$, find $z_1 + z_2$ and $z_1 - z_2$.

SOLUTION

$$z_1 + z_2 = (3 - 5i) + (-6 - 2i)$$
$$= -3 - 7i$$
$$z_1 - z_2 = (3 - 5i) - (-6 - 2i)$$
$$= 9 - 3i$$

POWERS OF i

If we assume the properties of exponents hold when the base is i, we can write any integer power of i as i, -1, $-i$, or 1. Using the fact that $i^2 = -1$, we have

$$i^1 = i$$

$$i^2 = -1$$

$$i^3 = i^2 \cdot i = -1(i) = -i$$

$$i^4 = i^2 \cdot i^2 = -1(-1) = 1$$

Since $i^4 = 1$, i^5 will simplify to i and we will begin repeating the sequence i, -1, $-i$, 1 as we increase our exponent by one each time.

$$i^5 = i^4 \cdot i = 1(i) = i$$

$$i^6 = i^4 \cdot i^2 = 1(-1) = -1$$

$$i^7 = i^4 \cdot i^3 = 1(-i) = -i$$

$$i^8 = i^4 \cdot i^4 = 1(1) = 1$$

We can simplify higher powers of i by writing them in terms of i^4 since i^4 is always 1.

 EXAMPLE 5   Simplify each power of i.

a. $i^{20} = (i^4)^5 = 1^5 = 1$

b. $i^{23} = (i^4)^5 \cdot i^3 = 1(-i) = -i$

c. $i^{30} = (i^4)^7 \cdot i^2 = 1(-1) = -1$

MULTIPLICATION AND DIVISION WITH COMPLEX NUMBERS

> **DEFINITION**
>
> If $z_1 = a_1 + b_1i$ and $z_2 = a_2 + b_2i$ are complex numbers, then their product is defined as follows:
>
> $$z_1 z_2 = (a_1 + b_1i)(a_2 + b_2i)$$
>
> $$= (a_1a_2 - b_1b_2) + (a_1b_2 + a_2b_1)i$$

This formula is simply a result of binomial multiplication and is much less complicated than it looks. Since complex numbers have the form of binomials with i as the variable, we can multiply two complex numbers using the same methods we use to multiply binomials and not have another formula to memorize.

 **EXAMPLE 6** Multiply $(3 - 4i)(2 - 5i)$.

SOLUTION   Multiplying as if these were two binomials, we have

$$(3 - 4i)(2 - 5i) = 3 \cdot 2 - 3 \cdot 5i - 2 \cdot 4i + 4i \cdot 5i$$
$$= 6 - 15i - 8i + 20i^2$$
$$= 6 - 23i + 20i^2$$

Now, since $i^2 = -1$, we can simplify further.

$$= 6 - 23i + 20(-1)$$
$$= 6 - 23i - 20$$
$$= -14 - 23i$$

EXAMPLE 7 Multiply $(4 - 5i)(4 + 5i)$.

SOLUTION   This product has the form $(a - b)(a + b)$, which we know results in the difference of two squares, $a^2 - b^2$.

$$(4 - 5i)(4 + 5i) = 4^2 - (5i)^2$$
$$= 16 - 25i^2$$
$$= 16 - 25(-1)$$
$$= 16 + 25$$
$$= 41$$

The product of the two complex numbers $4 - 5i$ and $4 + 5i$ is the real number 41. This fact is very useful and leads to the following definition.

DEFINITION

The complex numbers $a + bi$ and $a - bi$ are called *complex conjugates*. Their product is the real number $a^2 + b^2$. Here's why:

$$(a + bi)(a - bi) = a^2 - (bi)^2$$
$$= a^2 - b^2i^2$$
$$= a^2 - b^2(-1)$$
$$= a^2 + b^2$$

The fact that the product of two complex conjugates is a real number is the key to division with complex numbers.

 **EXAMPLE 8** Divide $\dfrac{5i}{2-3i}$.

SOLUTION We want to find a complex number in standard form that is equivalent to the quotient $5i/(2-3i)$. To do so, we need to replace the denominator with a real number. We can accomplish this by multiplying both the numerator and the denominator by $2+3i$, which is the conjugate of $2-3i$.

$$\frac{5i}{2-3i} = \frac{5i}{2-3i} \cdot \frac{(2+3i)}{(2+3i)}$$

$$= \frac{5i(2+3i)}{(2-3i)(2+3i)}$$

$$= \frac{10i + 15i^2}{4 - 9i^2}$$

$$= \frac{10i + 15(-1)}{4 - 9(-1)}$$

$$= \frac{-15 + 10i}{13}$$

$$= -\frac{15}{13} + \frac{10}{13}i$$

Notice that we have written our answer in standard form. The real part is $-15/13$ and the imaginary part is $10/13$.

CALCULATOR NOTE Some graphing calculators have the ability to perform operations with complex numbers. Check to see if your model has a key for entering the number i. Figure 1 shows how Examples 4 and 6 might look when solved on a graphing calculator. Example 8 is shown in Figure 2.

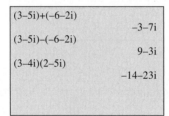

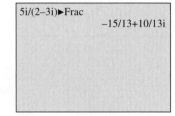

Figure 1   **Figure 2**

 **EXAMPLE 9** In the introduction to this section, we found that $x = 5 + \sqrt{-15}$ is one part of a solution to the system of equations

$$x + y = 10$$

$$xy = 40$$

Find the value of y that accompanies this value of x. Then show that together they form a solution to the system of equations that describe Cardan's problem.

SOLUTION   First we write the number $x = 5 + \sqrt{-15}$ with our complex number notation as $x = 5 + i\sqrt{15}$. Now, to find the value of y that accompanies this value of x, we substitute $5 + i\sqrt{15}$ for x in the equation $x + y = 10$ and solve for y:

$$\text{When} \qquad x = 5 + i\sqrt{15}$$

$$\text{then} \qquad x + y = 10$$

$$\text{becomes} \qquad 5 + i\sqrt{15} + y = 10$$

$$y = 5 - i\sqrt{15}$$

All we have left to do is show that these values of x and y satisfy the second equation, $xy = 40$. We do so by substitution:

$$\text{If} \qquad x = 5 + i\sqrt{15} \qquad \text{and} \qquad y = 5 - i\sqrt{15}$$

$$\text{then the equation} \qquad xy = 40$$

$$\text{becomes} \qquad (5 + i\sqrt{15})(5 - i\sqrt{15}) = 40$$

$$5^2 - (i\sqrt{15})^2 = 40$$

$$25 - i^2(\sqrt{15})^2 = 40$$

$$25 - (-1)(15) = 40$$

$$25 + 15 = 40$$

$$40 = 40$$

Since our last statement is a true statement, our two expressions for x and y together form a solution to the equation. ▮

GETTING READY FOR CLASS

After reading through the preceding section, respond in your own words and in complete sentences.

a. Give a definition for complex numbers.

b. How do you add two complex numbers?

c. What is a complex conjugate?

d. How do you divide two complex numbers?

PROBLEM SET 8.1

Write each expression in terms of i.

1. $\sqrt{-16}$ **2.** $\sqrt{-49}$ **3.** $\sqrt{-121}$ **4.** $\sqrt{-400}$
5. $\sqrt{-18}$ **6.** $\sqrt{-45}$ **7.** $\sqrt{-8}$ **8.** $\sqrt{-20}$

Write in terms of i and then simplify.

9. $\sqrt{-4} \cdot \sqrt{-9}$ **10.** $\sqrt{-25} \cdot \sqrt{-1}$ **11.** $\sqrt{-1} \cdot \sqrt{-9}$ **12.** $\sqrt{-16} \cdot \sqrt{-4}$

Find x and y so that each of the following equations is true:

13. $4 + 7i = 6x - 14yi$ **14.** $2 - 5i = -x + 10yi$
15. $(5x + 2) - 7i = 4 + (2y + 1)i$ **16.** $(7x - 1) + 4i = 2 + (5y + 2)i$
17. $(x^2 - 6) + 9i = x + y^2 i$ **18.** $(x^2 - 2x) + y^2 i = 8 + (2y - 1)i$

Find all x and y $(0 \le x, y < 2\pi)$ so that each of the following equations is true:

19. $\cos x + i \sin y = \sin x + i$ **20.** $\sin x + i \cos y = -\cos x - i$

21. $(\sin^2 x + 1) + i \tan y = 2 \sin x + i$ **22.** $(\cos^2 x + 1) + i \tan y = 2 \cos x - i$

Combine the following complex numbers:

23. $(7 + 2i) + (3 - 4i)$ **24.** $(3 - 5i) + (2 + 4i)$

25. $(6 + 7i) - (4 + i)$ **26.** $(5 + 2i) - (3 + 6i)$

27. $(7 - 3i) - (4 + 10i)$ **28.** $(11 - 6i) - (2 - 4i)$

29. $(3 \cos x + 4i \sin y) + (2 \cos x - 7i \sin y)$

30. $(2 \cos x - 3i \sin y) + (3 \cos x - 2i \sin y)$

31. $[(3 + 2i) - (6 + i)] + (5 + i)$ **32.** $[(4 - 5i) - (2 + i)] + (2 + 5i)$

33. $(7 - 4i) - [(-2 + i) - (3 + 7i)]$ **34.** $(10 - 2i) - [(2 + i) - (3 - i)]$

Simplify each power of i.

35. i^{12} **36.** i^{13} **37.** i^{14} **38.** i^{15}

39. i^{32} **40.** i^{34} **41.** i^{33} **42.** i^{35}

Find the following products:

43. $-6i(3 - 8i)$ **44.** $6i(3 + 8i)$

45. $(2 - 4i)(3 + i)$ **46.** $(2 + 4i)(3 - i)$

47. $(3 + 2i)^2$ **48.** $(3 - 2i)^2$

49. $(5 + 4i)(5 - 4i)$ **50.** $(4 + 5i)(4 - 5i)$

51. $(7 + 2i)(7 - 2i)$ **52.** $(2 + 7i)(2 - 7i)$

53. $2i(3 + i)(2 + 4i)$ **54.** $3i(1 + 2i)(3 + i)$

55. $3i(1 + i)^2$ **56.** $4i(1 - i)^2$

Find the following quotients. Write all answers in standard form for complex numbers.

57. $\dfrac{2i}{3 + i}$ **58.** $\dfrac{3i}{2 + i}$ **59.** $\dfrac{2 + 3i}{2 - 3i}$ **60.** $\dfrac{3 + 2i}{3 - 2i}$

61. $\dfrac{5 - 2i}{i}$ **62.** $\dfrac{5 - 2i}{-i}$ **63.** $\dfrac{2 + i}{5 - 6i}$ **64.** $\dfrac{5 + 4i}{3 + 6i}$

Let $z_1 = 2 + 3i$, $z_2 = 2 - 3i$, and $z_3 = 4 + 5i$, and find

65. $z_1 z_2$ **66.** $z_2 z_1$ **67.** $z_1 z_3$ **68.** $z_3 z_1$

69. $2z_1 + 3z_2$ **70.** $3z_1 + 2z_2$ **71.** $z_3(z_1 + z_2)$ **72.** $z_3(z_1 - z_2)$

73. Assume x represents a real number and multiply $(x + 3i)(x - 3i)$.

74. Assume x represents a real number and multiply $(x - 4i)(x + 4i)$.

75. Show that $x = 2 + 3i$ is a solution to the equation $x^2 - 4x + 13 = 0$.

76. Show that $x = 3 + 2i$ is a solution to the equation $x^2 - 6x + 13 = 0$.

77. Show that $x = a + bi$ is a solution to the equation $x^2 - 2ax + (a^2 + b^2) = 0$.

78. Show that $x = a - bi$ is a solution to the equation $x^2 - 2ax + (a^2 + b^2) = 0$.

Use the method shown in the introduction to this section to solve each system of equations.

79. $x + y = 8$ **80.** $x - y = 10$

       $xy = 20$ $xy = -40$

81. $2x + y = 4$ **82.** $3x + y = 6$

       $xy = 8$ $xy = 9$

83. If z is a complex number, show that the product of z and its conjugate is a real number.

84. If z is a complex number, show that the sum of z and its conjugate is a real number.

85. Is addition of complex numbers a commutative operation? That is, if z_1 and z_2 are two complex numbers, is it always true that $z_1 + z_2 = z_2 + z_1$?

86. Is subtraction with complex numbers a commutative operation?

REVIEW PROBLEMS

The problems that follow review material we covered in Sections 1.3, 3.1, and Chapter 7. Reviewing these problems will help you with some of the material in the next section.

Find $\sin \theta$ and $\cos \theta$ if the given point lies on the terminal side of θ.

87. $(3, -4)$ **88.** $(-5, 12)$ **89.** (a, b) **90.** $(1, -1)$

Find θ between $0°$ and $360°$ if

91. $\sin \theta = \dfrac{1}{\sqrt{2}}$ and $\cos \theta = -\dfrac{1}{\sqrt{2}}$ **92.** $\sin \theta = \dfrac{1}{2}$ and θ terminates in QII

Solve triangle ABC if

93. $A = 73.1°$, $b = 243$ cm, and $c = 157$ cm

94. $B = 24.2°$, $C = 63.8°$, and $b = 5.92$ inches

95. $a = 42.1$ m, $b = 56.8$ m, and $c = 63.4$ m

96. $B = 32.8°$, $a = 625$ ft, and $b = 521$ ft

SECTION 8.2 | TRIGONOMETRIC FORM FOR COMPLEX NUMBERS

Cardan

As you know, the quadratic formula, $x = \dfrac{-b \pm \sqrt{b^2 - 4ac}}{2a}$, can be used to solve any quadratic equation in the form $ax^2 + bx + c = 0$. In his book *Ars Magna*, Jerome Cardan gives a similar formula that can be used to solve certain cubic equations. Here it is in our notation:

If $\qquad x^3 = ax + b$

then $\qquad x = \sqrt[3]{\dfrac{b}{2} + \sqrt{\left(\dfrac{b}{2}\right)^2 - \left(\dfrac{a}{3}\right)^3}} + \sqrt[3]{\dfrac{b}{2} - \sqrt{\left(\dfrac{b}{2}\right)^2 - \left(\dfrac{a}{3}\right)^3}}$

This formula is known as Cardan's formula. In his book, Cardan attempts to use his formula to solve the equation

$$x^3 = 15x + 4$$

This equation has the form $x^3 = ax + b$, where $a = 15$ and $b = 4$. Substituting these values for a and b in Cardan's formula, we have

$$x = \sqrt[3]{\dfrac{4}{2} + \sqrt{\left(\dfrac{4}{2}\right)^2 - \left(\dfrac{15}{3}\right)^3}} + \sqrt[3]{\dfrac{4}{2} - \sqrt{\left(\dfrac{4}{2}\right)^2 - \left(\dfrac{15}{3}\right)^3}}$$

$$= \sqrt[3]{2 + \sqrt{4 - 125}} + \sqrt[3]{2 - \sqrt{4 - 125}}$$

$$= \sqrt[3]{2 + \sqrt{-121}} + \sqrt[3]{2 - \sqrt{-121}}$$

Cardan couldn't go any further than this because he didn't know what to do with $\sqrt{-121}$. In fact, he wrote:

I have sent to enquire after the solution to various problems for which you have given me no answer, one of which concerns the cube equal to an unknown plus a number. I have certainly grasped this rule, but when the cube of one-third of the coefficient of the unknown is greater in value than the square of one-half of the number, then, it appears, I cannot make it fit into the equation.

Notice in his formula that if $(a/3)^3 > (b/2)^2$, then the result will be a negative number inside the square root.

In this section, we will take the first step in finding cube roots of complex numbers by learning how to write complex numbers in what is called *trigonometric form*. Before we do so, let's look at a definition that will give us a visual representation for complex numbers.

DEFINITION

The graph of the complex number $x + yi$ is a vector (arrow) that extends from the origin out to the point (x, y).

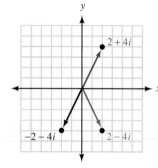

 EXAMPLE 1 Graph each complex number: $2 + 4i$, $-2 - 4i$, and $2 - 4i$.

SOLUTION The graphs are shown in Figure 1. Notice how the graphs of $2 + 4i$ and $2 - 4i$, which are conjugates, have symmetry about the *x*-axis. Note also that the graphs of $2 + 4i$ and $-2 - 4i$, which are opposites, have symmetry about the origin.

Figure 1

 **EXAMPLE 2** Graph the complex numbers 1, i, -1, and $-i$.

SOLUTION Here are the four complex numbers written in standard form.

$$1 = 1 + 0i \qquad i = 0 + i$$
$$-1 = -1 + 0i \qquad -i = 0 - i$$

The graph of each is shown in Figure 2.

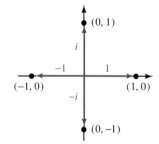

Figure 2

If we write a real number as a complex number in standard form, its graph will fall on the *x*-axis. Therefore, we call the *x*-axis the *real axis* when we are graphing complex numbers. Likewise, because the imaginary numbers i and $-i$ fall on the *y*-axis, we call the *y*-axis the *imaginary axis* when we are graphing complex numbers.

DEFINITION

The *absolute value* or *modulus* of the complex number $z = x + yi$ is the distance from the origin to the point (x, y). If this distance is denoted by r, then

$$r = |z| = |x + yi| = \sqrt{x^2 + y^2}$$

EXAMPLE 3 Find the modulus of each of the complex numbers $5i$, 7, and $3 + 4i$.

SOLUTION Writing each number in standard form and then applying the definition of modulus, we have

For $z = 5i = 0 + 5i$, $r = |z| = |0 + 5i| = \sqrt{0^2 + 5^2} = 5$

For $z = 7 = 7 + 0i$, $r = |z| = |7 + 0i| = \sqrt{7^2 + 0^2} = 7$

For $z = 3 + 4i$, $r = |z| = |3 + 4i| = \sqrt{3^2 + 4^2} = 5$

DEFINITION

The *argument* of the complex number $z = x + yi$ is the smallest positive angle θ from the positive real axis to the graph of z.

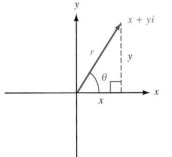

Figure 3

Figure 3 illustrates the relationships between the complex number $z = x + yi$, its graph, and the modulus r and argument θ of z. From Figure 3 we see that

$$\cos \theta = \frac{x}{r} \quad \text{or} \quad x = r \cos \theta$$

and

$$\sin \theta = \frac{y}{r} \quad \text{or} \quad y = r \sin \theta$$

We can use this information to write z in terms of r and θ.

$$z = x + yi$$
$$= r \cos \theta + (r \sin \theta)i$$
$$= r \cos \theta + ri \sin \theta$$
$$= r (\cos \theta + i \sin \theta)$$

This last expression is called the *trigonometric form* for z, which can be abbreviated as r cis θ. The formal definition follows.

DEFINITION

If $z = x + yi$ is a complex number in standard form, then the *trigonometric form* for z is given by

$$z = r(\cos \theta + i \sin \theta) = r \text{ cis } \theta$$

where r is the modulus of z and θ is the argument of z.

We can convert back and forth between standard form and trigonometric form by using the relationships that follow:

For $z = x + yi = r(\cos \theta + i \sin \theta) = r \text{ cis } \theta$

$$r = \sqrt{x^2 + y^2} \quad \text{and } \theta \text{ is such that}$$

$$\cos \theta = \frac{x}{r}, \quad \sin \theta = \frac{y}{r}, \quad \text{and} \quad \tan \theta = \frac{y}{x}$$

$-1 + i = \sqrt{2}$ cis 135°

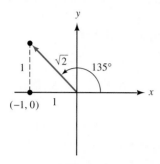

Figure 4

EXAMPLE 4 Write $z = -1 + i$ in trigonometric form.

SOLUTION We have $x = -1$ and $y = 1$; therefore,

$$r = \sqrt{(-1)^2 + 1^2} = \sqrt{2}$$

Angle θ is the smallest positive angle for which

$$\cos \theta = \frac{x}{r} = -\frac{1}{\sqrt{2}} \qquad \text{and} \qquad \sin \theta = \frac{y}{r} = \frac{1}{\sqrt{2}}$$

Therefore, θ must be 135°.

Using these values of r and θ in the formula for trigonometric form, we have

$$z = r(\cos \theta + i \sin \theta)$$
$$= \sqrt{2}(\cos 135° + i \sin 135°)$$
$$= \sqrt{2} \text{ cis } 135°$$

The graph of z is shown in Figure 4. ▪

2 cis 60° $= 1 + i\sqrt{3}$

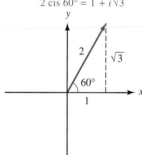

Figure 5

EXAMPLE 5 Write $z = 2$ cis 60° in standard form.

SOLUTION Using exact values for cos 60° and sin 60°, we have

$$z = 2 \text{ cis } 60°$$
$$= 2(\cos 60° + i \sin 60°)$$
$$= 2\left(\frac{1}{2} + i\frac{\sqrt{3}}{2}\right)$$
$$= 1 + i\sqrt{3}$$

The graph of z is shown in Figure 5. ▪

As you can see, converting from trigonometric form to standard form is usually more direct than converting from standard form to trigonometric form.

USING TECHNOLOGY

CONVERTING TO TRIGONOMETRIC FORM

Most graphing calculators are able to convert a complex number between standard form (sometimes called *rectangular form*) and trigonometric form (also called *polar form*). For example, on a TI-83 we could solve Example 4 using the **abs** and **angle** commands found in the MATH CPX menu, as shown in Figure 6. On a TI-86, the conversion commands are found in the CPLX menu. Figure 7 shows how these commands could be used to work Examples 4 and 5. The complex number $-1 + i$ is entered as an ordered pair $(-1, 1)$, and 2 cis 60° is entered as $(2\angle 60°)$. In both cases, we have set the calculator to degree mode.

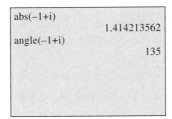

Figure 6

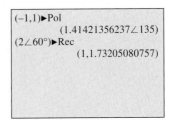

Figure 7

Check the manual for your model to see if your calculator is able to convert complex numbers to trigonometric form. You may find the commands listed in the index under polar coordinates (we will explore polar coordinates later in this chapter).

EXAMPLE 6   In the introduction to this section, we mentioned two complex numbers that Jerome Cardan had difficulty with: $2 + \sqrt{-121}$ and $2 - \sqrt{-121}$. Write each of these numbers in trigonometric form.

SOLUTION   First we write them as complex numbers

$$2 + \sqrt{-121} = 2 + 11i$$

$$2 - \sqrt{-121} = 2 - 11i$$

The modulus of each is

$$r = \sqrt{2^2 + 11^2} = \sqrt{125} = 5\sqrt{5}$$

For $2 + 11i$, we have $\tan \theta = \dfrac{11}{2}$. Using a calculator and rounding to the nearest hundredth of a degree, we find that $\theta = \tan^{-1}(5.5) = 79.70°$. Therefore,

$$2 + 11i = 5\sqrt{5}(\cos 79.70° + i \sin 79.70°)$$

$$= 5\sqrt{5} \text{ cis } 79.70°$$

For $2 - 11i$, we have $\tan \theta = -\dfrac{11}{2}$, giving us $\theta = 280.30°$ to the nearest hundredth of a degree. Therefore,

$$2 - 11i = 5\sqrt{5}(\cos 280.30° + i \sin 280.30°)$$

$$= 5\sqrt{5} \text{ cis } 280.30°$$

Note that if we were to take the trigonometric form of this last number and simplify it using decimals and a calculator, the result would not be exactly the same as the original number $2 - 11i$. That is,

$$5\sqrt{5}\,(\cos 280.30° + i \sin 280.30°) = 11.18(0.18 - i\,0.98)$$

$$= 2.01 - 10.96i$$

The difference is due to the rounding we did to obtain $280.30°$ and the decimal approximation to $5\sqrt{5}$. ■

GETTING READY FOR CLASS

After reading through the preceding section, respond in your own words and in complete sentences.

a. How do you draw the graph of a complex number?

b. Define the absolute value of a complex number.

c. What is the argument of a complex number?

d. What is the first step in writing the complex number $-1 + i$ in trigonometric form?

PROBLEM SET 8.2

Graph each complex number. In each case, give the absolute value of the number.

1. $3 + 4i$ **2.** $3 - 4i$ **3.** $1 + i$ **4.** $1 - i$

5. $-5i$ **6.** $4i$ **7.** 2 **8.** -4

9. $-4 - 3i$ **10.** $-3 - 4i$

Graph each complex number along with its opposite and conjugate.

11. $2 - i$ **12.** $2 + i$ **13.** $4i$ **14.** $-3i$

15. -3 **16.** 5 **17.** $-5 - 2i$ **18.** $-2 - 5i$

Write each complex number in standard form.

19. $2(\cos 30° + i \sin 30°)$ **20.** $4(\cos 30° + i \sin 30°)$

21. $4(\cos 120° + i \sin 120°)$ **22.** $8(\cos 120° + i \sin 120°)$

23. $1 \operatorname{cis} 210°$ **24.** $1 \operatorname{cis} 240°$

25. $1 \operatorname{cis} 315°$ **26.** $\sqrt{2} \operatorname{cis} 315°$

Use a calculator to help write each complex number in standard form. Round the numbers in your answers to the nearest hundredth.

27. $10(\cos 12° + i \sin 12°)$ **28.** $100(\cos 70° + i \sin 70°)$

29. $100(\cos 143° + i \sin 143°)$ **30.** $100(\cos 171° + i \sin 171°)$

31. $1 \operatorname{cis} 205°$ **32.** $1 \operatorname{cis} 261°$

33. $10 \operatorname{cis} 342°$ **34.** $10 \operatorname{cis} 318°$

Write each complex number in trigonometric form. In each case, begin by sketching the graph to help find the argument θ.

35. $-1 + i$ **36.** $1 + i$ **37.** $1 - i$ **38.** $-1 - i$

39. $3 + 3i$ **40.** $5 + 5i$ **41.** $8i$ **42.** $-8i$

43. -9 **44.** 2 **45.** $-2 + 2i\sqrt{3}$ **46.** $-2\sqrt{3} + 2i$

Write each complex number in trigonometric form. Round all angles to the nearest hundredth of a degree.

47. $3 + 4i$ **48.** $3 - 4i$ **49.** $20 + 21i$ **50.** $21 - 20i$

51. $7 - 24i$ **52.** $8 - 15i$ **53.** $11 + 2i$ **54.** $11 - 2i$

Use your graphing calculator to convert the complex number to trigonometric form in the problems below.

55. Problem 47   **56.** Problem 48   **57.** Problem 49   **58.** Problem 50

59. Problem 51   **60.** Problem 52   **61.** Problem 53   **62.** Problem 54

63. We know that $2i \cdot 3i = 6i^2 = -6$. Change $2i$ and $3i$ to trigonometric form, and then show that their product in trigonometric form is still -6.

64. Change $4i$ and 2 to trigonometric form and then multiply. Show that this product is $8i$.

65. Show that 2 cis 30° and 2 cis (−30°) are conjugates.

66. Show that 2 cis 60° and 2 cis (−60°) are conjugates.

67. Show that if $z = \cos \theta + i \sin \theta$, then $\left| z \right| = 1$.

68. Show that if $z = \cos \theta - i \sin \theta$, then $\left| z \right| = 1$.

REVIEW PROBLEMS

The problems that follow review material we covered in Sections 5.2 and 7.2. Reviewing the problems from 5.2 will help you understand the next section.

69. Use the formula for cos $(A + B)$ to find the exact value of cos 75°.

70. Use the formula for sin $(A + B)$ to find the exact value of sin 75°.

Let $\sin A = 3/5$ with A in QI and $\sin B = 5/13$ with B in QI and find

71. sin $(A + B)$ **72.** cos $(A + B)$

Simplify each expression to a single trigonometric function.

73. sin 30° cos 90° + cos 30° sin 90°   **74.** cos 30° cos 90° − sin 30° sin 90°

75. cos 18° cos 32° − sin 18° sin 32°   **76.** sin 18° cos 32° + cos 18° sin 32°

In triangle ABC, $A = 45.6°$ and $b = 567$ inches. Find B for each value of a below. (You may get one or two values of B, or you may find that no triangle fits the given description.)

77. $a = 234$ inches   **78.** $a = 678$ inches

79. $a = 456$ inches   **80.** $a = 789$ inches

EXTENDING THE CONCEPTS

81. With the publication of *Ars Magna,* a dispute arose between Cardan and another mathematician named Tartaglia. What was that dispute? In your opinion who was at fault?

Tartaglia

SECTION 8.3 | PRODUCTS AND QUOTIENTS IN TRIGONOMETRIC FORM

Multiplication and division with complex numbers becomes a very simple process when the numbers are written in trigonometric form. Let's state the rule for finding the product of two complex numbers written in trigonometric form as a theorem and then prove the theorem.

THEOREM (MULTIPLICATION)

If

$$z_1 = r_1(\cos \theta_1 + i \sin \theta_1) = r_1 \text{ cis } \theta_1$$

and

$$z_2 = r_2(\cos \theta_2 + i \sin \theta_2) = r_2 \text{ cis } \theta_2$$

are two complex numbers in trigonometric form, then their product, $z_1 z_2$, is

$$z_1 z_2 = [r_1(\cos \theta_1 + i \sin \theta_1)][r_2(\cos \theta_2 + i \sin \theta_2)]$$
$$= r_1 r_2[\cos (\theta_1 + \theta_2) + i \sin (\theta_1 + \theta_2)]$$
$$= r_1 r_2 \text{ cis } (\theta_1 + \theta_2)$$

In Words: To multiply two complex numbers in trigonometric form, multiply absolute values and add angles.

PROOF

We begin by multiplying algebraically. Then we simplify our product by using the sum formulas we introduced in Section 5.2.

$$z_1 z_2 = [r_1(\cos \theta_1 + i \sin \theta_1)][r_2(\cos \theta_2 + i \sin \theta_2)]$$

$$= r_1 r_2 (\cos \theta_1 + i \sin \theta_1)(\cos \theta_2 + i \sin \theta_2)$$

$$= r_1 r_2 (\cos \theta_1 \cos \theta_2 + i \cos \theta_1 \sin \theta_2 + i \sin \theta_1 \cos \theta_2 + i^2 \sin \theta_1 \sin \theta_2)$$

$$= r_1 r_2 [\cos \theta_1 \cos \theta_2 + i(\cos \theta_1 \sin \theta_2 + \sin \theta_1 \cos \theta_2) - \sin \theta_1 \sin \theta_2]$$

$$= r_1 r_2 [(\cos \theta_1 \cos \theta_2 - \sin \theta_1 \sin \theta_2) + i(\sin \theta_1 \cos \theta_2 + \cos \theta_1 \sin \theta_2)]$$

$$= r_1 r_2 [\cos (\theta_1 + \theta_2) + i \sin (\theta_1 + \theta_2)]$$

This completes our proof. As you can see, to multiply two complex numbers in trigonometric form, we multiply absolute values, $r_1 r_2$, and add angles, $\theta_1 + \theta_2$.

 EXAMPLE 1 Find the product of 3 cis 40° and 5 cis 10°.

SOLUTION Applying the formula from our theorem on products, we have

$$(3 \text{ cis } 40°)(5 \text{ cis } 10°) = 3 \cdot 5 \text{ cis } (40° + 10°) = 15 \text{ cis } (50°)$$

 EXAMPLE 2 Find the product of $z_1 = 1 + i\sqrt{3}$ and $z_2 = -\sqrt{3} + i$ in standard form, and then write z_1 and z_2 in trigonometric form and find their product again.

SOLUTION Leaving each complex number in standard form and multiplying we have

$$z_1 z_2 = (1 + i\sqrt{3})(-\sqrt{3} + i)$$

$$= -\sqrt{3} + i - 3i + i^2\sqrt{3}$$

$$= -2\sqrt{3} - 2i$$

Changing z_1 and z_2 to trigonometric form and multiplying looks like this:

$$z_1 = 1 + i\sqrt{3} = 2(\cos 60° + i \sin 60°)$$

$$z_2 = -\sqrt{3} + i = 2(\cos 150° + i \sin 150°)$$

$$z_1 z_2 = [2(\cos 60° + i \sin 60°)][2(\cos 150° + i \sin 150°)]$$

$$= 4(\cos 210° + i \sin 210°)$$

To compare our two products, we convert our product in trigonometric form to standard form.

$$4(\cos 210° + i \sin 210°) = 4\left(-\frac{\sqrt{3}}{2} - \frac{1}{2}i\right)$$

$$= -2\sqrt{3} - 2i$$

As you can see, both methods of multiplying complex numbers produce the same result.

The next theorem is an extension of the work we have done so far with multiplication. We will not give a formal proof of the theorem.

DEMOIVRE'S THEOREM

If $z = r(\cos \theta + i \sin \theta) = r \text{ cis } \theta$ is a complex number in trigonometric form and n is an integer, then

$$z^n = [r(\cos \theta + i \sin \theta)]^n$$
$$= r^n(\cos n\theta + i \sin n\theta)$$
$$= r^n \text{ cis } n\theta$$

The theorem seems reasonable after the work we have done with multiplication. For example, if n is 2,

$$[r(\cos \theta + i \sin \theta)]^2 = r(\cos \theta + i \sin \theta) \cdot r(\cos \theta + i \sin \theta)$$
$$= r \cdot r[\cos(\theta + \theta) + i \sin(\theta + \theta)]$$
$$= r^2(\cos 2\theta + i \sin 2\theta)$$

 EXAMPLE 3   Find $(1 + i)^{10}$.

SOLUTION   First we write $1 + i$ in trigonometric form:
$$1 + i = \sqrt{2}(\cos 45° + i \sin 45°)$$

Then we use DeMoivre's Theorem to raise this expression to the 10th power.
$$(1 + i)^{10} = [\sqrt{2}(\cos 45° + i \sin 45°)]^{10}$$
$$= (\sqrt{2})^{10}(\cos 10 \cdot 45° + i \sin 10 \cdot 45°)$$
$$= 32(\cos 450° + i \sin 450°)$$

which we can simplify to
$$= 32(\cos 90° + i \sin 90°)$$

since 90° and 450° are coterminal. In standard form, our result is
$$= 32(0 + i)$$
$$= 32i$$

That is,
$$(1 + i)^{10} = 32i$$

Since multiplication with complex numbers in trigonometric form is accomplished by multiplying absolute values and adding angles, we should expect that division is accomplished by dividing absolute values and subtracting angles.

THEOREM (DIVISION)

If

$$z_1 = r_1(\cos \theta_1 + i \sin \theta_1) = r_1 \text{ cis } \theta_1$$

and

$$z_2 = r_2(\cos \theta_2 + i \sin \theta_2) = r_2 \text{ cis } \theta_2$$

are two complex numbers in trigonometric form, then their quotient, z_1/z_2, is

$$\frac{z_1}{z_2} = \frac{r_1(\cos \theta_1 + i \sin \theta_1)}{r_2(\cos \theta_2 + i \sin \theta_2)}$$

$$= \frac{r_1}{r_2} [\cos(\theta_1 - \theta_2) + i \sin (\theta_1 - \theta_2)]$$

$$= \frac{r_1}{r_2} \text{ cis } (\theta_1 - \theta_2)$$

PROOF

As was the case with division of complex numbers in standard form, the major step in this proof is multiplying the numerator and denominator of our quotient by the conjugate of the denominator.

$$\frac{r_1(\cos \theta_1 + i \sin \theta_1)}{r_2(\cos \theta_2 + i \sin \theta_2)}$$

$$= \frac{r_1(\cos \theta_1 + i \sin \theta_1)}{r_2(\cos \theta_2 + i \sin \theta_2)} \cdot \frac{(\cos \theta_2 - i \sin \theta_2)}{(\cos \theta_2 - i \sin \theta_2)}$$

$$= \frac{r_1(\cos \theta_1 + i \sin \theta_1)(\cos \theta_2 - i \sin \theta_2)}{r_2(\cos^2 \theta_2 + \sin^2 \theta_2)}$$

$$= \frac{r_1}{r_2} (\cos \theta_1 \, \cos \theta_2 - i \cos \theta_1 \sin \theta_2 + i \sin \theta_1 \cos \theta_2 - i^2 \sin \theta_1 \sin \theta_2)$$

$$= \frac{r_1}{r_2} [(\cos \theta_1 \, \cos \theta_2 + \sin \theta_1 \sin \theta_2) + i(\sin \theta_1 \cos \theta_2 - \cos \theta_1 \sin \theta_2)]$$

$$= \frac{r_1}{r_2} [\cos (\theta_1 - \theta_2) + i \sin (\theta_1 - \theta_2)] \quad ■$$

 **EXAMPLE 4** Find the quotient when $20(\cos 75° + i \sin 75°)$ is divided by $4(\cos 40° + i \sin 40°)$.

SOLUTION We divide according to the formula given in our theorem on division.

$$\frac{20(\cos 75° + i \sin 75°)}{4(\cos 40° + i \sin 40°)} = \frac{20}{4} [\cos (75° - 40°) + i \sin (75° - 40°)]$$

$$= 5(\cos 35° + i \sin 35°) \quad ■$$

 **EXAMPLE 5** Divide $z_1 = 1 + i\sqrt{3}$ by $z_2 = \sqrt{3} + i$ and leave the answer in standard form. Then change each to trigonometric form and divide again.

SOLUTION Dividing in standard form, we have

$$\frac{z_1}{z_2} = \frac{1 + i\sqrt{3}}{\sqrt{3} + i}$$

$$= \frac{1 + i\sqrt{3}}{\sqrt{3} + i} \cdot \frac{\sqrt{3} - i}{\sqrt{3} - i}$$

$$= \frac{\sqrt{3} - i + 3i - i^2\sqrt{3}}{3 + 1}$$

$$= \frac{2\sqrt{3} + 2i}{4}$$

$$= \frac{\sqrt{3}}{2} + \frac{1}{2}i$$

Changing z_1 and z_2 to trigonometric form and dividing again, we have

$$z_1 = 1 + i\sqrt{3} = 2 \text{ cis } 60°$$

$$z_2 = \sqrt{3} + i = 2 \text{ cis } 30°$$

$$\frac{z_1}{z_2} = \frac{2 \text{ cis } 60°}{2 \text{ cis } 30°}$$

$$= \frac{2}{2} \text{ cis } (60° - 30°)$$

$$= 1 \text{ cis } 30°$$

which, in standard form, is

$$\frac{\sqrt{3}}{2} + \frac{1}{2}i \quad$$

GETTING READY FOR CLASS

After reading through the preceding section, respond in your own words and in complete sentences.

a. How do you multiply two complex numbers written in trigonometric form?

b. How do you divide two complex numbers written in trigonometric form?

c. What is the argument of the complex number that results when $\sqrt{2} (\cos 45° + i \sin 45°)$ is raised to the 10th power?

d. What is the modulus of the complex number that results when $\sqrt{2} (\cos 45° + i \sin 45°)$ is raised to the 10th power?

PROBLEM SET 8.3

Multiply. Leave all answers in trigonometric form.

1. $3(\cos 20° + i \sin 20°) \cdot 4(\cos 30° + i \sin 30°)$
2. $5(\cos 15° + i \sin 15°) \cdot 2(\cos 25° + i \sin 25°)$
3. $7(\cos 110° + i \sin 110°) \cdot 8(\cos 47° + i \sin 47°)$
4. $9(\cos 115° + i \sin 115°) \cdot 4(\cos 51° + i \sin 51°)$
5. $2 \operatorname{cis} 135° \cdot 2 \operatorname{cis} 45°$
6. $2 \operatorname{cis} 120° \cdot 4 \operatorname{cis} 30°$

Find the product z_1z_2 in standard form. Then write z_1 and z_2 in trigonometric form and find their product again. Finally, convert the answer that is in trigonometric form to standard form to show that the two products are equal.

7. $z_1 = 1 + i, z_2 = -1 + i$
8. $z_1 = 1 + i, z_2 = 2 + 2i$
9. $z_1 = 1 + i\sqrt{3}, z_2 = -\sqrt{3} + i$
10. $z_1 = -1 + i\sqrt{3}, z_2 = \sqrt{3} + i$
11. $z_1 = 3i, z_2 = -4i$
12. $z_1 = 2i, z_2 = -5i$
13. $z_1 = 1 + i, z_2 = 4i$
14. $z_1 = 1 + i, z_2 = 3i$
15. $z_1 = -5, z_2 = 1 + i\sqrt{3}$
16. $z_1 = -3, z_2 = \sqrt{3} + i$

Use DeMoivre's Theorem to find each of the following. Write your answer in standard form.

17. $[2(\cos 10° + i \sin 10°)]^6$
18. $[4(\cos 15° + i \sin 15°)]^3$
19. $(\cos 12° + i \sin 12°)^{10}$
20. $(\cos 18° + i \sin 18°)^{10}$
21. $(3 \operatorname{cis} 60°)^4$
22. $(3 \operatorname{cis} 30°)^4$
23. $(\sqrt{2} \operatorname{cis} 45°)^{10}$
24. $(\sqrt{2} \operatorname{cis} 70°)^6$
25. $(1 + i)^4$
26. $(1 + i)^5$
27. $(-\sqrt{3} + i)^4$
28. $(\sqrt{3} + i)^4$
29. $(1 + i)^6$
30. $(-1 + i)^8$
31. $(-2 + 2i)^3$
32. $(-2 - 2i)^3$

Divide. Leave your answers in trigonometric form.

33. $\dfrac{20(\cos 75° + i \sin 75°)}{5(\cos 40° + i \sin 40°)}$
34. $\dfrac{30(\cos 80° + i \sin 80°)}{10(\cos 30° + i \sin 30°)}$

35. $\dfrac{18(\cos 51° + i \sin 51°)}{12(\cos 32° + i \sin 32°)}$
36. $\dfrac{21(\cos 63° + i \sin 63°)}{14(\cos 44° + i \sin 44°)}$

37. $\dfrac{4 \operatorname{cis} 90°}{8 \operatorname{cis} 30°}$
38. $\dfrac{6 \operatorname{cis} 120°}{8 \operatorname{cis} 90°}$

Find the quotient z_1/z_2 in standard form. Then write z_1 and z_2 in trigonometric form and find their quotient again. Finally, convert the answer that is in trigonometric form to standard form to show that the two quotients are equal.

39. $z_1 = 2 + 2i, z_2 = 1 + i$
40. $z_1 = 2 - 2i, z_2 = 1 - i$
41. $z_1 = \sqrt{3} + i, z_2 = 2i$
42. $z_1 = 1 + i\sqrt{3}, z_2 = 2i$
43. $z_1 = 4 + 4i, z_2 = 2 - 2i$
44. $z_1 = 6 + 6i, z_2 = -3 - 3i$
45. $z_1 = 8, z_2 = -4$
46. $z_1 = -6, z_2 = 3$

Convert all complex numbers to trigonometric form and then simplify each expression. Write all answers in standard form.

47. $\dfrac{(1 + i)^4(2i)^2}{-2 + 2i}$
48. $\dfrac{(\sqrt{3} + i)^4(2i)^5}{(1 + i)^{10}}$

49. $\dfrac{(1 + i\sqrt{3})^4(\sqrt{3} - i)^2}{(1 - i\sqrt{3})^3}$
50. $\dfrac{(2 + 2i)^5(-3 + 3i)^3}{(\sqrt{3} + i)^{10}}$

51. Show that $x = 2(\cos 60° + i \sin 60°)$ is a solution to the quadratic equation $x^2 - 2x + 4 = 0$ by replacing x with $2(\cos 60° + i \sin 60°)$ and simplifying.

52. Show that $x = 2(\cos 300° + i \sin 300°)$ is a solution to the quadratic equation $x^2 - 2x + 4 = 0$.

53. Show that $w = 2(\cos 15° + i \sin 15°)$ is a fourth root of $z = 8 + 8i\sqrt{3}$ by raising w to the fourth power and simplifying to get z. (The number w is a fourth root of z, $w = z^{1/4}$, if the fourth power of w is z, $w^4 = z$.)

54. Show that $x = 1/2 + (\sqrt{3}/2)i$ is a cube root of -1.

DeMoivre's Theorem can be used to find reciprocals of complex numbers. Recall from algebra that the reciprocal of x is $1/x$, which can be expressed as x^{-1}. Use this fact, along with DeMoivre's Theorem, to find the reciprocal of each number below.

55. $1 + i$ **56.** $1 - i$ **57.** $\sqrt{3} - i$ **58.** $\sqrt{3} + i$

REVIEW PROBLEMS

The problems that follow review material we covered in Sections 5.3, 5.4, 7.1, and 7.3.

If $\cos A = -1/3$ and A is between $90°$ and $180°$, find

59. $\cos 2A$ **60.** $\sin 2A$ **61.** $\sin \dfrac{A}{2}$ **62.** $\cos \dfrac{A}{2}$

63. $\csc \dfrac{A}{2}$ **64.** $\sec \dfrac{A}{2}$ **65.** $\tan 2A$ **66.** $\tan \dfrac{A}{2}$

67. Distance and Bearing A crew member on a fishing boat traveling due north off the coast of California observes that the bearing of Morro Rock from the boat is N 35° E. After sailing another 9.2 miles, the crew member looks back to find that the bearing of Morro Rock from the ship is S 27° E. At that time, how far is the boat from Morro Rock?

68. Length A tent pole is held in place by two guy wires on opposite sides of the pole. One of the guy wires makes an angle of 43.2° with the ground and is 10.1 feet long. How long is the other guy wire if it makes an angle of 34.5° with the ground?

69. True Course and Speed A plane is flying with an airspeed of 170 miles per hour with a heading of 112°. The wind currents are a constant 28 miles per hour in the direction of due north. Find the true direction and ground speed of the plane.

70. Geometry If a parallelogram has sides of 33 centimeters and 22 centimeters that meet at an angle of 111°, how long is the longer diagonal?

SECTION 8.4   ROOTS OF A COMPLEX NUMBER

What is it about mathematics that draws some people toward it? In many cases, it is the fact that we can describe the world around us with mathematics; for some people, mathematics gives a clearer picture of the world in which we live. In other cases, the attraction is within mathematics itself. That is, for some people, mathematics itself is attractive, regardless of its connection to the real world. What you will find in this section is something in this second category. It is a property that real and complex numbers contain that is, by itself, surprising and attractive for people who enjoy mathematics. It has to do with roots of real and complex numbers.

If we solve the equation $x^2 = 25$, our solutions will be square roots of 25. Likewise, if we solve $x^3 = 8$, our solutions will be cube roots of 8. Further, the solutions to $x^4 = 81$ will be fourth roots of 81.

Without showing the work involved, here are the solutions to these three equations:

Equations	Solutions
$x^2 = 25$	-5 and 5
$x^3 = 8$	2, $-1 + i\sqrt{3}$, and $-1 - i\sqrt{3}$
$x^4 = 81$	-3, 3, $-3i$, and $3i$

The number 25 has two square roots, 8 has three cube roots, and 81 has four fourth roots. As you will see as we progress through this section, the numbers we used in the equations given are unimportant; every real (or complex) number has exactly two square roots, three cube roots, four fourth roots. In fact, every real (or complex) number has exactly n distinct nth roots, a surprising and attractive fact about numbers. The key to finding these roots is trigonometric form for complex numbers.

Suppose that z and w are complex numbers such that w is the nth root of z. That is,

$$w = \sqrt[n]{z}$$

If we raise both sides of this last equation to the nth power, we have

$$w^n = z$$

Now suppose $z = r(\cos \theta + i \sin \theta)$ and $w = s(\cos \alpha + i \sin \alpha)$. Substituting these expressions into the equation $w^n = z$, we have

$$[s(\cos \alpha + i \sin \alpha)]^n = r(\cos \theta + i \sin \theta)$$

We can rewrite the left side of this last equation using DeMoivre's Theorem.

$$s^n(\cos n\alpha + i \sin n\alpha) = r(\cos \theta + i \sin \theta)$$

The only way these two expressions can be equal is if their absolute values are equal and their angles are coterminal.

Absolute Values Equal	*Angles Coterminal*	
$s^n = r$	$n\alpha = \theta + 360°k$	$k = $ an integer

Solving for s and α, we have

$$s = r^{1/n} \qquad \alpha = \frac{\theta + 360°k}{n} = \frac{\theta}{n} + \frac{360°}{n} \cdot k$$

To summarize, we find the nth roots of a complex number by first finding the real nth root of the absolute value (the modulus). The angle for the first root will be θ/n, and the angles for the remaining roots are found by adding multiples of $360°/n$. The values for k will range from 0 to $n - 1$, because when $k = n$ we would be adding

$$\frac{360°}{n} \cdot n = 360°$$

which would be coterminal with the first angle.

THEOREM (ROOTS)

The nth roots of the complex number

$$z = r(\cos \theta + i \sin \theta) = r \text{ cis } \theta$$

are given by

$$w_k = r^{1/n}\left[\cos\left(\frac{\theta}{n} + \frac{360°}{n}k\right) + i \sin\left(\frac{\theta}{n} + \frac{360°}{n}k\right)\right]$$

$$= r^{1/n} \text{ cis }\left(\frac{\theta}{n} + \frac{360°}{n}k\right)$$

where $k = 0, 1, 2, \ldots, n - 1$.

 EXAMPLE 1   Find the 4 fourth roots of $z = 16(\cos 60° + i \sin 60°)$.

SOLUTION   According to the formula given in our theorem on roots, the 4 fourth roots will be

$$w_k = 16^{1/4}\left[\cos\left(\frac{60°}{4} + \frac{360°}{4}k\right) + i \sin\left(\frac{60°}{4} + \frac{360°}{4}k\right)\right] \qquad k = 0, 1, 2, 3$$

$$= 2[\cos(15° + 90°k) + i \sin(15° + 90°k)]$$

Replacing k with 0, 1, 2, and 3, we have

$$w_0 = 2(\cos 15° + i \sin 15°) = 2 \text{ cis } 15° \qquad \text{when } k = 0$$
$$w_1 = 2(\cos 105° + i \sin 105°) = 2 \text{ cis } 105° \qquad \text{when } k = 1$$
$$w_2 = 2(\cos 195° + i \sin 195°) = 2 \text{ cis } 195° \qquad \text{when } k = 2$$
$$w_3 = 2(\cos 285° + i \sin 285°) = 2 \text{ cis } 285° \qquad \text{when } k = 3$$

It is interesting to note the graphical relationship among these four roots, as illustrated in Figure 1. Because the modulus for each root is 2, the terminal point of the vector used to represent each root lies on a circle of radius 2. The vector representing the first root makes an angle of $\theta/n = 15°$ with the positive x-axis, then each following root has a vector that is rotated an additional $360°/n = 90°$ in the counterclockwise direction from the previous root. Because the fraction $360°/n$ represents one complete revolution divided into n equal parts, the nth roots of a complex number will always be evenly distributed around a circle of radius $r^{1/n}$.

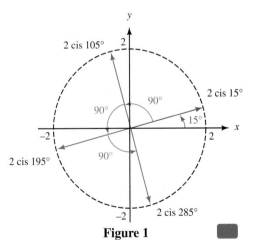

Figure 1

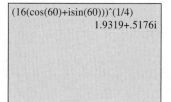

Figure 2

CALCULATOR NOTE If we use a graphing calculator to solve Example 1, we will only get one of the 4 fourth roots. Figure 2 shows the result with the values rounded to four decimal places. The complex number given by the calculator is an approximation of 2 cis 15°.

One application of finding roots is in solving certain equations, as our next examples illustrate.

EXAMPLE 2 Solve $x^3 + 1 = 0$.

SOLUTION Adding -1 to each side of the equation, we have

$$x^3 = -1$$

The solutions to this equation are the cube roots of -1. We already know that one of the cube roots of -1 is -1. There are two other complex cube roots as well. To find them, we write -1 in trigonometric form and then apply the formula from our theorem on roots. Writing -1 in trigonometric form, we have

$$-1 = 1(\cos 180° + i \sin 180°)$$

The 3 cube roots are given by

$$w_k = 1^{1/3}\left[\cos\left(\frac{180°}{3} + \frac{360°}{3}k\right) + i\sin\left(\frac{180°}{3} + \frac{360°}{3}k\right)\right]$$

$$= \cos(60° + 120°k) + i\sin(60° + 120°k)$$

where $k = 0$, 1, and 2. Replacing k with 0, 1, and 2 and then simplifying each result, we have

$$w_0 = \cos 60° + i \sin 60° = \frac{1}{2} + \frac{\sqrt{3}}{2}i \qquad \text{when } k = 0$$

$$w_1 = \cos 180° + i \sin 180° = -1 \qquad \text{when } k = 1$$

$$w_2 = \cos 300° + i \sin 300° = \frac{1}{2} - \frac{\sqrt{3}}{2}i \qquad \text{when } k = 2$$

The graphs of these three roots, which are evenly spaced around the unit circle, are shown in Figure 3.

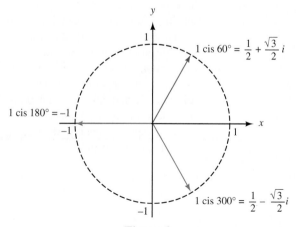

Figure 3

Note that the two complex roots are conjugates. Let's check root w_0 by cubing it.

$$w_0{}^3 = (\cos 60° + i \sin 60°)^3$$

$$= \cos 3 \cdot 60° + i \sin 3 \cdot 60°$$

$$= \cos 180° + i \sin 180°$$

$$= -1 \quad \blacksquare$$

You may have noticed that the equation in Example 2 can be solved by algebraic methods. Let's solve $x^3 + 1 = 0$ by factoring and compare our results to those in Example 2. [Recall the formula for factoring the sum of two cubes as follows: $a^3 + b^3 = (a + b)(a^2 - ab + b^2)$.]

$$x^3 + 1 = 0$$

$$(x + 1)(x^2 - x + 1) = 0 \qquad \text{Factor the left side}$$

$$x + 1 = 0 \qquad x^2 - x + 1 = 0 \qquad \text{Set factors equal to 0}$$

The first equation gives us $x = -1$ for a solution. To solve the second equation, we use the quadratic formula.

$$x = \frac{-(-1) \pm \sqrt{(-1)^2 - 4(1)(1)}}{2(1)}$$

$$= \frac{1 \pm \sqrt{-3}}{2}$$

$$= \frac{1 \pm i\sqrt{3}}{2}$$

This last expression gives us two solutions; they are

$$\frac{1}{2} + \frac{\sqrt{3}}{2}i \quad \text{and} \quad \frac{1}{2} - \frac{\sqrt{3}}{2}i$$

As you can see, the three solutions we found in Example 2 using trigonometry match the three solutions found using algebra. Notice that the algebraic method of solving the equation $x^3 + 1 = 0$ depends on our ability to factor the left side of the equation. The advantage to the trigonometric method is that it is independent of our ability to factor.

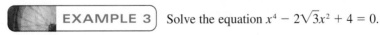 **EXAMPLE 3**   Solve the equation $x^4 - 2\sqrt{3}x^2 + 4 = 0$.

SOLUTION   The equation is quadratic in x^2. We can solve for x^2 by applying the quadratic formula.

$$x^2 = \frac{2\sqrt{3} \pm \sqrt{12 - 4(1)(4)}}{2}$$

$$= \frac{2\sqrt{3} \pm \sqrt{-4}}{2}$$

$$= \frac{2\sqrt{3} \pm 2i}{2}$$

$$= \sqrt{3} \pm i$$

The two solutions for x^2 are $\sqrt{3} + i$ and $\sqrt{3} - i$, which we write in trigonometric form as follows:

$$x^2 = \sqrt{3} + i \qquad\qquad \text{or} \qquad x^2 = \sqrt{3} - i$$
$$= 2(\cos 30° + i \sin 30°) \qquad\qquad = 2(\cos 330° + i \sin 330°)$$

Now each of these expressions has two square roots, each of which is a solution to our original equation.

When $x^2 = 2(\cos 30° + i \sin 30°)$

$$x = 2^{1/2} \left[\cos \left(\frac{30°}{2} + \frac{360°}{2}k \right) + i \sin \left(\frac{30°}{2} + \frac{360°}{2}k \right) \right] \quad \text{for } k = 0 \text{ and } 1$$

When $k = 0$, we have $x = \sqrt{2} \text{ cis } 15°$

When $k = 1$, we have $x = \sqrt{2} \text{ cis } 195°$

When $x^2 = 2(\cos 330° + i \sin 330°)$

$$x = 2^{1/2} \left[\cos \left(\frac{330°}{2} + \frac{360°}{2}k \right) + i \sin \left(\frac{330°}{2} + \frac{360°}{2}k \right) \right] \text{for } k = 0 \text{ and } 1$$

When $k = 0$, we have $x = \sqrt{2} \text{ cis } 165°$

When $k = 1$, we have $x = \sqrt{2} \text{ cis } 345°$

Using a calculator and rounding to the nearest hundredth, we can write decimal approximations to each of these four solutions.

SOLUTIONS		
Trigonometric form		**Decimal approximation**
$\sqrt{2} \text{ cis } 15°$	=	$1.37 + 0.37i$
$\sqrt{2} \text{ cis } 165°$	=	$-1.37 + 0.37i$
$\sqrt{2} \text{ cis } 195°$	=	$-1.37 - 0.37i$
$\sqrt{2} \text{ cis } 345°$	=	$1.37 - 0.37i$

GETTING READY FOR CLASS

After reading through the preceding section, respond in your own words and in complete sentences.

a. How many nth roots does a complex number contain?

b. How many degrees are there between the arguments of the four fourth roots of a number?

c. How many complex solutions are there to the equation $x^3 + 1 = 0$?

d. What is the first step in finding the three cube roots of -1?

PROBLEM SET 8.4

Find two square roots for each of the following complex numbers. Leave your answers in trigonometric form. In each case, graph the two roots.

1. $4(\cos 30° + i \sin 30°)$ **2.** $16(\cos 30° + i \sin 30°)$

3. $25(\cos 210° + i \sin 210°)$ **4.** $9(\cos 310° + i \sin 310°)$

5. $49 \text{ cis } 180°$ **6.** $81 \text{ cis } 150°$

Find two square roots for each of the following complex numbers. Write your answers in standard form.

7. $2 + 2i\sqrt{3}$ **8.** $-2 + 2i\sqrt{3}$ **9.** $4i$ **10.** $-4i$

11. -25 **12.** 25 **13.** $1 + i\sqrt{3}$ **14.** $1 - i\sqrt{3}$

Find three cube roots for each of the following complex numbers. Leave your answers in trigonometric form.

15. $8(\cos 210° + i \sin 210°)$ **16.** $27(\cos 303° + i \sin 303°)$

17. $4\sqrt{3} + 4i$ **18.** $-4\sqrt{3} + 4i$

19. -27 **20.** 8

21. $64i$ **22.** $-64i$

Solve each equation.

23. $x^3 - 27 = 0$ **24.** $x^3 + 8 = 0$ **25.** $x^4 - 16 = 0$ **26.** $x^4 + 81 = 0$

27. Find 4 fourth roots of $z = 16(\cos 120° + i \sin 120°)$. Write each root in standard form.

28. Find 4 fourth roots of $z = \cos 240° + i \sin 240°$. Leave your answers in trigonometric form.

29. Find 5 fifth roots of $z = 10^5 \text{ cis } 15°$. Write each root in trigonometric form and then give a decimal approximation, accurate to the nearest hundredth, for each one.

30. Find 5 fifth roots of $z = 10^{10} \text{ cis } 75°$. Write each root in trigonometric form and then give a decimal approximation, accurate to the nearest hundredth, for each one.

31. Find 6 sixth roots of $z = -1$. Leave your answers in trigonometric form. Graph all six roots on the same coordinate system.

32. Find 6 sixth roots of $z = 1$. Leave your answers in trigonometric form. Graph all six roots on the same coordinate system.

Solve each of the following equations. Leave your solutions in trigonometric form.

33. $x^4 - 2x^2 + 4 = 0$ **34.** $x^4 + 2x^2 + 4 = 0$

35. $x^4 + 2x^2 + 2 = 0$ **36.** $x^4 - 2x^2 + 2 = 0$

REVIEW PROBLEMS

The problems that follow review material we covered in Sections 4.2, 4.3, and 7.4.

Graph each equation on the given interval.

37. $y = -2 \sin (-3x), 0 \le x \le 2\pi$ **38.** $y = -2 \cos (-3x), 0 \le x \le 2\pi$

Graph one complete cycle of each of the following:

39. $y = -\cos \left(2x + \dfrac{\pi}{2} \right)$ **40.** $y = -\cos \left(2x - \dfrac{\pi}{2} \right)$

41. $y = 3 \sin \left(\dfrac{\pi}{3} x - \dfrac{\pi}{3} \right)$ **42.** $y = 3 \cos \left(\dfrac{\pi}{3} x - \dfrac{\pi}{3} \right)$

Find the area of triangle ABC if

43. $A = 56.2°$, $b = 2.65$ cm, and $c = 3.84$ cm
44. $B = 21.8°$, $a = 44.4$ cm, and $c = 22.2$ cm
45. $a = 2.3$ ft, $b = 3.4$ ft, and $c = 4.5$ ft
46. $a = 5.4$ ft, $b = 4.3$ ft, and $c = 3.2$ ft

EXTENDING THE CONCEPTS

47. Recall from the introduction to Section 8.2 that Jerome Cardan's solutions to the equation $x^3 = 15x + 4$ were given by

$$x = \sqrt[3]{2 + 11i} + \sqrt[3]{2 - 11i}$$

Let's assume that the cube roots shown above are complex conjugates. If they are, then we can simplify our work by noticing that

$$x = \sqrt[3]{2 + 11i} + \sqrt[3]{2 - 11i} = a + bi + a - bi = 2a$$

which means that we simply double the real part of each cube root of $2 + 11i$ to find the solutions to $x^3 = 15x + 4$. Now, to end our work with Cardan, find the 3 cube roots of $2 + 11i$. Then, noting the discussion above, use the 3 cube roots to solve the equation $x^3 = 15x + 4$. Write your answers accurate to the nearest thousandth.

Cardan

SECTION 8.5 | POLAR COORDINATES

Up to this point in our study of trigonometry, whenever we have given the position of a point in the plane, we have used rectangular coordinates. That is, we give the position of points in the plane relative to a set of perpendicular axes. In a way, the rectangular coordinate system is a map that tells us how to get anywhere in the plane by traveling so far horizontally and so far vertically relative to the origin of our coordinate system. For example, to reach the point whose address is (2, 1), we start at the origin and travel 2 units right and then 1 unit up. In this section, we will see that there are other ways to get to the point (2, 1). In particular, we can travel $\sqrt{5}$ units on the terminal side of an angle in standard position. This type of map is the basis of what we call *polar coordinates:* The address of each point in the plane is given by an ordered pair (r, θ), where r is a directed distance on the terminal side of standard position angle θ.

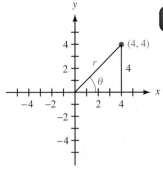

Figure 1

EXAMPLE 1 A point lies at (4, 4) on a rectangular coordinate system. Give its address in polar coordinates (r, θ).

SOLUTION We want to reach the same point by traveling r units on the terminal side of a standard position angle θ. Figure 1 shows the point (4, 4), along with the distance r and angle θ.

The triangle formed is a 45°–45°–90° right triangle. Therefore, r is $4\sqrt{2}$, and θ is 45°. In rectangular coordinates, the address of our point is (4, 4). In polar coordinates, the address is $(4\sqrt{2}, 45°)$.

Now let's formalize our ideas about polar coordinates.

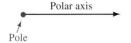

Figure 2

The foundation of the polar coordinate system is a ray called the *polar axis,* whose initial point is called the *pole* (Figure 2). In the polar coordinate system, points are named by ordered pairs (r, θ) in which r is the directed distance from the pole on the terminal side of an angle θ, the initial side of which is the polar axis. If the terminal side of θ has been rotated in a counterclockwise direction from the polar axis, then θ is a positive angle. Otherwise, θ is a negative angle. For example, the point $(5, 30°)$ is 5 units from the pole along a ray that has been rotated $30°$ from the polar axis, as shown in Figure 3.

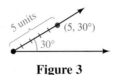

Figure 3

To simplify matters, in this book we place the pole at the origin of our rectangular coordinate system and take the polar axis to be the positive x-axis.

The circular grids used in Example 2 are helpful when graphing points given in polar coordinates. The lines on the grid are multiples of $15°$, and the circles have radii of 1, 2, 3, 4, 5, and 6.

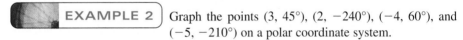

EXAMPLE 2   Graph the points $(3, 45°)$, $(2, -240°)$, $(-4, 60°)$, and $(-5, -210°)$ on a polar coordinate system.

SOLUTION   To graph $(3, 45°)$, we locate the point that is 3 units from the origin along the terminal side of $45°$, as shown in Figure 4. The point $(2, -240°)$ is 2 units out on the terminal side of $-240°$, as Figure 5 indicates.

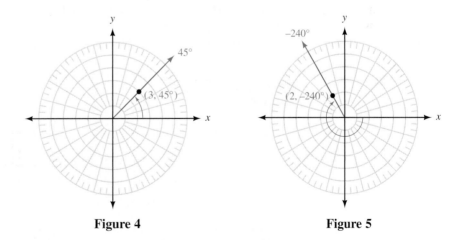

Figure 4   **Figure 5**

As you can see from Figures 4 and 5, if r is positive, we locate the point (r, θ) along the terminal side of θ. The next two points we will graph have negative values of r.

To graph a point (r, θ) in which r is negative, we look for the point that is r units in the *opposite* direction indicated by the terminal side of θ. That is, we extend a ray from the pole that is directly opposite the terminal side of θ and locate the point r units along this ray.

To graph $(-4, 60°)$, we first extend the terminal side of θ through the origin to create a ray in the opposite direction. Then we locate the point that is 4 units from the origin along this ray, as shown in Figure 6. The terminal side of θ has been drawn in blue, and the ray pointing in the opposite direction is shown in red.

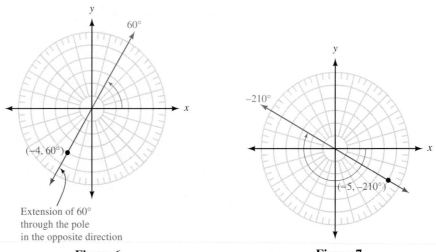

Figure 6   **Figure 7**

To graph $(-5, -210°)$, we look for the point that is 5 units from the origin along the ray in the opposite direction of $-210°$ (Figure 7).

In rectangular coordinates, each point in the plane is named by a unique ordered pair (x, y). That is, no point can be named by two different ordered pairs. The same is not true of points named by polar coordinates. As Example 3 illustrates, the polar coordinates of a point are not unique.

EXAMPLE 3 Give three other ordered pairs that name the same point as $(3, 60°)$.

SOLUTION As Figure 8 illustrates, the points $(-3, 240°)$, $(-3, -120°)$, and $(3, -300°)$ all name the point $(3, 60°)$. There are actually an infinite number of ordered pairs that name the point $(3, 60°)$. Any angle that is coterminal with $60°$ will have its terminal side pass through $(3, 60°)$. Therefore, all points of the form

$$(3, 60° + 360°k) \qquad k = \text{an integer}$$

will name the point $(3, 60°)$. Also, any angle that is coterminal with $240°$ can be used with a value of $r = -3$, so all points of the form

$$(-3, 240° + 360°k) \qquad k = \text{an integer}$$

will name the point $(3, 60°)$.

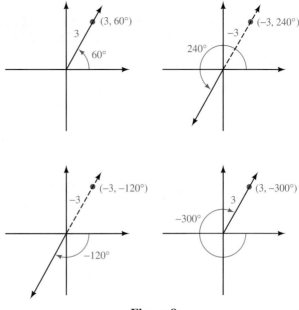

Figure 8

NOTE In Example 3, compare the ordered pairs (3, 60°) and (3, −300°). Since 60° − 360° = −300°, we can see that if 360° is subtracted from θ, r will be of the same sign. The same will be true if 360° is added to θ. However, if 180° is added or subtracted, r will be of the opposite sign. This can be seen with the ordered pairs (−3, 240°) and (−3, −120°), since 60° + 180° = 240° and 60° − 180° = −120°.

POLAR COORDINATES AND RECTANGULAR COORDINATES

To derive the relationship between polar coordinates and rectangular coordinates, we consider a point P with rectangular coordinates (x, y) and polar coordinates (r, θ).

To convert back and forth between polar and rectangular coordinates, we simply use the relationships that exist among x, y, r, and θ in Figure 9.

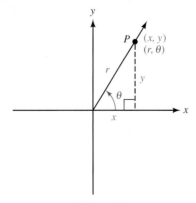

Figure 9

TO CONVERT RECTANGULAR COORDINATES TO POLAR COORDINATES

Let

$$r = \pm\sqrt{x^2 + y^2} \qquad \text{and} \qquad \tan\theta = \frac{y}{x}$$

where the sign of r and the choice of θ place the point (r, θ) in the same quadrant as (x, y).

TO CONVERT POLAR COORDINATES TO RECTANGULAR COORDINATES

Let

$$x = r\cos\theta \qquad \text{and} \qquad y = r\sin\theta$$

The process of converting to rectangular coordinates is simply a matter of substituting r and θ into the equations given above. To convert to polar coordinates, we have to choose θ and the sign of r so the point (r, θ) is in the same quadrant as the point (x, y).

 EXAMPLE 4   Convert to rectangular coordinates.

a. $(4, 30°)$ **b.** $(-\sqrt{2}, 135°)$ **c.** $(3, 270°)$

SOLUTION   To convert from polar coordinates to rectangular coordinates, we substitute the given values of r and θ into the equations

$$x = r\cos\theta \qquad \text{and} \qquad y = r\sin\theta$$

Here are the conversions for each point along with the graphs in both rectangular and polar coordinates.

a. $x = 4\cos 30° = 4\left(\dfrac{\sqrt{3}}{2}\right) = 2\sqrt{3}$

$y = 4\sin 30° = 4\left(\dfrac{1}{2}\right) = 2$

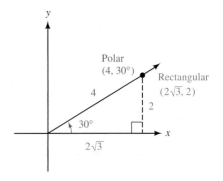

Figure 10

The point $(2\sqrt{3}, 2)$ in rectangular coordinates is equivalent to $(4, 30°)$ in polar coordinates. Figure 10 illustrates.

b. $x = -\sqrt{2} \cos 135° = -\sqrt{2}\left(-\dfrac{1}{\sqrt{2}}\right) = 1$

$y = -\sqrt{2} \sin 135° = -\sqrt{2}\left(\dfrac{1}{\sqrt{2}}\right) = -1$

The point $(1, -1)$ in rectangular coordinates is equivalent to $(-\sqrt{2}, 135°)$ in polar coordinates. Figure 11 illustrates.

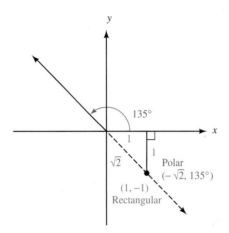

Figure 11

c. $x = 3 \cos 270° = 3(0) = 0$

$y = 3 \sin 270° = 3(-1) = -3$

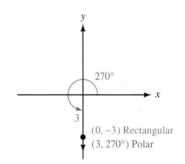

Figure 12

The point $(0, -3)$ in rectangular coordinates is equivalent to $(3, 270°)$ in polar coordinates. Figure 12 illustrates. ▨

EXAMPLE 5 Convert to polar coordinates.

a. $(3, 3)$ **b.** $(-2, 0)$ **c.** $(-1, \sqrt{3})$

SOLUTION
a. Since x is 3 and y is 3, we have

$$r = \pm\sqrt{9 + 9} = \pm 3\sqrt{2} \qquad \text{and} \qquad \tan \theta = \dfrac{3}{3} = 1$$

Since $(3, 3)$ is in quadrant I, we can choose $r = 3\sqrt{2}$ and $\theta = 45°$ (Figure 13). Remember, there are an infinite number of ordered pairs in polar coordinates that name the point $(3, 3)$. The point $(3\sqrt{2}, 45°)$ is just one of them. Generally, we choose r and θ so that r is positive and θ is between $0°$ and $360°$.

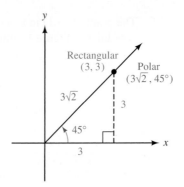

Figure 13

b. We have $x = -2$ and $y = 0$, so

$$r = \pm\sqrt{4 + 0} = \pm 2 \qquad \text{and} \qquad \tan \theta = \frac{0}{-2} = 0$$

Since $(-2, 0)$ is on the negative x-axis, we can choose $r = 2$ and $\theta = 180°$ to get the point $(2, 180°)$. Figure 14 illustrates.

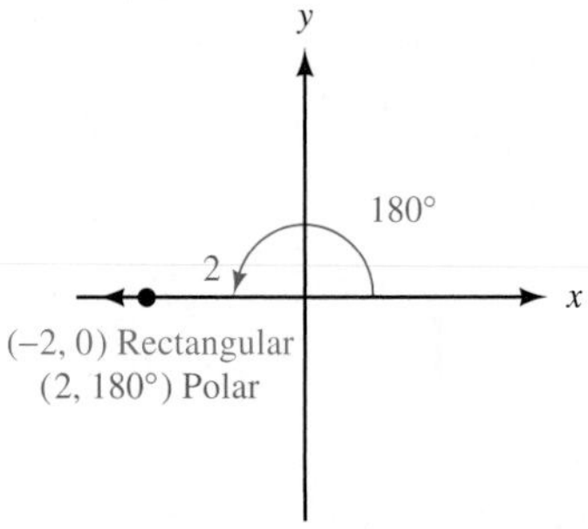

c. Since $x = -1$ and $y = \sqrt{3}$, we have

$$r = \pm\sqrt{1 + 3} = \pm 2 \qquad \text{and} \qquad \tan \theta = \frac{\sqrt{3}}{-1}$$

Since $(-1, \sqrt{3})$ is in quadrant II, we can let $r = 2$ and $\theta = 120°$. In polar coordinates, the point is $(2, 120°)$. Figure 15 illustrates.

Figure 14

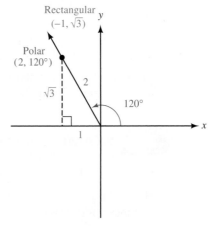

Figure 15

CALCULATOR NOTE If we use the $\boxed{\tan^{-1}}$ key on a calculator to find the angle θ in Example 5c, we will get

$$\theta = \tan^{-1}(-\sqrt{3}) = -60^\circ$$

This would place the terminal side of θ in quadrant IV. To locate our point in quadrant II, we could use a negative value of r. The polar coordinates would then be $(-2, -60^\circ)$.

USING TECHNOLOGY

CONVERTING BETWEEN POLAR AND RECTANGULAR COORDINATES

In Section 8.2 we mentioned that most graphing calculators are able to convert a complex number to trigonometric form. The same is true for converting ordered pairs from rectangular coordinates to polar coordinates. In fact, sometimes the very same commands are used.

In Figure 16 we used a TI-83 to solve Example 4a. The value 3.4641 is an approximation of $2\sqrt{3}$. Example 5a is shown in Figure 17, where 4.2426 is an approximation of $3\sqrt{2}$. For a TI-83, the conversion commands are located in the ANGLE menu.

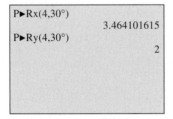

```
P►Rx(4,30°)
                3.464101615
P►Ry(4,30°)
                          2
```

Figure 16

```
R►Pr(3,3)
                4.242640687
R►Pθ(3,3)
                         45
```

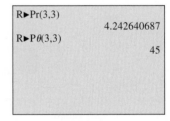

Figure 17

These same examples are solved in Figure 18 with a TI-86. The commands are found in the CPLX menu. Check the manual for your model to see if your calculator is able to convert rectangular coordinates to polar coordinates.

```
(4∠30°)►Rec
          (3.4641016154,2)
(3,3)►Pol
      (4.24264068712∠45)
```

Figure 18

EQUATIONS IN POLAR COORDINATES

Equations in polar coordinates have variables r and θ instead of x and y. The conversions we used to change ordered pairs from polar coordinates to rectangular coordinates and from rectangular coordinates to polar coordinates are the same ones we use to convert back and forth between equations given in polar coordinates and those in rectangular coordinates.

 EXAMPLE 6 Change $r^2 = 9 \sin 2\theta$ to rectangular coordinates.

SOLUTION Before we substitute to clear the equation of r and θ, we must use a double-angle identity to write $\sin 2\theta$ in terms of $\sin \theta$ and $\cos \theta$.

$$r^2 = 9 \sin 2\theta$$

$$r^2 = 9 \cdot 2 \sin \theta \cos \theta \qquad \text{Double-angle identity}$$

$$r^2 = 18 \cdot \frac{y}{r} \cdot \frac{x}{r} \qquad \text{Substitute } y/r \text{ for } \sin \theta \text{ and } x/r \text{ for } \cos \theta$$

$$r^2 = \frac{18xy}{r^2} \qquad \text{Multiply}$$

$$r^4 = 18xy \qquad \text{Multiply both sides by } r^2$$

$$(x^2 + y^2)^2 = 18xy \qquad \text{Substitute } x^2 + y^2 \text{ for } r^2 \quad \blacksquare$$

 EXAMPLE 7 Change $x + y = 4$ to polar coordinates.

SOLUTION Since $x = r \cos \theta$ and $y = r \sin \theta$, we have

$$r \cos \theta + r \sin \theta = 4$$

$$r(\cos \theta + \sin \theta) = 4 \qquad \text{Factor out } r$$

$$r = \frac{4}{\cos \theta + \sin \theta} \qquad \text{Divide both sides by } \cos \theta + \sin \theta$$

The last equation gives us r in terms of θ.

 GETTING READY FOR CLASS

After reading through the preceding section, respond in your own words and in complete sentences.

a. How do you plot the point whose polar coordinates are $(3, 45°)$?

b. Why do points in the coordinate plane have more than one representation in polar coordinates?

c. If you convert $(4, 30°)$ to rectangular coordinates, how do you find the x-coordinate?

d. What are the rectangular coordinates of the point $(3, 270°)$?

PROBLEM SET 8.5

Graph each ordered pair on a polar coordinate system.

1. $(2, 45°)$ **2.** $(3, 60°)$ **3.** $(3, 150°)$ **4.** $(4, 135°)$

5. $(1, -225°)$ **6.** $(2, -240°)$ **7.** $(-3, 45°)$ **8.** $(-4, 60°)$

9. $(-4, -210°)$ **10.** $(-5, -225°)$ **11.** $(-2, 0°)$ **12.** $(-2, 270°)$

For each ordered pair, give three other ordered pairs with θ between $-360°$ and $360°$ that name the same point.

13. $(2, 60°)$ **14.** $(1, 30°)$ **15.** $(5, 135°)$ **16.** $(3, 120°)$

17. $(-3, 30°)$ **18.** $(-2, 45°)$

Convert to rectangular coordinates. Use exact values.

19. $(2, 60°)$ **20.** $(-2, 60°)$ **21.** $(3, 270°)$ **22.** $(1, 180°)$

23. $(\sqrt{2}, -135°)$ **24.** $(\sqrt{2}, -225°)$ **25.** $(-4\sqrt{3}, 30°)$ **26.** $(4\sqrt{3}, -30°)$

Use your graphing calculator to convert to rectangular coordinates. Round all values to four significant digits.

27. $(2, 19°)$ **28.** $(3, 124°)$ **29.** $(-3, 293°)$ **30.** $(-4, 261°)$

Convert to polar coordinates with $r \geq 0$ and $0° \leq \theta < 360°$.

31. $(-3, 3)$ **32.** $(-3, -3)$ **33.** $(-2\sqrt{3}, 2)$ **34.** $(2, -2\sqrt{3})$

35. $(2, 0)$ **36.** $(-2, 0)$ **37.** $(-\sqrt{3}, -1)$ **38.** $(-1, -\sqrt{3})$

Convert to polar coordinates. Use a calculator to find θ to the nearest tenth of a degree. Keep r positive and θ between $0°$ and $360°$.

39. $(3, 4)$ **40.** $(4, 3)$ **41.** $(-1, 2)$ **42.** $(1, -2)$

43. $(-2, -3)$ **44.** $(-3, -2)$

Use your graphing calculator to convert to polar coordinates. Round all values to four significant digits.

45. $(5, 8)$ **46.** $(-2, 9)$ **47.** $(-1, -6)$ **48.** $(7, -3)$

Write each equation with rectangular coordinates.

49. $r^2 = 9$ **50.** $r^2 = 4$

51. $r = 6 \sin \theta$ **52.** $r = 6 \cos \theta$

53. $r^2 = 4 \sin 2\theta$ **54.** $r^2 = 4 \cos 2\theta$

55. $r(\cos \theta + \sin \theta) = 3$ **56.** $r(\cos \theta - \sin \theta) = 2$

Write each equation in polar coordinates.

57. $x - y = 5$ **58.** $x + y = 5$

59. $x^2 + y^2 = 4$ **60.** $x^2 + y^2 = 9$

61. $x^2 + y^2 = 6x$ **62.** $x^2 + y^2 = 4x$

63. $y = x$ **64.** $y = -x$

REVIEW PROBLEMS

The problems that follow review material we covered in Sections 4.1 and 4.2. Reviewing these problems will help you with the next section.

Graph one complete cycle of each equation.

65. $y = 6 \sin x$ **66.** $y = 6 \cos x$

67. $y = 4 \sin 2x$ **68.** $y = 2 \sin 4x$

69. $y = 4 + 2 \sin x$ **70.** $y = 4 + 2 \cos x$

SECTION 8.6 | EQUATIONS IN POLAR COORDINATES AND THEIR GRAPHS

Archimedes

Over 2,000 years ago, Archimedes described a curve as starting at a point and moving out along a half-line at a constant rate. The endpoint of the half-line was anchored at the initial point and was rotating about it at a constant rate also. The curve, called the spiral of Archimedes, is shown in Figure 1, with a rectangular coordinate system superimposed on it so that the origin of the coordinate system coincides with the initial point on the curve.

In rectangular coordinates, the equation that describes this curve is

$$\sqrt{x^2 + y^2} = 3 \tan^{-1} \frac{y}{x}$$

If we were to superimpose a polar coordinate system on the curve in Figure 1, instead of the rectangular coordinate system, then the equation in polar coordinates that describes the curve would be simply

$$r = 3\theta$$

As you can see, the equations for some curves are best given in terms of polar coordinates.

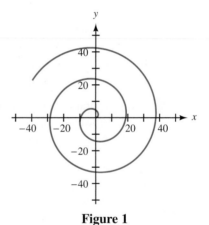

Figure 1

In this section, we will consider the graphs of polar equations. The solutions to these equations are ordered pairs (r, θ), where r and θ are the polar coordinates we defined in Section 8.5.

 **EXAMPLE 1** Sketch the graph of $r = 6 \sin \theta$.

SOLUTION We can find ordered pairs (r, θ) that satisfy the equation by making a table. The table is a little different from the ones we made for rectangular coordinates. With polar coordinates, we substitute convenient values for θ and then use the equation to find corresponding values of r. Let's use multiples of 30° and 45° for θ.

TABLE 1

θ	$r = 6 \sin \theta$	r	(r, θ)
0°	$r = 6 \sin 0° = 0$	0	$(0, 0°)$
30°	$r = 6 \sin 30° = 3$	3	$(3, 30°)$
45°	$r = 6 \sin 45° = 4.2$	4.2	$(4.2, 45°)$
60°	$r = 6 \sin 60° = 5.2$	5.2	$(5.2, 60°)$
90°	$r = 6 \sin 90° = 6$	6	$(6, 90°)$
120°	$r = 6 \sin 120° = 5.2$	5.2	$(5.2, 120°)$
135°	$r = 6 \sin 135° = 4.2$	4.2	$(4.2, 135°)$
150°	$r = 6 \sin 150° = 3$	3	$(3, 150°)$
180°	$r = 6 \sin 180° = 0$	0	$(0, 180°)$
210°	$r = 6 \sin 210° = -3$	-3	$(-3, 210°)$
225°	$r = 6 \sin 225° = -4.2$	-4.2	$(-4.2, 225°)$
240°	$r = 6 \sin 240° = -5.2$	-5.2	$(-5.2, 240°)$
270°	$r = 6 \sin 270° = -6$	-6	$(-6, 270°)$
300°	$r = 6 \sin 300° = -5.2$	-5.2	$(-5.2, 300°)$
315°	$r = 6 \sin 315° = -4.2$	-4.2	$(-4.2, 315°)$
330°	$r = 6 \sin 330° = -3$	-3	$(-3, 330°)$
360°	$r = 6 \sin 360° = 0$	0	$(0, 360°)$

NOTE   If we were to continue past 360° with values of θ, we would simply start to repeat the values of r we have already obtained, since $\sin \theta$ is periodic with period 360°.

Plotting each point on a polar coordinate system and then drawing a smooth curve through them, we have the graph in Figure 2.

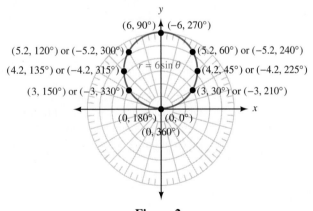

Figure 2

USING TECHNOLOGY

CREATING TABLES FOR POLAR EQUATIONS

A graphing calculator may be used to create a table of values for a polar equation. For example, to make a table for the equation in Example 1, first set the calculator to degree mode and polar mode. Define the function $r1 = 6 \sin \theta$. Set up your table to automatically generate values starting at $\theta = 0$ and using an increment of 15° (Figure 3). On some calculators, this is done by setting the independent variable to Auto.

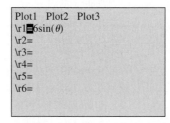

```
Plot1  Plot2  Plot3
\r1■6sin(θ)
\r2=
\r3=
\r4=
\r5=
\r6=
```

```
TABLE SETUP
 TblStart=0
 ΔTbl=15
Indpnt: Auto Ask
Depend: Auto Ask
```

Figure 3

Now display the table and you should see values similar to those shown in Figure 4. Scroll down the table to see additional pairs of values (Figure 5).

θ	r1	
0	0	
15	1.5529	
30	3	
45	4.2426	
60	5.1962	
75	5.7956	
90	6	
θ=0		

θ	r1	
105	5.7956	
120	5.1962	
135	4.2426	
150	3	
165	1.5529	
180	0	
195	—1.553	
θ=195		

Figure 4   **Figure 5**

Could we have found the graph of $r = 6 \sin \theta$ in Example 1 without making a table? The answer is yes, there are a couple of other ways to do so. One way is to convert to rectangular coordinates and see if we recognize the graph from the rectangular equation. We begin by replacing $\sin \theta$ with y/r.

$$r = 6\frac{y}{r} \qquad \sin \theta = \frac{y}{r}$$

$$r^2 = 6y \qquad \text{Multiply both sides by } r$$

$$x^2 + y^2 = 6y \qquad r^2 = x^2 + y^2$$

The equation is now written in terms of rectangular coordinates. If we add $-6y$ to both sides and then complete the square on y, we will obtain the rectangular equation of a circle with center at $(0, 3)$ and a radius of 3.

$$x^2 + y^2 - 6y = 0 \qquad \text{Add } -6y \text{ to both sides}$$

$$x^2 + y^2 - 6y + 9 = 9 \qquad \text{Complete the square on } y \text{ by adding 9} \\ \text{to both sides}$$

$$x^2 + (y - 3)^2 = 3^2 \qquad \text{Standard form for the equation of a circle}$$

This method of graphing, by changing to rectangular coordinates, works well only in some cases. Many of the equations we will encounter in polar coordinates do not have graphs that are recognizable in rectangular form.

In Example 2, we will look at another method of graphing polar equations that does not depend on the use of a table.

EXAMPLE 2   Sketch the graph of $r = 4 \sin 2\theta$.

SOLUTION   One way to visualize the relationship between r and θ as given by the equation $r = 4 \sin 2\theta$ is to sketch the graph of $y = 4 \sin 2x$ on a rectangular coordinate system. (Since we have been using degree measure for our angles in polar coordinates, we will label the x-axis for the graph of $y = 4 \sin 2x$ in degrees rather than radians as we usually do.) The graph of $y = 4 \sin 2x$ will have an amplitude of 4 and a period of $360°/2 = 180°$. Figure 6 shows the graph of $y = 4 \sin 2x$ between $0°$ and $360°$.

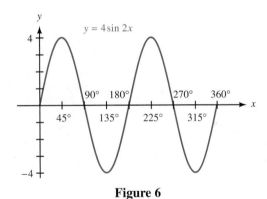

Figure 6

As you can see in Figure 6, as x goes from $0°$ to $45°$, y goes from 0 to 4. This means that, for the equation $r = 4 \sin 2\theta$, as θ goes from $0°$ to $45°$, r will go from 0 up to 4. A diagram of this is shown in Figure 7.

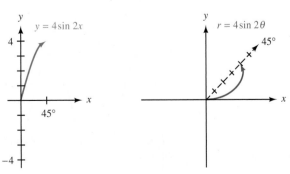

Figure 7

As x continues from $45°$ to $90°$, y decreases from 4 down to 0. Likewise, as θ rotates through $45°$ to $90°$, r will decrease from 4 down to 0. A diagram of this is shown in Figure 8. The numbers 1 and 2 in Figure 8 indicate the order in which those sections of the graph are drawn.

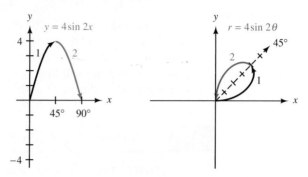

Figure 8

If we continue to reason in this manner, we will obtain a good sketch of the graph of $r = 4 \sin 2\theta$ by watching how y is affected by changes in x on the graph of $y = 4 \sin 2x$. Table 2 summarizes this information, and Figure 9 contains the graphs of both $y = 4 \sin 2x$ and $r = 4 \sin 2\theta$.

TABLE 2		
Reference number on graphs	**Variations in x (or θ)**	**Corresponding variations in y (or r)**
1	$0°$ to $45°$	0 to 4
2	$45°$ to $90°$	4 to 0
3	$90°$ to $135°$	0 to -4
4	$135°$ to $180°$	-4 to 0
5	$180°$ to $225°$	0 to 4
6	$225°$ to $270°$	4 to 0
7	$270°$ to $315°$	0 to -4
8	$315°$ to $360°$	-4 to 0

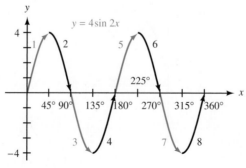

Figure 9

USING TECHNOLOGY

POLAR GRAPHS

If your graphing calculator can be set to polar mode, then it should be capable of producing polar graphs. To graph the equation $r = 4 \sin 2\theta$ from Example 2, set the calculator to degree mode and polar mode, and then define the function $r_1 = 4 \sin (2\theta)$. As shown in Example 2, we must allow θ to vary from $0°$ to $360°$ in order to obtain the complete graph. Set the window variables so that

$$0 \leq \theta \leq 360, \text{ step} = 7.5; \; -4.5 \leq x \leq 4.5; \; -4.5 \leq y \leq 4.5$$

Your graph should look similar to the one shown in Figure 10. For a more "true" representation of the graph, use your zoom-square command to remove the distortion caused by the rectangular screen (Figure 11).

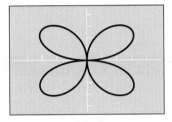

Figure 10

Figure 11

To find points on the graph, you can use the $\boxed{\text{TRACE}}$ key. Depending on your model of calculator, you may be able to obtain points in rectangular coordinates, polar coordinates, or both. (On the TI-83 and TI-86, select PolarGC in the FORMAT menu to see polar coordinates.) An added benefit of tracing the graph is that it reinforces the formation of the graph using the method explained in Example 2.

You may also have a command that allows you to find a point (x, y) on the graph directly for a given value of θ. Figure 12 shows how this command was used to find the point corresponding to $\theta = 112°$ in rectangular coordinates and then with polar coordinates.

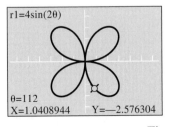

Figure 12

EXAMPLE 3 Sketch the graph of $r = 4 + 2 \sin \theta$.

SOLUTION The graph of $r = 4 + 2 \sin \theta$ (Figure 14) is obtained by first graphing $y = 4 + 2 \sin x$ (Figure 13) and then noticing the relationship between variations in x and the corresponding variations in y. These variations are equivalent to those that exist between θ and r.

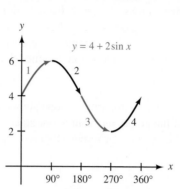

Figure 13

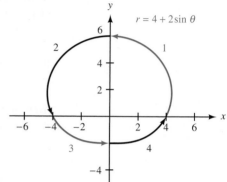

Figure 14

TABLE 3

Reference number on graphs	Variations in x (or θ)	Corresponding variations in y (or r)
1	0° to 90°	4 to 6
2	90° to 180°	6 to 4
3	180° to 270°	4 to 2
4	270° to 360°	2 to 4

Although the method of graphing presented in Examples 2 and 3 is sometimes difficult to comprehend at first, with a little practice it becomes much easier. In any case, the usual alternative is to make a table and plot points until the shape of the curve can be recognized. Probably the best way to graph these equations is to use a combination of both methods.

Here are some other common graphs in polar coordinates along with the equations that produce them (Figures 15–21). When you start graphing some of the equations in Problem Set 8.6, you may want to keep these graphs handy for reference. It is sometimes easier to get started when you can anticipate the general shape of the curve.

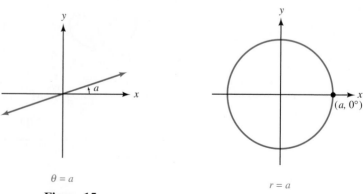

$\theta = a$

Figure 15

$r = a$

Figure 16

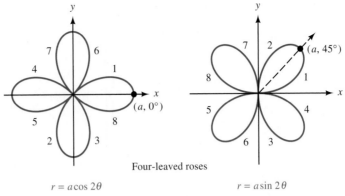

Four-leaved roses

$r = a\cos 2\theta$ $r = a\sin 2\theta$

Figure 17

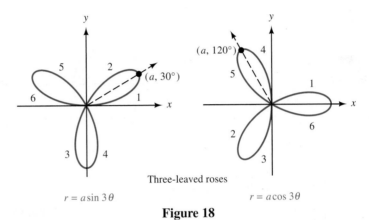

Three-leaved roses

$r = a\sin 3\theta$ $r = a\cos 3\theta$

Figure 18

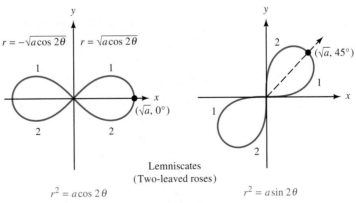

Lemniscates
(Two-leaved roses)

$r^2 = a\cos 2\theta$ $r^2 = a\sin 2\theta$

Figure 19

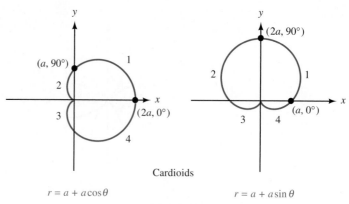

Cardioids

$$r = a + a\cos\theta \qquad\qquad r = a + a\sin\theta$$

Figure 20

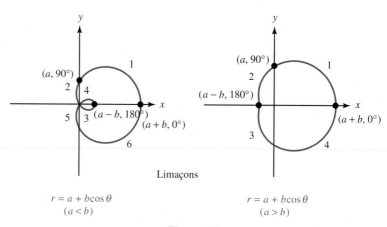

Limaçons

$$r = a + b\cos\theta \qquad\qquad r = a + b\cos\theta$$
$$(a < b) \qquad\qquad\qquad (a > b)$$

Figure 21

GETTING READY FOR CLASS

After reading through the preceding section, respond in your own words and in complete sentences.

a. What is one way to graph an equation written in polar coordinates?

b. What type of geometric object is the graph of $r = 6\sin\theta$?

c. What is the largest value of r that can be obtained from the equation $r = 6\sin\theta$?

d. What type of curve is the graph of $r = 4 + 2\sin\theta$?

PROBLEM SET 8.6

Sketch the graph of each equation by making a table using values of θ that are multiples of 45°.

1. $r = 6 \cos \theta$

2. $r = 4 \sin \theta$

3. $r = \sin 3\theta$

4. $r = \cos 3\theta$

 Use your graphing calculator in polar mode to generate a table for each equation using values of θ that are multiples of 15°. Sketch the graph of the equation using the values from your table.

5. $r = 2 \cos 2\theta$

6. $r = 2 \sin 2\theta$

7. $r = 3 + 3 \sin \theta$

8. $r = 3 + 3 \cos \theta$

Graph each equation.

9. $r = 3$

10. $r = 2$

11. $\theta = 45°$

12. $\theta = 135°$

13. $r = 3 \sin \theta$

14. $r = 3 \cos \theta$

15. $r = 4 + 2 \sin \theta$

16. $r = 4 + 2 \cos \theta$

17. $r = 2 + 4 \cos \theta$

18. $r = 2 + 4 \sin \theta$

19. $r = 2 + 2 \sin \theta$

20. $r = 2 + 2 \cos \theta$

21. $r^2 = 9 \sin 2\theta$

22. $r^2 = 4 \cos 2\theta$

23. $r = 2 \sin 2\theta$

24. $r = 2 \cos 2\theta$

25. $r = 4 \cos 3\theta$

26. $r = 4 \sin 3\theta$

Graph each equation using your graphing calculator in polar mode.

27. $r = 2 \cos 2\theta$

28. $r = 2 \sin 2\theta$

29. $r = 4 \sin 5\theta$

30. $r = 6 \cos 6\theta$

31. $r = 3 + 3 \sin \theta$

32. $r = 3 + 3 \cos \theta$

33. $r = 1 - 4 \cos \theta$

34. $r = 4 - 5 \sin \theta$

35. $r = 2 \cos 2\theta - 3 \sin \theta$

36. $r = 3 \sin 2\theta + 2 \cos \theta$

37. $r = 3 \sin 2\theta + \sin \theta$

38. $r = 2 \cos 2\theta - \cos \theta$

Convert each equation to polar coordinates and then sketch the graph.

39. $x^2 + y^2 = 16$

40. $x^2 + y^2 = 25$

41. $x^2 + y^2 = 6x$

42. $x^2 + y^2 = 6y$

43. $(x^2 + y^2)^2 = 2xy$

44. $(x^2 + y^2)^2 = x^2 - y^2$

Change each equation to rectangular coordinates and then graph.

45. $r(2 \cos \theta + 3 \sin \theta) = 6$

46. $r(3 \cos \theta - 2 \sin \theta) = 6$

47. $r(1 - \cos \theta) = 1$

48. $r(1 - \sin \theta) = 1$

49. $r = 4 \sin \theta$

50. $r = 6 \cos \theta$

51. Graph $r_1 = 2 \sin \theta$ and $r_2 = 2 \cos \theta$ and then name two points they have in common.

52. Graph $r_1 = 2 + 2 \cos \theta$ and $r_2 = 2 - 2 \cos \theta$ and name three points they have in common.

REVIEW PROBLEMS

The problems that follow review material we covered in Section 4.5. Graph each equation.

53. $y = \sin x - \cos x, \ 0 \le x \le 4\pi$

54. $y = \cos x - \sin x, \ 0 \le x \le 4\pi$

55. $y = x + \sin \pi x, \ 0 \le x \le 8$

56. $y = x + \cos \pi x, \ 0 \le x \le 8$

57. $y = 3 \sin x + \cos 2x, \ 0 \le x \le 4\pi$

58. $y = \sin x + \dfrac{1}{2} \cos 2x, \ 0 \le x \le 4\pi$

CHAPTER 8 SUMMARY

EXAMPLES

1. Each of the following is a complex number.

$$5 + 4i$$
$$-\sqrt{3} + i$$
$$7i$$
$$-8$$

The number $7i$ is complex because

$$7i = 0 + 7i$$

The number -8 is complex because

$$-8 = -8 + 0i$$

Definitions [8.1]

The number i is such that $i^2 = -1$. If $a > 0$, the expression $\sqrt{-a}$ can be written as $\sqrt{ai^2} = i\sqrt{a}$.

A *complex number* is any number that can be written in the form

$$a + bi$$

where a and b are real numbers and $i^2 = -1$. The number a is called the *real part* of the complex number, and b is called the *imaginary part*. The form $a + bi$ is called *standard form*.

All real numbers are also complex numbers since they can be put in the form $a + bi$, where $b = 0$. If $a = 0$ and $b \neq 0$, then $a + bi$ is called an *imaginary number*.

2. If $3x + 2i = 12 - 4yi$, then

$$3x = 12 \quad \text{and} \quad 2 = -4y$$
$$x = 4 \qquad \quad y = -1/2$$

Equality for Complex Numbers [8.1]

Two complex numbers are equal if and only if their real parts are equal and their imaginary parts are equal. That is,

$$a + bi = c + di \quad \text{if and only if} \quad a = c \quad \text{and} \quad b = d$$

3. If $z_1 = 2 - i$ and $z_2 = 4 + 3i$, then

Operations on Complex Numbers in Standard Form [8.1]

If $z_1 = a_1 + b_1i$ and $z_2 = a_2 + b_2i$ are two complex numbers in standard form, then the following definitions and operations apply.

$$z_1 + z_2 = 6 + 2i$$

Addition

$$z_1 + z_2 = (a_1 + a_2) + (b_1 + b_2)i$$

Add real parts; add imaginary parts.

$$z_1 - z_2 = -2 - 4i$$

Subtraction

$$z_1 - z_2 = (a_1 - a_2) + (b_1 - b_2)i$$

Subtract real parts; subtract imaginary parts.

$$z_1z_2 = (2 - i)(4 + 3i)$$
$$= 8 + 6i - 4i - 3i^2$$
$$= 11 + 2i$$

Multiplication

$$z_1z_2 = (a_1a_2 - b_1b_2) + (a_1b_2 + a_2b_1)i$$

In actual practice, simply multiply as you would multiply two binomials.

The conjugate of $4 + 3i$ is $4 - 3i$ and

$$(4 + 3i)(4 - 3i) = 16 + 9$$
$$= 25$$

Conjugates

The conjugate of $a + bi$ is $a - bi$. Their product is the real number $a^2 + b^2$.

$$\frac{z_1}{z_2} = \frac{2 - i}{4 + 3i} \cdot \frac{4 - 3i}{4 - 3i}$$
$$= \frac{5 - 10i}{25}$$
$$= \frac{1}{5} - \frac{2}{5}i$$

Division

Multiply the numerator and denominator of the quotient by the conjugate of the denominator.

4. $i^{20} = (i^4)^5 = 1$
$i^{21} = (i^4)^5 \cdot i = i$
$i^{22} = (i^4)^5 \cdot i^2 = -1$
$i^{23} = (i^4)^5 \cdot i^3 = -i$

Powers of i [8.1]

If n is an integer, then i^n can always be simplified to i, -1, $-i$, or 1.

5. The graph of $4 + 3i$ is

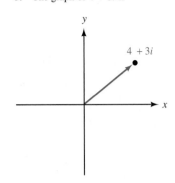

Graphing Complex Numbers [8.2]

The graph of the complex number $z = x + yi$ is the arrow (vector) that extends from the origin to the point (x, y).

6. If $z = \sqrt{3} + i$, then
$|z| = |\sqrt{3} + i| = \sqrt{3 + 1} = 2$

Absolute Value of a Complex Number [8.2]

The *absolute value* (or *modulus*) of the complex number $z = x + yi$ is the distance from the origin to the point (x, y). If this distance is denoted by r, then

$$r = |z| = |x + yi| = \sqrt{x^2 + y^2}$$

7. For $z = \sqrt{3} + i$, θ is the smallest positive angle for which

$$\sin \theta = \frac{1}{2} \text{ and } \cos \theta = \frac{\sqrt{3}}{2}$$

which means $\theta = 30°$.

Argument of a Complex Number [8.2]

The *argument* of the complex number $z = x + yi$ is the smallest positive angle from the positive x-axis to the graph of z. If the argument of z is denoted by θ, then

$$\sin \theta = \frac{y}{r}, \qquad \cos \theta = \frac{x}{r}, \qquad \text{and} \qquad \tan \theta = \frac{y}{x}$$

8. If $z = \sqrt{3} + i$, then in trigonometric form

$$z = 2(\cos 30° + i \sin 30°)$$
$$= 2 \text{ cis } 30°$$

Trigonometric Form of a Complex Number [8.2]

The complex number $z = x + yi$ is written in trigonometric form when it is written as

$$z = r(\cos \theta + i \sin \theta) = r \text{ cis } \theta$$

where r is the absolute value of z and θ is the argument of z.

9. If

$$z_1 = 8(\cos 40° + i \sin 40°) \text{ and}$$
$$z_2 = 4(\cos 10° + i \sin 10°),$$

then

$$z_1 z_2 = 32(\cos 50° + i \sin 50°)$$
$$\frac{z_1}{z_2} = 2(\cos 30° + i \sin 30°)$$

Products and Quotients in Trigonometric Form [8.3]

If $z_1 = r_1(\cos \theta_1 + i \sin \theta_1)$ and $z_2 = r_2(\cos \theta_2 + i \sin \theta_2)$ are two complex numbers in trigonometric form, then their product is

$$z_1 z_2 = r_1 r_2 [\cos (\theta_1 + \theta_2) + i \sin (\theta_1 + \theta_2)] = r_1 r_2 \text{ cis } (\theta_1 + \theta_2)$$

and their quotient is

$$\frac{z_1}{z_2} = \frac{r_1}{r_2} [\cos (\theta_1 - \theta_2) + i \sin (\theta_1 - \theta_2)] = \frac{r_1}{r_2} \text{ cis } (\theta_1 - \theta_2)$$

10. If $z = \sqrt{2} \text{ cis } 30°$, then

$$z^{10} = \sqrt{2}^{10} \text{ cis } (10 \cdot 30°)$$
$$= 32 \text{ cis } 300°$$

DeMoivre's Theorem [8.3]

If $z = r(\cos \theta + i \sin \theta)$ is a complex number in trigonometric form and n is an integer, then

$$z^n = r^n(\cos n\theta + i \sin n\theta) = r^n \text{ cis } (n\theta)$$

11. The 3 cube roots of $z = 8(\cos 60° + i \sin 60°)$ are given by

$$w_k = 8^{1/3} \text{ cis } \left(\frac{60°}{3} + \frac{360°}{3}k \right)$$
$$= 2 \text{ cis } (20° + 120°k)$$

where $k = 0, 1, 2$. That is,

$$w_0 = 2(\cos 20° + i \sin 20°)$$
$$w_1 = 2(\cos 140° + i \sin 140°)$$
$$w_2 = 2(\cos 260° + i \sin 260°)$$

Roots of a Complex Number [8.4]

The n nth roots of the complex number $z = r(\cos \theta + i \sin \theta)$ are given by the formula

$$w_k = r^{1/n} \left[\cos \left(\frac{\theta}{n} + \frac{360°}{n}k \right) + i \sin \left(\frac{\theta}{n} + \frac{360°}{n}k \right) \right]$$
$$= r^{1/n} \text{ cis } \left(\frac{\theta}{n} + \frac{360°}{n}k \right)$$

where $k = 0, 1, 2, \ldots, n - 1$.

12.

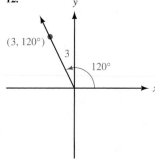

Polar Coordinates [8.5]

The ordered pair (r, θ) names the point that is r units from the origin along the terminal side of angle θ in standard position. The coordinates r and θ are said to be the *polar coordinates* of the point they name.

13. Convert $(-\sqrt{2}, 135°)$ to rectangular coordinates.

$$x = r \cos \theta$$
$$= -\sqrt{2} \cos 135°$$
$$= 1$$
$$y = r \sin \theta$$
$$= -\sqrt{2} \sin 135°$$
$$= -1$$

Polar Coordinates and Rectangular Coordinates [8.5]

To derive the relationship between polar coordinates and rectangular coordinates, we consider a point P with rectangular coordinates (x, y) and polar coordinates (r, θ).

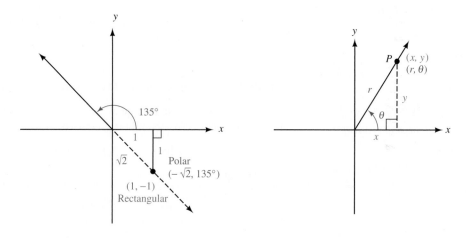

14. Change $x + y = 4$ to polar coordinates.

Since $x = r \cos \theta$ and $y = r \sin \theta$, we have

$$r \cos \theta + r \sin \theta = 4$$
$$r(\cos \theta + \sin \theta) = 4$$
$$r = \frac{4}{\cos \theta + \sin \theta}$$

The last equation gives us r in terms of θ.

Equations in Polar Coordinates [8.5, 8.6]

Equations in polar coordinates have variables r and θ instead of x and y. The conversions we use to change ordered pairs from polar coordinates to rectangular coordinates and from rectangular coordinates to polar coordinates are the same ones we use to convert back and forth between equations given in polar coordinates and those in rectangular coordinates.

CHAPTER 8 TEST

Write in terms of i.

1. $\sqrt{-25}$ **2.** $\sqrt{-12}$

Find x and y so that each of the following equations is true.

3. $7x - 6i = 14 - 3yi$ **4.** $(x^2 - 3x) + 16i = 10 + 8yi$

Combine the following complex numbers:

5. $(6 - 3i) + [(4 - 2i) - (3 + i)]$ **6.** $(7 + 3i) - [(2 + i) - (3 - 4i)]$

Simplify each power of i.

7. i^{16} **8.** i^{17}

Multiply. Leave your answer in standard form.

9. $(8 + 5i)(8 - 5i)$ **10.** $(3 + 5i)^2$

Divide. Write all answers in standard form.

11. $\dfrac{5 - 4i}{2i}$ **12.** $\dfrac{6 + 5i}{6 - 5i}$

For each of the following complex numbers give: (a) the absolute value; (b) the opposite; and (c) the conjugate.

13. $3 + 4i$ **14.** $3 - 4i$ **15.** $8i$ **16.** -4

Write each complex number in standard form.

17. $8(\cos 330° + i \sin 330°)$ **18.** $2 \operatorname{cis} 135°$

Write each complex number in trigonometric form.

19. $2 + 2i$ **20.** $-\sqrt{3} + i$ **21.** $5i$ **22.** -3

Multiply or divide as indicated. Leave your answers in trigonometric form.

23. $5(\cos 25° + i \sin 25°) \cdot 3(\cos 40° + i \sin 40°)$

24. $\dfrac{10(\cos 50° + i \sin 50°)}{2(\cos 20° + i \sin 20°)}$

25. $[2(\cos 10° + i \sin 10°)]^5$

26. $[3 \operatorname{cis} 20°]^4$

27. Find two square roots of $z = 49(\cos 50° + i \sin 50°)$. Leave your answer in trigonometric form.

28. Find 4 fourth roots for $z = 2 + 2i\sqrt{3}$. Leave your answer in trigonometric form.

Solve each equation. Write your solutions in trigonometric form.

29. $x^4 - 2\sqrt{3}x^2 + 4 = 0$ **30.** $x^3 = -1$

For each of the following points, name two other ordered pairs in polar coordinates that name the same point and then convert each of the original ordered pairs to rectangular coordinates.

31. $(4, 225°)$ **32.** $(-6, 60°)$

Convert to polar coordinates with r positive and θ between $0°$ and $360°$.

33. $(-3, 3)$ **34.** $(0, 5)$

Write each equation with rectangular coordinates.

35. $r = 6 \sin \theta$ **36.** $r = \sin 2\theta$

Write each equation in polar coordinates.

37. $x + y = 2$ **38.** $x^2 + y^2 = 8y$

Graph each equation.

39. $r = 4$ **40.** $\theta = 45°$
41. $r = 4 + 2 \cos \theta$ **42.** $r = \sin 2\theta$

 Graph each equation using your graphing calculator in polar mode.

43. $r = 4 - 4 \cos \theta$ **44.** $r = 6 \cos 3\theta$
45. $r = 3 \sin 2\theta - \sin \theta$ **46.** $r = 2 \cos 2\theta + 3 \sin \theta$

CHAPTER **8** GROUP PROJECT

INTERSECTIONS OF POLAR GRAPHS

Objective: To find the points of intersection of two polar graphs.

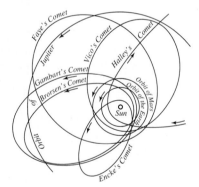

On July 16, 1994, Comet P/Shoemaker-Levy 9 collided with Jupiter (Figure 1). Potential collisions between planets and comets are found by identifying any intersection points in their orbits. These orbits are typically described using polar equations.

In Sections 8.5 and 8.6 we learned about the polar coordinate system and how to graph equations in polar coordinates. One of the things we might need to know about the graphs is if they have any points of intersection. In this project you will learn how to find these points of intersection.

First, we will find the points of intersection of the graphs of the polar equations $r = 1$ and $r = 1 + \cos \theta$.

Figure 1

1. We can approach the problem as a system of two equations in two variables. Using the substitution method, we can replace r in the second equation with the value 1 (from the first equation) to obtain an equation in just one variable:

$$1 = 1 + \cos \theta$$

Solve this equation for θ, if $0 \leq \theta < 2\pi$. Then use your values of θ to find the corresponding values of r. Write your results as ordered pairs (r, θ).

2. Verify your results from Question 1 graphically. Sketch the graph of both equations on the same polar coordinate system. How many points of intersection are there? Did you find them all using the substitution method in Question 1?

Next, we will find the points of intersection of the graphs of the polar equations $r = 3 \sin \theta$ and $r = 1 + \sin \theta$.

3. Use the substitution method to find any points of intersection for the two equations, if $0 \le \theta < 2\pi$. Write your results as ordered pairs (r, θ).

4. Verify your results graphically. Sketch the graph of both equations on the same polar coordinate system. How many points of intersection are there? Did you find them all using the substitution method in Question 3?

5. Show that the pole $r = 0$ does satisfy each equation for some value of θ.

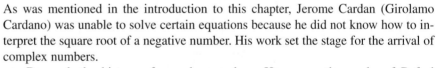 6. Use your graphing calculator to graph both equations. Make sure your calculator is set to polar mode, radians, and to graph equations *simultaneously*. Watch carefully as the graphs are drawn (at the same time) by your calculator. Did you notice any significant differences in how the intersection points occur? Write a small paragraph describing what you observed.

7. Based on your observations of the graphs, try to explain why you were not able to find one of the intersection points using the substitution method. You may find it helpful to keep in mind that each point in the plane has more than one different representation in polar coordinates. Write your explanation in a paragraph or two.

CHAPTER **8** RESEARCH PROJECT

COMPLEX NUMBERS

Girolamo Cardano

As was mentioned in the introduction to this chapter, Jerome Cardan (Girolamo Cardano) was unable to solve certain equations because he did not know how to interpret the square root of a negative number. His work set the stage for the arrival of complex numbers.

Research the history of complex numbers. How were the works of Rafael Bombelli, Jean Robert Argand, Leonhard Euler, and Abraham DeMoivre significant in the development of complex numbers? Write a paragraph or two about your findings.

REVIEW OF FUNCTIONS

Mathematics is not a careful march down a well-cleared high-way, but a journey into a strange wilderness, where the explorers often get lost.

W. S. Anglin

INTRODUCTION

In the United States, temperature is usually measured in terms of degrees Fahrenheit; however, most European countries prefer to measure temperature in degrees Celsius. Table 1 shows some temperatures measured using both systems.

TABLE 1	
Degrees Celsius (°C)	Degrees Fahrenheit (°F)
0	32
25	77
40	104
100	212

The relationship between these two systems of temperature measurement can be written in the form of an equation:

$$F = \frac{9}{5}C + 32$$

Every temperature measured in °C is associated with a unique measure in °F. We call this kind of relationship a function. In this appendix we will review the concept of a function, and see how in many cases a function may be "reversed."

A.1 | INTRODUCTION TO FUNCTIONS

The ad shown in the margin appeared in the help wanted section of the local newspaper the day we were writing this section of the book. We can use the information in the ad to start an informal discussion of our next topic: functions.

AN INFORMAL LOOK AT FUNCTIONS

To begin with, suppose you have a job that pays $7.50 per hour and that you work anywhere from 0 to 40 hours per week. The amount of money you make in one week depends on the number of hours you work that week. In mathematics we say that your weekly earnings are a *function* of the number of hours you work. If we let the variable x represent hours and the variable y represent the money you make, then the relationship between x and y can be written as

$$y = 7.5x \quad \text{for} \quad 0 \le x \le 40$$

 EXAMPLE 1   Construct a table and graph for the function

$$y = 7.5x \quad \text{for} \quad 0 \le x \le 40$$

SOLUTION   Table 1 gives some of the paired data that satisfy the equation $y = 7.5x$. Figure 1 is the graph of the equation with the restriction $0 \le x \le 40$.

TABLE 1 WEEKLY WAGES

Hours Worked	Rule	Pay
x	$y = 7.5x$	y
0	$y = 7.5(0)$	0
10	$y = 7.5(10)$	75
20	$y = 7.5(20)$	150
30	$y = 7.5(30)$	225
40	$y = 7.5(40)$	300

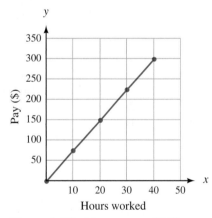

Figure 1 Weekly wages at $7.50 per hour.

The equation $y = 7.5x$ with the restriction $0 \le x \le 40$, Table 1, and Figure 1 are three ways to describe the same relationship between the number of hours you work in 1 week and your gross pay for that week. In all three, we *input* values of x, and then use the function rule to *output* values of y.

USING TECHNOLOGY

CREATING TABLES AND GRAPHS

```
Plot1  Plot2  Plot3
\Y1■7.5X
\Y2=
\Y3=
\Y4=
\Y5=
\Y6=
\Y7=
```

Figure 2

To create the table and graph for the function $y = 7.5x$ in Example 1 using a graphing calculator, first define the function as $Y_1 = 7.5x$ in the equation editor (Figure 2). Set up your table to automatically generate values starting at $x = 0$ and using an increment of 10 (Figure 3). On some calculators, this is done by setting the independent variable to Auto. Now display the table and you should see values similar to those shown in Figure 4.

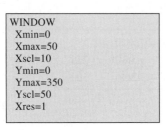

```
TABLE SETUP
 TblStart=0
 ΔTbl=10
Indpnt: Auto Ask
Depend: Auto Ask
```

Figure 3

X	Y1	
0	0	
10	75	
20	150	
30	225	
40	300	
50	375	
60	450	
X=0		

Figure 4

To create the graph of the function, set the window variables in agreement with the graph in Figure 1:

$$0 \le x \le 50, \text{ scale } = 10; \ 0 \le y \le 350, \text{ scale } = 50$$

By this, we mean that Xmin = 0, Xmax = 50, Xscl = 10, Ymin = 0, Ymax = 350, and Yscl = 50, as shown in Figure 5. Then graph the function. Your graph should be similar to the one shown in Figure 6.

```
WINDOW
 Xmin=0
 Xmax=50
 Xscl=10
 Ymin=0
 Ymax=350
 Yscl=50
 Xres=1
```

Figure 5

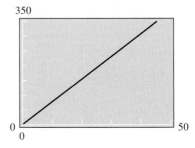

Figure 6

DOMAIN AND RANGE OF A FUNCTION

We began this discussion by saying that the number of hours worked during the week was from 0 to 40, so these are the values that x can assume. From the line graph in Figure 1, we see that the values of y range from 0 to 300. We call the complete set of values that x can assume the *domain* of the function. The values that are assigned to y are called the *range* of the function.

The Function Rule

Domain: The set of all inputs

Range: The set of all outputs

 EXAMPLE 2 State the domain and range for the function

$$y = 7.5x, \quad 0 \le x \le 40$$

SOLUTION   From the previous discussion we have

$$\text{Domain} = \{x \mid 0 \le x \le 40\}$$

$$\text{Range} = \{y \mid 0 \le y \le 300\}$$

A FORMAL LOOK AT FUNCTIONS

What is apparent from the preceding discussion is that we are working with paired data. The solutions to the equation $y = 7.5x$ are pairs of numbers, and the points on the line graph in Figure 1 come from paired data. We are now ready for the formal definition of a function.

DEFINITION

A *function* is a rule that pairs each element in one set, called the *domain,* with exactly one element from a second set, called the *range.*

In other words, a function is a rule for which each input is paired with exactly one output.

FUNCTIONS AS ORDERED PAIRS

The function rule $y = 7.5x$ from Example 1 produces ordered pairs of numbers (x, y). The same thing happens with all functions: The function rule produces ordered pairs of numbers. We use this result to write an alternative definition for a function

---

ALTERNATIVE DEFINITION

A *function* is a set of ordered pairs in which no two different ordered pairs have the same first coordinate. The set of all first coordinates is called the *domain* of the function. The set of all second coordinates is called the *range* of the function.

The restriction on first coordinates in the alternative definition keeps us from assigning a number in the domain to more than one number in the range.

A RELATIONSHIP THAT IS NOT A FUNCTION

You may be wondering if any sets of paired data fail to qualify as functions. The answer is yes, as the next example reveals.

 **EXAMPLE 3** Sketch the graph of $x = y^2$ and determine if this relationship is a function.

SOLUTION Without going into much detail, we graph the equation $x = y^2$ by finding a number of ordered pairs that satisfy the equation, plotting these points, then drawing a smooth curve that connects them. A table of values for x and y that satisfy the equation follows, along with the graph of $x = y^2$ shown in Figure 7.

x	y
0	0
1	1
1	−1
4	2
4	−2
9	3
9	−3

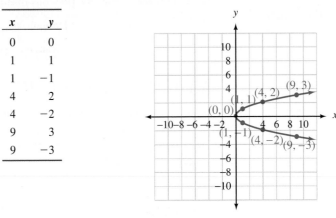

Figure 7

As you can see from looking at the table and the graph in Figure 7, several ordered pairs whose graphs lie on the curve have repeated first coordinates. For instance, (1, 1) and (1, −1), (4, 2) and (4, −2), and (9, 3) and (9, −3). The relationship is therefore not a function. ■

To classify all relationships specified by ordered pairs, whether they are functions or not, we include the following two definitions.

DEFINITION

A *relation* is a rule that pairs each element in one set, called the *domain*, with one or more elements from a second set, called the *range*.

ALTERNATIVE DEFINITION

A *relation* is a set of ordered pairs. The set of all first coordinates is the *domain* of the relation. The set of all second coordinates is the *range* of the relation.

Here are some facts that will help clarify the distinction between relations and functions.

1. Any rule that assigns numbers from one set to numbers in another set is a relation. If that rule makes the assignment so that no input has more than one output, then it is also a function.
2. Any set of ordered pairs is a relation. If none of the first coordinates of those ordered pairs is repeated, the set of ordered pairs is also a function.
3. Every function is a relation.
4. Not every relation is a function.

VERTICAL LINE TEST

Look at the graph shown in Figure 7. The reason this graph is the graph of a relation, but not of a function, is that some points on the graph have the same first coordinates—for example, the points $(4, 2)$ and $(4, -2)$. Furthermore, any time two points on a graph have the same first coordinates, those points must lie on a vertical line. This allows us to write the following test that uses the graph to determine whether a relation is also a function.

VERTICAL LINE TEST

If a vertical line crosses the graph of a relation in more than one place, the relation cannot be a function. If no vertical line can be found that crosses a graph in more than one place, then the graph is the graph of a function.

EXAMPLE 4 Graph $y = |x|$. Use the graph to determine whether we have the graph of a function. State the domain and range.

SOLUTION We let x take on values of $-4, -3, -2, -1, 0, 1, 2, 3$, and 4. The corresponding values of y are shown in the table. The graph is shown in Figure 8.

x	y
-4	4
-3	3
-2	2
-1	1
0	0
1	1
2	2
3	3
4	4

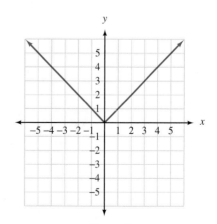

Figure 8

Because no vertical line can be found that crosses the graph in more than one place, $y = |x|$ is a function. The domain is all real numbers. The range is $\{y \mid y \geq 0\}$.

FUNCTION NOTATION

Let's return to the discussion that introduced us to functions. If a job pays \$7.50 per hour for working from 0 to 40 hours a week, then the amount of money y earned in 1 week is a function of the number of hours worked, x. The exact relationship between x and y is written

$$y = 7.5x \quad \text{for} \quad 0 \le x \le 40$$

Because the amount of money earned y depends on the number of hours worked x, we call y the *dependent variable* and x the *independent variable*. Furthermore, if we let f represent all the ordered pairs produced by the equation, then we can write

$$f = \{(x, y) \mid y = 7.5x \text{ and } 0 \le x \le 40\}$$

Once we have named a function with a letter, we can use an alternative notation to represent the dependent variable y. The alternative notation for y is $f(x)$. It is read "f of x" and can be used instead of the variable y when working with functions. The notation y and the notation $f(x)$ are equivalent—that is,

$$y = 7.5x \Leftrightarrow f(x) = 7.5x$$

When we use the notation $f(x)$ we are using *function notation*. The benefit of using function notation is that we can write more information with fewer symbols than we can by using just the variable y. For example, asking how much money a person will make for working 20 hours is simply a matter of asking for $f(20)$. Without function notation, we would have to say "find the value of y that corresponds to a value of $x = 20$." To illustrate further, using the variable y, we can say "y is 150 when x is 20." Using the notation $f(x)$, we simply say "$f(20) = 150$." Each expression indicates that you will earn \$150 for working 20 hours.

NOTE Some students like to think of functions as machines. Values of x are put into the machine, which transforms them into values of $f(x)$, which then are output by the machine.

Input x

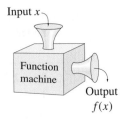

Function machine

Output
$f(x)$

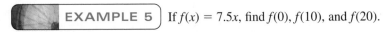

EXAMPLE 5 If $f(x) = 7.5x$, find $f(0)$, $f(10)$, and $f(20)$.

SOLUTION To find $f(0)$ we substitute 0 for x in the expression $7.5x$ and simplify. We find $f(10)$ and $f(20)$ in a similar manner—by substitution.

$$\text{If} \qquad f(x) = 7.5x$$

$$\text{then} \qquad f(\mathbf{0}) = 7.5(\mathbf{0}) = 0$$

$$f(\mathbf{10}) = 7.5(\mathbf{10}) = 75$$

$$f(\mathbf{20}) = 7.5(\mathbf{20}) = 150$$

FUNCTION NOTATION AND GRAPHS

We can visualize the relationship between x and $f(x)$ on the graph of the function. Figure 9 shows the graph of $f(x) = 7.5x$ along with two additional line segments. The horizontal line segment corresponds to $x = 20$, and the vertical line segment corresponds to $f(20)$. (Note that the domain is restricted to $0 \le x \le 40$.)

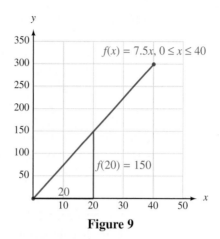

Figure 9

USING FUNCTION NOTATION

The remaining examples in this section show a variety of ways to use and interpret function notation.

 EXAMPLE 6 | A painting is purchased as an investment for $125. If its value increases continuously so that it doubles every 5 years, then its value is given by the function

$$V(t) = 125 \cdot 2^{t/5} \quad \text{for} \quad t \ge 0$$

where t is the number of years since the painting was purchased, and $V(t)$ is its value (in dollars) at time t. Find $V(5)$ and $V(10)$, and explain what they mean.

SOLUTION   The expression $V(5)$ is the value of the painting when $t = 5$ (5 years after it is purchased). We calculate $V(5)$ by substituting 5 for t in the equation $V(t) = 125 \cdot 2^{t/5}$. Here is our work:

$$V(\mathbf{5}) = 125 \cdot 2^{5/5} = 125 \cdot 2^1 = 125 \cdot 2 = 250$$

In words: After 5 years, the painting is worth $250.

The expression $V(10)$ is the value of the painting after 10 years. To find this number, we substitute 10 for t in the equation:

$$V(\mathbf{10}) = 125 \cdot 2^{10/5} = 125 \cdot 2^2 = 125 \cdot 4 = 500$$

In words: The value of the painting 10 years after it is purchased is $500.

The fact that $V(5) = 250$ means that the ordered pair $(5, 250)$ belongs to the function V. Likewise, the fact that $V(10) = 500$ tells us that the ordered pair $(10, 500)$ is a member of function V. ▪

We can generalize the discussion at the end of Example 6 this way:

$$(a, b) \in f \quad \text{if and only if} \quad f(a) = b$$

 EXAMPLE 7 If $f(x) = 3x^2 + 2x - 1$, find $f(0), f(3)$, and $f(-2)$.

SOLUTION Because $f(x) = 3x^2 + 2x - 1$, we have

$$f(\mathbf{0}) = 3(\mathbf{0})^2 + 2(\mathbf{0}) - 1 \quad = 0 + 0 - 1 = -1$$

$$f(\mathbf{3}) = 3(\mathbf{3})^2 + 2(\mathbf{3}) - 1 \quad = 27 + 6 - 1 = 32$$

$$f(-\mathbf{2}) = 3(-\mathbf{2})^2 + 2(-\mathbf{2}) - 1 = 12 - 4 - 1 = 7 \quad \blacksquare$$

In Example 7, the function f is defined by the equation $f(x) = 3x^2 + 2x - 1$. We could just as easily have said $y = 3x^2 + 2x - 1$; that is, $y = f(x)$. Saying $f(-2) = 7$ is exactly the same as saying y is 7 when x is -2.

 GETTING READY FOR CLASS

After reading through the preceding section, respond in your own words and in complete sentences.

a. What is a function?

b. What is the vertical line test?

c. Explain what you are calculating when you find $f(2)$ for a given function f.

d. If $f(2) = 3$ for a function f, what is the relationship between the numbers 2 and 3 and the graph of f?

PROBLEM SET A.1

For each of the following relations, give the domain and range, and indicate which are also functions.

1. $\{(1, 3), (2, 5), (4, 1)\}$

2. $\{(3, 1), (5, 7), (2, 3)\}$

3. $\{(-1, 3), (1, 3), (2, -5)\}$

4. $\{(3, -4), (-1, 5), (3, 2)\}$

5. $\{(7, -1), (3, -1), (7, 4)\}$

6. $\{(5, -2), (3, -2), (5, -1)\}$

State whether each of the following graphs represents a function.

7.

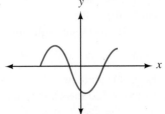

8.

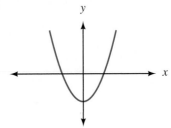

9.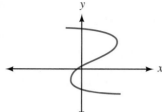

10.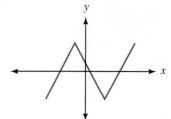

11.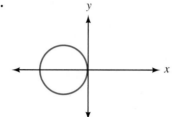

12.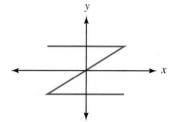

13.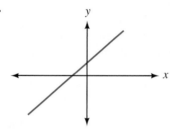

14.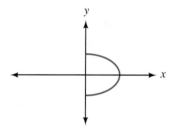

Graph each of the following relations. In each case, use the graph to find the domain and range, and indicate whether the graph is the graph of a function.

15. $y = x^2 - 1$ **16.** $y = x^2 + 4$ **17.** $x = y^2 + 4$
18. $x = y^2 - 9$ **19.** $y = |x - 2|$ **20.** $y = |x + 2|$
21. $y = |x| - 2$ **22.** $y = |x| + 2$

23. Suppose you have a job that pays \$8.50 per hour and you work anywhere from 10 to 40 hours per week.
    a. Write an equation, with a restriction on the variable x, that gives the amount of money, y, you will earn for working x hours in one week
    b. Use the function rule you have written in part (a) to complete Table 2.

TABLE 2 WEEKLY WAGES

Hours Worked x	Gross Pay (\$) y
10	
20	
30	
40	

    c. Construct a line graph from the information in Table 2.
    d. State the domain and range of this function.

24. The ad shown here was in the local newspaper. Suppose you are hired for the job described in the ad.

  a. If x is the number of hours you work per week and y is your weekly gross pay, write the equation for y. (Be sure to include any restrictions on the variable x that are given in the ad.)

  b. Use the function rule you have written in part (a) to complete Table 3.

TABLE 3 WEEKLY WAGES	
Hours Worked x	Gross Pay ($) y
15	
20	
25	
30	

  c. Construct a line graph from the information in Table 3.

  d. State the domain and range of this function.

  e. What is the minimum amount you can earn in a week with this job? What is the maximum amount?

 25. Tossing a Coin Hali is tossing a quarter into the air with an underhand motion. The distance the quarter is above her hand at any time is given by the function

$$h = 16t - 16t^2 \qquad \text{for} \qquad 0 \le t \le 1$$

where h is the height of the quarter in feet, and t is the time in seconds.

  a. Use the table feature of your graphing calculator to complete the table.

Time (sec) t	Distance (ft) h
0	
0.1	
0.2	
0.3	
0.4	
0.5	
0.6	
0.7	
0.8	
0.9	
1	

  b. State the domain and range of this function.

  c. Graph the function on your calculator using an appropriate window.

26. Intensity of Light The formula below gives the intensity of light that falls on a surface at various distances from a 100-watt light bulb:

$$I = \frac{120}{d^2} \quad \text{for} \quad d > 0$$

where I is the intensity of light (in lumens per square foot), and d is the distance (in feet) from the light bulb to the surface.

a. Use the table feature of your graphing calculator to complete the table.

Distance (ft) d	Intensity I
1	
2	
3	
4	
5	
6	

b. Graph the function on your calculator using an appropriate window.

27. Area of a Circle The formula for the area A of a circle with radius r is given by $A = \pi r^2$. The formula shows that A is a function of r.

a. Graph the function $A = \pi r^2$ for $0 \le r \le 3$. (On the graph, let the horizontal axis be the r-axis, and let the vertical axis be the A-axis.)

b. State the domain and range of the function $A = \pi r^2$, $0 \le r \le 3$.

28. Area and Perimeter of a Rectangle A rectangle is 2 inches longer than it is wide. Let $x =$ the width and $P =$ the perimeter.

a. Write an equation that will give the perimeter P in terms of the width x of the rectangle. Are there any restrictions on the values that x can assume?

b. Graph the relationship between P and x.

Let $f(x) = 2x - 5$ and $g(x) = x^2 + 3x + 4$. Evaluate the following.

29. $f(2)$ **30.** $f(3)$ **31.** $f(-3)$ **32.** $g(-2)$

33. $g(-1)$ **34.** $f(-4)$ **35.** $g(-3)$ **36.** $g(2)$

37. $g(4) + f(4)$ **38.** $f(2) - g(3)$ **39.** $f(3) - g(2)$ **40.** $g(-1) + f(-1)$

Let $f(x) = 3x^2 - 4x + 1$ and $g(x) = 2x - 1$. Evaluate the following.

41. $f(0)$ **42.** $g(0)$ **43.** $g(-4)$ **44.** $f(1)$

45. $f(-1)$ **46.** $g(-1)$ **47.** $g(10)$ **48.** $f(10)$

49. $f(3)$ **50.** $g(3)$ **51.** $g\left(\dfrac{1}{2}\right)$ **52.** $g\left(\dfrac{1}{4}\right)$

53. $f(a)$ **54.** $g(b)$

If $f = \{(1, 4), (-2, 0), \left(3, \frac{1}{2}\right), (\pi, 0)\}$ and $g = \{(1, 1), (-2, 2), \left(\frac{1}{2}, 0\right)\}$, find each of the following values of f and g.

55. $f(1)$ **56.** $g(1)$ **57.** $g\left(\dfrac{1}{2}\right)$ **58.** $f(3)$

59. $g(-2)$ **60.** $f(\pi)$

61. Graph the function $f(x) = \frac{1}{2}x + 2$. Then draw and label the line segments that represent $x = 4$ and $f(4)$.

62. Graph the function $f(x) = -\frac{1}{2}x + 6$. Then draw and label the line segments that represent $x = 4$ and $f(4)$.

63. Investing in Art A painting is purchased as an investment for $150. If its value increases continuously so that it doubles every 3 years, then its value is given by the function

$$V(t) = 150 \cdot 2^{t/3} \quad \text{for} \quad t \geq 0$$

where t is the number of years since the painting was purchased, and $V(t)$ is its value (in dollars) at time t. Find $V(3)$ and $V(6)$, and then explain what they mean.

64. Average Speed If it takes Minke t minutes to run a mile, then her average speed $s(t)$, in miles per hour, is given by the formula

$$s(t) = \frac{60}{t} \quad \text{for} \quad t > 0$$

Find $s(4)$ and $s(5)$, and then explain what they mean.

Area of a Circle The formula for the area A of a circle with radius r can be written with function notation as $A(r) = \pi r^2$.

65. Find $A(2)$, $A(5)$, and $A(10)$. (Use $\pi \approx 3.14$.)

66. Why doesn't it make sense to ask for $A(-10)$?

EXTENDING THE CONCEPTS

The graphs of two functions are shown in Figures 10 and 11. Use the graphs to find the following.

67. a. $f(2)$ **b.** $f(-4)$ **c.** $g(0)$ **d.** $g(3)$

68. a. $g(2) - f(2)$ **b.** $f(1) + g(1)$ **c.** $f[g(3)]$ **d.** $g[f(3)]$

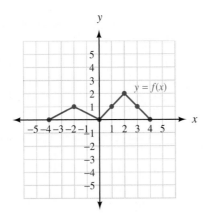

Figure 10

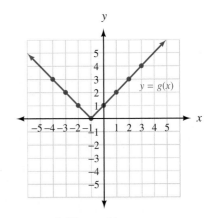

Figure 11

69. For the function $f(x) = \frac{1}{2}x + 2$, find the value of x for which $f(x) = x$.

70. For the function $f(x) = -\frac{1}{2}x + 6$, find the value of x for which $f(x) = x$.

A.2 | THE INVERSE OF A FUNCTION

The following diagram (Figure 1) shows the route Justin takes to school. He leaves his home and drives 3 miles east and then turns left and drives 2 miles north. When he leaves school to drive home, he drives the same two segments but in the reverse order and the opposite direction; that is, he drives 2 miles south, turns right, and drives 3 miles west. When he arrives home from school, he is right where he started. His route home "undoes" his route to school, leaving him where he began.

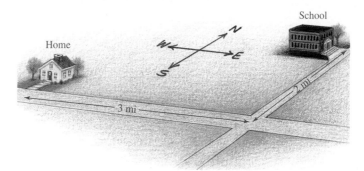

Figure 1

As you will see, the relationship between a function and its inverse function is similar to the relationship between Justin's route from home to school and his route from school to home.

Suppose the function f is given by

$$f = \{(1, 4), (2, 5), (3, 6), (4, 7)\}$$

The inverse of f is obtained by reversing the order of the coordinates in each ordered pair in f. The inverse of f is the relation given by

$$g = \{(4, 1), (5, 2), (6, 3), (7, 4)\}$$

It is obvious that the domain of f is now the range of g, and the range of f is now the domain of g. Every function (or relation) has an inverse that is obtained from the original function by interchanging the components of each ordered pair.

DEFINITION

The *inverse* of a relation is found by interchanging the coordinates in each ordered pair that is an element of the relation. That is, if (a, b) is an element of the relation, then (b, a) is an element of the inverse.

Suppose a function f is defined with an equation instead of a list of ordered pairs. We can obtain the equation of the inverse of f by interchanging the role of x and y in the equation for f.

EXAMPLE 1 If the function f is defined by $f(x) = 2x - 3$, find the equation that represents the inverse of f.

SOLUTION Since the inverse of f is obtained by interchanging the components of all the ordered pairs belonging to f, and each ordered pair in f satisfies the equation $y = 2x - 3$, we simply exchange x and y in the equation $y = 2x - 3$ to get the formula for the inverse of f:

$$x = 2y - 3$$

We now solve this equation for y in terms of x:

$$x + 3 = 2y$$

$$\frac{x + 3}{2} = y$$

$$y = \frac{x + 3}{2}$$

The last line gives the equation that defines the inverse of f. Let's compare the graphs of f and its inverse as given above. (See Figure 2.)

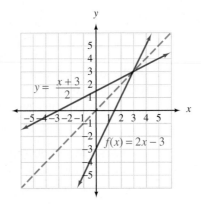

Figure 2

The graphs of f and its inverse have symmetry about the line $y = x$. We say that the graph of the inverse is a *reflection* of the graph of f about the line $y = x$. This is a reasonable result since the one function was obtained from the other by interchanging x and y in the equation. The ordered pairs (a, b) and (b, a) always have symmetry about the line $y = x$.

SYMMETRY PROPERTY OF INVERSES

The graph of the inverse of a relation (or function) will be a reflection of the graph of the original relation (or function) about the line $y = x$.

EXAMPLE 2 Graph the function $y = x^2 - 2$ and its inverse. Give the equation for the inverse.

SOLUTION We can obtain the graph of the inverse of $y = x^2 - 2$ by graphing $y = x^2 - 2$ by the usual methods and then reflecting the graph about the line $y = x$.

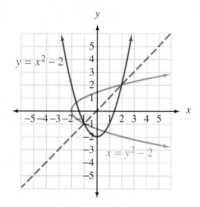

Figure 3

The equation that corresponds to the inverse of $y = x^2 - 2$ is obtained by interchanging x and y to get $x = y^2 - 2$.

We can solve the equation $x = y^2 - 2$ for y in terms of x as follows:

$$x = y^2 - 2$$

$$x + 2 = y^2$$

$$y = \pm\sqrt{x + 2}$$ ∎

USING TECHNOLOGY

GRAPHING AN INVERSE

```
Plot1  Plot2  Plot3
\X1▪T
 Y1▪T²–2
\X2ₜ=
 Y2ₜ=
\X3ₜ=
 Y3ₜ=
\X4ₜ=
```

Figure 4

One way to graph a function and its inverse is to use parametric equations, which we cover in more detail in Section 6.4. To graph the function $y = x^2 - 2$ and its inverse from Example 2, first set your graphing calculator to parametric mode. Then define the following set of parametric equations (Figure 4).

$$X_1 = t; Y_1 = t^2 - 2$$

Set the window variables so that

$$-3 \le t \le 3, \text{ step} = 0.05; -4 \le x \le 4; -4 \le y \le 4$$

Graph the function using the zoom-square command. Your graph should look similar to Figure 5.

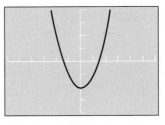

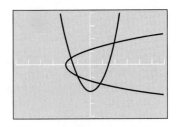

Figure 5   **Figure 6**

To graph the inverse, we need to interchange the roles of x and y for the original function. This is easily done by defining a new set of parametric equations that is just the reverse of the pair given above:

$$X2 = t^2 - 2, Y2 = t$$

Press $\boxed{\text{GRAPH}}$ again, and you should now see the graphs of the original function and its inverse (Figure 6). If you trace to any point on either graph, you can alternate between the two curves to see how the coordinates of the corresponding ordered pairs compare. As Figure 7 illustrates, the coordinates of a point on one graph are reversed for the other graph.

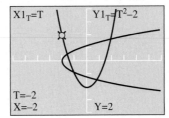

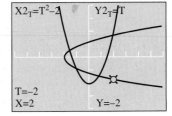

Figure 7

Comparing the graphs from Examples 1 and 2, we observe that the inverse of a function is not always a function. In Example 1, both f and its inverse have graphs that are nonvertical straight lines and therefore both represent functions. In Example 2, the inverse of function f is not a function since a vertical line crosses it in more than one place.

ONE-TO-ONE FUNCTIONS

We can distinguish between those functions with inverses that are also functions and those functions with inverses that are not functions with the following definition.

DEFINITION

A function is a *one-to-one function* if every element in the range comes from exactly one element in the domain.

This definition indicates that a one-to-one function will yield a set of ordered pairs in which no two different ordered pairs have the same second coordinates. For example, the function

$$f = \{(2, 3), (-1, 3), (5, 8)\}$$

is not one-to-one because the element 3 in the range comes from both 2 and -1 in the domain. On the other hand, the function

$$g = \{(5, 7), (3, -1), (4, 2)\}$$

is a one-to-one function because every element in the range comes from only one element in the domain.

HORIZONTAL LINE TEST

If we have the graph of a function, we can determine if the function is one-to-one with the following test. If a horizontal line crosses the graph of a function in more than one place, then the function is not a one-to-one function because the points at which the horizontal line crosses the graph will be points with the same y-coordinates but different x-coordinates. Therefore, the function will have an element in the range (the y-coordinate) that comes from more than one element in the domain (the x-coordinates).

FUNCTIONS WHOSE INVERSES ARE ALSO FUNCTIONS

Because one-to-one functions do not repeat second coordinates, when we reverse the order of the ordered pairs in a one-to-one function, we obtain a relation in which no two ordered pairs have the same first coordinate—by definition, this relation must be a function. In other words, every one-to-one function has an inverse that is itself a function. Because of this, we can use function notation to represent that inverse.

INVERSE FUNCTION NOTATION

If $y = f(x)$ is a one-to-one function, then the inverse of f is also a function and can be denoted by $y = f^{-1}(x)$.

To illustrate, in Example 1 we found the inverse of $f(x) = 2x - 3$ was the function $y = \dfrac{x + 3}{2}$. We can write this inverse function with inverse function notation as

$$f^{-1}(x) = \frac{x + 3}{2}$$

Note: The notation f^{-1} does not represent the reciprocal of f; that is, the -1 in this notation is not an exponent. The notation f^{-1} is defined as representing the inverse function for a one-to-one function.

However, the inverse of the function in Example 2 is not itself a function, so we do not use the notation $f^{-1}(x)$ to represent it.

EXAMPLE 3   Find the inverse of $g(x) = \dfrac{x-4}{x-2}$.

SOLUTION   To find the inverse for g, we begin by replacing $g(x)$ with y to obtain

$$y = \frac{x-4}{x-2} \qquad \text{The original function}$$

To find an equation for the inverse, we exchange x and y.

$$x = \frac{y-4}{y-2} \qquad \text{The inverse of the original function}$$

To solve for y, we first multiply each side by $y - 2$ to obtain

$$x(y - 2) = y - 4$$

$$xy - 2x = y - 4 \qquad \text{Distributive property}$$

$$xy - y = 2x - 4 \qquad \text{Collect all terms containing } y \text{ on the left side}$$

$$y(x - 1) = 2x - 4 \qquad \text{Factor } y \text{ from each term on the left side}$$

$$y = \frac{2x-4}{x-1} \qquad \text{Divide each side by } x - 1$$

Because our original function is one-to-one, as verified by the graph in Figure 8, its inverse is also a function. Therefore, we can use inverse function notation to write

$$g^{-1}(x) = \frac{2x-4}{x-1}$$

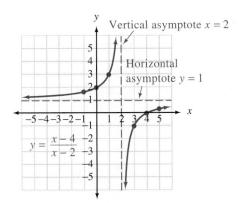

Figure 8

As we mentioned in the introduction to this section, one of the important relationships between a function and its inverse is that an inverse "undoes," or reverses, whatever actions were performed by the function. The next example illustrates this property of inverses.

EXAMPLE 4 Use the function $f(x) = \sqrt[3]{x - 2}$ and its inverse function $f^{-1}(x) = x^3 + 2$ to find the following:

a. $f(10)$
b. $f^{-1}[f(10)]$
c. $f^{-1}[f(x)]$

SOLUTION

a. $f(\mathbf{10}) = \sqrt[3]{\mathbf{10} - 2}$

$\qquad = \sqrt[3]{8}$

$\qquad = 2$

The function f takes an input of 10 and gives us a result of 2.

b. $f^{-1}[f(10)] = f^{-1}[\mathbf{2}]$ $\qquad$ Since $f(10) = 2$

$\qquad\qquad\quad = (\mathbf{2})^3 + 2$ $\qquad$ Replace x with 2 in $x^3 + 2$

$\qquad\qquad\quad = 8 + 2$

$\qquad\qquad\quad = 10$

The inverse f^{-1} takes the output of 2 from f and "undoes" the steps performed by f, with the end result that we return to our original input value of 10.

c. $f^{-1}[f(x)] = f^{-1}[\sqrt[3]{x - 2}]$

$\qquad\qquad\quad = (\sqrt[3]{x - 2})^3 + 2$

$\qquad\qquad\quad = (x - 2) + 2$

$\qquad\qquad\quad = x$

We see that when f^{-1} is applied to the output of f, the actions of f will be reversed, returning us to our original input x. ■

Example 4 illustrates another important property between a function and its inverse.

> If $y = f(x)$ is a one-to-one function with inverse $y = f^{-1}(x)$, then for all x in the domain of f,
>
> $$f^{-1}[f(x)] = x$$
>
> and for all x in the domain of f^{-1},
>
> $$f[f^{-1}(x)] = x$$

FUNCTIONS, RELATIONS, AND INVERSES— A SUMMARY

Here is a summary of some of the things we know about functions, relations, and their inverses:

1. Every function is a relation, but not every relation is a function.
2. Every function has an inverse, but only one-to-one functions have inverses that are also functions.

3. The domain of a function is the range of its inverse, and the range of a function is the domain of its inverse.
4. If $y = f(x)$ is a one-to-one function, then we can use the notation $y = f^{-1}(x)$ to represent its inverse function.
5. The graph of a function and its inverse have symmetry about the line $y = x$.
6. If (a, b) belongs to the function f, then the point (b, a) belongs to its inverse.
7. The inverse of f will "undo," or reverse, the actions performed by f.

GETTING READY FOR CLASS

After reading through the preceding section, respond in your own words and in complete sentences.

a. What is the inverse of a function?

b. What is the relationship between the graph of a function and the graph of its inverse?

c. Explain why only one-to-one functions have inverses that are also functions.

d. Describe the vertical line test, and explain the difference between the vertical line test and the horizontal line test.

PROBLEM SET A.2

For each of the following one-to-one functions, find the equation of the inverse. Write the inverse using the notation $f^{-1}(x)$.

1. $f(x) = 3x - 1$ **2.** $f(x) = 2x - 5$ **3.** $f(x) = x^3$

4. $f(x) = x^3 - 2$ **5.** $f(x) = \dfrac{x - 3}{x - 1}$ **6.** $f(x) = \dfrac{x - 2}{x - 3}$

7. $f(x) = \dfrac{x - 3}{4}$ **8.** $f(x) = \dfrac{x + 7}{2}$ **9.** $f(x) = \dfrac{1}{2}x - 3$

10. $f(x) = \dfrac{1}{3}x + 1$ **11.** $f(x) = \dfrac{2x + 1}{3x + 1}$ **12.** $f(x) = \dfrac{3x + 2}{5x + 1}$

For each of the following relations, sketch the graph of the relation and its inverse, and write an equation for the inverse.

13. $y = 2x - 1$ **14.** $y = 3x + 1$ **15.** $y = x^2 - 3$
16. $y = x^2 + 1$ **17.** $y = x^2 - 2x - 3$ **18.** $y = x^2 + 2x - 3$
19. $y = 4$ **20.** $y = -2$

21. $y = \dfrac{1}{2}x + 2$ **22.** $y = \dfrac{1}{3}x - 1$

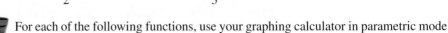 For each of the following functions, use your graphing calculator in parametric mode to graph the function and its inverse.

23. $y = \dfrac{1}{2}x^3$ **24.** $y = x^3 - 2$ **25.** $y = \sqrt{x + 2}$ **26.** $y = \sqrt{x} + 2$

27. Determine if the following functions are one-to-one.

a.

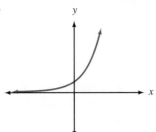

b.

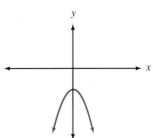

28. Could the following tables of values represent ordered pairs from one-to-one functions? Explain your answer.

a.

x	y
-2	5
-1	4
0	3
1	4
2	5

b.

x	y
1.5	0.1
2.0	0.2
2.5	0.3
3.0	0.4
3.5	0.5

29. If $f(x) = 3x - 2$, then $f^{-1}(x) = \dfrac{x+2}{3}$. Use these two functions to find

    a. $f(2)$ **b.** $f^{-1}(2)$ **c.** $f[f^{-1}(2)]$ **d.** $f^{-1}[f(2)]$

30. If $f(x) = \dfrac{1}{2}x + 5$, then $f^{-1}(x) = 2x - 10$. Use these two functions to find

    a. $f(-4)$ **b.** $f^{-1}(-4)$ **c.** $f[f^{-1}(-4)]$ **d.** $f^{-1}[f(-4)]$

31. Let $f(x) = \dfrac{1}{x}$, and find $f^{-1}(x)$.

32. Let $f(x) = \dfrac{a}{x}$, and find $f^{-1}(x)$. (a is a real number constant.)

33. Reading Tables Evaluate each of the following functions using the functions defined by Tables 1 and 2.

    a. $f[g(-3)]$ **b.** $g[f(-6)]$ **c.** $g[f(2)]$

    d. $f[g(3)]$ **e.** $f[g(-2)]$ **f.** $g[f(3)]$

    g. What can you conclude about the relationship between functions f and g?

34. Reading Tables Use the functions defined in Tables 1 and 2 in Problem 33 to answer the following questions.

    a. What are the domain and range of f?

    b. What are the domain and range of g?

    c. How are the domain and range of f related to the domain and range of g?

    d. Is f a one-to-one function?

    e. Is g a one-to-one function?

TABLE 1

x	$f(x)$
-6	3
2	-3
3	-2
6	4

TABLE 2

x	$g(x)$
-3	2
-2	3
3	-6
4	6

For each of the following functions, find $f^{-1}(x)$. Then show that $f[f^{-1}(x)] = x$.

35. $f(x) = 3x + 5$　　　　　　　　**36.** $f(x) = 6 - 8x$

37. $f(x) = x^3 + 1$　　　　　　　　**38.** $f(x) = x^3 - 8$

EXTENDING THE CONCEPTS

39. Inverse Functions in Words Inverses may also be found by *inverse reasoning*. For example, to find the inverse of $f(x) = 3x + 2$, first list, in order, the operations done to variable x:

a. Multiply by 3.

b. Add 2.

Then, to find the inverse, simply apply the inverse operations, in reverse order, to the variable x:

a. Subtract 2.

b. Divide by 3.

The inverse function then becomes $f^{-1}(x) = \dfrac{x - 2}{3}$. Use this method of "inverse reasoning" to find the inverse of the function $f(x) = \dfrac{x}{7} - 2$.

40. Inverse Functions in Words Refer to the method of *inverse reasoning* explained in Problem 39. Use *inverse reasoning* to find the following inverses:

a. $f(x) = 2x + 7$

b. $f(x) = \sqrt{x} - 9$

c. $f(x) = x^3 - 4$

d. $f(x) = \sqrt{x^3 - 4}$

41. The graphs of a function and its inverse are shown in the figure. Use the graphs to find the following:

a. $f(0)$　　　**b.** $f(1)$　　　**c.** $f(2)$　　　**d.** $f^{-1}(1)$

e. $f^{-1}(2)$　　**f.** $f^{-1}(5)$　　**g.** $f^{-1}[f(2)]$　　**h.** $f[f^{-1}(5)]$

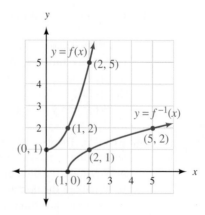

EXPONENTIAL AND LOGARITHMIC FUNCTIONS

Mathematics is the abstract key which turns the lock of the physical universe.

John Polkinghorne

INTRODUCTION

If you have had any problems with or had testing done on your thyroid gland, then you may have come in contact with radioactive iodine-131. Like all radioactive elements, iodine-131 decays naturally. The half-life of iodine-131 is 8 days, which means that every 8 days a sample of iodine-131 will decrease to half of its original amount. The following table and a graph show what happens to a 1,600-microgram sample of iodine-131 over time.

IODINE-131 AS A FUNCTION OF TIME	
t (Days)	*A* (Micrograms)
0	1,600
8	800
16	400
24	200
32	100

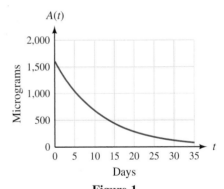

Figure 1

The function represented by the information in the table and Figure 1 is

$$A(t) = 1,600 \cdot 2^{-t/8}$$

It is one of the types of functions we will study in this appendix.

445

B.1 | EXPONENTIAL FUNCTIONS

To obtain an intuitive idea of how exponential functions behave, we can consider the heights attained by a bouncing ball. When a ball used in the game of racquetball is dropped from any height, the first bounce will reach a height that is $\frac{2}{3}$ of the original height. The second bounce will reach $\frac{2}{3}$ of the height of the first bounce, and so on, as shown in Figure 1.

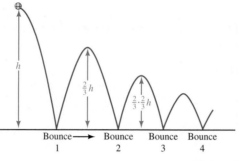
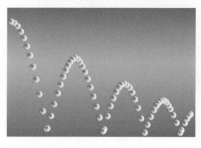

Figure 1

If the ball is dropped initially from a height of 1 meter, then during the first bounce it will reach a height of $\frac{2}{3}$ meter. The height of the second bounce will reach $\frac{2}{3}$ of the height reached on the first bounce. The maximum height of any bounce is $\frac{2}{3}$ of the height of the previous bounce.

Initial height: $h = 1$

Bounce 1: $h = \frac{2}{3}(1) = \frac{2}{3}$

Bounce 2: $h = \frac{2}{3}(\frac{2}{3}) = (\frac{2}{3})^2$

Bounce 3: $h = \frac{2}{3}(\frac{2}{3})^2 = (\frac{2}{3})^3$

Bounce 4: $h = \frac{2}{3}(\frac{2}{3})^3 = (\frac{2}{3})^4$

$\vdots$ $\vdots$

Bounce n: $h = \frac{2}{3}(\frac{2}{3})^{n-1} = (\frac{2}{3})^n$

This last equation is exponential in form. We classify all exponential functions together with the following definition.

DEFINITION

An *exponential function* is any function that can be written in the form

$$f(x) = b^x$$

where b is a positive real number other than 1.

Each of the following is an exponential function:

$$f(x) = 2^x \qquad y = 3^x \qquad f(x) = \left(\frac{1}{4}\right)^x$$

The first step in becoming familiar with exponential functions is to find some values for specific exponential functions.

 EXAMPLE 1 If the exponential functions f and g are defined by

$$f(x) = 2^x \quad \text{and} \quad g(x) = 3^x$$

then

$$f(0) = 2^0 = 1 \qquad\qquad g(0) = 3^0 = 1$$

$$f(1) = 2^1 = 2 \qquad\qquad g(1) = 3^1 = 3$$

$$f(2) = 2^2 = 4 \qquad\qquad g(2) = 3^2 = 9$$

$$f(3) = 2^3 = 8 \qquad\qquad g(3) = 3^3 = 27$$

$$f(-2) = 2^{-2} = \frac{1}{2^2} = \frac{1}{4} \qquad g(-2) = 3^{-2} = \frac{1}{3^2} = \frac{1}{9}$$

$$f(-3) = 2^{-3} = \frac{1}{2^3} = \frac{1}{8} \qquad g(-3) = 3^{-3} = \frac{1}{3^3} = \frac{1}{27}$$

In the introduction to this appendix we indicated that the half-life of iodine-131 is 8 days, which means that every 8 days a sample of iodine-131 will decrease to half of its original amount. If we start with A_0 micrograms of iodine-131, then after t days the sample will contain

$$A(t) = A_0 \cdot 2^{-t/8}$$

micrograms of iodine-131.

 EXAMPLE 2 A patient is administered a 1,200-microgram dose of iodine-131. How much iodine-131 will be in the patient's system after 10 days and after 16 days?

SOLUTION The initial amount of iodine-131 is $A_0 = 1,200$, so the function that gives the amount left in the patient's system after t days is

$$A(t) = 1,200 \cdot 2^{-t/8}$$

After 10 days, the amount left in the patient's system is

$$A(10) = 1,200 \cdot 2^{-10/8}$$

$$= 1,200 \cdot 2^{-1.25}$$

$$\approx 504.5 \text{ micrograms}$$

After 16 days, the amount left in the patient's system is

$$A(16) = 1,200 \cdot 2^{-16/8}$$

$$= 1,200 \cdot 2^{-2}$$

$$= 300 \text{ micrograms}$$

We will now turn our attention to the graphs of exponential functions. Since the notation y is easier to use when graphing, and $y = f(x)$, for convenience we will write the exponential functions as

$$y = b^x$$

EXAMPLE 3 Sketch the graph of the exponential function $y = 2^x$.

SOLUTION Using the results of Example 1, we have the following table. Graphing the ordered pairs given in the table and connecting them with a smooth curve, we have the graph of $y = 2^x$ shown in Figure 2.

x	y
-3	$\frac{1}{8}$
-2	$\frac{1}{4}$
-1	$\frac{1}{2}$
0	1
1	2
2	4
3	8

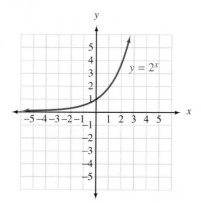

Figure 2

Notice that the graph does not cross the x-axis. It *approaches* the x-axis—in fact, we can get it as close to the x-axis as we want without it actually intersecting the x-axis. For the graph of $y = 2^x$ to intersect the x-axis, we would have to find a value of x that would make $2^x = 0$. Because no such value of x exists, the graph of $y = 2^x$ cannot intersect the x-axis. ■

EXAMPLE 4 Sketch the graph of $y = \left(\frac{1}{3}\right)^x$

SOLUTION The table shown here gives some ordered pairs that satisfy the equation. Using the ordered pairs from the table, we have the graph shown in Figure 3.

x	y
-3	27
-2	9
-1	3
0	1
1	$\frac{1}{3}$
2	$\frac{1}{9}$
3	$\frac{1}{27}$

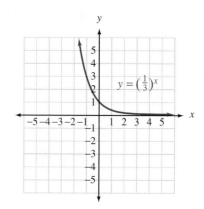

Figure 3 ■

The graphs of all exponential functions have two things in common: (1) each crosses the y-axis at $(0, 1)$ since $b^0 = 1$; and (2) none can cross the x-axis since $b^x = 0$ is impossible because of the restrictions on b.

Figures 4 and 5 show some families of exponential curves to help you become more familiar with them on an intuitive level.

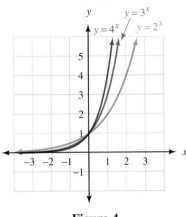

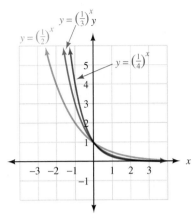

Figure 4   **Figure 5**

Among the many applications of exponential functions are the applications having to do with interest-bearing accounts. Here are the details.

COMPOUND INTEREST

If P dollars are deposited in an account with annual interest rate r, compounded n times per year, then the amount of money in the account after t years is given by the formula

$$A(t) = P\left(1 + \frac{r}{n}\right)^{nt}$$

 **EXAMPLE 5**   Suppose you deposit \$500 in an account with an annual interest rate of 8% compounded quarterly. Find an equation that gives the amount of money in the account after t years. Then find the amount of money in the account after 5 years.

SOLUTION   First we note that $P = 500$ and $r = 0.08$. Interest that is compounded quarterly is compounded 4 times a year, giving us $n = 4$. Substituting these numbers into the preceding formula, we have our function

$$A(t) = 500\left(1 + \frac{0.08}{4}\right)^{4t} = 500(1.02)^{4t}$$

To find the amount after 5 years, we let $t = 5$:

$$A(5) = 500(1.02)^{4 \cdot 5} = 500(1.02)^{20} \approx \$742.97$$

Our answer is found on a calculator and then rounded to the nearest cent.

THE NATURAL EXPONENTIAL FUNCTION

A commonly occurring exponential function is based on a special number we denote with the letter e. The number e is a number like π. It is irrational and occurs in many formulas that describe the world around us. Like π, it can be approximated with a decimal number. Whereas π is approximately 3.1416, e is approximately 2.7183. (If you have a calculator with a key labeled $\boxed{e^x}$, you can use it to find e^1 to find a more accurate approximation to e.) We cannot give a more precise definition of the number e without using some of the topics taught in calculus. For the work we are going to do with the number e, we only need to know that it is an irrational number that is approximately 2.7183.

Here are a table and graph for the natural exponential function.

$$y = f(x) = e^x$$

x	$f(x) = e^x$
-2	$f(-2) = e^{-2} = \dfrac{1}{e^2} \approx 0.135$
-1	$f(-1) = e^{-1} = \dfrac{1}{e} \approx 0.368$
0	$f(0) = e^0 = 1$
1	$f(1) = e^1 = e \approx 2.72$
2	$f(2) = e^2 \approx 7.39$
3	$f(3) = e^3 \approx 20.09$

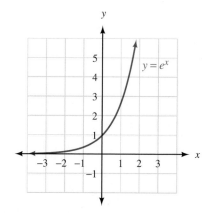

Figure 6

One common application of natural exponential functions is with interest-bearing accounts. In Example 5 we worked with the formula

$$A = P\left(1 + \frac{r}{n}\right)^{nt}$$

which gives the amount of money in an account if P dollars are deposited for t years at annual interest rate r, compounded n times per year. In Example 5 the number of compounding periods was 4. What would happen if we let the number of compounding periods become larger and larger, so that we compounded the interest every day, then every hour, then every second, and so on? If we take this as far as it can go, we end up compounding the interest every moment. When this happens, we have an account with interest that is compounded continuously, and the amount of money in such an account depends on the number e.

CONTINUOUSLY COMPOUNDED INTEREST

If P dollars are deposited in an account with annual interest rate r, compounded continuously, then the amount of money in the account after t years is given by the formula

$$A(t) = Pe^{rt}$$

EXAMPLE 6 Suppose you deposit $500 in an account with an annual interest rate of 8% compounded continuously. Find an equation that gives the amount of money in the account after t years. Then find the amount of money in the account after 5 years.

SOLUTION Since the interest is compounded continuously, we use the formula $A(t) = Pe^{rt}$. Substituting $P = 500$ and $r = 0.08$ into this formula we have

$$A(t) = 500e^{0.08t}$$

After 5 years, this account will contain

$$A(5) = 500e^{0.08 \cdot 5} = 500e^{0.4} \approx \$745.91$$

to the nearest cent. Compare this result with the answer to Example 5.   ■

GETTING READY FOR CLASS

After reading through the preceding section, respond in your own words and in complete sentences.

a. What is an exponential function?

b. In an exponential function, explain why the base b cannot equal 1. (What kind of function would you get if the base was equal to 1?)

c. Explain continuously compounded interest.

d. What characteristics do the graphs of $y = 2^x$ and $y = (\frac{1}{2})^x$ have in common?

PROBLEM SET B.1

Let $f(x) = 3^x$ and $g(x) = (\frac{1}{2})^x$, and evaluate each of the following:

1. $g(0)$ **2.** $f(0)$ **3.** $g(-1)$ **4.** $g(-4)$

5. $f(-3)$ **6.** $f(-1)$ **7.** $f(2) + g(-2)$ **8.** $f(2) - g(-2)$

Graph each of the following functions.

9. $y = 4^x$ **10.** $y = 2^{-x}$

11. $y = 3^{-x}$ **12.** $y = (\frac{1}{3})^{-x}$

13. $y = 2^{x+1}$ **14.** $y = 2^{x-3}$

15. $y = e^x$ **16.** $y = e^{-x}$

Graph each of the following functions on the same coordinate system for positive values of x only.

17. $y = 2x, y = x^2, y = 2^x$ **18.** $y = 3x, y = x^3, y = 3^x$

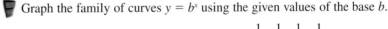 Graph the family of curves $y = b^x$ using the given values of the base b.

19. $b = 2, 4, 6, 8$ **20.** $b = \dfrac{1}{2}, \dfrac{1}{4}, \dfrac{1}{6}, \dfrac{1}{8}$

21. Bouncing Ball Suppose the ball mentioned in the introduction to this section is dropped from a height of 6 feet above the ground. Find an exponential equation that gives the height h the ball will attain during the nth bounce. How high will it bounce on the fifth bounce?

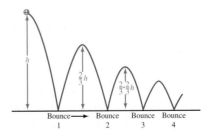

22. **Bouncing Ball** A golf ball is manufactured so that if it is dropped from A feet above the ground onto a hard surface, the maximum height of each bounce will be $\frac{1}{2}$ of the height of the previous bounce. Find an exponential equation that gives the height h the ball will attain during the nth bounce. If the ball is dropped from 10 feet above the ground onto a hard surface, how high will it bounce on the 8th bounce?

23. **Exponential Decay** The half-life of iodine-131 is 8 days. If a patient is administered a 1,400-microgram dose of iodine-131, how much iodine-131 will be in the patient's system after 8 days and after 11 days? (See Example 2.)

24. **Exponential Growth** Automobiles built before 1993 use Freon in their air conditioners. The federal government now prohibits the manufacture of Freon. Because the supply of Freon is decreasing, the price per pound is increasing exponentially. Current estimates put the formula for the price per pound of Freon at $p(t) = 1.89(1.25)^t$, where t is the number of years since 1990. Find the price of Freon in 2000 and 2005.

25. **Compound Interest** Suppose you deposit $1,200 in an account with an annual interest rate of 6% compounded quarterly.
   a. Find an equation that gives the amount of money in the account after t years.
   b. Find the amount of money in the account after 8 years.
   c. If the interest were compounded continuously, how much money would the account contain after 8 years?

26. **Compound Interest** Suppose you deposit $500 in an account with an annual interest rate of 8% compounded monthly.
   a. Find an equation that gives the amount of money in the account after t years.
   b. Find the amount of money in the account after 5 years.
   c. If the interest were compounded continuously, how much money would the account contain after 5 years?

27. **Compound Interest** If $5,000 is placed in an account with an annual interest rate of 12% compounded four times a year, how much money will be in the account 10 years later?

28. **Compound Interest** If $200 is placed in an account with an annual interest rate of 8% compounded twice a year, how much money will be in the account 10 years later?

29. **Bacteria Growth** Suppose it takes 1 day for a certain strain of bacteria to reproduce by dividing in half. If there are 100 bacteria present to begin with, then the total number present after x days will be $f(x) = 100 \cdot 2^x$. Find the total number present after 1 day, 2 days, 3 days, and 4 days.

30. **Bacteria Growth** Suppose it takes 12 hours for a certain strain of bacteria to reproduce by dividing in half. If there are 50 bacteria present to begin with, then the total number present after x days will be $f(x) = 50 \cdot 4^x$. Find the total number present after 1 day, 2 days, and 3 days.

31. **Health Care** In 1990, $699 billion were spent on health care expenditures. The amount of money, E, in billions spent on health care expenditures can be estimated using the function $E(t) = 78.16(1.11)^t$, where t is time in years since 1970 (U.S. Census Bureau).
   a. How close was the estimate determined by the function in estimating the actual amount of money spent on health care expenditures in 1990?
   b. What are the expected health care expenditures in 2005, 2006, and 2007? Round to the nearest billion.

32. Exponential Growth The cost of a can of Coca-Cola on January 1, 1960, was 10 cents. The function below gives the cost of a can of Coca-Cola t years after that.

$$C(t) = 0.10e^{0.0576t}$$

a. Use the function to fill in the table below. (Round to the nearest cent.)

Years Since 1960 t	Cost $C(t)$
0	$0.10
15	
40	
50	
90	

b. Use the table to find the cost of a can of Coca-Cola at the beginning of the year 2000.

33. Value of a Painting A painting is purchased as an investment for $150. If the painting's value doubles every 3 years, then its value is given by the function

$$V(t) = 150 \cdot 2^{t/3} \qquad \text{for } t \geq 0$$

where t is the number of years since it was purchased, and $V(t)$ is its value (in dollars) at that time. Graph this function.

34. Value of a Painting A painting is purchased as an investment for $125. If the painting's value doubles every 5 years, then its value is given by the function

$$V(t) = 125 \cdot 2^{t/5} \qquad \text{for} \qquad t \geq 0$$

where t is the number of years since it was purchased, and $V(t)$ is its value (in dollars) at that time. Graph this function.

35. Value of a Crane The function

$$V(t) = 450{,}000(1 - 0.30)^t$$

where V is value and t is time in years, can be used to find the value of a crane for the first 6 years of use.

a. What is the value of the crane after 3 years and 6 months?

b. State the domain of this function.

c. Sketch the graph of this function.

36. **Value of a Printing Press** The function $V(t) = 375{,}000(1 - 0.25)^t$, where V is value and t is time in years, can be used to find the value of a printing press during the first 7 years of use.
    a. What is the value of the printing press after 4 years and 9 months?
    b. State the domain of this function.
    c. Sketch the graph of this function.

EXTENDING THE CONCEPTS

37. **Drag Racing** A dragster is equipped with a computer. The table gives the speed of the dragster every second during one race at the 1993 Winternationals. Figure 7 is a line graph constructed from the data in the table.

SPEED OF A DRAGSTER	
Elapsed Time (sec)	**Speed (mi/hr)**
0	0.0
1	72.7
2	129.9
3	162.8
4	192.2
5	212.4
6	228.1

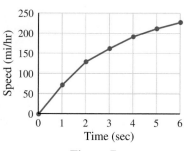

Figure 7

The graph of the following function contains the first point and the last point shown in Figure 7; that is, both (0, 0) and (6, 228.1) satisfy the function. Graph the function to see how close it comes to the other points in Figure 7.

$$s(t) = 250(1 - 1.5^{-t})$$

38. The graphs of two exponential functions are given in Figures 8 and 9. Use the graphs to find the following:
    a. $f(0)$ b. $f(-1)$ c. $f(1)$ d. $g(0)$ e. $g(1)$ f. $g(-1)$

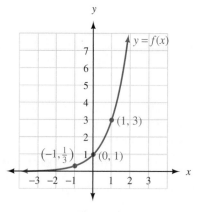

Figure 8

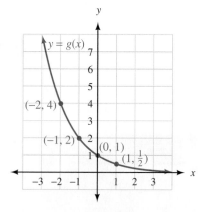

Figure 9

B.2 | LOGARITHMS ARE EXPONENTS

In January 1999, ABC News reported that an earthquake had occurred in Colombia, causing massive destruction. They reported the strength of the quake by indicating that it measured 6.0 on the Richter scale. For comparison, Table 1 gives the Richter magnitude of a number of other earthquakes.

TABLE 1 EARTHQUAKES

Year	Earthquake	Richter Magnitude
1971	Los Angeles	6.6
1985	Mexico City	8.1
1989	San Francisco	7.1
1992	Kobe, Japan	7.2
1994	Northridge	6.6
1999	Armenia, Colombia	6.0

Although the sizes of the numbers in the table do not seem to be very different, the intensity of the earthquakes they measure can be very different. For example, the 1989 San Francisco earthquake was more than 10 times stronger than the 1999 earthquake in Colombia. The reason behind this is that the Richter scale is a *logarithmic scale.* In this section, we start our work with logarithms, which will give you an understanding of the Richter scale. Let's begin.

As you know from your work in the previous section, equations of the form

$$y = b^x \qquad b > 0, b \neq 1$$

are called exponential functions. Any exponential function is a one-to-one function; therefore, it will have an inverse that is also a function. Because the equation of the inverse of a function can be obtained by exchanging x and y in the equation of the original function, the inverse of an exponential function must have the form

$$x = b^y \qquad b > 0, b \neq 1$$

Now, this last equation is actually the equation of a logarithmic function, as the following definition indicates:

DEFINITION

The equation $y = \log_b x$ is read "y is the logarithm to the base b of x" and is equivalent to the equation

$$x = b^y \qquad b > 0, b \neq 1$$

In words, we say "y is the number we raise b to in order to get x."

NOTATION   When an equation is in the form $x = b^y$, it is said to be in exponential form. On the other hand, if an equation is in the form $y = \log_b x$, it is said to be in logarithmic form.

Here are some equivalent statements written in both forms.

Exponential Form		Logarithmic Form
$8 = 2^3$	$\Leftrightarrow$	$\log_2 8 = 3$
$25 = 5^2$	$\Leftrightarrow$	$\log_5 25 = 2$
$0.1 = 10^{-1}$	$\Leftrightarrow$	$\log_{10} 0.1 = -1$
$\dfrac{1}{8} = 2^{-3}$	$\Leftrightarrow$	$\log_2 \dfrac{1}{8} = -3$
$r = z^s$	$\Leftrightarrow$	$\log_z r = s$

 EXAMPLE 1   Solve for x: $\log_3 x = -2$.

SOLUTION   In exponential form the equation looks like this:

$$x = 3^{-2}$$

$$\text{or} \quad x = \frac{1}{9}$$

The solution is $\frac{1}{9}$. ▪

 EXAMPLE 2   Solve $\log_x 4 = 3$.

SOLUTION   Again, we use the definition of logarithms to write the equation in exponential form:

$$4 = x^3$$

Taking the cube root of both sides, we have

$$\sqrt[3]{4} = \sqrt[3]{x^3}$$

$$x = \sqrt[3]{4}$$

The solution set is $\{\sqrt[3]{4}\}$. ▪

 EXAMPLE 3   Solve $\log_8 4 = x$.

SOLUTION   We write the equation again in exponential form:

$$4 = 8^x$$

Since both 4 and 8 can be written as powers of 2, we write them in terms of powers of 2:

$$2^2 = (2^3)^x$$

$$2^2 = 2^{3x}$$

The only way the left and right sides of this last line can be equal is if the exponents are equal—that is, if

$$2 = 3x$$

$$\text{or} \quad x = \frac{2}{3}$$

The solution is $\frac{2}{3}$. We check as follows:

$$\log_8 4 = \frac{2}{3} \Leftrightarrow 4 = 8^{2/3}$$

$$4 = (\sqrt[3]{8})^2$$

$$4 = 2^2$$

$$4 = 4$$

The solution checks when used in the original equation.

GRAPHING LOGARITHMIC FUNCTIONS

Graphing logarithmic functions can be done using the graphs of exponential functions and the fact that the graphs of inverse functions have symmetry about the line $y = x$. Here's an example to illustrate.

EXAMPLE 4 Graph the equation $y = \log_2 x$.

SOLUTION   The equation $y = \log_2 x$ is, by definition, equivalent to the exponential equation

$$x = 2^y$$

which is the equation of the inverse of the function

$$y = 2^x$$

We simply reflect the graph of $y = 2^x$ about the line $y = x$ to get the graph of $x = 2^y$, which is also the graph of $y = \log_2 x$. (See Figure 1.)

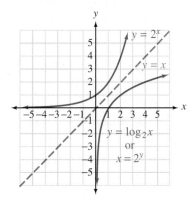

Figure 1

It is apparent from the graph that $y = \log_2 x$ is a function since no vertical line will cross its graph in more than one place. The same is true for all logarithmic equations of the form $y = \log_b x$, where b is a positive number other than 1. Note also that the graph of $y = \log_b x$ will always appear to the right of the y-axis, meaning that x will always be positive in the equation $y = \log_b x$.

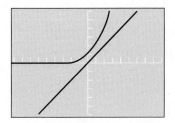

Figure 2

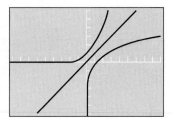

USING TECHNOLOGY →

GRAPHING LOGARITHMIC FUNCTIONS

As demonstrated in Example 4, we can graph the logarithmic function $y = \log_2 x$ as the inverse of the exponential function $y = 2^x$. Your graphing calculator most likely has a command to do this. First, define the exponential function as $Y_1 = 2^x$. To see the line of symmetry, define a second function $Y_2 = x$. Set the window variables so that

$$-6 \leq x \leq 6; \ -6 \leq y \leq 6$$

and use your zoom-square command to graph both functions. Your graph should look similar to the one shown in Figure 2. Now use the appropriate command to graph the inverse of the exponential function defined as Y_1 (Figure 3).

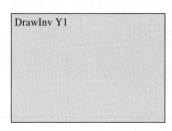

DrawInv Y1

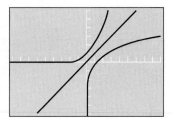

Figure 3

TWO SPECIAL IDENTITIES

If b is a positive real number other than 1, then each of the following is a consequence of the definition of a logarithm:

$$(1) \ \ b^{\log_b x} = x \qquad \text{and} \qquad (2) \ \ \log_b b^x = x$$

The justifications for these identities are similar. Let's consider only the first one. Consider the equation

$$y = \log_b x$$

By definition, it is equivalent to

$$x = b^y$$

Substituting $\log_b x$ for y in the last line gives us

$$x = b^{\log_b x}$$

The next examples in this section show how these two special properties can be used to simplify expressions involving logarithms.

 **EXAMPLE 5** Simplify $\log_2 8$.

SOLUTION Substitute 2^3 for 8:

$$\log_2 8 = \log_2 2^3$$

$$= 3$$

EXAMPLE 6   Simplify $\log_{10} 10{,}000$.

SOLUTION   10,000 can be written as 10^4:

$$\log_{10} 10{,}000 = \log_{10} 10^4$$
$$= 4 \quad \blacksquare$$

EXAMPLE 7   Simplify $\log_b b$ ($b > 0$, $b \neq 1$).

SOLUTION   Since $b^1 = b$, we have

$$\log_b b = \log_b b^1$$
$$= 1 \quad \blacksquare$$

EXAMPLE 8   Simplify $\log_b 1$ ($b > 0$, $b \neq 1$).

SOLUTION   Since $1 = b^0$, we have

$$\log_b 1 = \log_b b^0$$
$$= 0 \quad \blacksquare$$

EXAMPLE 9   Simplify $\log_4 (\log_5 5)$.

SOLUTION   Since $\log_5 5 = 1$,

$$\log_4 (\log_5 5) = \log_4 1$$
$$= 0 \quad \blacksquare$$

APPLICATION

As we mentioned in the introduction to this section, one application of logarithms is in measuring the magnitude of an earthquake. If an earthquake has a shock wave T times greater than the smallest shock wave that can be measured on a seismograph (Figure 4), then the magnitude M of the earthquake, as measured on the Richter scale, is given by the formula

$$M = \log_{10} T$$

(When we talk about the size of a shock wave, we are talking about its amplitude. The amplitude of a wave is half the difference between its highest point and its lowest point.)

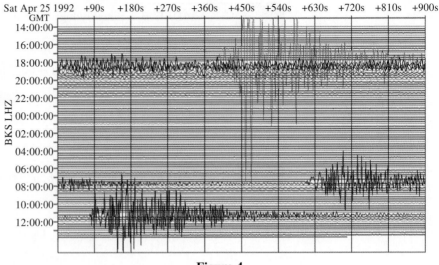

Figure 4

To illustrate the discussion, an earthquake that produces a shock wave that is 10,000 times greater than the smallest shock wave measurable on a seismograph will have a magnitude M on the Richter scale of

$$M = \log_{10} 10{,}000 = 4$$

 **EXAMPLE 10**   If an earthquake has a magnitude of $M = 5$ on the Richter scale, what can you say about the size of its shock wave?

SOLUTION   To answer this question, we put $M = 5$ into the formula $M = \log_{10} T$ to obtain

$$5 = \log_{10} T$$

Writing this equation in exponential form, we have

$$T = 10^5 = 100{,}000$$

We can say that an earthquake that measures 5 on the Richter scale has a shock wave 100,000 times greater than the smallest shock wave measurable on a seismograph.

From Example 10 and the discussion that preceded it, we find that an earthquake of magnitude 5 has a shock wave that is 10 times greater than an earthquake of magnitude 4 because 100,000 is 10 times 10,000.

GETTING READY FOR CLASS

After reading through the preceding section, respond in your own words and in complete sentences.

a. What is a logarithm?

b. What is the relationship between $y = 2^x$ and $y = \log_2 x$? How are their graphs related?

c. Will the graph of $y = \log_b x$ ever appear in the second or third quadrants? Explain why or why not.

d. Explain why $\log_2 0 = x$ has no solution for x.

PROBLEM SET B.2

Write each of the following equations in logarithmic form.

1. $2^4 = 16$ **2.** $3^2 = 9$ **3.** $125 = 5^3$ **4.** $16 = 4^2$

5. $0.01 = 10^{-2}$ **6.** $0.001 = 10^{-3}$ **7.** $2^{-5} = \dfrac{1}{32}$ **8.** $4^{-2} = \dfrac{1}{16}$

9. $\left(\dfrac{1}{2}\right)^{-3} = 8$ **10.** $\left(\dfrac{1}{3}\right)^{-2} = 9$ **11.** $27 = 3^3$ **12.** $81 = 3^4$

Write each of the following equations in exponential form.

13. $\log_{10} 100 = 2$ **14.** $\log_2 8 = 3$

15. $\log_2 64 = 6$ **16.** $\log_2 32 = 5$

17. $\log_8 1 = 0$ **18.** $\log_9 9 = 1$

19. $\log_{10} 0.001 = -3$ **20.** $\log_{10} 0.0001 = -4$

21. $\log_6 36 = 2$ **22.** $\log_7 49 = 2$

23. $\log_5 \dfrac{1}{25} = -2$ **24.** $\log_3 \dfrac{1}{81} = -4$

Solve each of the following equations for x.

25. $\log_3 x = 2$ **26.** $\log_4 x = 3$ **27.** $\log_5 x = -3$ **28.** $\log_2 x = -4$

29. $\log_2 16 = x$ **30.** $\log_3 27 = x$ **31.** $\log_8 2 = x$ **32.** $\log_{25} 5 = x$

33. $\log_x 4 = 2$ **34.** $\log_x 16 = 4$ **35.** $\log_x 5 = 3$ **36.** $\log_x 8 = 2$

Sketch the graph of each of the following logarithmic equations.

37. $y = \log_3 x$ **38.** $y = \log_{1/2} x$ **39.** $y = \log_{1/3} x$ **40.** $y = \log_4 x$

Use your graphing calculator to graph each exponential function with the zoom-square command. Then use the appropriate command to graph the inverse function, and write its equation in logarithmic form.

41. $y = 5^x$ **42.** $y = \left(\dfrac{1}{5}\right)^x$ **43.** $y = 10^x$ **44.** $y = e^x$

Simplify each of the following.

45. $\log_2 16$ **46.** $\log_3 9$ **47.** $\log_{25} 125$ **48.** $\log_9 27$

49. $\log_{10} 1{,}000$ **50.** $\log_{10} 10{,}000$ **51.** $\log_3 3$ **52.** $\log_4 4$

53. $\log_5 1$ **54.** $\log_{10} 1$ **55.** $\log_3 (\log_6 6)$ **56.** $\log_5 (\log_3 3)$

57. $\log_4 [\log_2 (\log_2 16)]$ **58.** $\log_4 [\log_3 (\log_2 8)]$

Measuring Acidity In chemistry, the pH of a solution is defined in terms of logarithms as $pH = -\log_{10} [H^+]$ where $[H^+]$ is the concentration of hydrogen ions in the solution. An acid solution has a pH below 7, and a basic solution has a pH higher than 7.

59. In distilled water, the concentration of hydrogen ions is $[H^+] = 10^{-7}$. What is the pH?

60. Find the pH of a bottle of vinegar if the concentration of hydrogen ions is $[H^+] = 10^{-3}$.

61. A hair conditioner has a pH of 6. Find the concentration of hydrogen ions, $[H^+]$, in the conditioner.

62. If a glass of orange juice has a pH of 4, what is the concentration of hydrogen ions, $[H^+]$, in the orange juice?

63. Magnitude of an Earthquake Find the magnitude M of an earthquake with a shock wave that measures $T = 100$ on a seismograph.

64. Magnitude of an Earthquake Find the magnitude M of an earthquake with a shock wave that measures $T = 100,000$ on a seismograph.

65. Shock Wave If an earthquake has a magnitude of 8 on the Richter scale, how many times greater is its shock wave than the smallest shock wave measurable on a seismograph?

66. Shock Wave If an earthquake has a magnitude of 6 on the Richter scale, how many times greater is its shock wave than the smallest shock wave measurable on a seismograph?

EXTENDING THE CONCEPTS

67. The graph of the exponential function $y = f(x) = b^x$ is shown here. Use the graph to complete parts a through d.

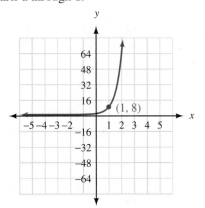

a. Fill in the table.

x	$f(x)$
-1	
0	
1	
2	

b. Fill in the table.

x	$f^{-1}(x)$
	-1
	0
	1
	2

c. Find the equation for $f(x)$.

d. Find the equation for $f^{-1}(x)$.

68. The graph of the exponential function $y = f(x) = b^x$ is shown here. Use the graph to complete parts a through d.

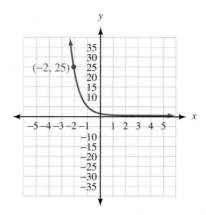

a. Fill in the table.

x	$f(x)$
-1	
0	
1	
2	

b. Fill in the table.

x	$f_-^{-1}(x)$
-1	
0	
1	
2	

c. Find the equation for $f(x)$.

d. Find the equation for $f^{-1}(x)$.

B.3 | PROPERTIES OF LOGARITHMS

If we search for the word *decibel* in *Microsoft Bookshelf 98,* we find the following definition:

A unit used to express relative difference in power or intensity, usually between two acoustic or electric signals, equal to ten times the common logarithm of the ratio of the two levels.

Decibels	Comparable to
10	A light whisper
20	Quiet conversation
30	Normal conversation
40	Light traffic
50	Typewriter, loud conversation
60	Noisy office
70	Normal traffic, quiet train
80	Rock music, subway
90	Heavy traffic, thunder
100	Jet plane at takeoff

The precise definition for a *decibel* is

$$D = 10 \log_{10} \left(\frac{I}{I_0} \right)$$

where I is the intensity of the sound being measured, and I_0 is the intensity of the least audible sound. (Sound intensity is related to the amplitude of the sound wave that models the sound and is given in units of watts per meter^2.) In this section, we will see that the preceding formula can also be written as

$$D = 10(\log_{10} I - \log_{10} I_0)$$

The rules we use to rewrite expressions containing logarithms are called the *properties of logarithms*. There are three of them.

For the following three properties, x, y, and b are all positive real numbers, $b \neq 1$, and r is any real number.

PROPERTY 1

$$\log_b (xy) = \log_b x + \log_b y$$

In words: The logarithm of a *product* is the *sum* of the logarithms.

PROPERTY 2

$$\log_b \left(\frac{x}{y} \right) = \log_b x - \log_b y$$

In words: The logarithm of a *quotient* is the *difference* of the logarithms.

PROPERTY 3

$$\log_b x^r = r \log_b x$$

In words: The logarithm of a number raised to a *power* is the *product* of the power and the logarithm of the number.

PROOF OF PROPERTY 1

To prove property 1, we simply apply the first identity for logarithms given in the preceding section:

$$b^{\log_b xy} = xy = (b^{\log_b x})(b^{\log_b y}) = b^{\log_b x + \log_b y}$$

Since the first and last expressions are equal and the bases are the same, the exponents $\log_b xy$ and $\log_b x + \log_b y$ must be equal. Therefore,

$$\log_b xy = \log_b x + \log_b y \quad \blacksquare$$

The proofs of properties 2 and 3 proceed in much the same manner, so we will omit them here. The examples that follow show how the three properties can be used.

 **EXAMPLE 1**   Expand, using the properties of logarithms: $\log_5 \dfrac{3xy}{z}$.

SOLUTION   Applying property 2, we can write the quotient of $3xy$ and z in terms of a difference:

$$\log_5 \frac{3xy}{z} = \log_5 3xy - \log_5 z$$

Applying property 1 to the product $3xy$, we write it in terms of addition:

$$\log_5 \frac{3xy}{z} = \log_5 3 + \log_5 x + \log_5 y - \log_5 z \quad \blacksquare$$

 EXAMPLE 2   Expand, using the properties of logarithms:

$$\log_2 \frac{x^4}{\sqrt{y} \cdot z^3}$$

SOLUTION   We write $\sqrt{y}$ as $y^{1/2}$ and apply the properties:

$$
\begin{aligned}
\log_2 \frac{x^4}{\sqrt{y} \cdot z^3} &= \log_2 \frac{x^4}{y^{1/2} z^3} & \sqrt{y} = y^{1/2} \\
&= \log_2 x^4 - \log_2 (y^{1/2} \cdot z^3) & \text{Property 2} \\
&= \log_2 x^4 - (\log_2 y^{1/2} + \log_2 z^3) & \text{Property 1} \\
&= \log_2 x^4 - \log_2 y^{1/2} - \log_2 z^3 & \text{Remove parentheses} \\
&= 4 \log_2 x - \frac{1}{2} \log_2 y - 3 \log_2 z & \text{Property 3} \quad \blacksquare
\end{aligned}
$$

We can also use the three properties to write an expression in expanded form as just one logarithm.

 EXAMPLE 3   Write as a single logarithm:

$$2 \log_{10} a + 3 \log_{10} b - \frac{1}{3} \log_{10} c$$

SOLUTION   We begin by applying property 3:

$$
\begin{aligned}
2 \log_{10} a &+ 3 \log_{10} b - \frac{1}{3} \log_{10} c \\
&= \log_{10} a^2 + \log_{10} b^3 - \log_{10} c^{1/3} & \text{Property 3} \\
&= \log_{10} (a^2 \cdot b^3) - \log_{10} c^{1/3} & \text{Property 1} \\
&= \log_{10} \frac{a^2 b^3}{c^{1/3}} & \text{Property 2} \\
&= \log_{10} \frac{a^2 b^3}{\sqrt[3]{c}} & c^{1/3} = \sqrt[3]{c} \quad \blacksquare
\end{aligned}
$$

The properties of logarithms along with the definition of logarithms are useful in solving equations that involve logarithms.

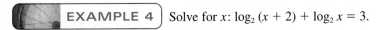 **EXAMPLE 4** Solve for x: $\log_2 (x + 2) + \log_2 x = 3$.

SOLUTION Applying property 1 to the left side of the equation allows us to write it as a single logarithm:

$$\log_2 (x + 2) + \log_2 x = 3$$

$$\log_2 [(x + 2)(x)] = 3$$

The last line can be written in exponential form using the definition of logarithms:

$$(x + 2)(x) = 2^3$$

Solve as usual:

$$x^2 + 2x = 8$$

$$x^2 + 2x - 8 = 0$$

$$(x + 4)(x - 2) = 0$$

$$x + 4 = 0 \quad \text{or} \quad x - 2 = 0$$

$$x = -4 \quad \text{or} \quad x = 2$$

In the previous section we noted the fact that x in the expression $y = \log_b x$ cannot be a negative number. Since substitution of $x = -4$ into the original equation gives

$$\log_2 (-2) + \log_2 (-4) = 3$$

which contains logarithms of negative numbers, we cannot use -4 as a solution. The solution set is $\{2\}$.

 GETTING READY FOR CLASS

After reading through the preceding section, respond in your own words and in complete sentences.

a. Explain why the following statement is false: "The logarithm of a product is the product of the logarithms."

b. Explain why the following statement is false: "The logarithm of a quotient is the quotient of the logarithms."

c. Explain the difference between $\log_b m + \log_b n$ and $\log_b (m + n)$. Are they equivalent?

d. Explain the difference between $\log_b (mn)$ and $(\log_b m)(\log_b n)$. Are they equivalent?

PROBLEM SET B.3

Use the three properties of logarithms given in this section to expand each expression as much as possible.

1. $\log_3 4x$

2. $\log_2 5x$

3. $\log_6 \dfrac{5}{x}$

4. $\log_3 \dfrac{x}{5}$

5. $\log_2 y^5$

6. $\log_7 y^3$

7. $\log_9 \sqrt[3]{z}$

8. $\log_8 \sqrt{z}$

9. $\log_6 x^2 y^4$

10. $\log_{10} x^2 y^4$

11. $\log_5 \sqrt{x} \cdot y^4$

12. $\log_8 \sqrt[3]{xy^6}$

13. $\log_b \dfrac{xy}{z}$

14. $\log_b \dfrac{3x}{y}$

15. $\log_{10} \dfrac{4}{xy}$

16. $\log_{10} \dfrac{5}{4y}$

17. $\log_{10} \dfrac{x^2 y}{\sqrt{z}}$

18. $\log_{10} \dfrac{\sqrt{x} \cdot y}{z^3}$

19. $\log_{10} \dfrac{x^3 \sqrt{y}}{z^4}$

20. $\log_{10} \dfrac{x^4 \sqrt[3]{y}}{\sqrt{z}}$

21. $\log_b \sqrt[3]{\dfrac{x^2 y}{z^4}}$

22. $\log_b \sqrt[4]{\dfrac{x^4 y^3}{z^5}}$

Write each expression as a single logarithm.

23. $\log_b x + \log_b z$

24. $\log_b x - \log_b z$

25. $2 \log_3 x - 3 \log_3 y$

26. $4 \log_2 x + 5 \log_2 y$

27. $\dfrac{1}{2} \log_{10} x + \dfrac{1}{3} \log_{10} y$

28. $\dfrac{1}{3} \log_{10} x - \dfrac{1}{4} \log_{10} y$

29. $3 \log_2 x + \dfrac{1}{2} \log_2 y - \log_2 z$

30. $2 \log_3 x + 3 \log_3 y - \log_3 z$

31. $\dfrac{1}{2} \log_2 x - 3 \log_2 y - 4 \log_2 z$

32. $3 \log_{10} x - \log_{10} y - \log_{10} z$

33. $\dfrac{3}{2} \log_{10} x - \dfrac{3}{4} \log_{10} y - \dfrac{4}{5} \log_{10} z$

34. $3 \log_{10} x - \dfrac{4}{3} \log_{10} y - 5 \log_{10} z$

Solve each of the following equations.

35. $\log_2 x + \log_2 3 = 1$

36. $\log_3 x + \log_3 3 = 1$

37. $\log_3 x - \log_3 2 = 2$

38. $\log_3 x + \log_3 2 = 2$

39. $\log_3 x + \log_3 (x - 2) = 1$

40. $\log_6 x + \log_6 (x - 1) = 1$

41. $\log_3 (x + 3) - \log_3 (x - 1) = 1$

42. $\log_4 (x - 2) - \log_4 (x + 1) = 1$

43. $\log_2 x + \log_2 (x - 2) = 3$

44. $\log_4 x + \log_4 (x + 6) = 2$

45. $\log_8 x + \log_8 (x - 3) = \dfrac{2}{3}$

46. $\log_{27} x + \log_{27} (x + 8) = \dfrac{2}{3}$

47. $\log_5 \sqrt{x} + \log_5 \sqrt{6x + 5} = 1$

48. $\log_2 \sqrt{x} + \log_2 \sqrt{6x + 5} = 1$

49. Food Processing The formula $M = 0.21(\log_{10} a - \log_{10} b)$ is used in the food processing industry to find the number of minutes M of heat processing a certain food should undergo at 250°F to reduce the probability of survival of *Clostridium botulinum* spores. The letter a represents the number of spores per can before heating, and b represents the number of spores per can after heating. Find M if $a = 1$ and $b = 10^{-12}$. Then find M using the same values for a and b in the formula

$$M = 0.21 \log_{10} \dfrac{a}{b}$$

50. **Acoustic Powers** The formula $N = \log_{10} \dfrac{P_1}{P_2}$ is used in radio electronics to find the ratio of the acoustic powers of two electric circuits in terms of their electric powers. Find N if P_1 is 100 and P_2 is 1. Then use the same two values of P_1 and P_2 to find N in the formula $N = \log_{10} P_1 - \log_{10} P_2$.

51. **Henderson–Hasselbalch Formula** Doctors use the Henderson–Hasselbalch formula to calculate the pH of a person's blood. pH is a measure of the acidity and/or the alkalinity of a solution. This formula is represented as

$$\text{pH} = 6.1 + \log_{10}\left(\frac{x}{y}\right)$$

where x is the base concentration and y is the acidic concentration. Rewrite the Henderson–Hasselbalch formula so that the logarithm of a quotient is not involved.

52. **Henderson–Hasselbalch Formula** Refer to the information in the preceding problem about the Henderson–Hasselbalch formula. If most people have a blood pH of 7.4, use the Henderson–Hasselbalch formula to find the ratio of x/y for an average person.

53. **Decibel Formula** Use the properties of logarithms to rewrite the decibel formula $D = 10 \log_{10}\left(\dfrac{I}{I_0}\right)$ so that the logarithm of a quotient is not involved.

54. **Decibel Formula** In the decibel formula $D = 10 \log_{10}\left(\dfrac{I}{I_0}\right)$, the threshold of hearing, I_0, is

$$I_0 = 10^{-12} \text{ watts/meter}^2$$

Substitute 10^{-12} for I_0 in the decibel formula, and then show that it simplifies to

$$D = 10(\log_{10} I + 12)$$

B.4 | COMMON LOGARITHMS AND NATURAL LOGARITHMS

Acid rain was first discovered in the 1960s by Gene Likens and his research team, who studied the damage caused by acid rain to Hubbard Brook in New Hampshire. Acid rain is rain with a pH of 5.6 and below. As you will see as you work your way through this section, pH is defined in terms of common logarithms—one of the topics we present in this section. So, when you are finished with this section, you will have a more detailed knowledge of pH and acid rain.

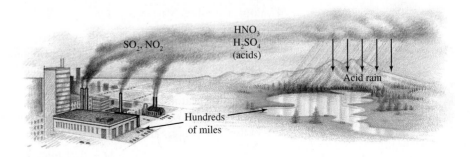

COMMON LOGARITHMS

Two kinds of logarithms occur more frequently than other logarithms. Logarithms with a base of 10 are very common because our number system is a base-10 number system. For this reason, we call base-10 logarithms *common logarithms.*

DEFINITION

A *common logarithm* is a logarithm with a base of 10. Because common logarithms are used so frequently, it is customary, in order to save time, to omit notating the base; that is,

$$\log_{10} x = \log x$$

When the base is not shown, it is assumed to be 10.

Common logarithms of powers of 10 are simple to evaluate. We need only recognize that $\log 10 = \log_{10} 10 = 1$ and apply the third property of logarithms: $\log_b x^r = r \log_b x$.

$$\log 1{,}000 = \log 10^3 = 3 \log 10 = 3(1) = 3$$
$$\log 100 = \log 10^2 = 2 \log 10 = 2(1) = 2$$
$$\log 10 = \log 10^1 = 1 \log 10 = 1(1) = 1$$
$$\log 1 = \log 10^0 = 0 \log 10 = 0(1) = 0$$
$$\log 0.1 = \log 10^{-1} = -1 \log 10 = -1(1) = -1$$
$$\log 0.01 = \log 10^{-2} = -2 \log 10 = -2(1) = -2$$
$$\log 0.001 = \log 10^{-3} = -3 \log 10 = -3(1) = -3$$

To find common logarithms of numbers that are not powers of 10, we use a calculator with a $\boxed{\log}$ key.

Check the following logarithms to be sure you know how to use your calculator. (These answers have been rounded to the nearest ten-thousandth.)

$$\log 7.02 \approx 0.8463$$
$$\log 1.39 \approx 0.1430$$
$$\log 6.00 \approx 0.7782$$
$$\log 9.99 \approx 0.9996$$

 EXAMPLE 1   Use a calculator to find log 2,760.

SOLUTION $\log 2{,}760 \approx 3.4409$

To work this problem on a scientific calculator, we simply enter the number 2,760 and press the key labeled $\boxed{\log}$. On a graphing calculator we press the $\boxed{\log}$ key first, then 2,760.

The 3 in the answer is called the *characteristic,* and the decimal part of the logarithm is called the *mantissa.* ◼

 **EXAMPLE 2** Find log 0.0391.

SOLUTION $\log 0.0391 \approx -1.4078$

 **EXAMPLE 3** Find log 0.00523.

SOLUTION $\log 0.00523 \approx -2.2815$ ▪

EXAMPLE 4 Find x if $\log x = 3.8774$.

SOLUTION We are looking for the number whose logarithm is 3.8774. On a scientific calculator, we enter 3.8774 and press the key labeled $\boxed{10^x}$. On a graphing calculator we press $\boxed{10^x}$ first, then 3.8774. The result is 7,540 to four significant digits. Here's why:

$$\text{If} \quad \log x = 3.8774$$
$$\text{then} \quad x = 10^{3.8774}$$
$$\approx 7,540$$

The number 7,540 is called the *antilogarithm* or just *antilog* of 3.8774; that is, 7,540 is the number whose logarithm is 3.8774. ▪

EXAMPLE 5 Find x if $\log x = -2.4179$.

SOLUTION Using the $\boxed{10^x}$ key, the result is 0.00382.

$$\text{If} \quad \log x = -2.4179$$
$$\text{then} \quad x = 10^{-2.4179}$$
$$\approx 0.00382$$

The antilog of -2.4179 is 0.00382; that is, the logarithm of 0.00382 is -2.4179. ▪

APPLICATIONS

Previously, we found that the magnitude M of an earthquake that produces a shock wave T times larger than the smallest shock wave that can be measured on a seismograph is given by the formula

$$M = \log_{10} T$$

We can rewrite this formula using our shorthand notation for common logarithms as

$$M = \log T$$

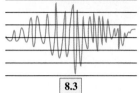

8.3

EXAMPLE 6 The San Francisco earthquake of 1906 is estimated to have measured 8.3 on the Richter scale. The San Fernando earthquake of 1971 measured 6.6 on the Richter scale. Find T for each earthquake, and then give some indication of how much stronger the 1906 earthquake was than the 1971 earthquake.

SOLUTION   For the 1906 earthquake:

$$\text{If } \log T = 8.3, \text{ then } T = 2.00 \times 10^8$$

For the 1971 earthquake:

$$\text{If } \log T = 6.6, \text{ then } T = 3.98 \times 10^6$$

Dividing the two values of T and rounding our answer to the nearest whole number, we have

$$\frac{2.00 \times 10^8}{3.98 \times 10^6} \approx 50$$

The shock wave for the 1906 earthquake was approximately 50 times stronger than the shock wave for the 1971 earthquake.

   In chemistry, the pH of a solution is the measure of the acidity of the solution. The definition for pH involves common logarithms. Here it is:

$$\text{pH} = -\log [\text{H}^+]$$

where $[\text{H}^+]$ is the concentration of hydrogen ions in moles per liter. The range for pH is from 0 to 14. Pure water, a neutral solution, has a pH of 7. An acidic solution, such as vinegar, will have a pH less than 7, and an alkaline solution, such as ammonia, has a pH above 7.

Increasingly alkaline ← NEUTRAL → Increasingly acidic

ACID RAIN

| 14 | 13 | 12 | 11 | 10 | 9 | 8 | 7 | 6 | 5 | 4 | 3 | 2 | 1 | 0 |

Lye — Ammonia — Milk of magnesia — Seawater, baking soda / Lake Ontario — Blood / NEUTRAL — Milk — Mean pH of Adirondack Lakes, 1930 / Upper limit at which some fish affected / "Clean" rain — Mean pH of Adirondack Lakes, 1975 / Average pH of Killarney Lakes, 1971 / Tomato juice / Average pH of rainfall, Toronto, Feb. 1979 — Apple juice / Vinegar / Lemon juice — Most acidic rainfall recorded in U.S. — Battery acid

THE ACID SCALE

EXAMPLE 7 Normal rainwater has a pH of 5.6. What is the concentration of hydrogen ions in normal rainwater?

SOLUTION   Substituting 5.6 for pH in the formula $\text{pH} = -\log [\text{H}^+]$, we have

$5.6 = -\log [\text{H}^+]$	Substitution
$\log [\text{H}^+] = -5.6$	Isolate the logarithm
$[\text{H}^+] = 10^{-5.6}$	Write in exponential form
$\approx 2.5 \times 10^{-6}$ moles per liter	Answer in scientific notation

EXAMPLE 8 The concentration of hydrogen ions in a sample of acid rain known to kill fish is 3.2×10^{-5} moles per liter. Find the pH of this acid rain to the nearest tenth.

SOLUTION Substituting 3.2×10^{-5} for $[H^+]$ in the formula pH $= -\log [H^+]$, we have

$$\text{pH} = -\log [3.2 \times 10^{-5}] \qquad \text{Substitution}$$

$$\text{pH} \approx -(-4.5) \qquad \text{Evaluate the logarithm}$$

$$= 4.5 \qquad \text{Simplify} \blacksquare$$

NATURAL LOGARITHMS

DEFINITION

A *natural logarithm* is a logarithm with a base of e. The natural logarithm of x is denoted by ln x; that is,
$$\ln x = \log_e x$$

The postage stamp shown here contains one of the two special identities we mentioned previously in this chapter, but stated in terms of natural logarithms.

We can assume that all our properties of exponents and logarithms hold for expressions with a base of e since e is a real number. Here are some examples intended to make you more familiar with the number e and natural logarithms.

EXAMPLE 9 Simplify each of the following expressions.

a. $e^0 = 1$
b. $e^1 = e$
c. $\ln e = 1$ In exponential form, $e^1 = e$
d. $\ln 1 = 0$ In exponential form, $e^0 = 1$
e. $\ln e^3 = 3$
f. $\ln e^{-4} = -4$
g. $\ln e^t = t$

EXAMPLE 10 Use the properties of logarithms to expand the expression $\ln Ae^{5t}$.

SOLUTION Since the properties of logarithms hold for natural logarithms, we have

$$\ln Ae^{5t} = \ln A + \ln e^{5t}$$

$$= \ln A + 5t \ln e$$

$$= \ln A + 5t \qquad \text{Because } \ln e = 1 \quad \blacksquare$$

 EXAMPLE 11 If $\ln 2 = 0.6931$ and $\ln 3 = 1.0986$, find

a. $\ln 6$ **b.** $\ln 0.5$ **c.** $\ln 8$

SOLUTION

a. Since $6 = 2 \cdot 3$, we have

$$\ln 6 = \ln (2 \cdot 3)$$

$$= \ln 2 + \ln 3$$

$$= 0.6931 + 1.0986$$

$$= 1.7917$$

b. Writing 0.5 as $\frac{1}{2}$ and applying property 2 for logarithms gives us

$$\ln 0.5 = \ln \frac{1}{2}$$

$$= \ln 1 - \ln 2$$

$$= 0 - 0.6931$$

$$= -0.6931$$

c. Writing 8 as 2^3 and applying property 3 for logarithms, we have

$$\ln 8 = \ln 2^3$$

$$= 3 \ln 2$$

$$= 3(0.6931)$$

$$= 2.0793 \quad \blacksquare$$

 GETTING READY FOR CLASS

After reading through the preceding section, respond in your own words and in complete sentences.

a. What is a common logarithm?

b. What is a natural logarithm?

c. Is e a rational number? Explain.

d. Find $\ln e$, and explain how you arrived at your answer.

PROBLEM SET B.4

Find the following logarithms.

1. log 378   **2.** log 426   **3.** log 37.8   **4.** log 42,600

5. log 3,780   **6.** log 0.4260   **7.** log 0.0378   **8.** log 0.0426

9. log 37,800   **10.** log 4,900   **11.** log 600   **12.** log 900

13. log 2,010   **14.** log 10,200   **15.** log 0.00971   **16.** log 0.0312

17. log 0.0314   **18.** log 0.00052   **19.** log 0.399   **20.** log 0.111

Find x in the following equations.

21. $\log x = 2.8802$ **22.** $\log x = 4.8802$ **23.** $\log x = -2.1198$

24. $\log x = -3.1198$ **25.** $\log x = 3.1553$ **26.** $\log x = 5.5911$

27. $\log x = -5.3497$ **28.** $\log x = -1.5670$

Find x.

29. $\log x = -7.0372$ **30.** $\log x = -4.2000$ **31.** $\log x = 10$

32. $\log x = -1$ **33.** $\log x = -10$ **34.** $\log x = 1$

35. $\log x = 20$ **36.** $\log x = -20$ **37.** $\log x = -2$

38. $\log x = 4$ **39.** $\log x = \log_2 8$ **40.** $\log x = \log_3 9$

Simplify each of the following expressions.

41. $\ln e$ **42.** $\ln 1$ **43.** $\ln e^5$ **44.** $\ln e^{-3}$ **45.** $\ln e^x$ **46.** $\ln e^y$

Use the properties of logarithms to expand each of the following expressions.

47. $\ln 10e^{3t}$ **48.** $\ln 10e^{4t}$ **49.** $\ln Ae^{-2t}$ **50.** $\ln Ae^{-3t}$

If $\ln 2 = 0.6931$, $\ln 3 = 1.0986$, and $\ln 5 = 1.6094$, find each of the following.

51. $\ln 15$ **52.** $\ln 10$ **53.** $\ln \dfrac{1}{3}$ **54.** $\ln \dfrac{1}{5}$

55. $\ln 9$ **56.** $\ln 25$ **57.** $\ln 16$ **58.** $\ln 81$

Measuring Acidity Previously we indicated that the pH of a solution is defined in terms of logarithms as

$$pH = -\log [H^+]$$

where $[H^+]$ is the concentration of hydrogen ions in that solution.

59. Find the pH of orange juice if the concentration of hydrogen ions in the juice is $[H^+] = 6.50 \times 10^{-4}$.

60. Find the pH of milk if the concentration of hydrogen ions in milk is $[H^+] = 1.88 \times 10^{-6}$.

61. Find the concentration of hydrogen ions in a glass of wine if the pH is 4.75.

62. Find the concentration of hydrogen ions in a bottle of vinegar if the pH is 5.75.

The Richter Scale Find the relative size T of the shock wave of earthquakes with the following magnitudes, as measured on the Richter scale.

63. 5.5   **64.** 6.6   **65.** 8.3   **66.** 8.7

67. Shock Wave How much larger is the shock wave of an earthquake that measures 6.5 on the Richter scale than one that measures 5.5 on the same scale?

68. Shock Wave How much larger is the shock wave of an earthquake that measures 8.5 on the Richter scale than one that measures 5.5 on the same scale?

69. Earthquake The chart below is a partial listing of earthquakes that were recorded in Canada in 2000. Complete the chart by computing the magnitude on the Richter scale, M, or the number of times the associated shock wave is larger than the smallest measurable shock wave, T.

Location	Date	Magnitude, M	Shock Wave, T
Moresby Island	January 23	4.0	
Vancouver Island	April 30		1.99×10^5
Quebec City	June 29	3.2	
Mould Bay	November 13	5.2	
St. Lawrence	December 14		5.01×10^3

Source: *National Resources Canada, National Earthquake Hazards Program.*

70. Earthquake On January 6, 2001, an earthquake with a magnitude of 7.7 on the Richter scale hit southern India (*National Earthquake Information Center*). By what factor was this earthquake's shock wave greater than the smallest measurable shock wave?

Depreciation The annual rate of depreciation r on a car that is purchased for P dollars and is worth W dollars t years later can be found from the formula

$$\log(1 - r) = \frac{1}{t} \log \frac{W}{P}$$

71. Find the annual rate of depreciation on a car that is purchased for $9,000 and sold 5 years later for $4,500.

72. Find the annual rate of depreciation on a car that is purchased for $9,000 and sold 4 years later for $3,000.

Two cars depreciate in value according to the following depreciation tables. In each case, find the annual rate of depreciation.

73.

Age in Years	Value in Dollars
new	7,550
5	5,750

74.

Age in Years	Value in Dollars
new	7,550
3	5,750

75. Getting Close to e Use a calculator to complete the following table.

x	$(1 + x)^{1/x}$
1	
0.5	
0.1	
0.01	
0.001	
0.0001	
0.00001	

What number does the expression $(1 + x)^{1/x}$ seem to approach as x gets closer and closer to zero?

76. Getting Close to *e* Use a calculator to complete the following table.

x	$\left(1 + \dfrac{1}{x}\right)^x$
1	
10	
50	
100	
500	
1,000	
10,000	
1,000,000	

What number does the expression $\left(1 + \dfrac{1}{x}\right)^x$ seem to approach as x gets larger and larger?

B.5 | EXPONENTIAL EQUATIONS AND CHANGE OF BASE

For items involved in exponential growth, the time it takes for a quantity to double is called the *doubling time*. For example, if you invest $5,000 in an account that pays 5% annual interest, compounded quarterly, you may want to know how long it will take for your money to double in value. You can find this doubling time if you can solve the equation

$$10,000 = 5,000 \, (1.0125)^{4t}$$

As you will see as you progress through this section, logarithms are the key to solving equations of this type.

Logarithms are very important in solving equations in which the variable appears as an exponent. The equation

$$5^x = 12$$

is an example of one such equation. Equations of this form are called *exponential equations*. Since the quantities 5^x and 12 are equal, so are their common logarithms. We begin our solution by taking the logarithm of both sides:

$$\log 5^x = \log 12$$

We now apply property 3 for logarithms, $\log x^r = r \log x$, to turn x from an exponent into a coefficient:

$$x \log 5 = \log 12$$

Dividing both sides by log 5 gives us

$$x = \frac{\log 12}{\log 5}$$

If we want a decimal approximation to the solution, we can find log 12 and log 5 on a calculator and divide:

$$x \approx \frac{1.0792}{0.6990}$$

$$\approx 1.5439$$

The complete problem looks like this:

$$5^x = 12$$

$$\log 5^x = \log 12$$

$$x \log 5 = \log 12$$

$$x = \frac{\log 12}{\log 5}$$

$$\approx \frac{1.0792}{0.6990}$$

$$\approx 1.5439$$

Here is another example of solving an exponential equation using logarithms.

 EXAMPLE 1   Solve for x: $25^{2x+1} = 15$.

SOLUTION   Taking the logarithm of both sides and then writing the exponent $(2x + 1)$ as a coefficient, we proceed as follows:

$$25^{2x+1} = 15$$

$$\log 25^{2x+1} = \log 15 \qquad \text{Take the log of both sides}$$

$$(2x + 1) \log 25 = \log 15 \qquad \text{Property 3}$$

$$2x + 1 = \frac{\log 15}{\log 25} \qquad \text{Divide by log 25}$$

$$2x = \frac{\log 15}{\log 25} - 1 \qquad \text{Add } -1 \text{ to both sides}$$

$$x = \frac{1}{2}\left(\frac{\log 15}{\log 25} - 1\right) \qquad \text{Multiply both sides by } \tfrac{1}{2}$$

Using a calculator, we can write a decimal approximation to the answer:

$$x \approx \frac{1}{2}\left(\frac{1.1761}{1.3979} - 1\right)$$

$$\approx \frac{1}{2}(0.8413 - 1)$$

$$\approx \frac{1}{2}(-0.1587)$$

$$\approx -0.0794 \quad \blacksquare$$

If you invest P dollars in an account with an annual interest rate r that is compounded n times a year, then t years later the amount of money in that account will be

$$A = P\left(1 + \frac{r}{n}\right)^{nt}$$

 EXAMPLE 2 How long does it take for \$5,000 to double if it is deposited in an account that yields 5% interest compounded once a year?

SOLUTION Substituting $P = 5{,}000$, $r = 0.05$, $n = 1$, and $A = 10{,}000$ into our formula, we have

$$10{,}000 = 5{,}000(1 + 0.05)^t$$

$$10{,}000 = 5{,}000(1.05)^t$$

$$2 = (1.05)^t \qquad \text{Divide by 5,000}$$

This is an exponential equation. We solve by taking the logarithm of both sides:

$$\log 2 = \log(1.05)^t$$

$$= t \log 1.05$$

Dividing both sides by $\log 1.05$, we have

$$t = \frac{\log 2}{\log 1.05}$$

$$\approx 14.2$$

It takes a little over 14 years for \$5,000 to double if it earns 5% interest per year, compounded once a year.

There is a fourth property of logarithms we have not yet considered. This last property allows us to change from one base to another and is therefore called the *change-of-base property*.

PROPERTY 4 (CHANGE OF BASE)

If a and b are both positive numbers other than 1, and if $x > 0$, then

$$\log_a x = \frac{\log_b x}{\log_b a}$$

$$\uparrow \qquad \uparrow$$

$$\text{Base } a \quad \text{Base } b$$

The logarithm on the left side has a base of a, and both logarithms on the right side have a base of b. This allows us to change from base a to any other base b that is a positive number other than 1. Here is a proof of property 4 for logarithms.

PROOF

We begin by writing the identity

$$a^{\log_a x} = x$$

Taking the logarithm base b of both sides and writing the exponent $\log_a x$ as a coefficient, we have

$$\log_b a^{\log_a x} = \log_b x$$

$$\log_a x \, \log_b a = \log_b x$$

Dividing both sides by $\log_b a$, we have the desired result:

$$\frac{\log_a x \, \log_b a}{\log_b a} = \frac{\log_b x}{\log_b a}$$

$$\log_a x = \frac{\log_b x}{\log_b a} \quad \blacksquare$$

We can use this property to find logarithms we could not otherwise compute on our calculators—that is, logarithms with bases other than 10 or e. The next example illustrates the use of this property.

 EXAMPLE 3 Find $\log_8 24$.

SOLUTION   Since we do not have base-8 logarithms on our calculators, we can change this expression to an equivalent expression that contains only base-10 logarithms:

$$\log_8 24 = \frac{\log 24}{\log 8} \qquad \text{Property 4}$$

Don't be confused. We did not just drop the base, we changed to base 10. We could have written the last line like this:

$$\log_8 24 = \frac{\log_{10} 24}{\log_{10} 8}$$

From our calculators, we write

$$\log_8 24 \approx \frac{1.3802}{0.9031}$$

$$\approx 1.5283 \quad \blacksquare$$

APPLICATION

 EXAMPLE 4 Suppose that the population in a small city is 32,000 in the beginning of 1994 and that the city council assumes that the population size t years later can be estimated by the equation

$$P = 32{,}000e^{0.05t}$$

Approximately when will the city have a population of 50,000?

SOLUTION   We substitute 50,000 for P in the equation and solve for t:

$$50{,}000 = 32{,}000e^{0.05t}$$

$$1.5625 = e^{0.05t} \qquad \text{Divide both sides by 32,000}$$

To solve this equation for t, we can take the natural logarithm of each side:

$$\ln 1.5625 = \ln e^{0.05t}$$

$$= 0.05t \ln e \qquad \text{Property 3 for logarithms}$$

$$= 0.05t \qquad \text{Because } \ln e = 1$$

$$t = \frac{\ln 1.5625}{0.05} \qquad \text{Divide each side by 0.05}$$

$$\approx 8.93 \text{ years}$$

We can estimate that the population will reach 50,000 toward the end of 2002.

USING TECHNOLOGY

SOLVING EXPONENTIAL EQUATIONS

We can solve the equation $50{,}000 = 32{,}000e^{0.05t}$ from Example 4 with a graphing calculator by defining the expression on each side of the equation as a function. The solution to the equation will be the x-value of the point where the two graphs intersect.

First define functions $Y_1 = 50000$ and $Y_2 = 32000e^{(0.05x)}$ as shown in Figure 1. Set your window variables so that

$$0 \le x \le 15; \ 0 \le y \le 70{,}000, \ \text{scale} = 10{,}000$$

Graph both functions, then use the appropriate command on your calculator to find the coordinates of the intersection point. From Figure 2 we see that the x-coordinate of this point is $x \approx 8.93$.

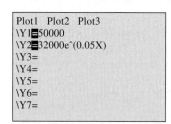

Plot1 Plot2 Plot3
\Y1■50000
\Y2■32000e^(0.05X)
\Y3=
\Y4=
\Y5=
\Y6=
\Y7=

Figure 1

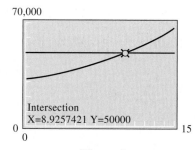

70,000

Intersection
X=8.9257421 Y=50000

0
0   15

Figure 2

GETTING READY FOR CLASS

After reading through the preceding section, respond in your own words and in complete sentences.

a. What is an exponential equation?

b. How do logarithms help you solve exponential equations?

c. What is the change-of-base property?

d. Write an application modeled by the equation $A = 10{,}000 \left(1 + \dfrac{0.08}{2}\right)^{2 \cdot 5}$.

PROBLEM SET B.5

Solve each exponential equation. Use a calculator to write the answer in decimal form.

1. $3^x = 5$	**2.** $4^x = 3$	**3.** $5^x = 3$	**4.** $3^x = 4$
5. $5^{-x} = 12$	**6.** $7^{-x} = 8$	**7.** $12^{-x} = 5$	**8.** $8^{-x} = 7$
9. $8^{x+1} = 4$	**10.** $9^{x+1} = 3$	**11.** $4^{x-1} = 4$	**12.** $3^{x-1} = 9$
13. $3^{2x+1} = 2$	**14.** $2^{2x+1} = 3$	**15.** $3^{1-2x} = 2$	**16.** $2^{1-2x} = 3$
17. $15^{3x-4} = 10$	**18.** $10^{3x-4} = 15$	**19.** $6^{5-2x} = 4$	**20.** $97^{-3x} = 5$

Use the change-of-base property and a calculator to find a decimal approximation to each of the following logarithms.

21. $\log_8 16$	**22.** $\log_9 27$	**23.** $\log_{16} 8$	**24.** $\log_{27} 9$
25. $\log_7 15$	**26.** $\log_3 12$	**27.** $\log_{15} 7$	**28.** $\log_{12} 3$
29. $\log_8 240$	**30.** $\log_6 180$	**31.** $\log_4 321$	**32.** $\log_5 462$

Find a decimal approximation to each of the following natural logarithms.

33. $\ln 345$	**34.** $\ln 3,450$	**35.** $\ln 0.345$	**36.** $\ln 0.0345$
37. $\ln 10$	**38.** $\ln 100$	**39.** $\ln 45,000$	**40.** $\ln 450,000$

41. Compound Interest How long will it take for $500 to double if it is invested at 6% annual interest compounded twice a year?

42. Compounded Interest How long will it take for $500 to double if it is invested at 6% annual interest compounded 12 times a year?

43. Compound Interest How long will it take for $1,000 to triple if it is invested at 12% annual interest compounded 6 times a year?

44. Compound Interest How long will it take for $1,000 to become $4,000 if it is invested at 12% annual interest compounded 6 times a year?

45. Doubling Time How long does it take for an amount of money P to double itself if it is invested at 8% interest compounded 4 times a year?

46. Tripling Time How long does it take for an amount of money P to triple itself if it is invested at 8% interest compounded 4 times a year?

47. Tripling Time If a $25 investment is worth $75 today, how long ago must that $25 have been invested at 6% interest computed twice a year?

48. Doubling Time If a $25 investment is worth $50 today, how long ago must that $25 have been invested at 6% interest computed twice a year?

Recall that if P dollars are invested in an account with annual interest rate r, compounded continuously, then the amount of money in the account after t years is given by the formula $A(t) = Pe^{rt}$.

49. Continuously Compounded Interest Repeat Problem 41 if the interest is compounded continuously.

50. Continuously Compounded Interest Repeat Problem 44 if the interest is compounded continuously.

51. Continuously Compounded Interest How long will it take $500 to triple if it is invested at 6% annual interest, compounded continuously?

52. Continuously Compounded Interest How long will it take $500 to triple if it is invested at 12% annual interest, compounded continuously?

53. Exponential Growth Suppose that the population in a small city is 32,000 at the beginning of 1994 and that the city council assumes that the population size t years later can be estimated by the equation

$$P(t) = 32{,}000e^{0.05t}$$

Approximately when will the city have a population of 64,000?

54. Exponential Growth Suppose the population of a city is given by the equation

$$P(t) = 100{,}000e^{0.05t}$$

where t is the number of years from the present time. How large is the population now? (Now corresponds to a certain value of t. Once you realize what that value of t is, the problem becomes very simple.)

▼ **55. Exponential Growth** Suppose the population of a city is given by the equation

$$P(t) = 15{,}000e^{0.04t}$$

where t is the number of years from the present time. How long will it take for the population to reach 45,000? Solve the problem using algebraic methods first, and then verify your solution with your graphing calculator using the intersection of graphs method.

▼ **56. Exponential Growth** Suppose the population of a city is given by the equation

$$P(t) = 15{,}000e^{0.08t}$$

where t is the number of years from the present time. How long will it take for the population to reach 45,000? Solve the problem using algebraic methods first, and then verify your solution with your graphing calculator using the intersection of graphs method.

EXTENDING THE CONCEPTS

57. Solve the formula $A = Pe^{rt}$ for t.
58. Solve the formula $A = Pe^{-rt}$ for t.
59. Solve the formula $A = P \cdot 2^{-kt}$ for t.
60. Solve the formula $A = P \cdot 2^{kt}$ for t.
61. Solve the formula $A = P(1 - r)^t$ for t.
62. Solve the formula $A = P(1 + r)^t$ for t.

ANSWERS TO ODD-NUMBERED EXERCISES AND CHAPTER TESTS

CHAPTER 1

PROBLEM SET 1.1

1. Acute, complement is 80°, supplement is 170°   **3.** Acute, complement is 45°, supplement is 135°   **5.** Obtuse, complement is $-30°$, supplement is 60°   **7.** We can't tell if x is acute or obtuse (or neither), complement is $90° - x$, supplement is $180° - x$
9. 60°   **11.** 45°   **13.** 50° (Look at it in terms of the big triangle ABC.)
15. Complementary   **17.** 65°   **19.** 180°   **21.** 120°   **23.** 60°
25. 5 (This triangle is called a 3–4–5 right triangle. You will see it again.)   **27.** 15
29. 5   **31.** $3\sqrt{2}$ (Note that this must be a 45°–45°–90° triangle.)
33. 4 (This is a 30°–60°–90° triangle.)   **35.** 1   **37.** $\sqrt{41}$ **39.** 6   **41.** 22.5 ft
43. Longest side 2, third side $\sqrt{3}$ **45.** Shortest side 4, third side $4\sqrt{3}$

47. Shortest side $\dfrac{6}{\sqrt{3}} = 2\sqrt{3}$, longest side $4\sqrt{3}$ **49.** 40 ft   **51.** 101.6 ft²

53. $\dfrac{4\sqrt{2}}{5}$ **55.** 8   **57.** $\dfrac{4}{\sqrt{2}} = 2\sqrt{2}$ **59.** 1,414 ft

61. a. $\sqrt{2}$ inches   **b.** $\sqrt{3}$ inches   **63. a.** $x\sqrt{2}$ **b.** $x\sqrt{3}$ **65.** $\sqrt{3}$ ft
67. See Figure 1 on page 1.

PROBLEM SET 1.2

1.–11. (odd)

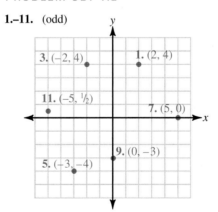

3. $(-2, 4)$ **1.** $(2, 4)$ **11.** $(-5, \frac{1}{2})$ **7.** $(5, 0)$ **9.** $(0, -3)$ **5.** $(-3, -4)$

13.

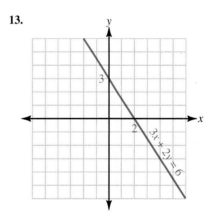

$3x + 2y = 6$

15.

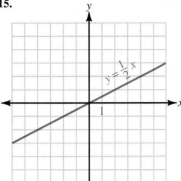

17.

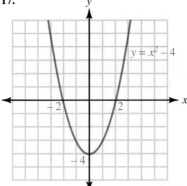

19.

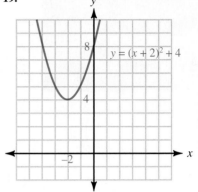

21.

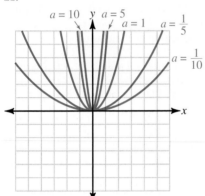

The parabola gets wider; the parabola gets narrower.

23.

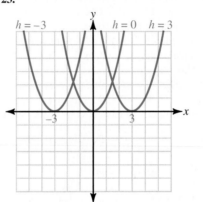

The parabola is shifted to the left; the parabola is shifted to the right.

25.

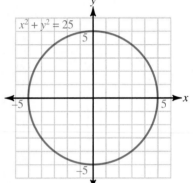

27.

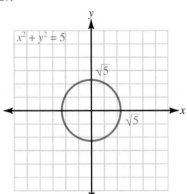

29. (0.5, 0.8660), (0.5, −0.8660)   **31.** (0.7071, 0.7071), (0.7071, −0.7071)   **33.** (−0.8660, 0.5), (−0.8660, −0.5)

35. (5, 0) and (0, 5)   **37.** $\left(-\dfrac{\sqrt{2}}{2}, -\dfrac{\sqrt{2}}{2}\right)$ and $\left(\dfrac{\sqrt{2}}{2}, \dfrac{\sqrt{2}}{2}\right)$ **39.** 5   **41.** 13   **43.** $\sqrt{130}$ **45.** 5

47. −1, 3   **49.** 1.3 mi   **51.** homeplate: (0, 0); first base: (60, 0); second base: (60, 60); third base: (0, 60)

53. QII and QIII   **55.** QIII

57.

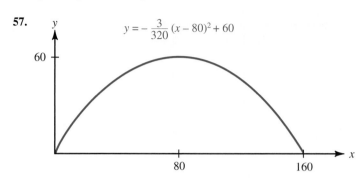

59. 45°   **61.** 30°   **63.** 60°   **65.** 90°   **67.** 300°   **69.** 150°

NOTE   For Problems 71 through 77, other answers are possible.

71.

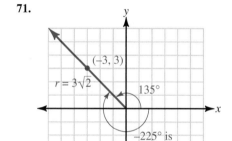

73.

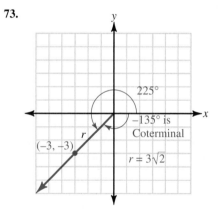

75.

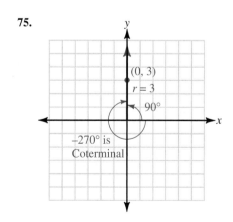

77.

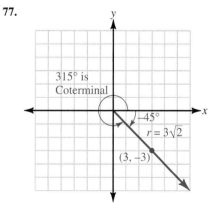

79.

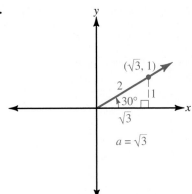

81.

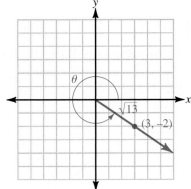

83.

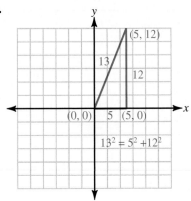

85. Answers will vary.

PROBLEM SET 1.3

	$\sin \theta$	$\cos \theta$	$\tan \theta$	$\cot \theta$	$\sec \theta$	$\csc \theta$
1.	$\dfrac{4}{5}$	$\dfrac{3}{5}$	$\dfrac{4}{3}$	$\dfrac{3}{4}$	$\dfrac{5}{3}$	$\dfrac{5}{4}$
3.	$\dfrac{4}{5}$	$-\dfrac{3}{5}$	$-\dfrac{4}{3}$	$-\dfrac{3}{4}$	$-\dfrac{5}{3}$	$\dfrac{5}{4}$
5.	$\dfrac{12}{13}$	$-\dfrac{5}{13}$	$-\dfrac{12}{5}$	$-\dfrac{5}{12}$	$-\dfrac{13}{5}$	$\dfrac{13}{12}$
7.	$-\dfrac{2}{\sqrt{5}}$	$-\dfrac{1}{\sqrt{5}}$	2	$\dfrac{1}{2}$	$-\sqrt{5}$	$-\dfrac{\sqrt{5}}{2}$
9.	$\dfrac{b}{\sqrt{a^2 + b^2}}$	$\dfrac{a}{\sqrt{a^2 + b^2}}$	$\dfrac{b}{a}$	$\dfrac{a}{b}$	$\dfrac{\sqrt{a^2 + b^2}}{a}$	$\dfrac{\sqrt{a^2 + b^2}}{b}$
11.	0	-1	0	undefined	-1	undefined
13.	$-\dfrac{1}{2}$	$\dfrac{\sqrt{3}}{2}$	$-\dfrac{1}{\sqrt{3}}$	$-\sqrt{3}$	$\dfrac{2}{\sqrt{3}}$	-2
15.	$\dfrac{2}{3}$	$-\dfrac{\sqrt{5}}{3}$	$-\dfrac{2}{\sqrt{5}}$	$-\dfrac{\sqrt{5}}{2}$	$-\dfrac{3}{\sqrt{5}}$	$\dfrac{3}{2}$
17.	$\dfrac{4}{5}$	$\dfrac{3}{5}$	$\dfrac{4}{3}$	$\dfrac{3}{4}$	$\dfrac{5}{3}$	$\dfrac{5}{4}$
19.	$-\dfrac{12}{13}$	$\dfrac{5}{13}$	$-\dfrac{12}{5}$	$-\dfrac{5}{12}$	$\dfrac{13}{5}$	$-\dfrac{13}{12}$

21. $\sin \theta = \dfrac{4}{5}$, $\cos \theta = \dfrac{3}{5}$, $\tan \theta = \dfrac{4}{3}$ **23.** $\sin \theta = \dfrac{5}{\sqrt{34}}$, $\cos \theta = \dfrac{3}{\sqrt{34}}$, $\tan \theta = \dfrac{5}{3}$ **25.** $\sin \theta = 0.6$, $\cos \theta = 0.8$

27.

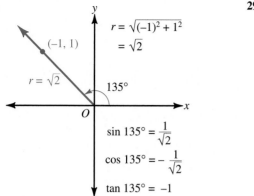

29.

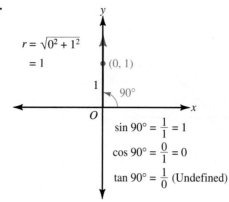

31.

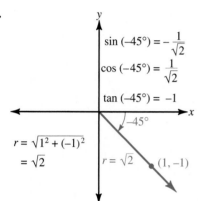

33.

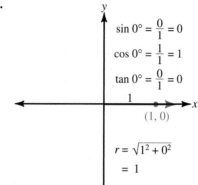

35. QI, QIV   **37.** QIII, QIV   **39.** QIII   **41.** QI (both positive), QIV (both negative)

	$\sin \theta$	$\cos \theta$	$\tan \theta$	$\cot \theta$	$\sec \theta$	$\csc \theta$
43.	$\dfrac{12}{13}$	$\dfrac{5}{13}$	$\dfrac{12}{5}$	$\dfrac{5}{12}$	$\dfrac{13}{5}$	$\dfrac{13}{12}$
45.	$-\dfrac{7}{25}$	$\dfrac{24}{25}$	$-\dfrac{7}{24}$	$-\dfrac{24}{7}$	$\dfrac{25}{24}$	$-\dfrac{25}{7}$
47.	$\dfrac{3}{5}$	$\dfrac{4}{5}$	$\dfrac{3}{4}$	$\dfrac{4}{3}$	$\dfrac{5}{4}$	$\dfrac{5}{3}$
49.	$-\dfrac{20}{29}$	$-\dfrac{21}{29}$	$\dfrac{20}{21}$	$\dfrac{21}{20}$	$-\dfrac{29}{21}$	$-\dfrac{29}{20}$
51.	$-\dfrac{12}{13}$	$\dfrac{5}{13}$	$-\dfrac{12}{5}$	$-\dfrac{5}{12}$	$\dfrac{13}{5}$	$-\dfrac{13}{12}$
53.	$\dfrac{2}{\sqrt{5}}$	$\dfrac{1}{\sqrt{5}}$	2	$\dfrac{1}{2}$	$\sqrt{5}$	$\dfrac{\sqrt{5}}{2}$
55.	$-\dfrac{1}{2}$	$\dfrac{\sqrt{3}}{2}$	$-\dfrac{1}{\sqrt{3}}$	$-\sqrt{3}$	$\dfrac{2}{\sqrt{3}}$	-2
57.	$\dfrac{1}{\sqrt{5}}$	$-\dfrac{2}{\sqrt{5}}$	$-\dfrac{1}{2}$	-2	$-\dfrac{\sqrt{5}}{2}$	$\sqrt{5}$
59.	$\dfrac{a}{\sqrt{a^2 + b^2}}$	$\dfrac{b}{\sqrt{a^2 + b^2}}$	$\dfrac{a}{b}$	$\dfrac{b}{a}$	$\dfrac{\sqrt{a^2 + b^2}}{b}$	$\dfrac{\sqrt{a^2 + b^2}}{a}$

61. $90°$ **63.** $225°$ **65.** $\sin \theta = \dfrac{2}{\sqrt{5}}$, $\cos \theta = \dfrac{1}{\sqrt{5}}$ **67.** $\sin \theta = \dfrac{3}{\sqrt{10}}$, $\tan \theta = -3$

69.

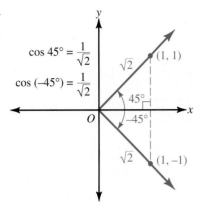

$$\cos 45° = \frac{1}{\sqrt{2}}$$

$$\cos (-45°) = \frac{1}{\sqrt{2}}$$

71. ± 4

PROBLEM SET 1.4

1. $\dfrac{1}{7}$ **3.** $-\dfrac{3}{2}$ **5.** $-\sqrt{2}$ **7.** $\dfrac{1}{x}$ $(x \neq 0)$ **9.** $\dfrac{5}{4}$ **11.** $-\dfrac{1}{2}$ **13.** $\dfrac{1}{a}$ **15.** $-\dfrac{3}{4}$ **17.** $\dfrac{12}{5}$

19. $\dfrac{1}{2}$ **21.** 8 **23.** $\dfrac{12}{5}$ **25.** $-\dfrac{13}{5}$ **27.** $\dfrac{4}{5}$ **29.** $-\dfrac{2\sqrt{2}}{3}$ **31.** $-\dfrac{3}{5}$ **33.** $\dfrac{1}{2}$ **35.** $-\dfrac{2}{\sqrt{5}}$

37. $\dfrac{1}{2\sqrt{2}}$ **39.** $-\dfrac{17}{15}$ **41.** $\dfrac{29}{20}$

	$\sin \theta$	$\cos \theta$	$\tan \theta$	$\cot \theta$	$\sec \theta$	$\csc \theta$
43.	$\dfrac{5}{13}$	$\dfrac{12}{13}$	$\dfrac{5}{12}$	$\dfrac{12}{5}$	$\dfrac{13}{12}$	$\dfrac{13}{5}$
45.	$-\dfrac{1}{2}$	$\dfrac{\sqrt{3}}{2}$	$-\dfrac{1}{\sqrt{3}}$	$-\sqrt{3}$	$\dfrac{2}{\sqrt{3}}$	-2
47.	$\dfrac{\sqrt{3}}{2}$	$\dfrac{1}{2}$	$\sqrt{3}$	$\dfrac{1}{\sqrt{3}}$	2	$\dfrac{2}{\sqrt{3}}$
49.	$-\dfrac{3}{\sqrt{13}}$	$\dfrac{2}{\sqrt{13}}$	$-\dfrac{3}{2}$	$-\dfrac{2}{3}$	$\dfrac{\sqrt{13}}{2}$	$-\dfrac{\sqrt{13}}{3}$
51.	$-\dfrac{2\sqrt{2}}{3}$	$-\dfrac{1}{3}$	$2\sqrt{2}$	$\dfrac{1}{2\sqrt{2}}$	-3	$-\dfrac{3}{2\sqrt{2}}$
53.	$\dfrac{1}{a}$	$\dfrac{\sqrt{a^2 - 1}}{a}$	$\dfrac{1}{\sqrt{a^2 - 1}}$	$\sqrt{a^2 - 1}$	$\dfrac{a}{\sqrt{a^2 - 1}}$	a
55.	0.23	0.97	0.24	4.17	1.03	4.35
57.	0.59	-0.81	-0.73	-1.37	-1.24	1.69

Your answers for 55 and 57 may differ from the answers here in the hundredths column if you found the reciprocal of an unrounded number.

NOTE   As Problems 59 and 61 indicate, the slope of a line through the origin is the same as the tangent of the angle the line makes with the positive *x*-axis.

59. 3 **61.** m

PROBLEM SET 1.5

1. $\pm\sqrt{1-\sin^2\theta}$ **3.** $\pm\dfrac{\sqrt{1-\sin^2\theta}}{\sin\theta}$ **5.** $\dfrac{1}{\cos\theta}$ **7.** $\pm\dfrac{\sqrt{1-\cos^2\theta}}{\cos\theta}$ **9.** $\dfrac{\cos\theta}{\sin^2\theta}$ **11.** $\dfrac{1}{\cos\theta}$

13. $\dfrac{\sin\theta}{\cos\theta}$ **15.** $\dfrac{1}{\sin\theta}$ **17.** $\dfrac{\sin^2\theta}{\cos^2\theta}$ **19.** $\sin^2\theta$ **21.** $\dfrac{\sin\theta+1}{\cos\theta}$ **23.** $2\cos\theta$ **25.** $\cos\theta$

27. $\dfrac{\sin^2\theta+\cos\theta}{\sin\theta\cos\theta}$ **29.** $\dfrac{\cos\theta-\sin\theta}{\sin\theta\cos\theta}$ **31.** $\dfrac{\sin\theta\cos\theta+1}{\cos\theta}$ **33.** $\dfrac{\cos^2\theta}{\sin\theta}$ **35.** $\sin^2\theta+7\sin\theta+12$

37. $8\cos^2\theta+2\cos\theta-15$ **39.** $\cos^2\theta$ **41.** $1-\tan^2\theta$ **43.** $1-2\sin\theta\cos\theta$ **45.** $\sin^2\theta-8\sin\theta+16$

47. $2|\sec\theta|$ **49.** $3|\cos\theta|$ **51.** $6|\tan\theta|$ **53.** $4|\sec\theta|$ **55.** $6|\cos\theta|$ **57.** $9|\tan\theta|$

For problems 59 through 91, a few selected solutions are given here. See the Solutions Manual for solutions to problems not shown.

59. $\cos\theta\tan\theta=\cos\theta\cdot\dfrac{\sin\theta}{\cos\theta}=\sin\theta$

63. $\dfrac{\sin\theta}{\csc\theta}=\dfrac{\sin\theta}{\dfrac{1}{\sin\theta}}$

$=\sin\theta\cdot\dfrac{\sin\theta}{1}$

$=\sin^2\theta$

71. $\sin\theta\tan\theta+\cos\theta=\sin\theta\cdot\dfrac{\sin\theta}{\cos\theta}+\cos\theta$

$=\dfrac{\sin^2\theta}{\cos\theta}+\cos\theta$

$=\dfrac{\sin^2\theta+\cos^2\theta}{\cos\theta}$

$=\dfrac{1}{\cos\theta}$

$=\sec\theta$

75. $\csc\theta-\sin\theta=\dfrac{1}{\sin\theta}-\sin\theta$

$=\dfrac{1}{\sin\theta}-\dfrac{\sin^2\theta}{\sin\theta}$

$=\dfrac{1-\sin^2\theta}{\sin\theta}$

$=\dfrac{\cos^2\theta}{\sin\theta}$

83. $\dfrac{\cos\theta}{\sec\theta}+\dfrac{\sin\theta}{\csc\theta}=\dfrac{\cos\theta}{\dfrac{1}{\cos\theta}}+\dfrac{\sin\theta}{\dfrac{1}{\sin\theta}}$

$=\cos^2\theta+\sin^2\theta$

$=1$

87. $\sin\theta(\sec\theta+\csc\theta)=\sin\theta\cdot\sec\theta+\sin\theta\cdot\csc\theta$

$=\sin\theta\cdot\dfrac{1}{\cos\theta}+\sin\theta\cdot\dfrac{1}{\sin\theta}$

$=\dfrac{\sin\theta}{\cos\theta}+\dfrac{\sin\theta}{\sin\theta}$

$=\tan\theta+1$

CHAPTER 1 TEST

1. $20°, 110°$ **2.** $3\sqrt{3}$ **3.** 3 **4.** $h=5\sqrt{3}, r=5\sqrt{6}, y=5, x=10$ **5.** $x=6, h=3\sqrt{3}, s=3\sqrt{3}, r=3\sqrt{6}$
6. $2\sqrt{13}$ **7.** $90°$ **8.** $5/2$ and $5\sqrt{3}/2$ **9.** $15\sqrt{2}$ ft **10.** $108°$ **11.** $120°$
12. **13.** 13 **14.** $\sqrt{a^2+b^2}$ **15.** $-5, 1$

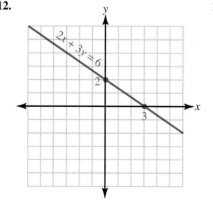

16.

$$y = -\frac{1}{98}(x - 70)^2 + 50$$

17. $\sin 90° = 1$, $\cos 90° = 0$, $\tan 90°$ is undefined

18. $\sin(-45°) = -1/\sqrt{2}$, $\cos(-45°) = 1/\sqrt{2}$, $\tan(-45°) = -1$ **19.** QIV   **20.** QII

	$\sin\theta$	$\cos\theta$	$\tan\theta$	$\cot\theta$	$\sec\theta$	$\csc\theta$
21.	$\dfrac{4}{5}$	$-\dfrac{3}{5}$	$-\dfrac{4}{3}$	$-\dfrac{3}{4}$	$-\dfrac{5}{3}$	$\dfrac{5}{4}$
22.	$-\dfrac{1}{\sqrt{10}}$	$-\dfrac{3}{\sqrt{10}}$	$\dfrac{1}{3}$	3	$-\dfrac{\sqrt{10}}{3}$	$-\sqrt{10}$
23.	$\dfrac{1}{2}$	$-\dfrac{\sqrt{3}}{2}$	$-\dfrac{1}{\sqrt{3}}$	$-\sqrt{3}$	$-\dfrac{2}{\sqrt{3}}$	2
24.	$-\dfrac{12}{13}$	$-\dfrac{5}{13}$	$\dfrac{12}{5}$	$\dfrac{5}{12}$	$-\dfrac{13}{5}$	$-\dfrac{13}{12}$

25. $\sin\theta = -2/\sqrt{5}$, $\cos\theta = 1/\sqrt{5}$ **26.** $-4/3$ **27.** $-1/2$ **28.** $1/27$ **29.** $\cos\theta = 1/3$, $\sin\theta = -2\sqrt{2}/3$,

$\tan\theta = -2\sqrt{2}$ **30.** $\cos\theta = \dfrac{\sqrt{a^2 - 1}}{a}$, $\csc\theta = a$, $\cot\theta = \sqrt{a^2 - 1}$ **31.** $\sin^2\theta - 4\sin\theta - 21$

32. $1 - 2\sin\theta\cos\theta$ **33.** $\cos^2\theta/\sin\theta$ **34.** $2|\cos\theta|$

For Problems 35 through 38, solutions are given in the Solutions Manual.

CHAPTER 2

PROBLEM SET 2.1

	$\sin A$	$\cos A$	$\tan A$	$\cot A$	$\sec A$	$\csc A$
1.	$\dfrac{4}{5}$	$\dfrac{3}{5}$	$\dfrac{4}{3}$	$\dfrac{3}{4}$	$\dfrac{5}{3}$	$\dfrac{5}{4}$
3.	$\dfrac{2}{\sqrt{5}}$	$\dfrac{1}{\sqrt{5}}$	2	$\dfrac{1}{2}$	$\sqrt{5}$	$\dfrac{\sqrt{5}}{2}$
5.	$\dfrac{2}{3}$	$\dfrac{\sqrt{5}}{3}$	$\dfrac{2}{\sqrt{5}}$	$\dfrac{\sqrt{5}}{2}$	$\dfrac{3}{\sqrt{5}}$	$\dfrac{3}{2}$

	$\sin A$	$\cos A$	$\tan A$	$\sin B$	$\cos B$	$\tan B$
7.	$\dfrac{5}{6}$	$\dfrac{\sqrt{11}}{6}$	$\dfrac{5}{\sqrt{11}}$	$\dfrac{\sqrt{11}}{6}$	$\dfrac{5}{6}$	$\dfrac{\sqrt{11}}{5}$
9.	$\dfrac{1}{\sqrt{2}}$	$\dfrac{1}{\sqrt{2}}$	1	$\dfrac{1}{\sqrt{2}}$	$\dfrac{1}{\sqrt{2}}$	1
11.	$\dfrac{3}{5}$	$\dfrac{4}{5}$	$\dfrac{3}{4}$	$\dfrac{4}{5}$	$\dfrac{3}{5}$	$\dfrac{4}{3}$
13.	$\dfrac{\sqrt{3}}{2}$	$\dfrac{1}{2}$	$\sqrt{3}$	$\dfrac{1}{2}$	$\dfrac{\sqrt{3}}{2}$	$\dfrac{1}{\sqrt{3}}$

15. (4, 3), $\sin A = \dfrac{3}{5}$, $\cos A = \dfrac{4}{5}$, $\tan A = \dfrac{3}{4}$ **17.** $80°$ **19.** $82°$ **21.** $90° - x$ **23.** x

25.

x	$\sin x$	$\csc x$
$0°$	0	undefined
$30°$	$\dfrac{1}{2}$	2
$45°$	$\dfrac{1}{\sqrt{2}}$	$\sqrt{2}$
$60°$	$\dfrac{\sqrt{3}}{2}$	$\dfrac{2}{\sqrt{3}}$
$90°$	1	1

27. 2 **29.** 3 **31.** $\dfrac{2 + \sqrt{3}}{2}$ **33.** 0 **35.** $4 + 2\sqrt{3}$ **37.** 1 **39.** $2\sqrt{3}$

41. $-\dfrac{3\sqrt{3}}{2}$ **43.** $\sqrt{2}$ **45.** $\dfrac{2}{\sqrt{3}}$ **47.** $\dfrac{2}{\sqrt{3}}$ **49.** 1 **51.** $\sqrt{2}$

	$\sin A$	$\cos A$	$\sin B$	$\cos B$
53.	0.60	0.80	0.80	0.60
55.	0.96	0.28	0.28	0.96

57. $\dfrac{1}{\sqrt{3}}, \dfrac{\sqrt{2}}{\sqrt{3}}$ **59.** $\dfrac{1}{\sqrt{3}}, \dfrac{\sqrt{2}}{\sqrt{3}}$ **61.** 5 **63.** $-1, 3$

65.

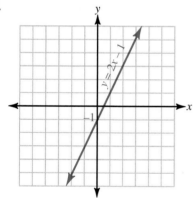

67. See Example 7, Section 1.2 **69.** $225°$ **71.** $60°$

PROBLEM SET 2.2

1. $64° \, 9'$ **3.** $89° \, 40'$ **5.** $106° \, 49'$ **7.** $55° \, 48'$ **9.** $59° \, 43'$ **11.** $53° \, 50'$ **13.** $39° \, 50'$ **15.** $35° \, 24'$
17. $16° \, 15'$ **19.** $92° \, 33'$ **21.** $19° \, 54'$ **23.** $45.2°$ **25.** $62.6°$ **27.** $17.33°$ **29.** $48.45°$ **31.** 0.4571
33. 0.9511 **35.** 21.3634 **37.** 1.6643 **39.** 1.5003 **41.** 4.0906 **43.** 0.9100 **45.** 0.9083 **47.** 0.8355
49. 1.4370

51.

X	sin X	csc X
0°	0	Error
30°	0.5	2
45°	0.7071	1.4142
60°	0.8660	1.1547
90°	1	1

53.

X	sin X	cos X	tan X
0°	0	1	0
15°	0.2588	0.9659	0.2679
30°	0.5	0.8660	0.5774
45°	0.7071	0.7071	1
60°	0.8660	0.5	1.7321
75°	0.9659	0.2588	3.7321
90°	1	0	Error

55. 12.3°   **57.** 34.5°   **59.** 78.9°   **61.** 11.1°   **63.** 33.3°   **65.** 55.5°   **67.** 44° 44′   **69.** 65° 43′   **71.** 10° 10′
73. 9° 9′   **75.** 0.3907   **77.** 1.2134   **79.** 0.0787   **81.** 1   **83.** 1   **85.** You should get an error message. The
sine of an angle can never exceed 1.   **87.** You get an error message; tan 90° is undefined.

89. a.

X	tan X
87°	19.1
87.5°	22.9
88°	28.6
88.5°	38.2
89°	57.3
89.5°	114.6
90°	undefined

b.

X	tan X
89.4°	95.5
89.5°	114.6
89.6°	143.2
89.7°	191.0
89.8°	286.5
89.9°	573.0
90°	undefined

91. $\sin \theta = -\dfrac{2}{\sqrt{13}}$, $\cos \theta = \dfrac{3}{\sqrt{13}}$, $\tan \theta = -\dfrac{2}{3}$ **93.** $\sin 90° = 1$, $\cos 90° = 0$, $\tan 90°$ is undefined

95. $\sin \theta = -\dfrac{12}{13}$, $\tan \theta = \dfrac{12}{5}$, $\cot \theta = \dfrac{5}{12}$, $\sec \theta = -\dfrac{13}{5}$, $\csc \theta = -\dfrac{13}{12}$ **97.** QII

PROBLEM SET 2.3

1. 10 ft   **3.** 39 m   **5.** 2.13 ft   **7.** 8.535 yd   **9.** 32°   **11.** 30°   **13.** 59.20°
15. $B = 65°$, $a = 10$ m, $b = 22$ m   **17.** $B = 57.4°$, $b = 67.9$ inches, $c = 80.6$ inches
19. $B = 79° 18′$, $a = 1.121$ cm, $c = 6.037$ cm   **21.** $A = 14°$, $a = 1.4$ ft, $b = 5.6$ ft
23. $A = 63° 30′$, $a = 650$ mm, $c = 726$ mm   **25.** $A = 66.55°$, $b = 2.356$ mi, $c = 5.921$ mi
27. $A = 23°$, $B = 67°$, $c = 95$ ft   **29.** $A = 42.8°$, $B = 47.2°$, $b = 2.97$ cm
31. $A = 61.42°$, $B = 28.58°$, $a = 22.41$ inches   **33.** 49°   **35.** 11   **37.** 36   **39.** $x = 79$, $h = 40$
41. 42°   **43.** $x = 7.4$, $y = 6.2$ **45.** $h = 18$, $x = 11$ **47.** 15   **49.** 35.3°   **51.** 35.3°   **53.** 30.7 ft

55. 201.5 ft   **57. a.** 159.8 ft   **b.** 195.8 ft   **c.** 40.8 ft   **59.** 1/4   **61.** $\dfrac{2\sqrt{2}}{3}$ **63.** $-\dfrac{\sqrt{21}}{5}$

	sin θ	cos θ	tan θ	cot θ	sec θ	csc θ
65.	$\dfrac{\sqrt{3}}{2}$	$-\dfrac{1}{2}$	$-\sqrt{3}$	$-\dfrac{1}{\sqrt{3}}$	-2	$\dfrac{2}{\sqrt{3}}$
67.	$-\dfrac{\sqrt{3}}{2}$	$-\dfrac{1}{2}$	$\sqrt{3}$	$\dfrac{1}{\sqrt{3}}$	-2	$-\dfrac{2}{\sqrt{3}}$

69. The angle is approximately 60°.

PROBLEM SET 2.4

1. 39 cm, 69°   **3.** 78.4°   **5.** 39 ft   **7.** 36.6°   **9.** 55.1°   **11.** 61 cm   **13. a.** 800 ft   **b.** 200 ft   **c.** 14.0°
15. 6.4 ft   **17.** 31 mi, N 78° E   **19.** 39 mi   **21.** 63.4 mi north, 48.0 mi west   **23.** 161 ft   **25.** 78.9 ft
27. 26 ft   **29.** 4,000 mi   **31.** 6.2 mi   **33.** $\theta_1 = 45.00°$, $\theta_2 = 35.26°$, $\theta_3 = 30.00°$ **35.** $1 - 2 \sin \theta \cos \theta$

For Problems 37 through 41, solutions are given in the Solutions Manual.
43. Answers will vary.

PROBLEM SET 2.5

1.

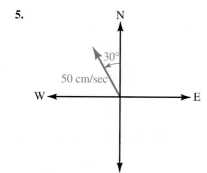

3.

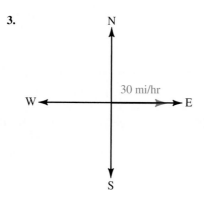

5.

7.

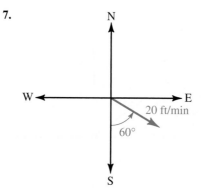

9. 49.6 mi, N 85.8° E   **11.** 6.80 mi   **13.** $\left| \mathbf{V}_x \right| = 12.6$, $\left| \mathbf{V}_y \right| = 5.66$ **15.** $\left| \mathbf{V}_x \right| = 339$, $\left| \mathbf{V}_y \right| = 248$
17. $\left| \mathbf{V}_x \right| = 64$, $\left| \mathbf{V}_y \right| = 0$ **19.** 43.6   **21.** 5.9   **23.** Both are 850 ft/sec   **25.** 2,550 ft
27. 38.1 ft/sec at an elevation of 23.2°   **29.** 100 km south, 90 km east   **31.** 250 mi north, 140 mi east
33. $\left| \mathbf{H} \right| = 42.0$ lb, $\left| \mathbf{T} \right| = 59.4$ lb   **35.** $\left| \mathbf{N} \right| = 9.7$ lb, $\left| \mathbf{F} \right| = 2.6$ lb   **37.** $\left| \mathbf{F} \right| = 33.1$ lb   **39.** 2,800 ft-lb

41. 8,200 ft-lb   **43.** $\sin 135° = \dfrac{1}{\sqrt{2}}$, $\cos 135° = -\dfrac{1}{\sqrt{2}}$, $\tan 135° = -1$ **45.** $\sin \theta = \dfrac{2}{\sqrt{5}}$, $\cos \theta = \dfrac{1}{\sqrt{5}}$

47. $x = \pm 6$

CHAPTER 2 TEST

	sin A	cos A	tan A	sin B	cos B	tan B
1.	$\dfrac{1}{\sqrt{5}}$	$\dfrac{2}{\sqrt{5}}$	$\dfrac{1}{2}$	$\dfrac{2}{\sqrt{5}}$	$\dfrac{1}{\sqrt{5}}$	2
2.	$\dfrac{\sqrt{3}}{2}$	$\dfrac{1}{2}$	$\sqrt{3}$	$\dfrac{1}{2}$	$\dfrac{\sqrt{3}}{2}$	$\dfrac{1}{\sqrt{3}}$
3.	$\dfrac{3}{5}$	$\dfrac{4}{5}$	$\dfrac{3}{4}$	$\dfrac{4}{5}$	$\dfrac{3}{5}$	$\dfrac{4}{3}$
4.	$\dfrac{5}{13}$	$\dfrac{12}{13}$	$\dfrac{5}{12}$	$\dfrac{12}{13}$	$\dfrac{5}{13}$	$\dfrac{12}{5}$

5. 76°   **6.** 17°   **7.** 5/4   **8.** 2   **9.** 0   **10.** $\dfrac{\sqrt{3}}{2}$ **11.** 73° 10′   **12.** 9° 43′   **13.** 73° 12′

14. 16° 27′   **15.** 2.8°   **16.** 79.5°   **17.** 0.4120   **18.** 0.7902   **19.** 2.0353   **20.** 0.3378   **21.** 4.7°
22. 58.7°   **23.** 71.2°   **24.** 13.3°   **25.** 4   **26.** 2   **27.** $A = 33.2°$, $B = 56.8°$, $c = 124$
28. $A = 30.3°$, $B = 59.7°$, $b = 41.5$ **29.** $A = 65.1°$, $a = 657$, $c = 724$ **30.** $B = 54° 30′$, $a = 0.268$, $b = 0.376$
31. 86 cm   **32.** 5.8 ft   **33.** 70 ft   **34.** $\left| \mathbf{V}_x \right| = 4.3$, $\left| \mathbf{V}_y \right| = 2.5$ **35.** 70°   **36.** $\left| \mathbf{V}_x \right| = 380$ ft/sec,
$\left| \mathbf{V}_y \right| = 710$ ft/sec   **37.** 60 mi south, 100 mi east   **38.** $\left| \mathbf{H} \right| = 45.6$ lb   **39.** $\left| \mathbf{F} \right| = 8.57$ lb   **40.** 2,100 ft-lb

CHAPTER 3

PROBLEM SET 3.1

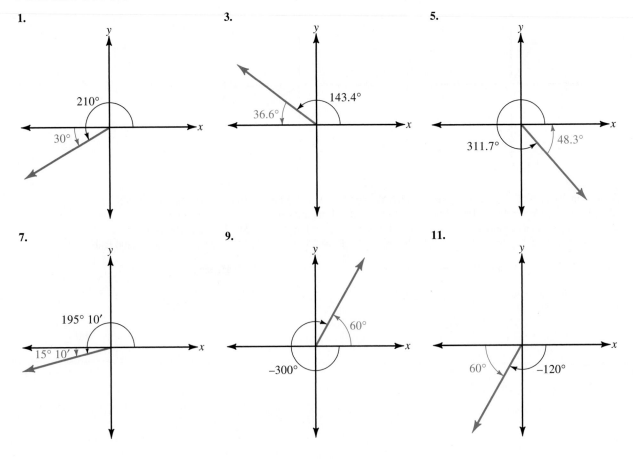

13. $-\dfrac{1}{\sqrt{2}}$ **15.** $\dfrac{\sqrt{3}}{2}$ **17.** -1 **19.** $-\dfrac{1}{2}$ **21.** -2 **23.** 2   **25.** $\dfrac{1}{2}$ **27.** $-\dfrac{1}{\sqrt{3}}$ **29.** 0.9744

31. -4.8901 **33.** -0.7427 **35.** 1.5032   **37.** 1.7321   **39.** -1.7263 **41.** 0.7071   **43.** 0.2711

45. 1.4225   **47.** -0.8660 **49.** 198.0°   **51.** 140.0°   **53.** 210.5°   **55.** 74.7°   **57.** 105.2°   **59.** 314.3°

61. 156.4°   **63.** 126.4°   **65.** 236.0°   **67.** 240°   **69.** 135°   **71.** 300°   **73.** 240°   **75.** 120°   **77.** 135°

79. 315°   **81.** Complement 20°, supplement 110°   **83.** Complement 90° − x, supplement 180° − x

85. Side opposite 30° is 5, side opposite 60° is $5\sqrt{3}$ **87.** $\dfrac{1}{4}$ **89.** 1

PROBLEM SET 3.2

1. 3   **3.** $\dfrac{1}{2}$ **5.** 3π **7.** 2   **9.** 0.1125 radian

11.

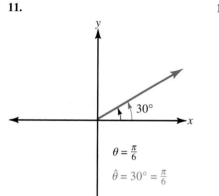

13.

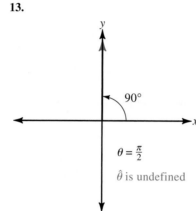

15.

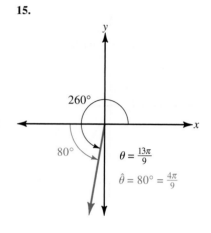

17.

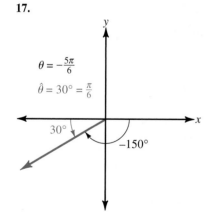

19.

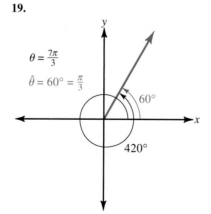

21.

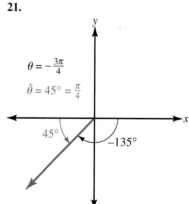

23. 2.11   **25.** 0.000291   **27.** 1.16 mi   **29.** $\dfrac{\pi}{6}$

31.

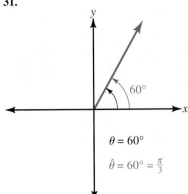

$\theta = 60°$

$\hat{\theta} = 60° = \frac{\pi}{3}$

33.

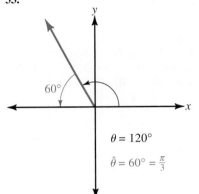

$\theta = 120°$

$\hat{\theta} = 60° = \frac{\pi}{3}$

35.

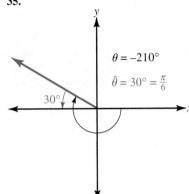

$\theta = -210°$

$\hat{\theta} = 30° = \frac{\pi}{6}$

37.

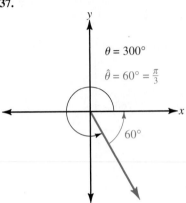

$\theta = 300°$

$\hat{\theta} = 60° = \frac{\pi}{3}$

39.

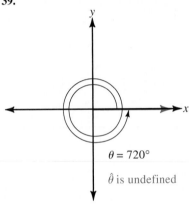

$\theta = 720°$

$\hat{\theta}$ is undefined

41.

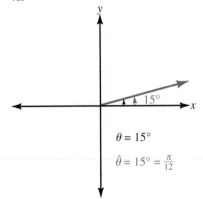

$\theta = 15°$

$\hat{\theta} = 15° = \frac{\pi}{12}$

43. $57.3°$ **45.** $74.5°$ **47.** $43.0°$ **49.** $286.5°$ **51.** $-\dfrac{\sqrt{3}}{2}$ **53.** $\dfrac{1}{\sqrt{3}}$ **55.** -2 **57.** 2 **59.** $-2\sqrt{2}$

61. $-\dfrac{1}{\sqrt{2}}$ **63.** $\sqrt{3}$ **65.** $\dfrac{\sqrt{3}}{2}$ **67.** 0 **69.** $\dfrac{\sqrt{3}}{2}$ **71.** -2 **73.** $(0,0), \left(\dfrac{\pi}{4}, \dfrac{1}{\sqrt{2}}\right), \left(\dfrac{\pi}{2}, 1\right), \left(\dfrac{3\pi}{4}, \dfrac{1}{\sqrt{2}}\right), (\pi, 0)$

75. $(0,0), \left(\dfrac{\pi}{2}, 2\right), (\pi, 0), \left(\dfrac{3\pi}{2}, -2\right), (2\pi, 0)$ **77.** $(0,0), \left(\dfrac{\pi}{4}, 1\right), \left(\dfrac{\pi}{2}, 0\right), \left(\dfrac{3\pi}{4}, -1\right), (\pi, 0)$

79. $\left(\dfrac{\pi}{2}, 0\right), (\pi, 1), \left(\dfrac{3\pi}{2}, 0\right), (2\pi, -1), \left(\dfrac{5\pi}{2}, 0\right)$ **81.** $\left(-\dfrac{\pi}{4}, 0\right), (0, 3), \left(\dfrac{\pi}{4}, 0\right), \left(\dfrac{\pi}{2}, -3\right), \left(\dfrac{3\pi}{4}, 0\right)$

	$\sin \theta$	$\cos \theta$	$\tan \theta$	$\cot \theta$	$\sec \theta$	$\csc \theta$	
85.	$-\dfrac{3}{\sqrt{10}}$	$\dfrac{1}{\sqrt{10}}$	-3	$-\dfrac{1}{3}$	$\sqrt{10}$	$-\dfrac{\sqrt{10}}{3}$	
87.	$\dfrac{n}{r}$	$\dfrac{m}{r}$	$\dfrac{n}{m}$	$\dfrac{m}{n}$	$\dfrac{r}{m}$	$\dfrac{r}{n}$	where $r = \sqrt{m^2 + n^2}$
89.	$\dfrac{1}{2}$	$-\dfrac{\sqrt{3}}{2}$	$-\dfrac{1}{\sqrt{3}}$	$-\sqrt{3}$	$-\dfrac{2}{\sqrt{3}}$	2	
91.	$\dfrac{2}{\sqrt{5}}$	$\dfrac{1}{\sqrt{5}}$	2	$\dfrac{1}{2}$	$\sqrt{5}$	$\dfrac{\sqrt{5}}{2}$	

PROBLEM SET 3.3

	sin θ	cos θ	tan θ	cot θ	sec θ	csc θ
1.	$\dfrac{1}{2}$	$-\dfrac{\sqrt{3}}{2}$	$-\dfrac{1}{\sqrt{3}}$	$-\sqrt{3}$	$-\dfrac{2}{\sqrt{3}}$	2
3.	$-\dfrac{1}{2}$	$\dfrac{\sqrt{3}}{2}$	$-\dfrac{1}{\sqrt{3}}$	$-\sqrt{3}$	$\dfrac{2}{\sqrt{3}}$	-2
5.	0	-1	0	undefined	-1	undefined
7.	$\dfrac{1}{\sqrt{2}}$	$-\dfrac{1}{\sqrt{2}}$	-1	-1	$-\sqrt{2}$	$\sqrt{2}$

9. $\dfrac{1}{2}$ **11.** $-\dfrac{\sqrt{3}}{2}$ **13.** $-\dfrac{1}{2}$ **15.** $-\dfrac{1}{\sqrt{2}}$ **17.** $\dfrac{\pi}{6}, \dfrac{5\pi}{6}$ **19.** $\dfrac{5\pi}{6}, \dfrac{7\pi}{6}$ **21.** $\dfrac{2\pi}{3}, \dfrac{5\pi}{3}$

23. $\sin\left(\dfrac{\pi}{4}\right) \approx 0.7071$, $\cos\left(\dfrac{\pi}{4}\right) \approx 0.7071$ **25.** $\sin\left(\dfrac{7\pi}{6}\right) = -0.5$, $\cos\left(\dfrac{7\pi}{6}\right) \approx -0.8660$ **27.** $0.5236, 2.6180$

29. 3.1416 **31.** $\sin(120°) \approx 0.8660$, $\cos(120°) = -0.5$ **33.** $\sin(75°) \approx 0.9659$, $\cos(75°) \approx 0.2588$

35. $210°, 330°$ **37.** $45°, 225°$ **39.** $\sin\theta = -2/\sqrt{5}$, $\cos\theta = 1/\sqrt{5}$, $\tan\theta = -2$ **41.** $\dfrac{1}{3}$ **43.** 1 **45.** $\dfrac{1}{2}$

47. 1 **49.** $\dfrac{1}{2}$ **51.**

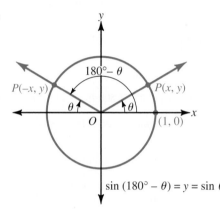

$\sin(180° - \theta) = y = \sin\theta$

For Problems 53 through 59, solutions are given in the Solutions Manual.

61.

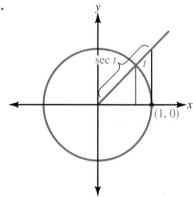

63. $B = 48°$, $a = 24$, $b = 27$ **65.** $A = 68°$, $a = 790$, $c = 850$ **67.** $A = 33.1°$, $B = 56.9°$, $c = 37.5$ **69.** $A = 44.7°$, $B = 45.3°$, $b = 4.41$

PROBLEM SET 3.4

1. 6 inches   **3.** 2.25 ft   **5.** 2π cm $\approx$ 6.28 cm   **7.** $\dfrac{4\pi}{3}$ mm $\approx$ 4.19 mm   **9.** $\dfrac{40\pi}{3}$ inches $\approx$ 41.9 inches

11. 5.03 cm   **13.** 4,400 mi   **15.** $\dfrac{4\pi}{9}$ ft $\approx$ 1.40 ft   **17.** 33.0 feet   **19.** 1.92 radians, 110°   **21.** 2,100 mi

23. 65.4 ft   **25.** 480 ft   **27. a.** 103 ft   **b.** 361 ft   **c.** 490 ft   **29.** 0.5 ft   **31.** 3 inches   **33.** 4 cm

35. 1 m   **37.** $\dfrac{16}{5\pi}$ km $\approx$ 1.02 km   **39.** 9 cm²   **41.** 19.2 inches²   **43.** $\dfrac{9\pi}{10}$ m² $\approx$ 2.83 m²

45. $\dfrac{25\pi}{24}$ m² $\approx$ 3.27 m²   **47.** 4 inches²   **49.** 2 cm   **51.** $\dfrac{4}{\sqrt{3}}$ inches $\approx$ 2.31 inches   **53.** 900π ft² $\approx$ 2,830 ft²

55. $\dfrac{175\pi}{2}$ mm $\approx$ 275 mm   **57.** 60.2°   **59.** 2.31 ft   **61.** 74.0°   **63.** 62.3 ft   **65.** 0.009 radian $\approx$ 0.518°

67. The sun is also about 400 times farther away from the earth.

PROBLEM SET 3.5

1. 1.5 ft/min   **3.** 3 cm/sec   **5.** 15 mi/hr   **7.** 80 ft   **9.** 22.5 mi   **11.** 7 mi   **13.** $\dfrac{2\pi}{15}$ rad/sec $\approx$ 0.419 rad/sec

15. 4 rad/min   **17.** $\dfrac{8}{3}$ rad/sec $\approx$ 2.67 rad/sec   **19.** 37.5π rad/hr $\approx$ 118 rad/hr

21. $d = 100\tan\dfrac{\pi}{2}t$; when $t = \dfrac{1}{2}$, $d = 100$ ft; when $t = \dfrac{3}{2}$, $d = -100$ ft; when $t = 1$, d is undefined because the light rays are parallel to the wall.   **23.** 40 inches   **25.** 180π m $\approx$ 565 m   **27.** 4,500 ft   **29.** 20π rad/min $\approx$ 62.8 rad/min

31. $\dfrac{200\pi}{3}$ rad/min $\approx$ 209 rad/min   **33.** 11.6π rad/min $\approx$ 36.4 rad/min   **35.** 10 inches/sec   **37.** 0.5 rad/sec

39. 80π ft/min $\approx$ 251 ft/min   **41.** $\dfrac{\pi}{12}$ rad/hr $\approx$ 0.262 rad/hr   **43.** 300π ft/min $\approx$ 942 ft/min   **45.** 9.50 mi/hr

47. 23.3 rpm   **49.** 5.65 ft/sec   **51.** 0.47 mi/hr   **53.** 889 rad/min (53,300 rad/hr)   **55.** 18.8 km/hr
57. 52.3 mm   **59.** 80.8 rpm   **61.** 10.4 mi/hr at N 24.1° W   **63.** $\left|\mathbf{V}_x\right| = 54.3$ ft/sec, $\left|\mathbf{V}_y\right| = 40.9$ ft/sec
65. 46.2 mi south, 71.9 mi west   **67.** See the Solutions Manual.

CHAPTER 3 TEST

1.

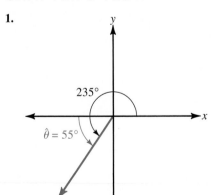

2.

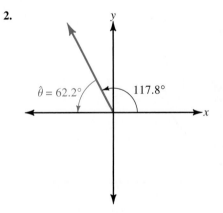

3.

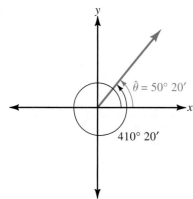

4.

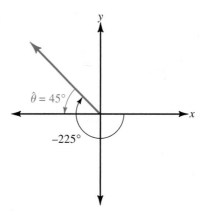

5. -1.1918 **6.** -2.1445 **7.** 1.1964 **8.** 1.2639 **9.** -1.2991 **10.** -6.5121 **11.** $174.0°$ **12.** $241.5°$

13. $226.0°$ **14.** $310.3°$ **15.** $-\dfrac{1}{\sqrt{2}}$ **16.** $-\dfrac{1}{\sqrt{2}}$ **17.** $-\dfrac{1}{\sqrt{3}}$ **18.** $\dfrac{2}{\sqrt{3}}$ **19.** $\dfrac{25\pi}{18}$ **20.** $-\dfrac{13\pi}{6}$

21. $240°$ **22.** $105°$ **23.** $\dfrac{\sqrt{3}}{2}$ **24.** $-\dfrac{1}{2}$ **25.** $-2\sqrt{2}$ **26.** 1 **27.** $-\dfrac{2}{\sqrt{3}}$ **28.** 2 **29.** 0 **30.** $2\sqrt{2}$

31. Begin by writing $\cot(-\theta) = \dfrac{\cos(-\theta)}{\sin(-\theta)}$ **32.** First use odd and even functions to write everything in terms of θ instead of $-\theta$.   **33.** 2π m ≈ 6.28 m   **34.** 2π ft ≈ 6.28 ft   **35.** 4 cm   **36.** 0.375 cm   **37.** 4π inches^2 ≈ 12.6 inches^2
38. 10.8 cm^2 **39.** 2π cm ≈ 6.28 cm   **40.** 10π ft ≈ 31.4 ft   **41.** 8 inches^2 **42.** 90 ft   **43.** $3,960$ ft   **44.** 72 inches
45. 120π ft ≈ 377 ft   **46.** 12π rad/min ≈ 37.7 rad/min   **47.** 4π rad/min ≈ 12.6 rad/min   **48.** 0.5 rad/sec
49. $\dfrac{5}{3}$ rad/sec   **50.** 80π ft/min ≈ 251 ft/min   **51.** 20π ft/min ≈ 62.8 ft/min   **52.** 4 rad/sec for the 6-cm pulley and
3 rad/sec for the 8-cm pulley   **53.** $2,700\pi$ ft/min $\approx 8,480$ ft/min   **54.** 900π inches/min $\approx 2,830$ inches/min
55. 17.5 mi/hr, 26.2 mi/hr   **56.** 22.0 rpm   **57.** 41.6 km/hr

CHAPTER 4

PROBLEM SET 4.1

If you want graphs that look similar to the ones here, use graph paper on which each square is 1/8 inch on each side. Then let two squares equal one unit. This way you can let the number π be approximately 6 squares. Here is an example:

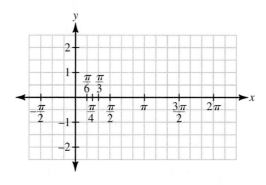

1.

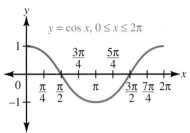

3.

5.

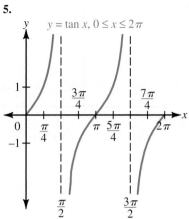

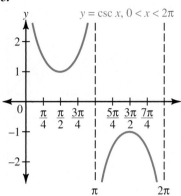

7.

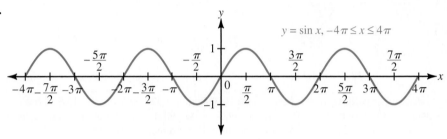

9.

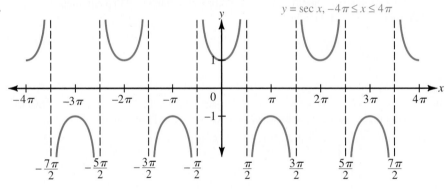

11.

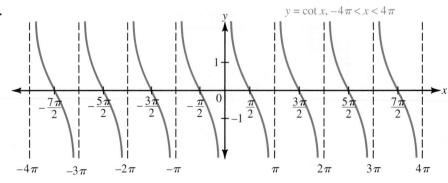

13. $\dfrac{\pi}{2} + k\pi$ **15.** $\dfrac{\pi}{2} + 2k\pi$ **17.** $k\pi$ **19.** $\dfrac{\pi}{2} + 2k\pi$ **21.** $\dfrac{\pi}{2} + k\pi$ **23.** $k\pi$

25. Amplitude = 3, period = π **27.** Amplitude = 2, period = 2   **29.** Amplitude = 3, period = π
31. After you have tried the problem yourself, look at Example 3.

33. After you have tried the problem yourself, look at Example 1 in Section 4.2.

35.

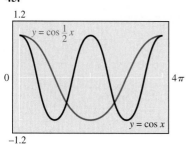

37.

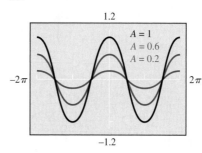

39.

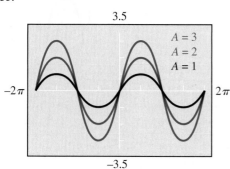

41.

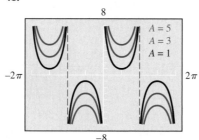

43.

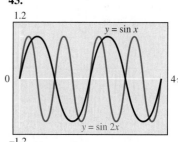

45.

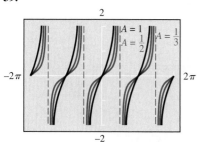

47.

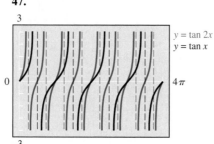

49.

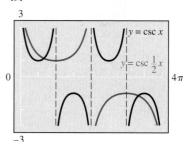

51.

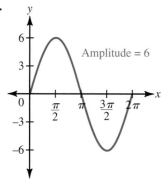

53.

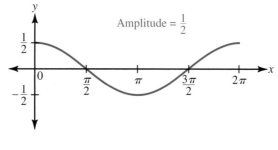

55.

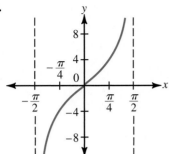

57.

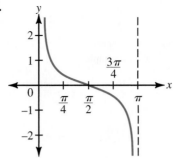

59.

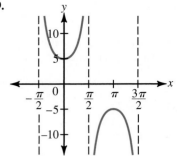

61.

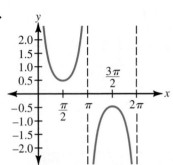

63.

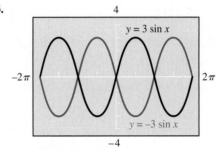

65.

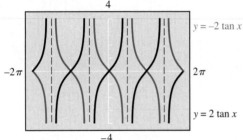

67.

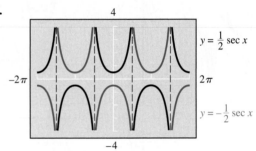

For Problems 69 through 73, solutions are given in the Solutions Manual.

75. 60°   **77.** 45°   **79.** 120°   **81.** 330°

PROBLEM SET 4.2

1.

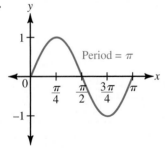

3.

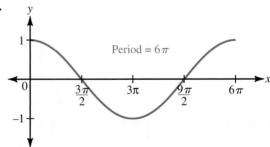

5.

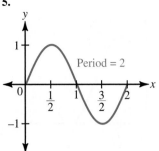

Period = 2

7.

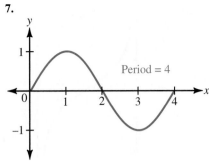

Period = 4

9.

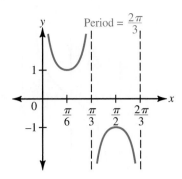

Period = $\frac{2\pi}{3}$

11.

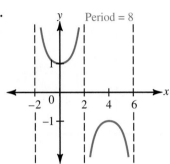

Period = 8

13.

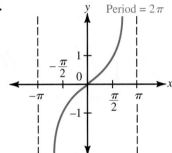

Period = 2π

15.

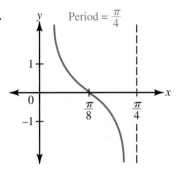

Period = $\frac{\pi}{4}$

17.

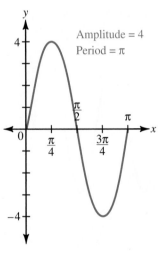

Amplitude = 4
Period = π

19.

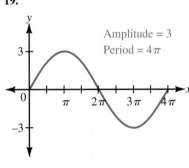

Amplitude = 3
Period = 4π

21.

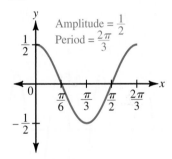

Amplitude = $\frac{1}{2}$
Period = $\frac{2\pi}{3}$

23.

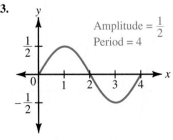

Amplitude = $\frac{1}{2}$
Period = 4

25.

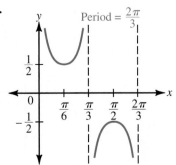

Period = $\frac{2\pi}{3}$

27.

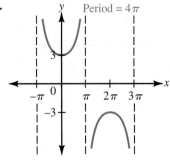

Period = 4π

29.

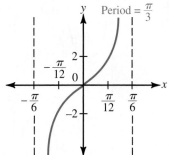

Period = $\frac{\pi}{3}$

31.

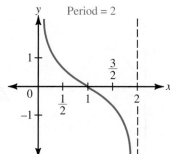

Period = 2

33.

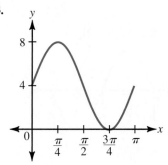

35.

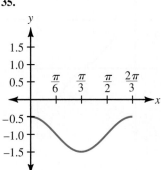

37.

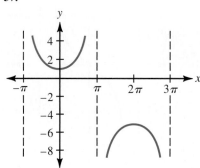

39.

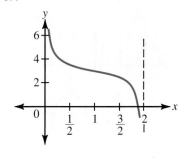

41.

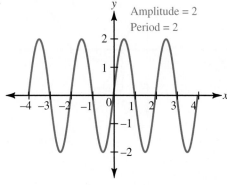

Amplitude = 2
Period = 2

43.

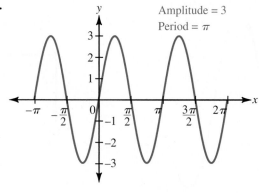

Amplitude = 3
Period = π

45.

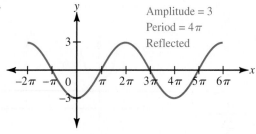

Amplitude = 3
Period = 4π
Reflected

47.

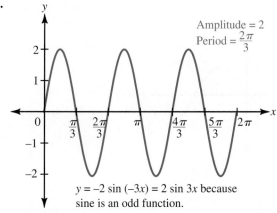

Amplitude = 2
Period = $\frac{2\pi}{3}$

$y = -2 \sin(-3x) = 2 \sin 3x$ because
sine is an odd function.

49.

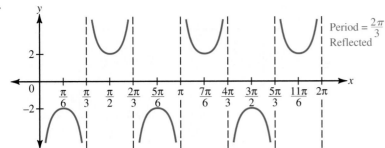

Period = $\frac{2\pi}{3}$
Reflected

51.

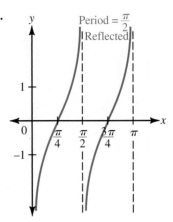

Period = $\frac{\pi}{2}$
Reflected

53. Maximum value of I is 20 amperes; one complete cycle takes 1/60 second.

55.

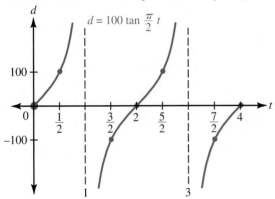

$d = 100 \tan \frac{\pi}{2} t$

57. a.

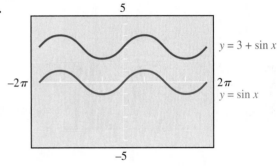

$y = 3 + \sin x$

$y = \sin x$

b.

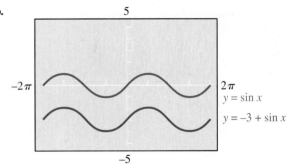

$y = \sin x$

$y = -3 + \sin x$

c.

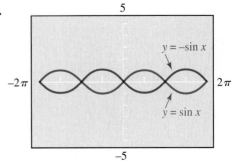

$y = -\sin x$

$y = \sin x$

59. a.

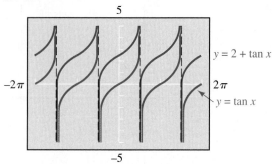

b.

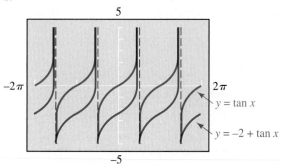

c.

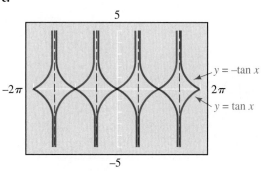

61. a.

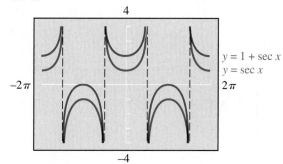

b.

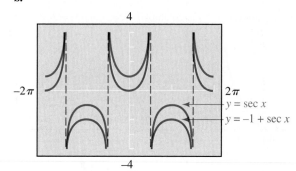

c.

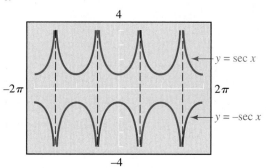

63. a.

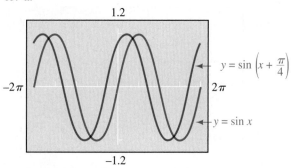

b.

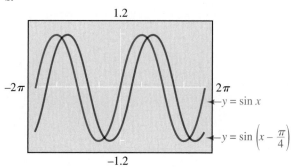

65. a.

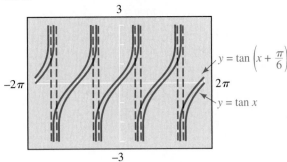

b.

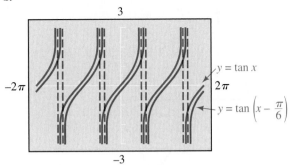

67. 0   **69.** 1   **71.** $\dfrac{\sqrt{3}}{2}$ **73.** $\dfrac{3}{2}$ **75.** $\dfrac{\pi}{4}$ **77.** $\dfrac{\pi}{3}$ **79.** $\dfrac{5\pi}{6}$ **81.** $\dfrac{5\pi}{4}$

PROBLEM SET 4.3

1.

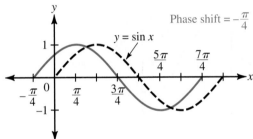

Phase shift $= -\dfrac{\pi}{4}$

3.

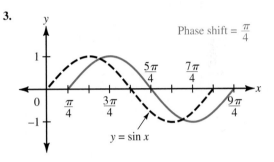

Phase shift $= \dfrac{\pi}{4}$

5.

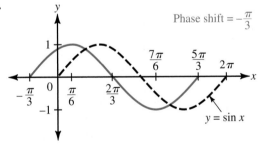

Phase shift $= -\dfrac{\pi}{3}$

7.

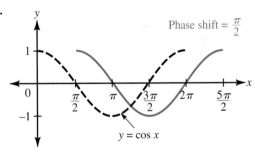

Phase shift $= \dfrac{\pi}{2}$

9.

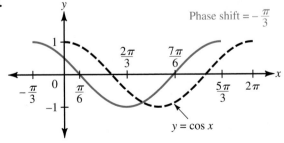

Phase shift $= -\dfrac{\pi}{3}$

11.

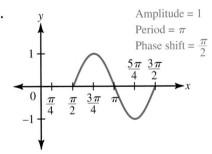

Amplitude $= 1$
Period $= \pi$
Phase shift $= \dfrac{\pi}{2}$

13.

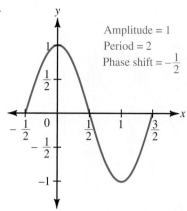

Amplitude = 1
Period = 2
Phase shift = $-\dfrac{1}{2}$

15.

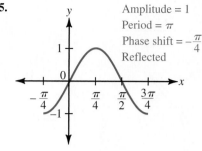

Amplitude = 1
Period = π
Phase shift = $-\dfrac{\pi}{4}$
Reflected

17.

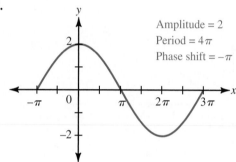

Amplitude = 2
Period = 4π
Phase shift = $-\pi$

19.

Amplitude = $\dfrac{1}{2}$
Period = $\dfrac{2\pi}{3}$
Phase shift = $\dfrac{\pi}{6}$

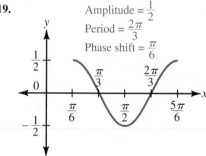

21.

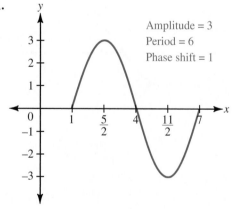

Amplitude = 3
Period = 6
Phase shift = 1

23.

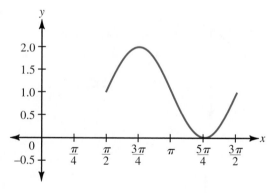

25.

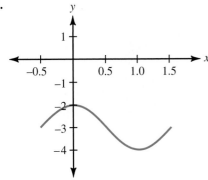

27.

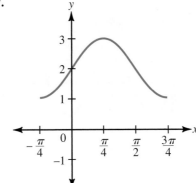

29.

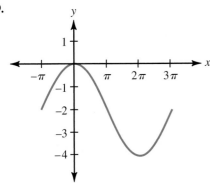

31.

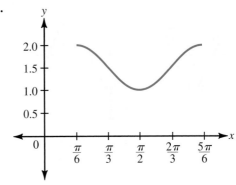

33.

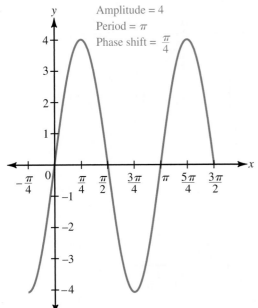

Amplitude = 4
Period = π
Phase shift = $\frac{\pi}{4}$

35.

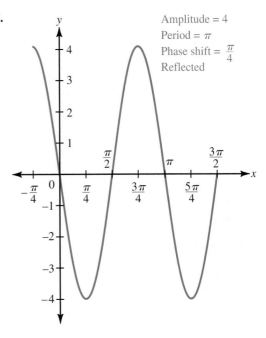

Amplitude = 4
Period = π
Phase shift = $\frac{\pi}{4}$
Reflected

37.

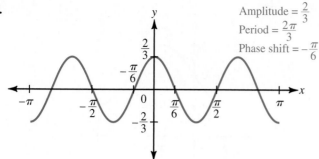

Amplitude = $\frac{2}{3}$

Period = $\frac{2\pi}{3}$

Phase shift = $-\frac{\pi}{6}$

39.

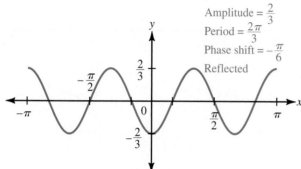

Amplitude = $\frac{2}{3}$

Period = $\frac{2\pi}{3}$

Phase shift = $-\frac{\pi}{6}$

Reflected

41.

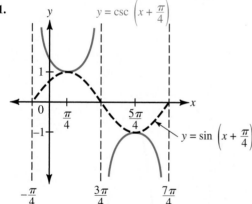

$y = \csc\left(x + \frac{\pi}{4}\right)$

$y = \sin\left(x + \frac{\pi}{4}\right)$

43.

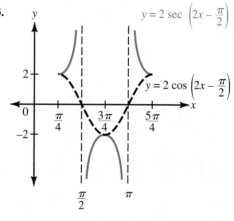

$y = 2\sec\left(2x - \frac{\pi}{2}\right)$

$y = 2\cos\left(2x - \frac{\pi}{2}\right)$

45.

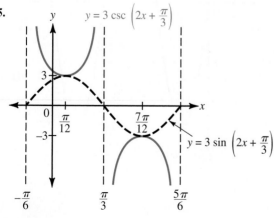

$y = 3\csc\left(2x + \frac{\pi}{3}\right)$

$y = 3\sin\left(2x + \frac{\pi}{3}\right)$

47.

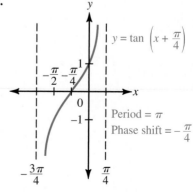

$y = \tan\left(x + \frac{\pi}{4}\right)$

Period = π

Phase shift = $-\frac{\pi}{4}$

49.

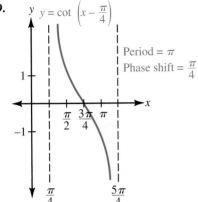

$y = \cot\left(x - \dfrac{\pi}{4}\right)$

Period = π
Phase shift = $\dfrac{\pi}{4}$

51.

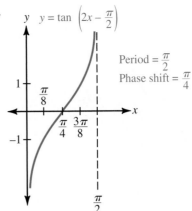

$y = \tan\left(2x - \dfrac{\pi}{2}\right)$

Period = $\dfrac{\pi}{2}$
Phase shift = $\dfrac{\pi}{4}$

53. $\dfrac{5\pi}{3}$ cm   **55.** $2.6\pi \approx 8.2$ cm   **57.** $\dfrac{2}{3}$ ft or 8 inches

PROBLEM SET 4.4

1. $y = \dfrac{1}{2}x + 1$ **3.** $y = 2x - 3$ **5.** $y = \sin x$ **7.** $y = 3\cos x$ **9.** $y = -3\cos x$ **11.** $y = \sin 3x$

13. $y = \sin \dfrac{1}{3}x$ **15.** $y = 2\cos 3x$ **17.** $y = 4\sin \pi x$ **19.** $y = -4\sin \pi x$ **21.** $y = 2 - 4\sin \pi x$

23. $y = 3\cos\left(2x + \dfrac{\pi}{2}\right)$ **25.** $y = -2\cos\left(3x + \dfrac{\pi}{2}\right)$ **27.** $y = -3 + 3\cos\left(2x + \dfrac{\pi}{2}\right)$

29. $y = 2 - 2\cos\left(3x + \dfrac{\pi}{2}\right)$

31.

t (min)	h (ft)
0	12
1.875	40.8
3.75	110.5
5.625	180.2
7.5	209
9.375	180.2
11.25	110.5
13.125	40.8
15	12

$h = 110.5 - 98.5\cos\dfrac{2\pi}{15}t$

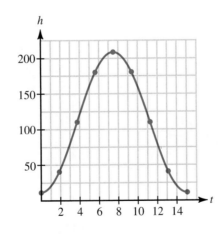

33. $39°$ **35.** $56°$ **37.** $84°$ **39.** $131.8°$ **41.** $205.5°$ **43.** $281.8°$

PROBLEM SET 4.5

1.

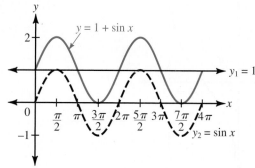

3.

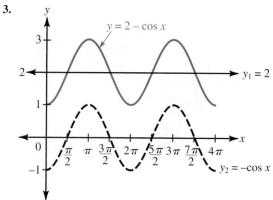

5.

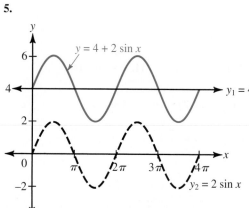

7.

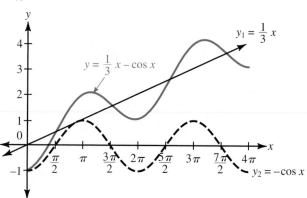

9.

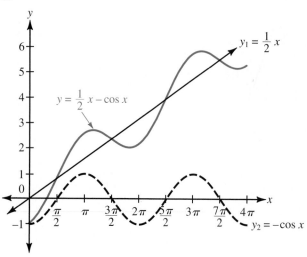

11.

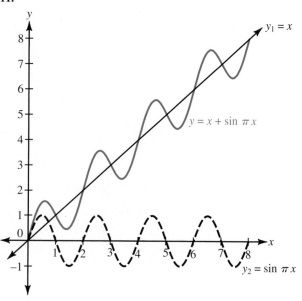

13.

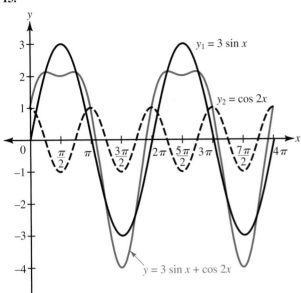

15.

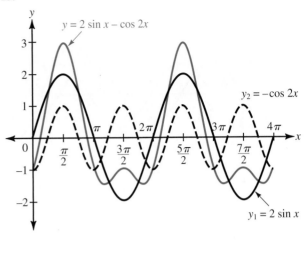

17.

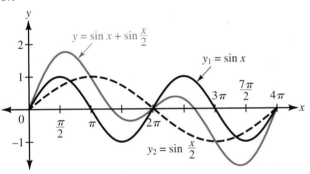

19.

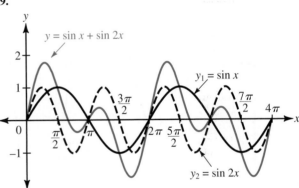

21.

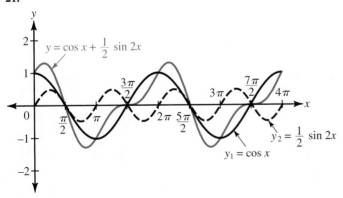

23.

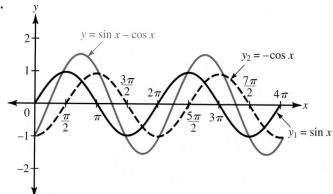

25.

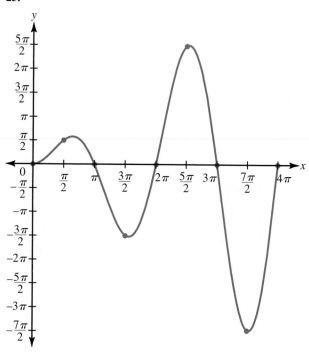

27.

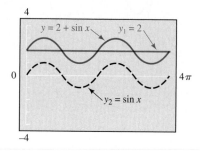

29.

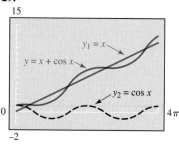

31.

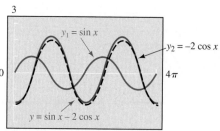

33.

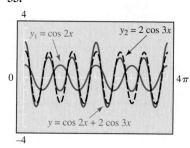

35. a.

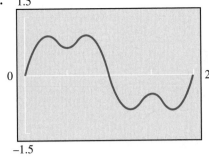

b.

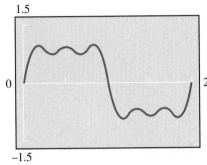

c.

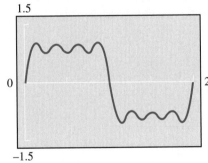

d.

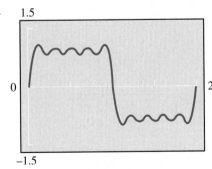

37. $\dfrac{1}{4}$ ft/sec   **39.** 1,200 ft   **41.** 180 m   **43.** 4π rad/sec   **45.** 40π cm

PROBLEM SET 4.6

1.

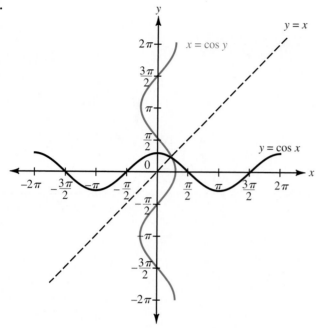

3. We have not included the graph of $y = \tan x$ with the graph of $x = \tan y$ because placing them both on the same coordinate system makes the diagram too complicated. It is best to graph $y = \tan x$ lightly in pencil and then reflect the graph about the line $y = x$ to get the graph of $x = \tan y$.

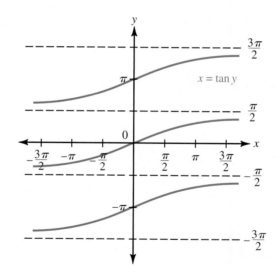

5. $\dfrac{\pi}{3}$ **7.** π **9.** $\dfrac{\pi}{4}$ **11.** $\dfrac{3\pi}{4}$ **13.** $-\dfrac{\pi}{6}$ **15.** $\dfrac{\pi}{3}$ **17.** 0 **19.** $-\dfrac{\pi}{6}$ **21.** $\dfrac{2\pi}{3}$ **23.** $\dfrac{\pi}{6}$

25. $9.8°$ **27.** $147.4°$ **29.** $20.8°$ **31.** $74.3°$ **33.** $117.8°$ **35.** $-70.0°$ **37.** $-50.0°$

39. a. $-30°$ **b.** $60°$ **c.** $-45°$ **41. a.** $-45°$ **b.** $60°$ **c.** $-30°$ **43.** $2 \sin \theta$

45. $\dfrac{3}{5}$ **47.** $\dfrac{1}{2}$ **49.** $\dfrac{1}{2}$ **51.** $-45°$ **53.** $\dfrac{\pi}{3}$ **55.** $120°$ **57.** $\dfrac{\pi}{4}$ **59.** $45°$ **61.** $-\dfrac{\pi}{6}$

63. $\dfrac{4}{5}$ **65.** $\dfrac{3}{4}$ **67.** $\sqrt{5}$ **69.** $\dfrac{\sqrt{3}}{2}$ **71.** 2   **73.** x **75.** x **77.** $\sqrt{1-x^2}$ **79.** $\dfrac{x}{\sqrt{x^2+1}}$

81. $\dfrac{\sqrt{x^2-1}}{x}$ **83.** x

85.

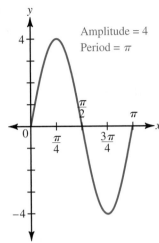

Amplitude = 4
Period = π

87.

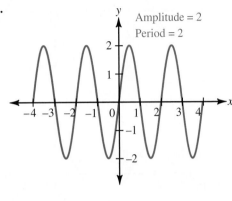

Amplitude = 2
Period = 2

89.

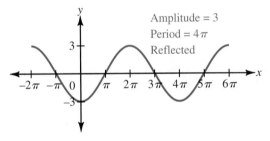

Amplitude = 3
Period = 4π
Reflected

91.

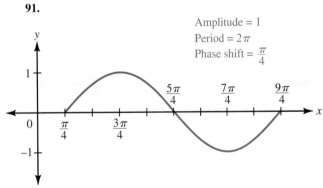

Amplitude = 1
Period = 2π
Phase shift = $\dfrac{\pi}{4}$

93.

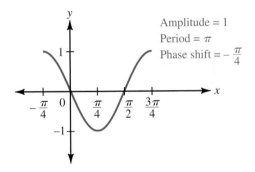

Amplitude = 1
Period = π
Phase shift = $-\dfrac{\pi}{4}$

95.

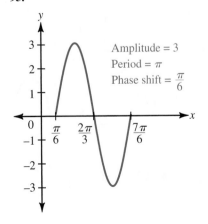

Amplitude = 3
Period = π
Phase shift = $\dfrac{\pi}{6}$

CHAPTER 4 TEST

1.

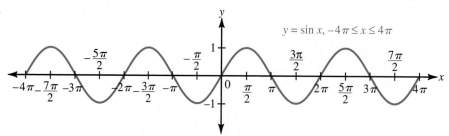

2.

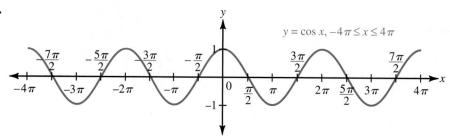

3.

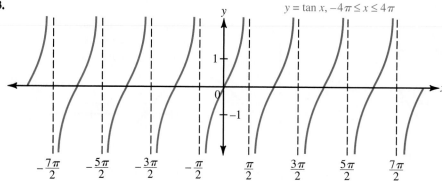

4.

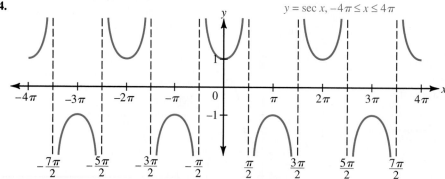

5. 4   **6.** 8   **7.** $-3\pi, -\pi, \pi, 3\pi$ **8.** $-\dfrac{7\pi}{2}, -\dfrac{5\pi}{2}, -\dfrac{3\pi}{2}, -\dfrac{\pi}{2}, \dfrac{\pi}{2}, \dfrac{3\pi}{2}, \dfrac{5\pi}{2}, \dfrac{7\pi}{2}$

9.

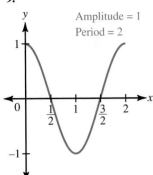

Amplitude = 1
Period = 2

10.

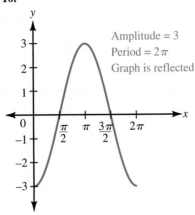

Amplitude = 3
Period = 2π
Graph is reflected

11.

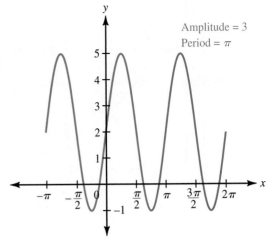

Amplitude = 3
Period = π

12.

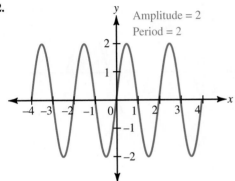

Amplitude = 2
Period = 2

13.

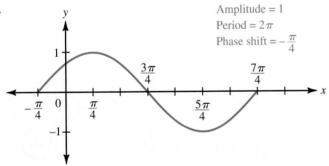

Amplitude = 1
Period = 2π
Phase shift = $-\dfrac{\pi}{4}$

14.

Amplitude = 1
Period = 2π
Phase shift = $\dfrac{\pi}{2}$

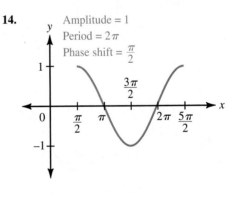

15.

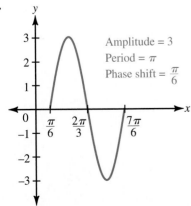

Amplitude = 3
Period = π
Phase shift = $\dfrac{\pi}{6}$

16.

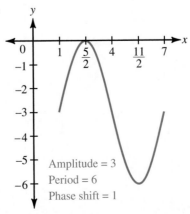

Amplitude = 3
Period = 6
Phase shift = 1

17.

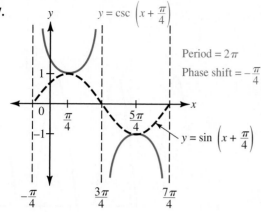

$$y = \csc\left(x + \frac{\pi}{4}\right)$$

$$\text{Period} = 2\pi$$
$$\text{Phase shift} = -\frac{\pi}{4}$$

$$y = \sin\left(x + \frac{\pi}{4}\right)$$

18.

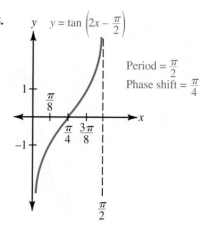

$$y = \tan\left(2x - \frac{\pi}{2}\right)$$

$$\text{Period} = \frac{\pi}{2}$$
$$\text{Phase shift} = \frac{\pi}{4}$$

19.

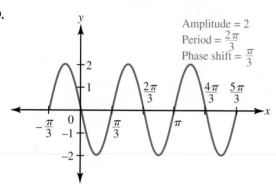

$$\text{Amplitude} = 2$$
$$\text{Period} = \frac{2\pi}{3}$$
$$\text{Phase shift} = \frac{\pi}{3}$$

20.

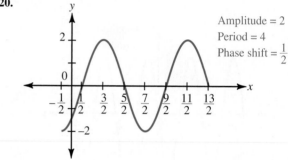

$$\text{Amplitude} = 2$$
$$\text{Period} = 4$$
$$\text{Phase shift} = \frac{1}{2}$$

21. $y = 2\sin\left(\dfrac{1}{2}x + \dfrac{\pi}{2}\right)$ **22.** $y = \dfrac{1}{2} + \dfrac{1}{2}\sin\dfrac{\pi}{2}x$

23.

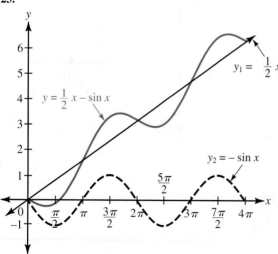

$$y_1 = \frac{1}{2}x$$

$$y = \frac{1}{2}x - \sin x$$

$$y_2 = -\sin x$$

24.

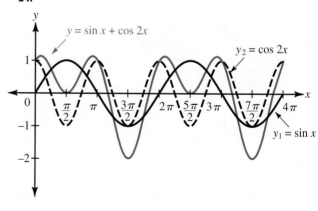

$$y = \sin x + \cos 2x$$

$$y_2 = \cos 2x$$

$$y_1 = \sin x$$

25.

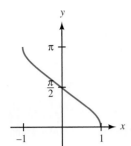

26.

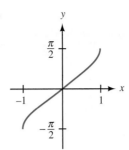

27. $\dfrac{\pi}{6}$ **28.** $\dfrac{5\pi}{6}$ **29.** $-\dfrac{\pi}{4}$ **30.** $\dfrac{\pi}{2}$ **31.** $36.4°$ **32.** $-39.7°$ **33.** $134.3°$ **34.** $-12.5°$ **35.** $\dfrac{\sqrt{5}}{2}$

36. $\dfrac{3}{\sqrt{13}}$ **37.** $30°$ **38.** $\dfrac{\pi}{6}$ **39.** $\sqrt{1-x^2}$ **40.** $\dfrac{x}{\sqrt{1-x^2}}$

CHAPTER 5

PROBLEM SET 5.1

For some of the problems in the beginning of this problem set we will give the complete proof. Remember, however, that there is often more than one way to prove an identity. You may have a correct proof even if it doesn't match the one you find here. As the problem set progresses, we will give hints on how to begin the proof instead of the complete proof. Solutions to problems not shown are given in the Solutions Manual.

1. $\cos\theta \tan\theta = \cos\theta \cdot \dfrac{\sin\theta}{\cos\theta}$

$\qquad\qquad\quad = \sin\theta$

9. $\cos x(\csc x + \tan x) = \cos x \csc x + \cos x \tan x$

$\qquad\qquad = \cos x \cdot \dfrac{1}{\sin x} + \cos x \cdot \dfrac{\sin x}{\cos x}$

$\qquad\qquad = \dfrac{\cos x}{\sin x} + \sin x$

$\qquad\qquad = \cot x + \sin x$

17. $\dfrac{\cos^4 t - \sin^4 t}{\sin^2 t} = \dfrac{(\cos^2 t + \sin^2 t)(\cos^2 t - \sin^2 t)}{\sin^2 t}$

$\qquad\qquad = \dfrac{\cos^2 t - \sin^2 t}{\sin^2 t}$

$\qquad\qquad = \dfrac{\cos^2 t}{\sin^2 t} - \dfrac{\sin^2 t}{\sin^2 t}$

$\qquad\qquad = \cot^2 t - 1$

19. Write the numerator on the right side as $1 - \sin^2\theta$ and then factor it.   **25.** Factor the left side and then write it in terms of sines and cosines.   **27.** Change the left side to sines and cosines and then add the resulting fractions.   **33.** See Example 6 in this section.   **37.** Rewrite the left side in terms of cosine and then simplify.   **67.** Is an identity   **69.** Not an identity
71. Not an identity   **73.** Is an identity

75. $-\dfrac{\pi}{3}$ is one possible answer.   **77.** 0 is one possible answer.   **79.** $\dfrac{\pi}{4}$ is one possible answer.

81. See the Solutions Manual.   **83.** $\cos A = \dfrac{4}{5}$, $\tan A = \dfrac{3}{4}$ **85.** $\dfrac{\sqrt{3}}{2}$ **87.** $\dfrac{\sqrt{3}}{2}$ **89.** $15°$ **91.** $105°$

PROBLEM SET 5.2

1. $\dfrac{\sqrt{6} - \sqrt{2}}{4}$ **3.** $\dfrac{\sqrt{6} - \sqrt{2}}{\sqrt{6} + \sqrt{2}}$ or $\dfrac{\sqrt{3} - 1}{\sqrt{3} + 1}$ **5.** $\dfrac{\sqrt{6} + \sqrt{2}}{4}$ **7.** $\dfrac{\sqrt{2} - \sqrt{6}}{4}$

9. $\sin (x + 2\pi) = \sin x \cos 2\pi + \cos x \sin 2\pi$
$ = \sin x(1) + \cos x(0)$
$ = \sin x$

For problems 11–19, proceed as in Problem 9. Expand the left side and simplify. Solutions are given in the Solutions Manual.

21. $\sin 5x$ **23.** $\cos 6x$ **25.** $\cos 90° = 0$

27.

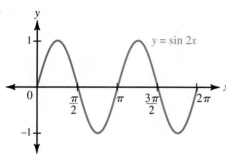

29.

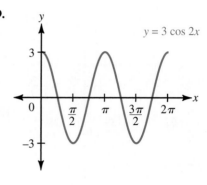

31.

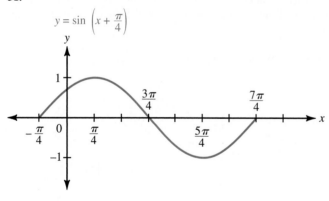

33.

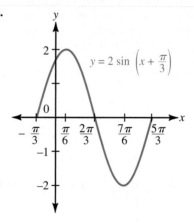

35. $-\dfrac{16}{65}, \dfrac{63}{65}, -\dfrac{16}{63}$, QIV   **37.** $2, \dfrac{1}{2}$, QI   **39.** 1

41. $\sin 2x = 2 \sin x \cos x$

For Problems 43–57, solutions are given in the Solutions Manual.

NOTE For Problems 59–63, when the equation is an identity, the proof is given in the Solutions Manual.

59. Is an identity   **61.** Not an identity   **63.** Is an identity

65.

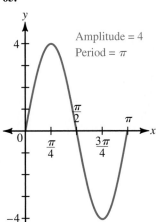

Amplitude = 4
Period = π

67.

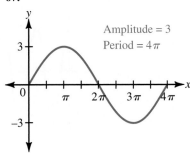

Amplitude = 3
Period = 4π

69.

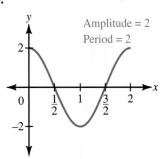

Amplitude = 2
Period = 2

71. See the solution to Problem 9 in Problem Set 4.2.

73.

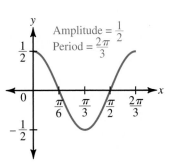

Amplitude = $\dfrac{1}{2}$
Period = $\dfrac{2\pi}{3}$

75.

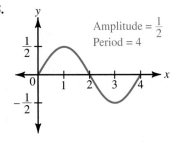

Amplitude = $\dfrac{1}{2}$
Period = 4

PROBLEM SET 5.3

1. $\dfrac{24}{25}$ **3.** $\dfrac{24}{7}$ **5.** $-\dfrac{4}{5}$ **7.** $\dfrac{4}{3}$ **9.** $\dfrac{120}{169}$ **11.** $\dfrac{169}{120}$ **13.** $\dfrac{3}{5}$ **15.** $\dfrac{5}{3}$

17. $y = 4 - 8 \sin^2 x$
$\quad = 4(1 - 2 \sin^2 x)$
$\quad = 4 \cos 2x$

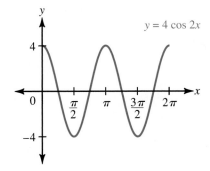

$y = 4 \cos 2x$

19.

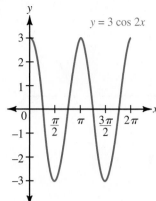

21.

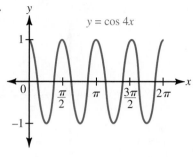

For Problems 23 and 25, see the Solutions Manual.

27. $\dfrac{24}{7}$ **29.** $\dfrac{1}{2}$ **31.** $-\dfrac{\sqrt{3}}{2}$ **33.** $\dfrac{1}{4}$ **35.** $\dfrac{1}{2}$

For Problems 37–53, solutions are given in the Solutions Manual.

NOTE For Problems 55–59, when the equation is an identity, the proof is given in the Solutions Manual.

55. Is an identity   **57.** Not an identity   **59.** Is an identity

61. $\dfrac{1}{2}\left(\tan^{-1}\dfrac{x}{5} - \dfrac{5x}{x^2 + 25}\right)$ **63.** $\dfrac{1}{2}\left(\sin^{-1}\dfrac{x}{3} - \dfrac{x\sqrt{9 - x^2}}{9}\right)$

65.

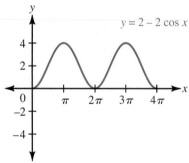

67.

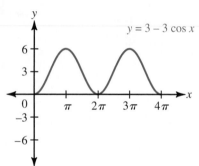

69. See the solution to Problem 21 in Problem Set 4.5.

71.

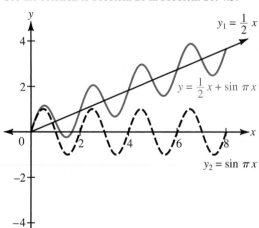

PROBLEM SET 5.4

1. $\dfrac{1}{2}$ **3.** 2   **5.** $-\dfrac{1}{\sqrt{10}}$ **7.** $-\sqrt{10}$ **9.** $\sqrt{\dfrac{3 + 2\sqrt{2}}{6}}$ **11.** $-\sqrt{\dfrac{3 - 2\sqrt{2}}{6}}$ **13.** $-3 - 2\sqrt{2}$

15. $\dfrac{2}{\sqrt{5}}$ **17.** $-\dfrac{7}{25}$ **19.** $-\dfrac{25}{7}$ **21.** $\dfrac{3}{\sqrt{10}}$ **23.** $\dfrac{7}{25}$ **25.** 0

27. See the solution to Problem 65 in Problem Set 5.3.

29.

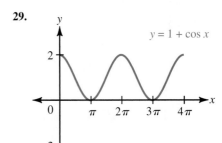

31. $\dfrac{\sqrt{2 + \sqrt{3}}}{2}$ **33.** $\dfrac{\sqrt{2 + \sqrt{3}}}{2}$ **35.** $-\dfrac{\sqrt{2 - \sqrt{3}}}{2}$

For Problems 37–47, see the Solutions Manual.

49. $\dfrac{3}{5}$ **51.** $\dfrac{1}{\sqrt{5}}$ **53.** $\dfrac{x}{\sqrt{x^2 + 1}}$ **55.** $\dfrac{x}{\sqrt{1 - x^2}}$

57. See graph on page 222.

PROBLEM SET 5.5

1. $-\dfrac{1}{\sqrt{5}}$ **3.** $\dfrac{2\sqrt{3} - 1}{2\sqrt{5}}$ **5.** $\dfrac{4}{5}$ **7.** $\dfrac{x}{\sqrt{1 - x^2}}$ **9.** $2x\sqrt{1 - x^2}$ **11.** $2x^2 - 1$

13. See the Solutions Manual.   **15.** $5(\sin 8x + \sin 2x)$ **17.** $\dfrac{1}{2}(\cos 10x + \cos 6x)$ **19.** $\dfrac{1}{2}(\sin 90° + \sin 30°) = \dfrac{3}{4}$

21. $\dfrac{1}{2}(\cos 2\pi - \cos 6\pi) = 0$ **23.** See the Solutions Manual.   **25.** $2 \sin 5x \cos 2x$

27. $2 \cos 30° \cos 15° = \sqrt{3} \cos 15°$ **29.** $2 \cos \dfrac{\pi}{3} \sin \dfrac{\pi}{4} = \dfrac{1}{\sqrt{2}}$

For Problems 31–35, see the Solutions Manual.

37.

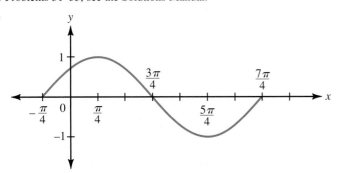

39.

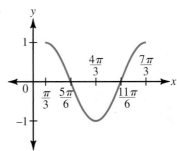

41.

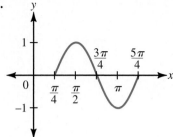

43.

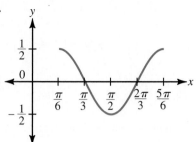

45.

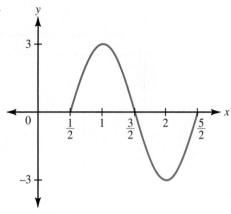

CHAPTER 5 TEST

For Problems 1–12, see the Solutions Manual.

NOTE For Problems 13–18, when the equation is an identity, the proof is given in the Solutions Manual.

13. Is an identity **14.** Is an identity **15.** Not an identity **16.** Not an identity **17.** Is an identity

18. Is an identity **19.** $\dfrac{63}{65}$ **20.** $-\dfrac{56}{65}$ **21.** $-\dfrac{119}{169}$ **22.** $-\dfrac{120}{169}$ **23.** $\dfrac{1}{\sqrt{10}}$ **24.** $-\dfrac{3}{\sqrt{10}}$

NOTE For Problems 25–28, other answers are possible depending on the identity used.

25. $\dfrac{\sqrt{6}+\sqrt{2}}{4}$ **26.** $\dfrac{\sqrt{6}+\sqrt{2}}{4}$ **27.** $\dfrac{\sqrt{3}-1}{\sqrt{3}+1}$ **28.** $\dfrac{\sqrt{3}+1}{\sqrt{3}-1}$ **29.** $\cos 9x$ **30.** $\sin 90° = 1$

31. $\dfrac{3}{5}, -\sqrt{\dfrac{5-2\sqrt{5}}{10}}$ **32.** $\dfrac{3}{5}, \sqrt{\dfrac{10-\sqrt{10}}{20}}$ **33.** 1 **34.** $\pm\dfrac{\sqrt{3}}{2}$ **35.** $\dfrac{11}{5\sqrt{5}}$ **36.** $\dfrac{11}{5\sqrt{5}}$ **37.** $1-2x^2$

38. $2x\sqrt{1-x^2}$ **39.** $\dfrac{1}{2}(\cos 2x - \cos 10x)$ **40.** $2\cos 45° \cos(-30°) = \dfrac{\sqrt{6}}{2}$

CHAPTER 6

PROBLEM SET 6.1

1. 30°, 150° **3.** 30°, 330° **5.** 135°, 315° **7.** $\pi/3, 2\pi/3$ **9.** $\dfrac{\pi}{6}, \dfrac{11\pi}{6}$ **11.** $\dfrac{3\pi}{2}$ **13.** 48.6°, 131.4°

15. $\varnothing$ **17.** 228.6°, 311.4° **19.** $\dfrac{\pi}{2}, \dfrac{\pi}{6}, \dfrac{5\pi}{6}$ **21.** $0, \dfrac{\pi}{4}, \pi, \dfrac{5\pi}{4}$ **23.** $0, \dfrac{2\pi}{3}, \pi, \dfrac{4\pi}{3}$ **25.** $\dfrac{\pi}{2}, \dfrac{7\pi}{6}, \dfrac{11\pi}{6}$

27. 120°, 150°, 210°, 240° **29.** 0°, 60°, 180°, 120° **31.** 120°, 240° **33.** 201.5°, 338.5° **35.** 51.8°, 308.2°

37. $17.0°, 163.0°$ **39.** $30° + 360°k, 150° + 360°k$ **41.** $\dfrac{\pi}{3} + 2k\pi, \dfrac{2\pi}{3} + 2k\pi$ **43.** $\dfrac{3\pi}{2} + 2k\pi$

45. $48.6° + 360°k, 131.4° + 360°k$ **47.** $40° + 180°k, 190° + 180°k$ **49.** $120°k, 40° + 120°k$

51. $35° + 90°k, 65° + 90°k$ **53.** $42° + 72°k, 60° + 72°k$

For Problems 55–73, see the answer for the corresponding problem.

75. $h = -16t^2 + 750t$ **77.** $1{,}436 \text{ ft}$ **79.** $15.7°$ **81.** $\sin 2A = 2 \sin A \cos A$ **83.** $\cos 2A = 2 \cos^2 A - 1$

85. $\dfrac{1}{\sqrt{2}} \sin \theta + \dfrac{1}{\sqrt{2}} \cos \theta$ **87.** $\dfrac{\sqrt{6} + \sqrt{2}}{4}$ **89.** See the Solutions Manual.

PROBLEM SET 6.2

1. $30°, 330°$ **3.** $225°, 315°$ **5.** $45°, 135°, 225°, 315°$ **7.** $30°, 150°$ **9.** $30°, 90°, 150°, 270°$

11. $60°, 180°, 300°$ **13.** $\dfrac{7\pi}{6}, \dfrac{3\pi}{2}, \dfrac{11\pi}{6}$ **15.** $0, \dfrac{2\pi}{3}, \dfrac{4\pi}{3}$ **17.** $\dfrac{\pi}{2}, \dfrac{7\pi}{6}, \dfrac{11\pi}{6}$ **19.** $\dfrac{\pi}{3}, \dfrac{5\pi}{3}$

21. $\dfrac{2\pi}{3}, \dfrac{4\pi}{3}$ **23.** $\dfrac{\pi}{4}$ **25.** $30°, 90°$ **27.** $60°, 180°$ **29.** $60°, 300°$ **31.** $120°, 180°$ **33.** $210°, 330°$

35. $36.9°, 48.2°, 311.8°, 323.1°$ **37.** $36.9°, 143.1°, 216.9°, 323.1°$ **39.** $225° + 360°k, 315° + 360°k$

41. $\dfrac{\pi}{4} + 2k\pi$ **43.** $120° + 360°k, 180° + 360°k$ **45.** See the Solutions Manual.   **47.** $68.5°, 291.5°$

49. $218.2°, 321.8°$ **51.** $73.0°, 287.0°$ **53.** $0.3630, 2.1351$ **55.** $3.4492, 5.9756$ **57.** $0.3166, 1.9917$

59. $\sqrt{\dfrac{3 - \sqrt{5}}{6}}$ **61.** $\sqrt{\dfrac{6}{3 - \sqrt{5}}}$ **63.** $\sqrt{\dfrac{3 - \sqrt{5}}{3 + \sqrt{5}}}$ or $\dfrac{3 - \sqrt{5}}{2}$

65. See the solution to Problem 65 in Problem Set 5.3.   **67.** $\dfrac{\sqrt{2 - \sqrt{2}}}{2}$

PROBLEM SET 6.3

1. $30°, 60°, 210°, 240°$ **3.** $67.5°, 157.5°, 247.5°, 337.5°$ **5.** $60°, 180°, 300°$ **7.** $\dfrac{\pi}{8}, \dfrac{3\pi}{8}, \dfrac{9\pi}{8}, \dfrac{11\pi}{8}$

9. $\dfrac{\pi}{3}, \pi, \dfrac{5\pi}{3}$ **11.** $\dfrac{\pi}{6}, \dfrac{2\pi}{3}, \dfrac{7\pi}{6}, \dfrac{5\pi}{3}$ **13.** $15° + 180°k, 75° + 180°k$ **15.** $30° + 120°k, 90° + 120°k$

17. $6° + 36°k, 12° + 36°k$ **19.** $112.5°, 157.5°, 292.5°, 337.5°$ **21.** $20°, 100°, 140°, 220°, 260°, 340°$

23. $15°, 105°, 195°, 285°$ **25.** $\dfrac{\pi}{18}, \dfrac{5\pi}{18}, \dfrac{13\pi}{18}, \dfrac{17\pi}{18}, \dfrac{25\pi}{18}, \dfrac{29\pi}{18}$ **27.** $\dfrac{5\pi}{18}, \dfrac{7\pi}{18}, \dfrac{17\pi}{18}, \dfrac{19\pi}{18}, \dfrac{29\pi}{18}, \dfrac{31\pi}{18}$

29. $\dfrac{\pi}{10} + \dfrac{2k\pi}{5}$ **31.** $\dfrac{\pi}{8} + \dfrac{k\pi}{2}, \dfrac{3\pi}{8} + \dfrac{k\pi}{2}$ **33.** $\dfrac{\pi}{5} + \dfrac{2k\pi}{5}$ **35.** $10° + 120°k, 50° + 120°k, 90° + 120°k$

37. $60° + 180°k, 90° + 180°k, 120° + 180°k$ **39.** $20° + 60°k, 40° + 60°k$ **41.** $0°, 270°$ **43.** $180°, 270°$

45. $96.8°, 173.2°, 276.8°, 353.2°$ **47.** $27.4°, 92.6°, 147.4°, 212.6°, 267.4°, 332.6°$

49. $50.4°, 84.6°, 140.4°, 174.6°, 230.4°, 264.6°, 320.4°, 354.6°$

51. $4.0 \text{ min and } 16.0 \text{ min}$ **53.** 6 **55.** $\dfrac{1}{4}$ second (and every second after that)   **57.** $\dfrac{1}{12}$

For Problems 59–63, see the Solutions Manual.

65. $-\dfrac{4\sqrt{2}}{9}$ **67.** $\sqrt{\dfrac{3 - 2\sqrt{2}}{6}}$ **69.** $\dfrac{4 - 6\sqrt{2}}{15}$ **71.** $\dfrac{15}{4 - 6\sqrt{2}}$

PROBLEM SET 6.4

1.

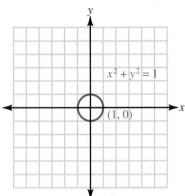

3.

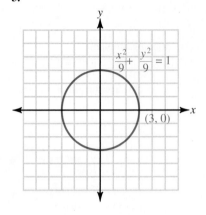

5.

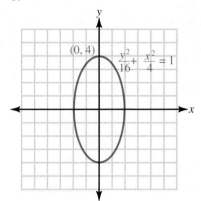

7.

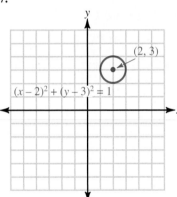

9.

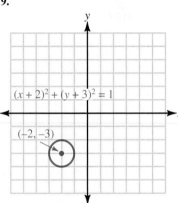

11.

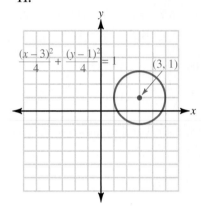

13.

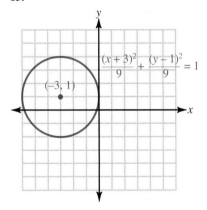

15. $x^2 - y^2 = 1$ **17.** $\dfrac{x^2}{9} - \dfrac{y^2}{9} = 1$ **19.** $\dfrac{(y-4)^2}{9} - \dfrac{(x-2)^2}{9} = 1$ **21.** $x = 1 - 2y^2$ **23.** $y = x$ **25.** $2x = 3y$

27. $x = 98.5 \cos\left(\dfrac{2\pi}{15}T - \dfrac{\pi}{2}\right)$

$y = 110.5 + 98.5 \sin\left(\dfrac{2\pi}{15}T - \dfrac{\pi}{2}\right)$

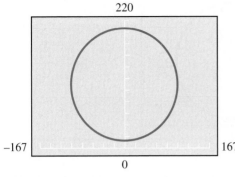

29. $\dfrac{\sqrt{3}}{2}$ **31.** $\dfrac{\sqrt{15}+3}{4\sqrt{10}}$ **33.** $\sqrt{1-x^2}$ **35.** $\dfrac{1-x^2}{1+x^2}$ **37.** $4(\sin 5x + \sin x)$

39. To three significant digits:
   maximum height $= 63.0$ ft
   maximum distance $= 145$ ft
   time he hits the net $= 3.97$ sec
   The graph is shown in Figure 17

CHAPTER 6 TEST

1. $30°, 150°$ **2.** $150°, 330°$ **3.** $30°, 90°, 150°, 270°$ **4.** $0°, 60°, 180°, 300°$ **5.** $45°, 135°, 225°, 315°$
6. $90°, 210°, 330°$ **7.** $180°$ **8.** $0°, 240°$ **9.** $48.6°, 131.4°, 210°, 330°$
10. $95° + 120°k, 115° + 120°k$ where $k = 0, 1, 2$ **11.** $0°, 90°$ **12.** $90°, 180°$ **13.** $40°, 80°, 160°, 200°, 280°, 320°$

14. $22.5°, 112.5°, 202.5°, 292.5°$ **15.** $2k\pi, \dfrac{\pi}{3} + 2k\pi, \dfrac{5\pi}{3} + 2k\pi$ **16.** $\dfrac{\pi}{6} + 2k\pi, \dfrac{7\pi}{6} + 2k\pi$ **17.** $\dfrac{\pi}{2} + \dfrac{2k\pi}{3}$

18. $\dfrac{\pi}{8} + \dfrac{k\pi}{2}$ **19.** $90°, 203.6°, 336.4°$ **20.** $111.5°, 248.5°$ **21.** $0.7297, 2.4119$ **22.** $1.0598, 2.5717$

23. $0.3076, 2.8340$ **24.** $0.3218, 1.2490, 3.4633, 4.3906$ **25.** 5.3 min and 14.7 min
26.   **27.**   **28.**

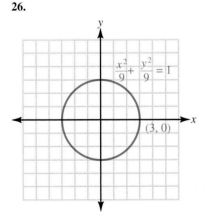
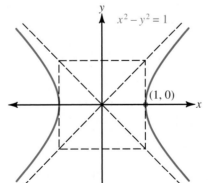
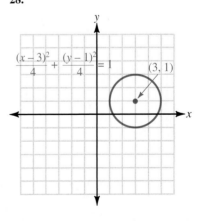

29.

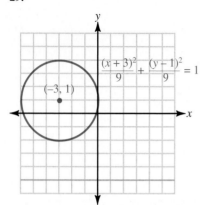

$$\frac{(x+3)^2}{9} + \frac{(y-1)^2}{9} = 1$$

30. $x = 90 \cos\left(\frac{2\pi}{3}t - \frac{\pi}{2}\right)$, $y = 98 + 90 \sin\left(\frac{2\pi}{3}t - \frac{\pi}{2}\right)$

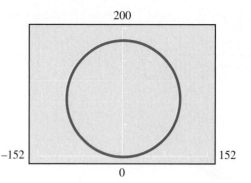

CHAPTER 7

PROBLEM SET 7.1

1. 16 cm   **3.** 71 inches   **5.** 140 yd   **7.** $C = 70°$, $c = 44$ km   **9.** $C = 80°$, $a = 11$ cm
11. $C = 66.1°$, $b = 302$ inches, $c = 291$ inches   **13.** $C = 39°$, $a = 7.8$ m, $b = 11$ m
15. $B = 16°$, $b = 1.39$ ft, $c = 4.36$ ft   **17.** $A = 141.8°$, $b = 118$ cm, $c = 214$ cm   **19.** $\sin B = 5$, which is impossible
21. 11   **23.** 20   **25.** 209 ft   **27.** 273 ft   **29.** 5,900 ft   **31.** 42 ft   **33.** 14 mi, 9.3 mi
35. $|\mathbf{CB}| = 341$ lb, $|\mathbf{CA}| = 345$ lb   **37.** $|\mathbf{CA}| = 3,240$ lb, $|\mathbf{CB}| = 3,190$ lb   **39.** 45°, 135°   **41.** 90°, 270°

43. 30°, 90°, 150°   **45.** 54.1°, 305.9°   **47.** $\frac{\pi}{6} + 2k\pi$, $\frac{3\pi}{2} + 2k\pi$, $\frac{5\pi}{6} + 2k\pi$ **49.** 47.6°, 132.4°

51. 75.2°, 104.8°

PROBLEM SET 7.2

1. $\sin B = 2$ is impossible   **3.** $B = 35°$ is the only possibility for B **5.** $B = 77°$ or $B' = 103°$
7. $B = 54°$, $C = 88°$, $c = 67$ ft; $B' = 126°$, $C' = 16°$, c' = 18 ft   **9.** $B = 28.1°$, $C = 39.7°$, $c = 30.2$ cm
11. $B = 34° \ 50'$, $A = 117° \ 20'$, $a = 660$ m; $B' = 145° \ 10'$, $A' = 7°$, $a' = 90.6$ m
13. $C = 26° \ 20'$, $A = 108° \ 30'$, $a = 2.39$ inches   **15.** no solution   **17.** no solution
19. $B = 26.8°$, $A = 126.4°$, $a = 65.7$ km   **21.** 15 ft or 38 ft

23.

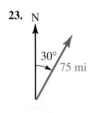

25.

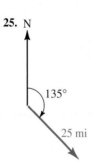

27. 310 mi/hr or 360 mi/hr   **29.** $6\sqrt{3}$ mi/hr ≈ 10 mi/hr   **31.** Yes, it makes an angle of 88° with the ground.
33. 30°, 150°, 210°, 330°   **35.** 90°, 270°   **37.** 41.8°, 48.6°, 131.4°, 138.2°

39. $\frac{7\pi}{6} + 2k\pi$, $\frac{11\pi}{6} + 2k\pi$ **41.** $\frac{3\pi}{4} + 2k\pi$, $\frac{7\pi}{4} + 2k\pi$

PROBLEM SET 7.3

1. 100 inches   **3.** $C = 69°$ **5.** 9.4 m   **7.** $A = 128°$

For Problems 9–15, answers may vary depending on the order in which the angles are found. Your answers may be slightly different but still be correct.

9. $A = 15.6°$, $C = 12.9°$, $b = 727$ m   **11.** $A = 39°$, $B = 57°$, $C = 84°$ **13.** $B = 114° \, 10'$, $C = 22° \, 30'$, $a = 0.694$ km
15. $A = 55.4°$, $B = 45.5°$, $C = 79.1°$ **17.** See the Solutions Manual.   **19.** 24 inches   **21.** 130 mi   **23.** 462 mi
25. 190 mi/hr (two significant digits) with heading 153°   **27.** 18.5 mi/hr at 81.3° from due north   **29.** 59.5 cm

31. 61.2 cm, 55.6°   **33.** $\dfrac{\pi}{18} + \dfrac{2k\pi}{3}, \dfrac{5\pi}{18} + \dfrac{2k\pi}{3}$ **35.** $\dfrac{\pi}{12} + \dfrac{k\pi}{3}, \dfrac{\pi}{4} + \dfrac{k\pi}{3}$ **37.** $20° + 120°k$, $100° + 120°k$

39. $45° + 60°k$ **41.** $0°, 90°$ **43.** 182 mi/hr at 54.5° from due north   **45.** 198 mi/hr at N 20.5° E

PROBLEM SET 7.4

1. 1,520 cm²   **3.** 342 m²   **5.** 0.123 km²   **7.** 26.3 m²   **9.** 28,300 in²   **11.** 2.09 ft²   **13.** 1,410 in²
15. 15.0 yd²   **17.** 8.15 ft²   **19.** 156 in²   **21.** 14.3 cm   **23.** See the solution to Problem 1 in Problem Set 6.4.
25. See the solution to Problem 11 in Problem Set 6.4.

27. $\dfrac{y^2}{9} - \dfrac{x^2}{4} = 1$ **29.** $y = 1 - 2x^2$

PROBLEM SET 7.5

1.

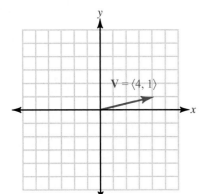

3.

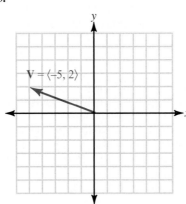

5.

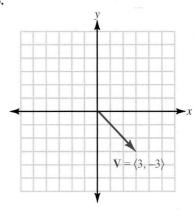

7.

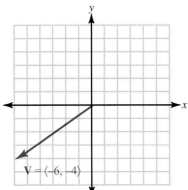

9.

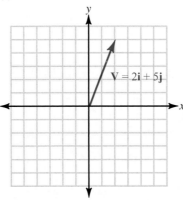

11.

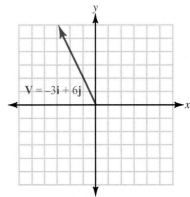

13.

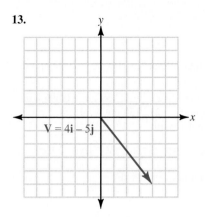

15.

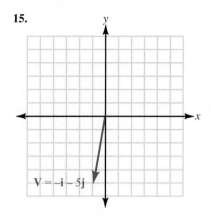

17. $\sqrt{61}$ **19.** 2   **21.** $\sqrt{29}$ **23.** $|\mathbf{V}| = 5$ **25.** $|\mathbf{U}| = 13$ **27.** $|\mathbf{W}| = \sqrt{5}$ **29.** $\langle 8, 0 \rangle, \langle 0, 8 \rangle, \langle -4, 20 \rangle$
31. $\langle 2, -7 \rangle, \langle 2, 7 \rangle, \langle 4, 21 \rangle$ **33.** $\langle -1, 3 \rangle, \langle 9, -1 \rangle, \langle 23, -4 \rangle$ **35.** $2\mathbf{i}, 2\mathbf{j}, 5\mathbf{i} + \mathbf{j}$ **37.** $6\mathbf{i} - 8\mathbf{j}, 6\mathbf{i} + 8\mathbf{j}, 18\mathbf{i} - 16\mathbf{j}$
39. $7\mathbf{i} + 7\mathbf{j}, -3\mathbf{i} + 3\mathbf{j}, 16\mathbf{i} + 19\mathbf{j}$ **41.** $\mathbf{V} = \langle 14, 12 \rangle = 14\mathbf{i} + 12\mathbf{j}$ to 2 significant figures
43. $\mathbf{W} = \langle -5.1, -6.1 \rangle = -5.1\mathbf{i} - 6.1\mathbf{j}$ to 2 significant figures   **45.** $|\mathbf{U}| = 3\sqrt{2}, 45°$ **47.** $|\mathbf{W}| = 2, 240°$
49. $|\mathbf{N}| = 9.7$ lb, $|\mathbf{F}| = 2.6$ lb   **51.** $|\mathbf{H}| = 45.6$ lb   **53.** $|\mathbf{T_1}| = 16$ lb, $|\mathbf{T_2}| = 20$ lb

PROBLEM SET 7.6

1. 48   **3.** −369   **5.** 0   **7.** 0   **9.** 20   **11.** $\theta = 90.0°$ **13.** $\theta = 81.1°$ **15.** $\theta = 111.3°$
17. $\langle 1, 0 \rangle \cdot \langle 0, 1 \rangle = 0$ **19.** $\langle -1, 0 \rangle \cdot \langle 0, 1 \rangle = 0$ **21.** $\langle a, b \rangle \cdot \langle -b, a \rangle = 0$ **23.** 696 ft-lb   **25.** 5,340 ft-lb
27. 510 ft-lb   **29.** 0 ft-lb   **31.** See the Solutions Manual.   **33.** 2,800 ft-lb   **35.** 8,200 ft-lb

CHAPTER 7 TEST

1. 6.7 inches   **2.** 4.3 inches   **3.** $C = 78.4°, a = 26.5$ cm, $b = 38.3$ cm   **4.** $B = 49.2°, a = 18.8$ cm, $c = 43.2$ cm
5. $\sin B = 3.0311$, which is impossible   **6.** $B = 29°$ is the only possibility for B
7. $B = 71°, C = 58°, c = 7.1$ ft; $B' = 109°, C' = 20°, c' = 2.9$ ft
8. $B = 59°, C = 95°, c = 11$ ft; $B' = 121°, C' = 33°, c' = 6.0$ ft   **9.** 11 cm   **10.** 19 cm   **11.** 95.7°   **12.** 69.5°
13. $A = 43°, B = 18°, c = 8.1$ m   **14.** $B = 34°, C = 111°, a = 3.8$ m   **15.** 498 cm²   **16.** 307 cm²   **17.** 52 cm²
18. 52 cm²   **19.** 17 km²   **20.** 52 km²   **21.** 51°   **22.** 59 ft   **23.** 410 ft   **24.** 14.1 m   **25.** 142 mi
26. 4.2 mi, S 75° W   **27.** 90 ft   **28.** 300 mi/hr or 388 mi/hr   **29.** 65 ft   **30.** 260 mi/hr at 88.9° from due north
31. 13   **32.** $-5\mathbf{i} + 41\mathbf{j}$ **33.** $35\mathbf{i} + 31\mathbf{j}$ **34.** $\sqrt{117}$ **35.** −8   **36.** 98.6°   **37.** See the Solutions Manual.
38. $\mathbf{V} \cdot \mathbf{W} = 0$ **39.** $b = -\dfrac{5}{3}$ **40.** 1,808

CHAPTER 8

PROBLEM SET 8.1

1. $4i$ **3.** $11i$ **5.** $3i\sqrt{2}$ **7.** $2i\sqrt{2}$ **9.** −6   **11.** −3   **13.** $x = \dfrac{2}{3}, y = -\dfrac{1}{2}$ **15.** $x = \dfrac{2}{5}, y = -4$

17. $x = -2$ or $3, y = \pm 3$ **19.** $x = \dfrac{\pi}{4}$ or $\dfrac{5\pi}{4}, y = \dfrac{\pi}{2}$ **21.** $x = \dfrac{\pi}{2}, y = \dfrac{\pi}{4}$ or $\dfrac{5\pi}{4}$ **23.** $10 - 2i$ **25.** $2 + 6i$

27. $3 - 13i$ **29.** $5 \cos x - 3i \sin y$ **31.** $2 + 2i$ **33.** $12 + 2i$ **35.** 1   **37.** −1   **39.** 1   **41.** i

43. $-48 - 18i$ **45.** $10 - 10i$ **47.** $5 + 12i$ **49.** 41   **51.** 53   **53.** $-28 + 4i$ **55.** −6   **57.** $\dfrac{1}{5} + \dfrac{3}{5}i$

59. $-\dfrac{5}{13} + \dfrac{12}{13}i$ **61.** $-2 - 5i$ **63.** $\dfrac{4}{61} + \dfrac{17}{61}i$ **65.** 13   **67.** $-7 + 22i$ **69.** $10 - 3i$ **71.** $16 + 20i$

73. $x^2 + 9$ **75.** See the Solutions Manual.   **77.** See the Solutions Manual.
79. $x = 4 + 2i, y = 4 - 2i; x = 4 - 2i, y = 4 + 2i$

81. $x = 1 + i\sqrt{3}$, $y = 2 - 2i\sqrt{3}$; $x = 1 - i\sqrt{3}$, $y = 2 + 2i\sqrt{3}$ **83.** See the Solutions Manual.   **85.** Yes

87. $\sin\theta = -\dfrac{4}{5}$, $\cos\theta = \dfrac{3}{5}$ **89.** $\sin\theta = \dfrac{b}{\sqrt{a^2 + b^2}}$, $\cos\theta = \dfrac{a}{\sqrt{a^2 + b^2}}$ **91.** $135°$

93. $B = 69.6°$, $C = 37.3°$, $a = 248$ cm   **95.** $A = 40.5°$, $B = 61.3°$, $C = 78.2°$

PROBLEM SET 8.2

1.

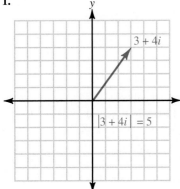

3.

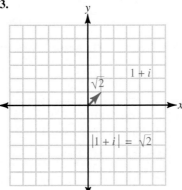

5.

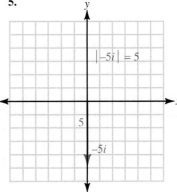

7.

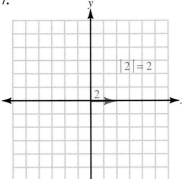

9.

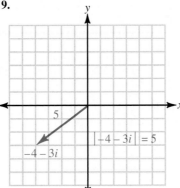

11.

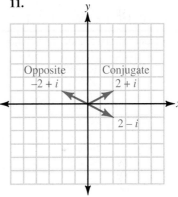

13.

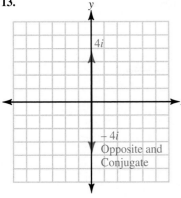

15.

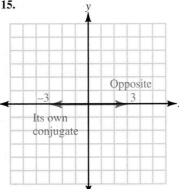

17.

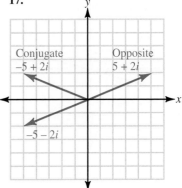

19. $\sqrt{3} + i$ **21.** $-2 + 2i\sqrt{3}$ **23.** $-\dfrac{\sqrt{3}}{2} - \dfrac{1}{2}i$ **25.** $\dfrac{\sqrt{2}}{2} - \dfrac{\sqrt{2}}{2}i$ **27.** $9.78 + 2.08i$

29. $-79.86 + 60.18i$ **31.** $-0.91 - 0.42i$ **33.** $9.51 - 3.09i$ **35.** $\sqrt{2}(\cos 135° + i \sin 135°) = \sqrt{2}$ cis $135°$

37. $\sqrt{2}(\cos 315° + i \sin 315°) = \sqrt{2}$ cis $315°$ **39.** $3\sqrt{2}(\cos 45° + i \sin 45°) = 3\sqrt{2}$ cis $45°$

41. $8(\cos 90° + i \sin 90°) = 8$ cis $90°$ **43.** $9(\cos 180° + i \sin 180°) = 9$ cis $180°$

45. $4(\cos 120° + i \sin 120°) = 4 \text{ cis } 120°$ **47.** $5(\cos 53.13° + i \sin 53.13°) = 5 \text{ cis } 53.13°$
49. $29(\cos 46.40° + i \sin 46.40°) = 29 \text{ cis } 46.40°$ **51.** $25(\cos 286.26° + i \sin 286.26°) = 25 \text{ cis } 286.26°$
53. $5\sqrt{5}(\cos 10.30° + i \sin 10.30°) = 5\sqrt{5} \text{ cis } 10.30°$

For Problems 55–61, see the answer to the corresponding problem.
63. $(2i)(3i) = [2(\cos 90° + i \sin 90°)][3(\cos 90° + i \sin 90°)]$
$= -6$
65. $2(\cos 30° + i \sin 30°) = \sqrt{3} + i;\ 2[\cos(-30°) + i \sin(-30°)] = \sqrt{3} - i$
67. $|z| = \sqrt{\cos^2 \theta + \sin^2 \theta} = 1$ **69.** $\dfrac{\sqrt{6} - \sqrt{2}}{4}$ **71.** $\dfrac{56}{65}$ **73.** $\sin 120° = \dfrac{\sqrt{3}}{2}$ **75.** $\cos 50°$

77. No triangle exists.   **79.** $B = 62.7°$ or $B = 117.3°$ **81.** Answers will vary.

PROBLEM SET 8.3

1. $12(\cos 50° + i \sin 50°)$ **3.** $56(\cos 157° + i \sin 157°)$ **5.** $4(\cos 180° + i \sin 180°) = 4 \text{ cis } 180°$
7. $2(\cos 180° + i \sin 180°) = -2$ **9.** $4(\cos 210° + i \sin 210°) = -2\sqrt{3} - 2i$ **11.** $12(\cos 360° + i \sin 360°) = 12$
13. $4\sqrt{2}(\cos 135° + i \sin 135°) = -4 + 4i$ **15.** $10(\cos 240° + i \sin 240°) = -5 - 5i\sqrt{3}$ **17.** $32 + 32i\sqrt{3}$
19. $-\dfrac{1}{2} + \dfrac{\sqrt{3}}{2}i$ **21.** $-\dfrac{81}{2} - \dfrac{81\sqrt{3}}{2}i$ **23.** $32i$ **25.** -4 **27.** $-8 - 8i\sqrt{3}$ **29.** $-8i$
31. $16 + 16i$ **33.** $4(\cos 35° + i \sin 35°)$ **35.** $1.5(\cos 19° + i \sin 19°)$ **37.** $0.5(\cos 60° + i \sin 60°) = 0.5 \text{ cis } 60°$
39. $2(\cos 0° + i \sin 0°) = 2$ **41.** $\cos(-60°) + i \sin(-60°) = \dfrac{1}{2} - \dfrac{\sqrt{3}}{2}i$ **43.** $2[\cos(-270°) + i \sin(-270°)] = 2i$
45. $2[\cos(-180°) + i \sin(-180°)] = -2$ **47.** $-4 - 4i$ **49.** 8

For Problems 51 and 53, see the Solutions Manual.

55. $(1 + i)^{-1} = [\sqrt{2}(\cos 45° + i \sin 45°)]^{-1} = \dfrac{1}{2} - \dfrac{1}{2}i$ **57.** $\dfrac{\sqrt{3}}{4} + \dfrac{1}{4}i$ **59.** $-\dfrac{7}{9}$ **61.** $\dfrac{\sqrt{6}}{3}$ **63.** $\dfrac{\sqrt{6}}{2}$
65. $\dfrac{4\sqrt{2}}{7}$ **67.** 6.0 mi   **69.** 103° at 160 mi/hr

PROBLEM SET 8.4

1.

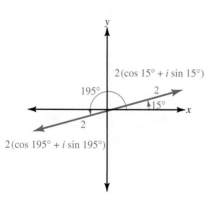

3.

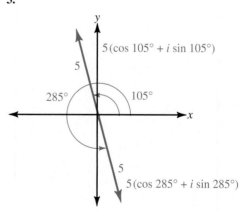

5.

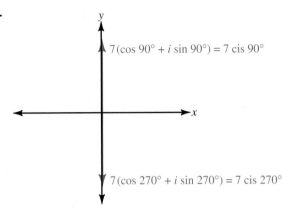

$7(\cos 90° + i \sin 90°) = 7 \text{ cis } 90°$

$7(\cos 270° + i \sin 270°) = 7 \text{ cis } 270°$

7. $\sqrt{3} + i, -\sqrt{3} - i$ **9.** $\sqrt{2} + i\sqrt{2}, -\sqrt{2} - i\sqrt{2}$ **11.** $5i, -5i$ **13.** $\dfrac{\sqrt{6}}{2} + \dfrac{\sqrt{2}}{2}i, -\dfrac{\sqrt{6}}{2} - \dfrac{\sqrt{2}}{2}i$

15. $2(\cos 70° + i \sin 70°), 2(\cos 190° + i \sin 190°), 2(\cos 310° + i \sin 310°)$
17. $2(\cos 10° + i \sin 10°), 2(\cos 130° + i \sin 130°), 2(\cos 250° + i \sin 250°)$
19. $3(\cos 60° + i \sin 60°), 3(\cos 180° + i \sin 180°), 3(\cos 300° + i \sin 300°)$
21. $4(\cos 30° + i \sin 30°), 4(\cos 150° + i \sin 150°), 4(\cos 270° + i \sin 270°)$

23. $3, -\dfrac{3}{2} + \dfrac{3\sqrt{3}}{2}i, -\dfrac{3}{2} - \dfrac{3\sqrt{3}}{2}i$ **25.** $2, -2, 2i, -2i$ **27.** $\sqrt{3} + i, -1 + i\sqrt{3}, -\sqrt{3} - i, 1 - i\sqrt{3}$

29. $10(\cos 3° + i \sin 3°) \approx 9.99 + 0.52i$
$10(\cos 75° + i \sin 75°) \approx 2.59 + 9.66i$
$10(\cos 147° + i \sin 147°) \approx -8.39 + 5.45i$
$10(\cos 219° + i \sin 219°) \approx -7.77 - 6.29i$
$10(\cos 291° + i \sin 291°) \approx 3.58 - 9.34i$

31.

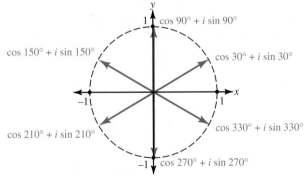

$\cos 90° + i \sin 90°$

$\cos 150° + i \sin 150°$

$\cos 30° + i \sin 30°$

$\cos 210° + i \sin 210°$

$\cos 330° + i \sin 330°$

$\cos 270° + i \sin 270°$

33. $\sqrt{2}(\cos \theta + i \sin \theta)$ where $\theta = 30°, 150°, 210°, 330°$
35. $\sqrt[4]{2}(\cos \theta + i \sin \theta)$ where $\theta = 67.5°, 112.5°, 247.5°, 292.5°$

37.

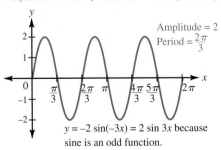

Amplitude $= 2$
Period $= \dfrac{2\pi}{3}$

$y = -2 \sin(-3x) = 2 \sin 3x$ because
sine is an odd function.

39.

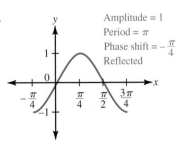

Amplitude $= 1$
Period $= \pi$
Phase shift $= -\dfrac{\pi}{4}$
Reflected

41.

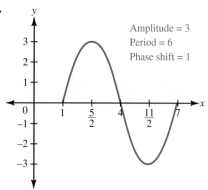

Amplitude = 3
Period = 6
Phase shift = 1

43. 4.23 cm²   **45.** 3.8 ft²   **47.** −3.732, −0.268, 4.000

PROBLEM SET 8.5

1.–11. (odd)

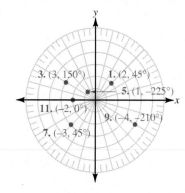

13. $(2, -300°), (-2, 240°), (-2, -120°)$ **15.** $(5, -225°), (-5, 315°), (-5, -45°)$
17. $(-3, -330°), (3, -150°), (3, 210°)$ **19.** $(1, \sqrt{3})$ **21.** $(0, -3)$ **23.** $(-1, -1)$ **25.** $(-6, -2\sqrt{3})$
27. $(1.891, 0.6511)$ **29.** $(-1.172, 2.762)$ **31.** $(3\sqrt{2}, 135°)$ **33.** $(4, 150°)$ **35.** $(2, 0°)$ **37.** $(2, 210°)$
39. $(5, 53.1°)$ **41.** $(\sqrt{5}, 116.6°)$ **43.** $(\sqrt{13}, 236.3°)$ **45.** $(9.434, 57.99°)$ **47.** $(6.083, -99.46°)$
49. $x^2 + y^2 = 9$ **51.** $x^2 + y^2 = 6y$ **53.** $(x^2 + y^2)^2 = 8xy$ **55.** $x + y = 3$ **57.** $r(\cos \theta - \sin \theta) = 5$
59. $r^2 = 4$ **61.** $r = 6 \cos \theta$ **63.** $\theta = 45°$ or $\cos \theta = \sin \theta$

65.

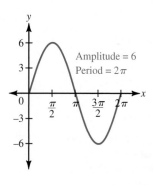

Amplitude = 6
Period = 2π

67.

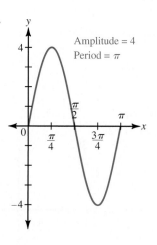

Amplitude = 4
Period = π

69.

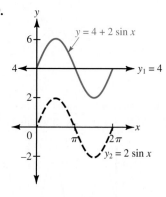

$y = 4 + 2 \sin x$
$y_1 = 4$
$y_2 = 2 \sin x$

PROBLEM SET 8.6

1.

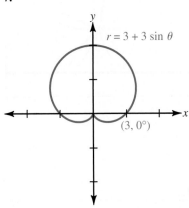

3.

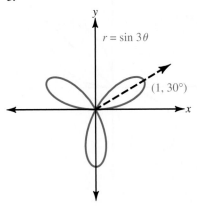

5.

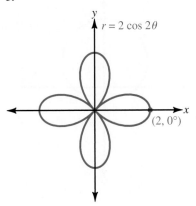

7.

9.

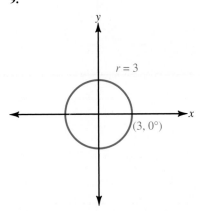

11.

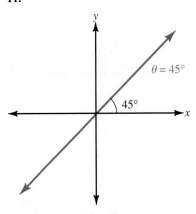

13.

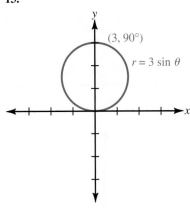

15.

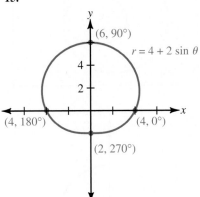

17.

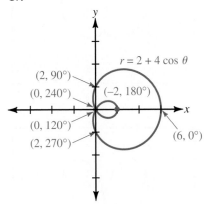

19.

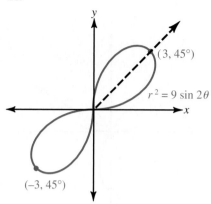

$(4, 90°)$
$r = 2 + 2 \sin \theta$
$(2, 180°)$
$(2, 0°)$
$(0, 270°)$

21.

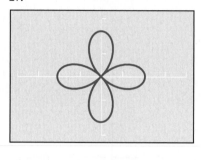

$(3, 45°)$
$r^2 = 9 \sin 2\theta$
$(-3, 45°)$

23.

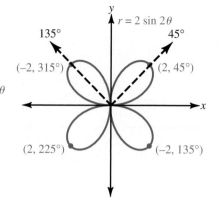

$r = 2 \sin 2\theta$
$135°$
$45°$
$(-2, 315°)$
$(2, 45°)$
$(2, 225°)$
$(-2, 135°)$

25.

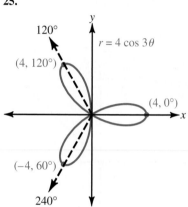

$120°$
$r = 4 \cos 3\theta$
$(4, 120°)$
$(4, 0°)$
$(-4, 60°)$
$240°$

27.

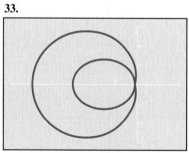

29.

31.

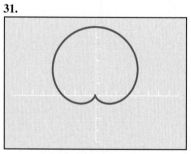

33.

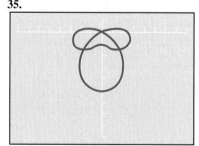

35.

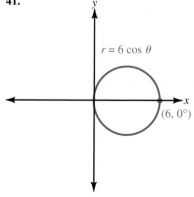

37.

39.

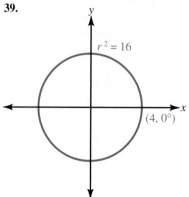

$r^2 = 16$
$(4, 0°)$

41.

$r = 6 \cos \theta$
$(6, 0°)$

43.

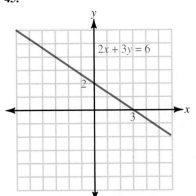

$r^2 = \sin 2\theta$

$(1, 45°)$

$(-1, 45°)$

45.

$2x + 3y = 6$

47.

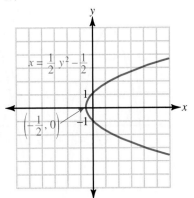

$x = \frac{1}{2} y^2 - \frac{1}{2}$

$\left(-\frac{1}{2}, 0\right)$

-1

49.

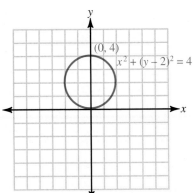

$(0, 4)$

$x^2 + (y - 2)^2 = 4$

51.

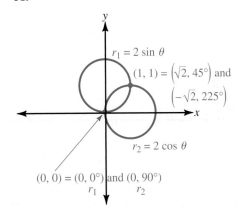

$r_1 = 2 \sin \theta$

$(1, 1) = \left(\sqrt{2}, 45°\right)$ and $\left(-\sqrt{2}, 225°\right)$

$r_2 = 2 \cos \theta$

$(0, 0) = (0, 0°)$ and $(0, 90°)$

r_1 r_2

53.

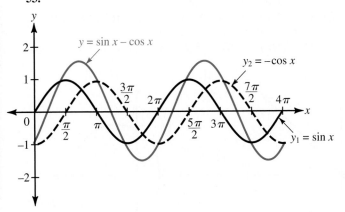

$y = \sin x - \cos x$

$y_2 = -\cos x$

$\frac{3\pi}{2}$ 2π $\frac{7\pi}{2}$ 4π

$\frac{\pi}{2}$ π $\frac{5\pi}{2}$ 3π

$y_1 = \sin x$

55.

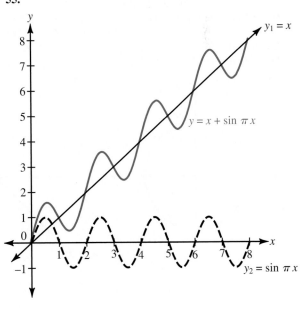

$y_1 = x$

$y = x + \sin \pi x$

$y_2 = \sin \pi x$

57.

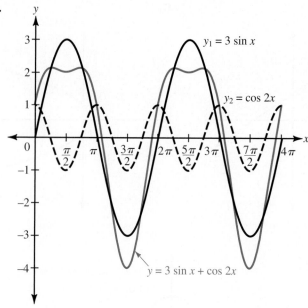

$y_1 = 3 \sin x$

$y_2 = \cos 2x$

$y = 3 \sin x + \cos 2x$

CHAPTER 8 TEST

1. $5i$ **2.** $2i\sqrt{3}$ **3.** $x = 2, y = 2$ **4.** $x = -2$ or $5, y = 2$ **5.** $7 - 6i$ **6.** $8 - 2i$ **7.** 1 **8.** i **9.** 89

10. $-16 + 30i$ **11.** $-2 - \dfrac{5}{2}i$ **12.** $\dfrac{11}{61} + \dfrac{60}{61}i$ **13. a.** 5 **b.** $-3 - 4i$ **c.** $3 - 4i$ **14. a.** 5

b. $-3 + 4i$ **c.** $3 + 4i$ **15. a.** 8 **b.** $-8i$ **c.** $-8i$ **16. a.** 4 **b.** 4 **c.** -4 **17.** $4\sqrt{3} - 4i$

18. $-\sqrt{2} + i\sqrt{2}$ **19.** $2\sqrt{2}(\cos 45° + i \sin 45°)$ **20.** $2(\cos 150° + i \sin 150°)$ **21.** $5(\cos 90° + i \sin 90°)$

22. $3(\cos 180° + i \sin 180°)$ **23.** $15(\cos 65° + i \sin 65°)$ **24.** $5(\cos 30° + i \sin 30°)$ **25.** $32(\cos 50° + i \sin 50°)$

26. $81(\cos 80° + i \sin 80°) = 81 \text{ cis } 80°$ **27.** $7(\cos 25° + i \sin 25°), 7(\cos 205° + i \sin 205°)$

28. $\sqrt{2}(\cos \theta + i \sin \theta)$ where $\theta = 15°, 105°, 195°, 285°$ **29.** $x = \sqrt{2}(\cos \theta + i \sin \theta)$ where $\theta = 15°, 165°, 195°, 345°$

30. $x = \cos \theta + i \sin \theta$ where $\theta = 60°, 180°, 300°$ **31.** $(-4, 45°), (4, -135°); (-2\sqrt{2}, -2\sqrt{2})$

32. $(6, 240°), (6, -120°); (-3, -3\sqrt{3})$ **33.** $(3\sqrt{2}, 135°)$ **34.** $(5, 90°)$ **35.** $x^2 + y^2 = 6y$ **36.** $(x^2 + y^2)^{3/2} = 2xy$

37. $r(\cos \theta + \sin \theta) = 2$ **38.** $r = 8 \sin \theta$

39.   **40.**   **41.**

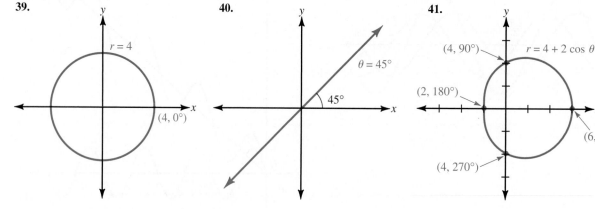

$r = 4$

$(4, 0°)$

$\theta = 45°$

$45°$

$(4, 90°)$ $r = 4 + 2 \cos \theta$

$(2, 180°)$

$(6, 0°)$

$(4, 270°)$

42.

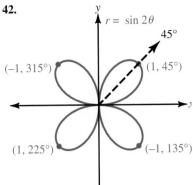

43.

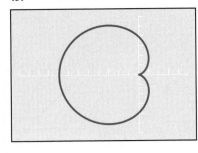

44.

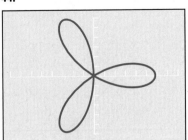

45.

46.

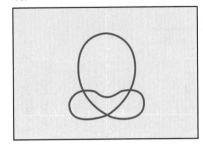

APPENDIX A

PROBLEM SET A.1

1. Domain = $\{1, 2, 4\}$; range = $\{3, 5, 1\}$; a function
3. Domain = $\{-1, 1, 2\}$; range = $\{3, -5\}$; a function
5. Domain = $\{7, 3\}$; range = $\{-1, 4\}$; not a function
7. Yes   **9.** No   **11.** No   **13.** Yes
15. Domain = all real numbers;
range = $\{y \mid y \geq -1\}$; a function
17. Domain = $\{x \mid x \geq 4\}$;
range = all real numbers; not a function

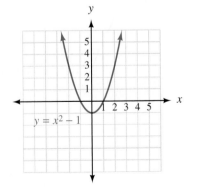

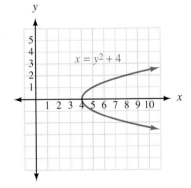

19. Domain = all real numbers;
range = $\{y \mid y \geq 0\}$; a function

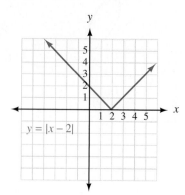

21. Domain = all real numbers;
range = $\{y \mid y \geq -2\}$; a function

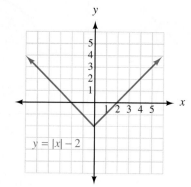

23. a. $y = 8.5x$ for $10 \leq x \leq 40$

b.

Hours Worked	Gross Pay ($)
x	y
10	85
20	170
30	255
40	340

c.

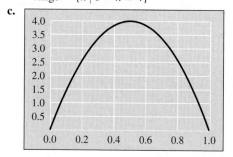

d. Domain = $\{x \mid 10 \leq x \leq 40\}$; range = $\{y \mid 85 \leq y \leq 340\}$

25. a.

Time (sec)	Distance (ft)
t	h
0	0
0.1	1.44
0.2	2.56
0.3	3.36
0.4	3.84
0.5	4
0.6	3.84
0.7	3.36
0.8	2.56
0.9	1.44
1	0

b. Domain = $\{t \mid 0 \leq t \leq 1\}$;
range = $\{h \mid 0 \leq h \leq 4\}$

c.

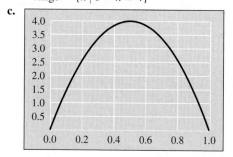

27. a.

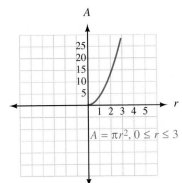

$A = \pi r^2, 0 \leq r \leq 3$

b. Domain $= \{r \mid 0 \leq r \leq 3\}$;
range $= \{A \mid 0 \leq A \leq 9\pi\}$

29. -1 **31.** -11 **33.** 2   **35.** 4   **37.** 35   **39.** -13 **41.** 1   **43.** -9 **45.** 8   **47.** 19   **49.** 16

51. 0   **53.** $3a^2 - 4a + 1$ **55.** 4   **57.** 0   **59.** 2

61.

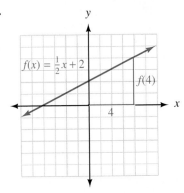

$f(x) = \frac{1}{2}x + 2$

$f(4)$

4

63. $V(3) = 300$, the painting is worth \$300 in 3 years; $V(6) = 600$, the painting is worth \$600 in 6 years.
65. $A(2) = 12.56$; $A(5) = 78.5$; $A(10) = 314$
67. a. 2   **b.** 0   **c.** 1   **d.** 4   **69.** $x = 4$

PROBLEM SET A.2

1. $f^{-1}(x) = \dfrac{x + 1}{3}$ **3.** $f^{-1}(x) = \sqrt[3]{x}$ **5.** $f^{-1}(x) = \dfrac{x - 3}{x - 1}$ **7.** $f^{-1}(x) = 4x + 3$ **9.** $f^{-1}(x) = 2x + 6$

11. $f^{-1}(x) = \dfrac{1 - x}{3x - 2}$

13.

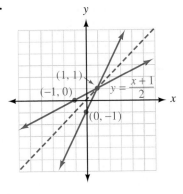

$(1, 1)$
$(-1, 0)$
$y = \dfrac{x + 1}{2}$
$(0, -1)$

15.

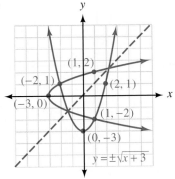

$(1, 2)$
$(-2, 1)$ $(2, 1)$
$(-3, 0)$
$(1, -2)$
$(0, -3)$
$y = \pm\sqrt{x + 3}$

17.

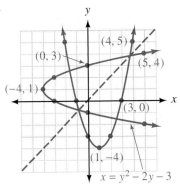

$(4, 5)$
$(0, 3)$ $(5, 4)$
$(-4, 1)$
$(3, 0)$
$(1, -4)$
$x = y^2 - 2y - 3$

19.

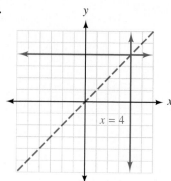

21.

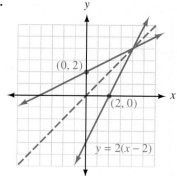

23.

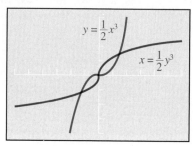

25.

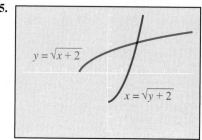

27. a. Yes   **b.** No   **29. a.** 4   **b.** $\dfrac{4}{3}$ **c.** 2   **d.** 2   **31.** $f^{-1}(x) = \dfrac{1}{x}$

33. a. -3 **b.** -6 **c.** 2   **d.** 3   **e.** -2 **f.** 3   **g.** They are inverses of each other.   **35.** $f^{-1}(x) = \dfrac{x-5}{3}$

37. $f^{-1}(x) = \sqrt[3]{x-1}$ **39.** $f^{-1}(x) = 7(x+2)$ **41. a.** 1   **b.** 2   **c.** 5   **d.** 0   **e.** 1   **f.** 2   **g.** 2   **h.** 5

APPENDIX B

PROBLEM SET B.1

1. 1   **3.** 2   **5.** $\dfrac{1}{27}$ **7.** 13

9.

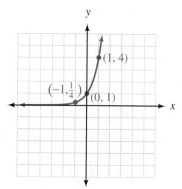

11.

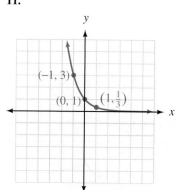

13.

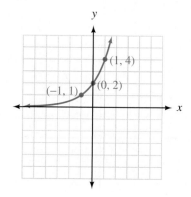

15.

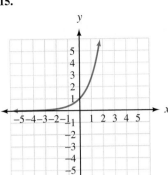

17.

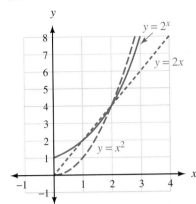

19.

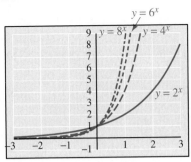

21. $h = 6 \cdot \left(\dfrac{2}{3}\right)^n$; 5th bounce: $6\left(\dfrac{2}{3}\right)^5 \approx 0.79$ feet

23. After 8 days, 700 micrograms; after 11 days, $1400 \cdot 2^{-11/8} \approx 539.8$ micrograms

25. a. $A(t) = 1200\left(1 + \dfrac{.06}{4}\right)^{4t}$ **b.** $1932.39   **c.** $1939.29   **27.** $16,310.19

29. 200, 400, 800, 1,600   **31. a.** The function underestimated the expenditures by $69 billion.
 b. $3,015 billion
   $3,347 billion
   $3,715 billion

33.

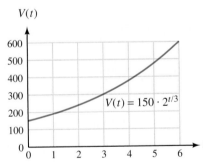

35. a. $129,138.48
b. $0 \le t \le 6$
c.

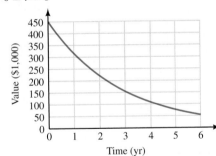

37.

t	$s(t)$
0	0
1	83.3
2	138.9
3	175.9
4	200.6
5	217.1
6	228.1

PROBLEM SET B.2

1. $\log_2 16 = 4$ **3.** $\log_5 125 = 3$ **5.** $\log_{10} 0.01 = -2$ **7.** $\log_2 \dfrac{1}{32} = -5$ **9.** $\log_{1/2} 8 = -3$ **11.** $\log_3 27 = 3$

13. $10^2 = 100$ **15.** $2^6 = 64$ **17.** $8^0 = 1$ **19.** $10^{-3} = 0.001$ **21.** $6^2 = 36$ **23.** $5^{-2} = \dfrac{1}{25}$ **25.** 9

27. $\dfrac{1}{125}$ **29.** 4   **31.** $\dfrac{1}{3}$ **33.** 2   **35.** $\sqrt[3]{5}$

37.

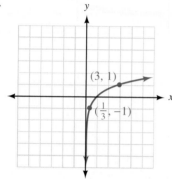

39.

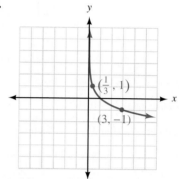

41.

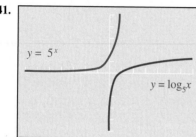

43.

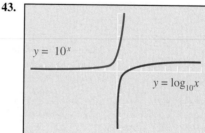

45. 4   **47.** $\dfrac{3}{2}$ **49.** 3   **51.** 1   **53.** 0   **55.** 0   **57.** $\dfrac{1}{2}$ **59.** 7   **61.** 10^{-6} **63.** 2   **65.** 10^8 times as large

67. a.

x	$f(x)$
-1	$\dfrac{1}{8}$
0	1
1	8
2	64

b.

x	$f^{-1}(x)$
$\dfrac{1}{8}$	-1
1	0
8	1
64	2

c. $f(x) = 8^x$ **d.** $f^{-1}(x) = \log_8 x$

PROBLEM SET B.3

1. $\log_3 4 + \log_3 x$ **3.** $\log_6 5 - \log_6 x$ **5.** $5 \log_2 y$ **7.** $\dfrac{1}{3} \log_9 z$ **9.** $2 \log_6 x + 4 \log_6 y$ **11.** $\dfrac{1}{2} \log_5 x + 4 \log_5 y$

13. $\log_b x + \log_b y - \log_b z$ **15.** $\log_{10} 4 - \log_{10} x - \log_{10} y$ **17.** $2 \log_{10} x + \log_{10} y - \dfrac{1}{2} \log_{10} z$

19. $3 \log_{10} x + \dfrac{1}{2} \log_{10} y - 4 \log_{10} z$ **21.** $\dfrac{2}{3} \log_b x + \dfrac{1}{3} \log_b y - \dfrac{4}{3} \log_b z$ **23.** $\log_b xz$ **25.** $\log_3 \dfrac{x^2}{y^3}$

27. $\log_{10} \sqrt{x} \sqrt[3]{y}$ **29.** $\log_2 \dfrac{x^3\sqrt{y}}{z}$ **31.** $\log_2 \dfrac{\sqrt{x}}{y^3 z^4}$ **33.** $\log_{10} \dfrac{x^{3/2}}{y^{3/4} z^{4/5}}$ **35.** $\dfrac{2}{3}$ **37.** 18

39. Possible solutions -1 and 3; only 3 checks; 3   **41.** 3   **43.** Possible solutions -2 and 4; only 4 checks; 4

45. Possible solutions -1 and 4; only 4 checks; 4   **47.** Possible solutions $-\dfrac{5}{2}$ and $\dfrac{5}{3}$, only $\dfrac{5}{3}$ checks; $\dfrac{5}{3}$

49. 2.52 min   **51.** $pH = 6.1 + \log_{10} x - \log_{10} y$ **53.** $D = 10(\log_{10} I - \log_{10} I_0)$

PROBLEM SET B.4

1. 2.5775   **3.** 1.5775   **5.** 3.5775   **7.** -1.4225 **9.** 4.5775   **11.** 2.7782   **13.** 3.3032   **15.** -2.0128
17. -1.5031 **19.** -0.3990 **21.** 759   **23.** 0.00759   **25.** 1,430   **27.** 0.00000447   **29.** 0.0000000918

31. 10^{10} **33.** 10^{-10} **35.** 10^{20} **37.** $\dfrac{1}{100}$ **39.** 1,000   **41.** 1   **43.** 5   **45.** x **47.** $\ln 10 + 3t$

49. $\ln A - 2t$ **51.** 2.7080   **53.** -1.0986 **55.** 2.1972   **57.** 2.7724   **59.** 3.19   **61.** 1.78×10^{-5}
63. 3.16×10^5 **65.** 2.00×10^8 **67.** 10 times larger

69.

Location	Date	Magnitude (M)	Shock Wave (T)
Moresby Island	January 23	4.0	1.00×10^4
Vancouver Island	April 30	5.3	1.99×10^5
Quebec City	June 29	3.2	1.58×10^3
Mould Bay	November 13	5.2	1.58×10^5
St. Lawrence	December 14	3.7	5.01×10^3

71. 12.9%   **73.** 5.3%

75.

x	$(1 + x)^{1/x}$	; the number e
1	2.0000	
0.5	2.2500	
0.1	2.5937	
0.01	2.7048	
0.001	2.7169	
0.0001	2.7181	
0.00001	2.7183	

PROBLEM SET B.5

1. 1.4650   **3.** 0.6826   **5.** -1.5440 **7.** -0.6477 **9.** -0.3333 **11.** 2.0000   **13.** -0.1845 **15.** 0.1845
17. 1.6168   **19.** 2.1131   **21.** 1.3333   **23.** 0.7500   **25.** 1.3917   **27.** 0.7186   **29.** 2.6356   **31.** 4.1632
33. 5.8435   **35.** -1.0642 **37.** 2.3026   **39.** 10.7144   **41.** 11.7 years   **43.** 9.25 years   **45.** 8.75 years
47. 18.6 years   **49.** 11.6 years   **51.** 18.3 years   **53.** 13.9 years later, or toward the end of 2007   **55.** 27.5 years

57. $t = \dfrac{1}{r} \ln \dfrac{A}{P}$ **59.** $t = \dfrac{1}{k} \dfrac{\log P - \log A}{\log 2}$ **61.** $t = \dfrac{\log A - \log P}{\log (1 - r)}$

PHOTO CREDITS

INDEX

Operations on Complex Numbers in Standard Form [8.1]

If $z_1 = a_1 + b_1 i$ and $z_2 = a_2 + b_2 i$ are two complex numbers in standard form, then the following definitions and operations apply.

Addition
$$z_1 + z_2 = (a_1 + a_2) + (b_1 + b_2)i$$
Add real parts; add imaginary parts.

Subtraction
$$z_1 - z_2 = (a_1 - a_2) + (b_1 - b_2)i$$
Subtract real parts; subtract imaginary parts.

Multiplication
$$z_1 z_2 = (a_1 a_2 - b_1 b_2) + (a_1 b_2 + a_2 b_1)i$$
In actual practice, simply multiply as you would multiply two binomials.

Conjugates
The conjugate of $a + bi$ is $a - bi$. Their product is the real number $a^2 + b^2$.

Division
Multiply the numerator and denominator of the quotient by the conjugate of the denominator.

Graphing Complex Numbers [8.2]

The graph of the complex number $z = x + yi$ is the arrow (vector) that extends from the origin to the point (x, y).

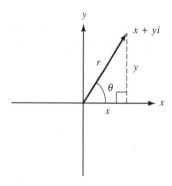

Absolute Value of a Complex Number [8.2]

The *absolute value* (or *modulus*) of the complex number $z = x + yi$ is the distance from the origin to the point (x, y). If this distance is denoted by r, then

$$r = |z| = |x + yi| = \sqrt{x^2 + y^2}$$

Argument of a Complex Number [8.2]

The *argument* of the complex number $z = x + yi$ is the smallest positive angle from the positive x-axis to the graph of z. If the argument of z is denoted by θ, then

$$\sin \theta = \frac{y}{r}, \qquad \cos \theta = \frac{x}{r}, \qquad \text{and} \qquad \tan \theta = \frac{y}{x}$$

Trigonometric Form of a Complex Number [8.2]

The complex number $z = x + yi$ is written in trigonometric form when it written as

$$z = r(\cos \theta + i \sin \theta) = r \operatorname{cis} \theta$$

where r is the absolute value of z and θ is the argument of z.

Products and Quotients in Trigonometric Form [8.3]

If $z_1 = r_1(\cos \theta_1 + i \sin \theta_1)$ and $z_2 = r_2(\cos \theta_2 + i \sin \theta_2)$ then

$$z_1 z_2 = r_1 r_2 [\cos (\theta_1 + \theta_2) + i \sin (\theta_1 + \theta_2)]$$

$$\frac{z_1}{z_2} = \frac{r_1}{r_2} [\cos (\theta_1 - \theta_2) + i \sin (\theta_1 - \theta_2)]$$

DeMoivre's Theorem [8.3]

If $z = r(\cos \theta + i \sin \theta)$ is a complex number in trigonometric form and n an integer, then

$$z^n = r^n(\cos n\theta + i \sin n\theta) = r^n \operatorname{cis} (n\theta)$$

Roots of a Complex Number [8.4]

The nth roots of the complex number

$$z = r(\cos \theta + i \sin \theta) = r \operatorname{cis} \theta$$

are given by

$$w_k = r^{1/n} \left[\cos \left(\frac{\theta}{n} + \frac{360°}{n} k \right) + i \sin \left(\frac{\theta}{n} + \frac{360°}{n} k \right) \right]$$

$$= r^{1/n} \operatorname{cis} \left(\frac{\theta}{n} + \frac{360°}{n} k \right)$$

where $k = 0, 1, 2, \ldots, n - 1$.